RADIATION IN THE ATMOSPHERE
A Course in Theoretical Meteorology

This book presents the theory and applications of radiative transfer in the atmosphere. It is written for graduate students and researchers in the fields of meteorology and related sciences.

The book begins with important basic definitions of the radiative transfer theory. It presents the hydrodynamic derivation of the radiative transfer equation and the principles of invariance. The authors examine in detail various quasi-exact solutions of the radiative transfer equation, such as the matrix operator method, the discrete ordinate method, and the Monte Carlo method. A thorough treatment of the radiative perturbation theory is given. The book also presents various two-stream methods for the approximate solution of the radiative transfer equation. The interaction of radiation with matter is discussed as well as the transmission in individual spectral lines and in bands of lines. It formulates the theory of gaseous absorption and analyzes the normal vibrations of linear and non-linear molecules. The book presents the Schrödinger equation and describes the computation of transition probabilities, before examining the mathematical formulation of spectral line intensities. A rigorous treatment of Mie scattering is given, including Rayleigh scattering as a special case, and the important efficiency factors for extinction, scattering and absorption are derived. Polarization effects are introduced with the help of Stokes parameters. The fundamentals of remote sensing applications of radiative transfer are presented.

Problems of varying degrees of difficulty are included at the end of each chapter, so readers can further their understanding of the materials covered in the book.

WILFORD ZDUNKOWSKI is Professor Emeritus at the Institute of Atmospheric Physics in the University of Mainz, Germany.

THOMAS TRAUTMANN is Head of the Unit for Atmospheric Processors at the Remote Sensing Technology Institute, German Aerospace Center (DLR) Oberpfaffenhofen, Germany, and Professor for Meteorology at the University of Leipzig, Germany.

ANDREAS BOTT is a university professor for Theoretical Meteorology at the Meteorological Institute of the University of Bonn, Germany.

Together, the authors have a wealth of experience in teaching atmospheric physics, including courses on atmospheric thermodynamics, atmospheric dynamics, cloud microphysics, atmospheric chemistry, and numerical modeling. Professors Zdunkowski and Bott co-wrote *Dynamics of the Atmosphere* and *Thermodynamics of the Atmosphere* (Cambridge University Press 2003 and 2004).

RADIATION IN THE ATMOSPHERE

A Course in Theoretical Meteorology

WILFORD ZDUNKOWSKI
Formerly University of Mainz, Germany

THOMAS TRAUTMANN
German Aerospace Center (DLR), Germany

ANDREAS BOTT
University of Bonn, Germany

CAMBRIDGE
UNIVERSITY PRESS

CAMBRIDGE
UNIVERSITY PRESS

University Printing House, Cambridge CB2 8BS, United Kingdom

One Liberty Plaza, 20th Floor, New York, NY 10006, USA

477 Williamstown Road, Port Melbourne, VIC 3207, Australia

314-321, 3rd Floor, Plot 3, Splendor Forum, Jasola District Centre, New Delhi - 110025, India

79 Anson Road, #06-04/06, Singapore 079906

Cambridge University Press is part of the University of Cambridge.

It furthers the University's mission by disseminating knowledge in the pursuit of education, learning and research at the highest international levels of excellence.

www.cambridge.org
Information on this title: www.cambridge.org/9781108462723

First published 2007
First paperback edition 2018

A catalogue record for this publication is available from the British Library

ISBN 978-0-521-87107-5 Hardback
ISBN 978-1-108-46272-3 Paperback

This book is dedicated to the memory of
Professor Fritz Möller (1906–1983) and
Professor J. Vern Hales (1917–1997).
We have profited directly and indirectly
from their lectures and research.

Contents

Preface

Radiation in the Atmosphere is the third volume in the series *A Course in Theoretical Meteorology*. The first two volumes entitled *Dynamics of the Atmosphere* and *Thermodynamics of the Atmosphere* were first published in the years 2003 and 2004.

The present textbook is written for graduate students and researchers in the field of meteorology and related sciences. Radiative transfer theory has reached a high point of development and is still a vastly expanding subject. Kourganoff (1952) in the postscript of his well-known book on radiative transfer speaks of the three olympians named completeness, up-to-date-ness and clarity. We have not made any attempt to be complete, but we have tried to be reasonably up-to-date, if this is possible at all with the many articles on radiative transfer appearing in various monthly journals. Moreover, we have tried very hard to present a coherent and consistent development of radiative transfer theory as it applies to the atmosphere. We have given principle allegiance to the olympian clarity and sincerely hope that we have succeeded.

In the selection of topics we have resisted temptation to include various additional themes which traditionally belong to the fields of physical meteorology and physical climatology. Had we included these topics, our book, indeed, would be very bulky, and furthermore, we would not have been able to cover these subjects in the required depth. Neither have we made any attempt to include radiative transfer theory as it pertains to the ocean, a subject well treated by Thomas and Stamnes (1999) in their book *Radiative Transfer in the Atmosphere and Ocean*.

As in the previous books of the series, we were guided by the principle to make the book as self-contained as possible. As far as space would permit, all but the most trivial mathematical steps have been included in the presentation of the theory to encourage students to follow the various developments. Nevertheless, here and there students may find it difficult to follow our reasoning. In this case, we encourage them not to get stuck with a particular detail but to continue with the subject. Additional details given later may clarify any questions. Moreover, on a second reading everything will become much clearer.

We will now give a brief description of the various chapters and topics treated in this book. Chapter 1 gives the general introduction to the book. Various important definitions such as the radiance and the net flux density are given to describe the radiation field. The interaction of radiation with matter is briefly discussed by introducing the concepts of absorption and scattering. To get an overall view of the mean global radiation budget of the system Earth–atmosphere, it is shown that the incoming and outgoing energy at the top of the atmosphere are balanced.

In Chapter 2 the hydrodynamic derivation of the radiative transfer equation (RTE) is worked out; this is in fact the budget equation for photons. The radiatively induced temperature change is formulated with the help of the first law of thermodynamics. Some basic formulas from spherical harmonics, which are needed to evaluate certain transfer integrals, are presented. Various special cases are discussed.

Chapter 3 presents the principle of invariance which, loosely speaking, is a collection of common sense statements about the exact mathematical structure of the radiation field. At first glance the mathematical formalism looks much worse than it really is. A systematic study of the mathematical and physical principles of invariance it quite rewarding.

Quasi-exact solutions of the RTE, such as the matrix operator method together with the doubling algorithm are presented in Chapter 4. Various other prominent solutions such as the successive order of scattering and the Monte Carlo methods are discussed in some detail.

Chapter 5 presents the radiative perturbation theory. The concept of the adjoint formulation of the RTE is introduced, and it is shown that in the adjoint formulation certain radiative effects can be evaluated with much higher numerical efficiency than with the so-called forward mode methods.

For many practical purposes in connection with numerical weather prediction it is sufficient to obtain fast approximate solutions of the RTE. These are known as two-stream methods and are treated in Chapter 6. Partial cloudiness is introduced in the solution scheme on the basis of two differing assumptions. The method allows fairly general situations to be handled.

In Chapter 7, the theory of individual spectral lines and band models is treated in some detail. In those cases in which scattering effects can be ignored, formulas are worked out to describe the mean absorption of homogeneous atmospheric layers. A technique is introduced which makes it possible to replace the transmission through an inhomogeneous atmosphere by a nearly equivalent homogeneous layer.

The theory of gaseous absorption is formulated in Chapter 8. The analysis of normal vibrations of linear and nonlinear molecules is introduced. The Schrödinger equation is presented and the computation of transition probabilities is described, which finally leads to the mathematical formulation of spectral line intensities. Simple but instructive analytic solutions of Schrödinger's equation are obtained leading, for example, to the description of the vibration–rotation spectrum of diatomic molecules.

Not only atmospheric gaseous absorbers influence the radiation field but also aerosol particles and cloud droplets. Chapter 9 gives a rigorous treatment of Mie scattering which includes Rayleigh scattering as a special case. The important efficiency factors for extinction, scattering and absorption are derived. The mathematical analysis requires the mathematical skill which the graduate student has acquired in various mathematics and physics courses. The effects of nonspherical particles are not treated in this book.

So far polarization has not been included in the RTE, which is usually satisfactory for energy considerations but may not be sufficient for optical applications. To give a complete description of the radiation field the polarization effects are introduced in Chapter 10 with the help of the Stokes parameters. This finally leads to the most general vector formulation of the RTE in terms of the phase matrix while the phase function is sufficient if polarization may be ignored.

Chapter 11 introduces remote sensing applications of radiative transfer. After the general description of some basic ideas, the RTE is presented in a form which is suitable to recover the atmospheric temperature profile by special inversion techniques. The chapter closes with a description of the way in which the atmospheric ozone profile can be retrieved using radiative perturbation theory.

The book closes with Chapter 12 in which a simple and brief account of the influence of clouds on climate is given. The student will be exposed to concepts such as cloud forcing and cloud radiative feedback.

Problems of various degrees of difficulty are included at the end of each chapter. Some of the included problems are almost trivial. They serve the purpose of making students familiar with new concepts and terminologies. Other problems are more demanding. Where necessary answers to problems are given at the end of the book.

One of the problems that any author of a physical science textbook is confronted with, is the selection of proper symbols. Inspection of the book shows that many times the same symbol is used to label several quite different physical entities. It would be ideal to represent each physical quantity by a unique symbol which is not used again in some other context. Consider, for example, the letter k. For the Boltzmann constant we could have written k_B, for Hooke's constant k_H, for the wave number k_w, and k_s for the climate sensitivity constant. It would have been possible, in addition to using the Greek alphabet, to also employ the letters of another alphabet, e.g. Hebrew, to label physical quantities in order to obtain uniqueness in notation. Since usually confusion is unlikely, we have tried to use standard notation even if the same symbol is used more than once. For example, the climate sensitivity parameter k appears in Chapter 12, Hooke's constant in Chapter 8 and Boltzmann's constant in Chapter 1.

The book concludes with a list of frequently used symbols and a list of constants.

We would like to give recognition to the excellent textbooks *Radiative Transfer* by the late S. Chandrasekhar (1960), to *Atmospheric Radiation* by R. M. Goody (1964) and the updated version of this book by Goody and Yung (1989). These books have been an invaluable guidance to us in research and teaching.

We would like to give special recognition to Dr W. G. Panhans for his splendid cooperation in organizing and conducting our exercise classes. Recognition is due to Dr Jochen Landgraf for discussions related to the perturbation theory and to ozone retrieval. Moreover, we will be indebted to Sebastian Otto for carrying out the transfer calculations presented in Section 7.5. We also wish to express our gratitude to many colleagues and graduate students for helpful comments while preparing the text. Last but not least we wish to thank our families for their patience and encouragement during the preparation of this book.

It seems to be one of the unfortunate facts of life that no book as technical as this one can be written free of error. However, each author takes comfort in the thought that any errors appearing in this book are due to one of the other two. To remove such errors, we will be grateful for anyone pointing these out to us.

1

Introduction

1.1 The atmospheric radiation field

The theory presented in this book applies to the lower 50 km of the Earth's atmosphere, that is to the troposphere and to the stratosphere. In this part of the atmosphere the so-called *local thermodynamic equilibrium* is observed.

In general, the condition of *thermodynamic equilibrium* is described by the state of matter and radiation inside a constant temperature enclosure. The radiation inside the enclosure is known as *black body radiation*. The conditions describing thermodynamic equilibrium were first formulated by Kirchhoff (1882). He stated that within the enclosure the radiation field is:

(1) isotropic and unpolarized;
(2) independent of the nature and shape of the cavity walls;
(3) dependent only on the temperature.

The existence of local thermodynamic equilibrium in the atmosphere implies that a local temperature can be assigned everywhere. In this case the *thermal radiation* emitted by each atmospheric layer can be described by *Planck's radiation law*. This results in a relatively simple treatment of the thermal radiation transport in the lower sections of the atmosphere.

Kirchhoff's and Planck's laws, fundamental in radiative transfer theory, will be described in the following chapters. While the derivation of Planck's law requires a detailed microscopic picture, Kirchhoff's law may be obtained by using purely thermodynamic arguments. The derivation of Kirchhoff's law is presented in numerous textbooks such as in *Thermodynamics of the Atmosphere* by Zdunkowski and Bott (2004).[1]

---

[1] Whenever we make reference to this book, henceforth we simply refer to THD (2004).

The atmosphere, some sort of an open system, is not in thermodynamic equilibrium since the temperature and the radiation field vary in space and in time. Nevertheless, in the troposphere and within the stratosphere the *emission* of thermal radiation is still governed by Kirchhoff's law at the local temperature. The reason for this is that in these atmospheric regions the density of the air is sufficiently high so that the mean time between molecular collisions is much smaller than the mean lifetime of an excited state of a radiating molecule. Hence, equilibrium conditions exist between *vibrational* and *rotational* and the *translational energy* of the molecule. At levels higher than 50 km, the two time scales become comparable resulting in a sufficiently strong deviation from thermodynamic equilibrium so that Kirchhoff's law cannot be applied anymore.

The breakdown of thermodynamic equilibrium in higher regions of the atmosphere also implies that Planck's law no longer adequately describes the *thermal emission* so that quantum theoretical arguments must be introduced to describe radiative transfer. Quantum theoretical considerations of this type will not be treated in this book. For a study of this situation we refer the reader to the textbook *Atmospheric Radiation* by Goody and Yung (1989).

The units usually employed to measure the wavelength of radiation are the micrometer (μm) with $1\,\mu\text{m} = 10^{-6}$ m or the nanometer (nm) with $1\,\text{nm} = 10^{-9}$ m and occasionally Ångströms (Å) where $1\,\text{Å} = 10^{-10}$ m. The thermal radiation spectrum of the Sun, also called the *solar radiation spectrum*, stretches from roughly 0.2–3.5 μm where practically all the thermal energy of the solar radiation is located. It consists of ultraviolet radiation ($< 0.4\,\mu$m), visible radiation (0.4–0.76 μm), and infrared radiation $> 0.76\,\mu$m. The thermal radiation spectrum of the Earth ranges from about 3.5–100 μm so that for all practical purposes the solar and the terrestrial radiation spectrum are separated. As will be seen later, this feature is of great importance facilitating the calculation of atmospheric radiative transfer. Due to the positions of the spectral regions of the solar and the terrestrial radiation we speak of *short-wave* and *long-wave radiation*. The terrestrial radiation spectrum is also called the *infrared radiation spectrum*.

Important applications of atmospheric radiative transfer are climate modeling and weather prediction which require the evaluation of a prognostic temperature equation. One important term in this equation, see e.g. Chapter 3 of THD (2004), is the divergence of the *net radiative flux density* whose evaluation is fairly involved, even for conditions of local thermodynamic equilibrium. Accurate numerical radiative transfer algorithms exist that can be used to evaluate the radiation part of the temperature prediction equation. In order to judiciously apply any such computer model, some detailed knowledge of radiative transfer is required.

There are other areas of application of radiative transfer such as remote sensing. In the concluding chapter of this textbook we will present various examples.

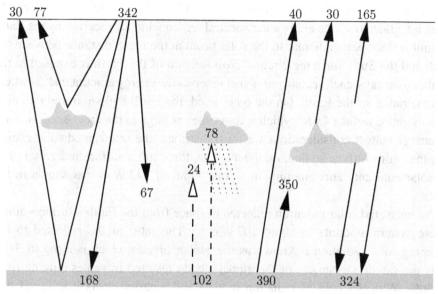

30   77   342   40   30   165

78

24

67

350

168   102   390   324

Fig. 1.1 The Earth's annual global mean energy budget, after Kiehl and Trenberth (1997), see also Houghton *et al.* (1996). Units are (W m^{-2}).

1.2 The mean global radiation budget of the Earth

Owing to the advanced satellite observational techniques now at our disposal, we are able to study with some confidence the Earth's annual mean global energy budget. Early meteorologists and climatologists have already understood the importance of this topic, but they did not have the observational basis to verify their results. A summary of pre-satellite investigations is given by Hunt *et al.* (1986). In the following we wish to briefly summarize the mean global radiation budget of the Earth according to Kiehl and Trenberth (1997). Here we have an instructive example showing in which way radiative transfer models can be applied to interpret observations.

The evaluation of the radiation model requires vertical distributions of absorbing gases, clouds, temperature, and pressure. For the major absorbing gases, namely water vapor and ozone, numerous observational data must be handled and supplemented with model atmospheres. In order to calculate the important influence of CO_2 on the infrared radiation budget, Kiehl and Trenberth specify a constant volume mixing ratio of about 350 ppmv. Moreover, it is necessary to employ distributions of the less important absorbing gases CH_4, N_2O, and of other trace gases. Using the best data presently available, they have provided the radiation budget as displayed in Figure 1.1.

The analysis employs a *solar constant* $S_0 = 1368\,\mathrm{W\,m^{-2}}$. This is the solar radiation, integrated over the entire solar spectral region which is received by the Earth per unit surface perpendicular to the solar beam at the mean distance between the Earth and the Sun. Since the circular cross-section of the Earth is exposed to the parallel solar rays, each second our planet receives the energy amount $\pi R^2 S_0$ where R is the radius of the Earth. On the other hand, the Earth emits infrared radiation from its entire surface $4\pi R^2$ which is four times as large as the cross-section. Thus for energy budget considerations we must distribute the intercepted solar energy over the entire surface so that, on the average, the Earth's surface receives $1/4$ of the solar constant. This amounts to a solar input of $342\,\mathrm{W\,m^{-2}}$ as shown in the figure.

The measured solar radiation reflected to space from the Earth's surface–atmosphere system amounts to about $107\,\mathrm{W\,m^{-2}}$. The ratio of the reflected to the incoming solar radiation is known as the *global albedo* which is close to 31%. Early pre-satellite estimates of the global albedo resulted in values ranging from 40–50%. With the help of radiation models and measurements it is found that cloud reflection and scattering by atmospheric molecules and aerosol particles contribute $77\,\mathrm{W\,m^{-2}}$ while ground reflection contributes $30\,\mathrm{W\,m^{-2}}$. In order to have a balanced radiation budget at the top of the atmosphere, the net gain $342 - 107 = 235\,\mathrm{W\,m^{-2}}$ of the short-wave solar radiation must be balanced by emission of long-wave radiation to space. Indeed, this is verified by satellite measurements of the outgoing long-wave radiation.

Let us now briefly consider the radiation budget at the surface of the Earth, which can be determined only with the help of radiation models since sufficiently dense surface measurements are not available. Assuming that the ground emits black body radiation at the temperature of $15\,^\circ\mathrm{C}$, an amount of $390\,\mathrm{W\,m^{-2}}$ is lost by the ground. According to Figure 1.1 this energy loss is partly compensated by a short-wave gain of $168\,\mathrm{W\,m^{-2}}$ and by a long-wave gain of $324\,\mathrm{W\,m^{-2}}$ because of the thermal emission of the atmospheric *greenhouse gases* (H_2O, CO_2, O_3, CH_4, etc.) and clouds. Thus the total energy gain $168 + 324 = 492\,\mathrm{W\,m^{-2}}$ exceeds the long-wave loss of $390\,\mathrm{W\,m^{-2}}$ by $102\,\mathrm{W\,m^{-2}}$.

In order to have a balanced energy budget at the Earth's surface, other physical processes must be active since a continuous energy gain would result in an ever increasing temperature of the Earth's surface. From observations, Kiehl and Trenberth estimated a mean global precipitation rate of $2.69\,\mathrm{mm\,day^{-1}}$ enabling them to compute a surface energy loss due to evapotranspiration. Multiplying $2.69\,\mathrm{mm\,day^{-1}}$ by the density of water and by the latent heat of vaporization amounts to a latent heat flux density of $78\,\mathrm{W\,m^{-2}}$. Thus the surface budget is still unbalanced by $24\,\mathrm{W\,m^{-2}}$. Assigning a surface energy loss of $-24\,\mathrm{W\,m^{-2}}$ resulting from sensible heat fluxes yields a balanced energy budget at the Earth's surface. The individual losses due to turbulent surface fluxes are uncertain within several percent since it is very difficult to accurately assess the global amount of precipitation which implies that the estimated sensible heat flux density is also quite uncertain. Only the sum of the turbulent surface flux densities is reasonably certain.

Finally, we must study the budget of the atmosphere itself. Figure 1.1 reveals that the atmosphere gains 67 W m^{-2} by absorption of solar radiation, 102 W m^{-2} by turbulent surface fluxes, and additionally 350 W m^{-2} resulting from long-wave radiation emitted by the surface of the Earth and intercepted by atmospheric greenhouse gases and clouds. The total of 519 W m^{-2} must be re-emitted by the atmosphere. As shown in the figure, the atmospheric greenhouse gases and the clouds emit $165 + 30 = 195$ W m^{-2} to space and 324 W m^{-2} as back-radiation to the surface of the Earth just balancing the atmospheric energy gain.

We also see that from the 390 W m^{-2} emitted by the Earth's surface only 350 W m^{-2} are intercepted by the atmosphere. To account for the remaining 40 W m^{-2} we recognize that these escape more or less unimpeded to space in the so-called *spectral window region* as will be discussed later.

By considering the budget in Figure 1.1, we observe that only the reflected solar radiation and the long-wave radiation emitted to space are actually verified by measurements. However, the remaining budget components should also be taken seriously since nowadays radiation models are quite accurate. Nevertheless, the output of the models cannot be any more accurate than the input data. In future days further refinements and improvements of the global energy budget can be expected.

In order to calculate the global radiation budget, we must have some detailed information on the absorption behavior of atmospheric trace gases and the physical properties of aerosol and cloud particles. In a later chapter we will study the radiative characteristics of spherical particles by means of the solution of *Maxwell's equations* of electromagnetic theory. Here we will only qualitatively present the absorption spectrum of the most important greenhouse gases.

Figure 1.2 combines some important information regarding the solar spectrum. The upper curve labeled TOA (top of the atmosphere) shows the extraterrestrial incoming solar radiation after Coulson (1975). For wavelengths exceeding 1.4 μm this curve coincides closely with a Planckian black body curve of 6000 K. The lower curve depicts the total solar radiation reaching the Earth's surface for a solar zenith angle $\theta_0 = 60°$. The calculations were carried out with sufficiently high spectral resolution using the so-called Moderate Resolution Atmospheric Radiance and Transmittance Model (MODTRAN; version 3.5; Anderson, 1996; Kneizys *et al.*, 1996) program package. All relevant absorbing trace gases shown in the figure are included in the calculations. Not shown are the positions of the CO and CH$_4$ absorption bands which are located in the solar spectrum and in the near infared spectral region of

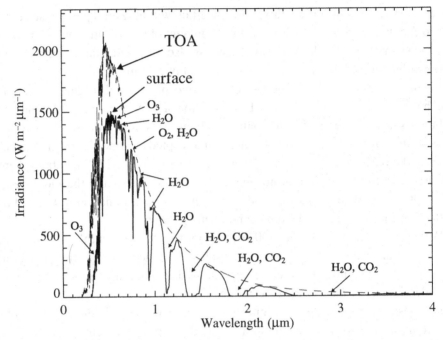

Fig. 1.2 Incoming solar flux density at the top of the atmosphere (TOA) and at ground level. The solar zenith angle is $\theta_0 = 60°$, ground albedo $A_g = 0$. The spectral positions of major absorption bands of the trace gases are shown.

thermal radiation. A tabulation of bands of these two trace gases is given, for example, in Goody (1964a). Since the radiation curve for ground level shows a high spectral variability, it was artificially smoothed for better display to a somewhat lower spectral resolution.

Figure 1.3 depicts the spectral distribution of the upwelling thermal radiance as a function of the wave number (to be defined later) at a height of 60 km. For comparison purposes the Planck black body radiance curves for several temperatures are shown also. The maximum of the 300 K black body curve is located at roughly $600 \, \text{cm}^{-1}$. The calculations were carried out with the same program package (MODTRAN) using a spectral resolution of $1 \, \text{cm}^{-1}$. All relevant absorbing and emitting gases have been accounted for. The widths of the major infrared absorption bands (H_2O, CO_2, O_3) are also shown in the figure.

Kiehl and Trenberth (1997) produced similar curves for the solar and infrared radiative fluxes per unit surface. However, in addition to the absorption by gases shown in Figures 1.2 and 1.3, they also included the effects of clouds in their calculations by assuming an effective droplet radius of $10 \, \mu\text{m}$ and suitable liquid water contents. Moreover, assumptions were made about the spatial distributions of clouds. Their results indicate that water vapor is the most important

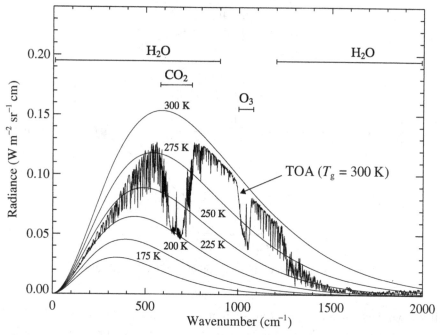

Fig. 1.3 Upwelling infrared radiance at a height of 60 km for a clear sky mid-latitude summer atmosphere.

gas absorbing 38 W m^{-2} of solar radiation which is followed by O_3 (15 W m^{-2}) and O_2 (2 W m^{-2}) while the effect of CO_2 may be ignored. Thus the greenhouse gases absorb 55 W m^{-2}. Figure 1.1, however, requires 67 W m^{-2}. The 12 W m^{-2} still missing must be attributed to partial cloudiness and to spectral overlap effects, i.e., cloud droplets and gases absorb at the same wavelength. Handling clouds in the radiative transfer problem is usually very difficult since in general water droplet size distributions are unknown.

Finally, let us consider the gaseous absorption bands of the infrared spectrum. In the calculations of Kiehl and Trenberth (1997) analogous to Figure 1.3, the surface is assumed to emit black body radiation with a temperature of 15°C. The major absorbing gases are H_2O, O_3, and CO_2. Of course, the same distribution of absorbing gases and clouds as for solar radiation is assumed. Integration of the infrared curve at the top of the atmosphere over the entire spectral region yields 235 W m^{-2} as required by Figure 1.1.

We conclude this section by considering a simple example to obtain the effective emission temperature of the system Earth's surface–atmosphere. As we have discussed above, the cross-section of the Earth intercepts the solar energy $\pi R^2 S_0$. Since the global albedo is 31%, the rate of absorption is $1368(1 - 0.31) = 944$ W m^{-2}. Assuming that the Earth emits black body radiation, we must apply the

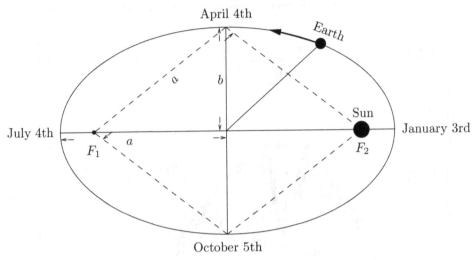

Fig. 1.4 Simplified elliptical geometry of the Earth's orbit.

well-known *Stefan–Boltzmann law* so that the Earth's surface emits $4\pi R^2 \sigma T^4$ where σ is the *Stefan–Boltzmann constant*. Assuming steady-state conditions, we have $\pi R^2 \times 944$ W m$^{-2} = 4\pi R^2 \sigma T^4$ from which we obtain the temperature $T = 254$ K which resembles the effective emission temperature of our planet.

1.3 Solar–terrestrial relations

To a high degree of accuracy the Earth's orbit around the Sun can be described by an ellipse with eccentricity $e = \sqrt{a^2 - b^2}/a = 0.01673$, where a and b are, respectively, the semi-major and semi-minor axis of the ellipse, see Figure 1.4. The Sun's position is located in one of the two elliptical foci (F_1, F_2). For demonstration purposes, the figure exaggerates the eccentricity of the elliptical orbit. The *perihelion*, that is the shortest distance r_{min} between Sun and Earth, occurs around January 3rd, while the *aphelion*, that is the largest distance r_{max} between Sun and Earth, is registered around July 4th. These times are not constant, but they vary from year to year. Often the mean distance between the Earth and the Sun is approximated by

$$a = \frac{r_{min} + r_{max}}{2} = 1.496 \times 10^8 \text{ km} \tag{1.1}$$

The distances r_{min} and r_{max} are related to a and e via

$$
\begin{aligned}
r_{min} &= a(1 - e) = 1.471 \times 10^8 \text{ km} \\
r_{max} &= a(1 + e) = 1.521 \times 10^8 \text{ km}
\end{aligned}
\tag{1.2}
$$

Beginning with January 1st, i.e. Julian day number 1 of the year, a normal year counts 365 days (for simplicity we will not take the occurrence of leap years into account). A particular day of the year is then labelled with its corresponding Julian day number J.

We introduce the rotation angle Γ of the Earth beginning with the 1st of January as

$$\Gamma = \frac{2\pi}{365}(J - 1) \tag{1.3}$$

where Γ is expressed in radians.

During the course of the year the angular distance Sun–Earth, the *solar declination* δ, and the so-called *equation of time ET* change in a more or less harmonic manner. In the following we will discuss simple expressions developed by Spencer (1971) which are accurate enough to evaluate the quantities $(a/r)^2$, δ, and ET, where r is the actual distance between Sun and Earth. The term $(a/r)^2$ is given by

$$\left(\frac{a}{r}\right)^2 = 1.000110 + 0.034221 \cos \Gamma + 0.001280 \sin \Gamma$$
$$+ 0.000719 \cos 2\Gamma + 0.000077 \sin 2\Gamma \tag{1.4}$$

with a maximum error of approximately 10^{-4}. If $S_0 = 1368 \, \text{W m}^{-2}$ is the *solar constant* for the mean distance between Sun and Earth, the actual solar constant varies as a function of J

$$S_0(J) = S_0 \left(\frac{a}{r(J)}\right)^2 \tag{1.5}$$

According to (1.4) the maximum change of $S_0(J)$ relative to S_0 has an amplitude of approximately 3.3%.

The solar declination δ is defined as the angle between the Earth's equatorial plane and the actual position of the Sun as seen from the center of the Earth. The Earth's rotational axis and the normal to the Earth's plane of the ecliptic make on average an angle of $\varepsilon = 23°27'$, δ amounts to $+23°27'$ and $-23°27'$ at summer solstice (around June 21st) and winter solstice (around December 22nd), respectively. These relations are illustrated in Figure 1.5 and in the three-dimensional view of the Sun–Earth geometry of Figure 1.6.

The *equinox points* are defined as the intersecting line (equinox line) between the Earth's plane of the ecliptic and the Sun's equatorial plane. A second line which is normal to the equinox line and which is located in the Earth's plane of the ecliptic intersects the Earth's orbit in the points WS (winter solstice) and SS (summer solstice). The perihelion P and the aphelion A, which both lie on the semi-major axis of the Earth's elliptical orbit, make an angle $\psi = 11°08'$ with the solstice line.

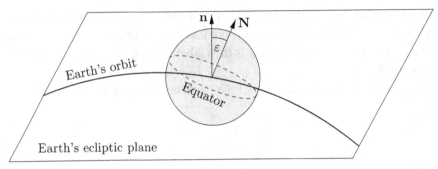

Fig. 1.5 Relation between the Earth's orbit, the normal vector **n** to the plane of the ecliptic, the Earth's rotational vector **N** and the angle of the ecliptic ε.

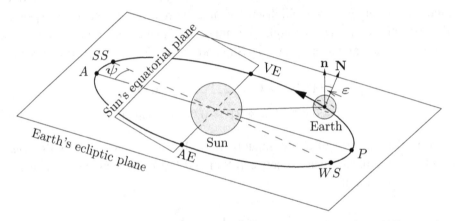

Fig. 1.6 Schematical view of the Sun–Earth geometry. P, perihelion; VE, vernal equinox; SS, summer solstice; A, aphelion; AE, autumnal equinox; WS, winter solstice; ε, angle of the ecliptic; ψ, angle between the distances (SS, WS) and (A, P); **N**, vector along the rotational axis of the Earth; **n**, normal unit vector with respect to the Earth's plane of the ecliptic.

It should be observed that the vector **N** is fixed in direction pointing to the polar star. At the solstices the vectors **N**, **n** and the line between the solstice points lie in the same plane so that $\delta = \pm 23°27'$. At the equinox points $(\delta = 0°)$ the line between the Earth and the Sun is at a right angle to the line (SS, WS).

The solar declination δ is a function of the Julian day number J. It can be expressed as

$$\delta = 0.006918 - 0.399912 \cos \Gamma + 0.070257 \sin \Gamma$$
$$- 0.006758 \cos 2\Gamma + 0.000907 \sin 2\Gamma \tag{1.6}$$

with δ expressed in radians. Due to Spencer (1971) this approximate formula has an error in δ less than $12'$. Figure 1.7 depicts a plot of δ versus J.

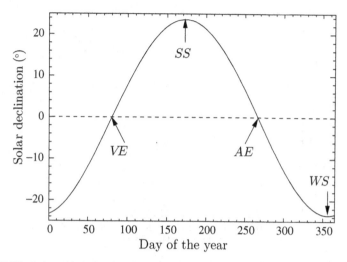

Fig. 1.7 Variation of the solar declination δ as a function of the Julian day J, see (1.6). VE, vernal equinox; SS, summer solstice; AE, autumnal equinox; WS, winter solstice.

1.3.1 The equation of time

In the following we assume that the period of the rotation of the Earth around the North Pole is constant. The time interval between two successive passages of a fixed star as seen from the local meridian of an observer on the Earth's surface is called a *sidereal day*. Due to the fact that the Earth moves around the Sun in an elliptical orbit, the time interval between two successive passages of the Sun in the local meridian, i.e. the so-called *solar day*, is about 4 min longer than the length of the sidereal day.

For a practical definition of time, one introduces the so-called *mean solar day* which is exactly divided into 24-h periods. Thus the local noon with respect to the local mean time (LMT) is defined by the passage of a mean fictitious Sun as registered from the Earth observer's local meridian. Clearly, depending on the Julian day J the real Sun appears somewhat earlier or later in the local meridian than the fictitious Sun. The time difference between the noon of the *true solar time* (TST) and the noon of the *local mean time* (LMT) is the so-called *equation of time ET*

$$\boxed{ET = TST - LMT} \tag{1.7}$$

Following the analysis of Spencer (1971), a functional fit expression can be derived for ET in the form

$$ET = \frac{1440}{2\pi}(0.000075 + 0.001868 \cos \Gamma - 0.032077 \sin \Gamma$$
$$- 0.014615 \cos 2\Gamma - 0.040849 \sin 2\Gamma) \tag{1.8}$$

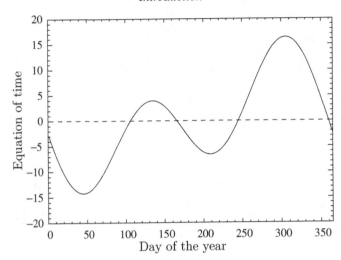

Fig. 1.8 Variation of the equation of time ET (in minutes) during the course of the year as given by (1.8).

where ET is expressed in minutes and 1440 is the number of minutes per day. The accuracy of this approximation is better than 35 s. The maximum time difference between TST and LMT amounts to less than about ± 15 min. Figure 1.8 depicts the variation of ET during the course of the year. Note that the irregularities of the Earth's orbit around the Sun lead to a complicated shape of the functional form of ET versus J.

Universal time UT, or *Greenwich mean time* GMT, is defined as the LMT at Greenwich's (UK) meridian at $0°$ in longitude. Since 24 h cover an entire rotation of the Earth, LMT increases by exactly 1 h per $15°$ in eastern longitude, i.e. 4 min per degree of eastern longitude. Similarly, LMT decreases by 4 min if one moves by one degree of longitude in the western direction. For the true solar time we thus obtain the relation

$$TST = UT + 4\lambda + ET \tag{1.9}$$

where TST, UT, and ET are given in minutes and the longitude λ is in units of degree ($-180° < \lambda \le 180°$).

The *hour angle of the Sun* H is defined as the angle between the local observer's meridian and the solar meridian, see Figure 1.9. If H is expressed in degrees longitude one obtains

$$H = 15(12 - TST) \tag{1.10}$$

where TST has to be inserted in hours. Note that $H > 0$ in the morning and $H < 0$ in the afternoon.

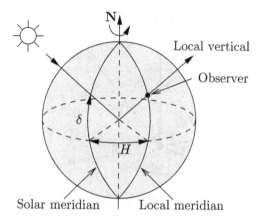

Fig. 1.9 Relation between hour angle H, solar declination δ, the solar and the local meridian.

The *local standard time LST* is defined as the local mean time for a given meridian being a multiple of 15° away from the Greenwich meridian (0°). Therefore, LST and UT differ by an integral number of hours. For particular countries, differences of 30 and 45 min relative to the standard time meridians have been introduced for convenience. Note also that for locations with daylight saving time, the local mean time differs by 1 to 2 h relative to LST.

1.3.2 Geographical coordinates and the solar position

A particular point P on the Earth's surface is identified by the pair of *geographical coordinates* (λ, ϕ), where λ is the longitude and ϕ is the latitude. Note that ϕ is counted positive in the northern hemisphere and negative in the southern hemisphere. The coordinates of the Sun relative to P are defined by the *solar zenith angle* ϑ_0 and the solar *azimuth angle* φ_0. If the Sun is at the zenith we have $\vartheta_0 = 180°$, and $\vartheta_0 = 90°$ if it is at the horizon, see Figure 1.10. The *solar height h* is given by $h = \vartheta_0 - \pi/2$. The solar azimuth φ_0 is defined as the angle between the solar vertical plane and a vertical plane of reference which is aligned with the north–south direction. Here, $\varphi_0 = 0°$ if the Sun is exactly over the southern direction and φ_0 is counted positive in the eastward direction. Figure 1.11 depicts the apparent track of the Sun during the day.

The position angles (ϑ_0, φ_0) of the Sun are usually not measured directly and must be determined from other known angles. Utilizing the laws of spherical trigonometry it can be shown that the following relations are valid

$$
\begin{aligned}
\text{(a)} \quad & \cos(\pi - \vartheta_0) = \sin\varphi \sin\delta + \cos\varphi \cos\delta \cos H \\
\text{(b)} \quad & \cos\varphi_0 = \frac{\cos(\pi - \vartheta_0)\sin\varphi - \sin\delta}{\sin(\pi - \vartheta_0)\cos\varphi}
\end{aligned}
\tag{1.11}
$$

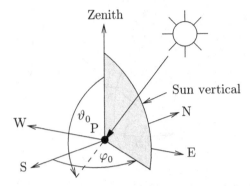

Fig. 1.10 Coordinates defining the position of the Sun.

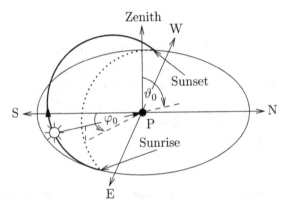

Fig. 1.11 Apparent solar track during the course of the day. The dotted curve marks
the projection of the solar path onto the horizontal plane.

At solar noon at any latitude we have $H = 0$. In this case we obtain from (1.11a)
$(\pi - \vartheta_0) = \varphi - \delta$. At sunrise or sunset at any latitude $\vartheta_0 = 90°$ and $H = D_h$. The
term D_h is also called the *half-day length* since it is half the time interval between
sunrise and sunset. Excepting the poles we find from (1.11a)

$$\cos D_h = -\tan \varphi \tan \delta \tag{1.12}$$

At the equator on all days and at the equinoxes ($\delta = 0$) at all latitudes (with
$\varphi \neq \pm 90°$) we find $D_h = 90°$ or 6 h. The latitude of the *polar night* is found by
setting in (1.12) $D_h = 0$ so that $\tan \varphi = -\cot \delta$ (with $\delta \neq 0$) and φ(polar night) $=$
$90° - |\delta|$ in the winter hemisphere.

The daily total solar radiation Q_s incident on a horizontal surface at the top of
the atmosphere is found by integrating the incoming solar radiation over the length

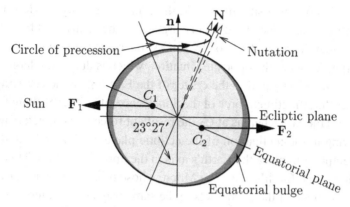

Fig. 1.12 Precession and nutation of the Earth.

of the day. Thus from (1.5) we find

$$Q_s = S_0 \left(\frac{a}{r(J)} \right)^2 \int_{-D_h}^{D_h} \cos(\pi - \vartheta_0) dt \qquad (1.13)$$

Since the angular velocity of the Earth can be written as $\Omega = dH/dt = 2\pi$ day^{-1} we obtain from (1.11a) after some simple integration

$$Q_s = S_0 \left(\frac{a}{r(J)} \right)^2 \frac{86400}{\pi} (D_h \sin \varphi \sin \delta + \cos \varphi \cos \delta \sin D_h) \; J\,m^{-2}day^{-1}$$

$$(1.14)$$

In the first term D_h must be expressed in radians. The expression $(a/r(J))^2$ never departs by more than about 3% from unity. Graphical representations of this formula are given in various texts, for example in Sellers (1965) where additional details may be found.

1.3.3 Long-term variations of the Earth's orbital parameters

For completeness we briefly discuss the most important variations of the Earth's orbit around the Sun. The *eccentricity e* of the Earth's elliptical orbit varies irregularly between 0 and 0.05 with its current value $e = 0.01673$. The period of this oscillation is approximately 100 000 years. The Earth's rotational axis N precesses around the normal of the ecliptic plane n with an angle of 23°27'. The reason for the precession of the Earth is that it is not an ideal sphere, but it has the shape of a geoid, that is, the poles are flattened and an *equatorial bulge* of about 21 km is observed, see Figure 1.12. First we investigate the influence of the Sun on the geoidal form of the Earth. At the center of the Earth the gravitational attraction of the Sun and the centrifugal force due to the revolving motion of the Earth around the Sun are equal but opposite in sign. At the center of gravity C_1 (left half of the geoid) the attractional force of the Sun is larger than the centrifugal force, which is due to the smaller distance of C_1 to the Sun. At the center of gravity C_2 (right half of the geoid) we observe the opposite situation, which is due to the larger distance of C_2 to the Sun as compared to the center of the Earth; here the centrifugal force preponderates the attractional force of the Sun. Hence, at C_1 the resultant force $\mathbf{F}_1$ is directed toward the Sun whereas at C_2 the resultant force $\mathbf{F}_2$ is directed away from the Sun. Owing to the inclination of the ecliptic plane the forces $\mathbf{F}_1$ and $\mathbf{F}_2$ form a couple attempting to place the Earth's axis in the upright position. This results in the precession of the Earth's axis. The Moon, whose orbital plane nearly coincides with the orbital plane of the Earth, acts in the same way but even more effectively. Here, the small mass of the Moon in comparison with the mass of the Sun is over-compensated by the small distance between Moon and Earth. As a result of these forces, $\mathbf{N}$ revolves on the mantle of a cone as shown in Figure 1.12. The time for a full rotation around the *circle of precession* amounts to about 25 780 years.

Apart from the Sun and the Moon the other planets of the solar system also exert an influence on the inclination of the ecliptic leading to changes in ε between $21°55'$ and $24°18'$ having a period of about 40 000 years. Finally, in addition to the precession, the Earth's rotational axis exhibits also a nodding motion. This effect is caused by the fact that the Moon's gravitational influence varies in time. This *nutation* leads to a small variation of the Earth's axis inclination and has a period of about 18.6 years.

1.4 Basic definitions of radiative quantities

In this section we will present some basic definitions and the terminologies used in this book. The *photon* is considered to be an idealized infinitesimally small particle with zero rest mass carrying the energy

$$e(\nu) = h\nu \tag{1.15}$$

where $h = 6.626196 \times 10^{-34}$ J s is *Planck's constant*, and ν is the frequency of the electromagnetic radiation. Frequency units are Hertz (Hz) where 1 Hz is 1 cycle s^{-1}. Considering a single photon one may attribute to it a *momentum* $\mathbf{p}(\nu)$ with magnitude

$$p(\nu) = |\mathbf{p}(\nu)| = \frac{e(\nu)}{c} \tag{1.16}$$

where $c = 2.997925 \times 10^8$ m s^{-1} is the *vacuum speed of light*. Photons may travel in an arbitrary direction specified by the unit vector $\mathbf{\Omega}$. Therefore, the vectorial

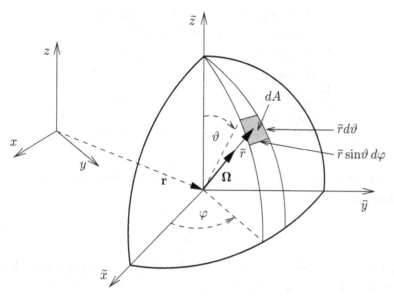

Fig. 1.13 Definition of the local spherical $(\tilde{r}, \vartheta, \varphi)$-coordinate system and the direction Ω.

notation of the photon's momentum can be expressed as

$$\mathbf{p}(\nu) = \frac{e(\nu)}{c}\Omega = \frac{h\nu}{c}\Omega \qquad (1.17)$$

As soon as the photon interferes with matter, various types of interactions between the atoms of the material and the photon may occur. A single interaction may be an *absorption* or a *scattering process*. Between any two scattering interactions the photon is assumed to travel in a straight line with the speed of light c. We will also assume that during a scattering process the photon suffers no change in frequency. In this case one speaks of *elastic scattering*.

In some situations *inelastic scattering* might be of importance where in addition to the change of flight direction a shift in the photon's frequency occurs. One important example for atmospheric applications is *Raman scattering*. *Rayleigh scattering* and *Mie scattering* to be discussed later are examples of elastic scattering processes. Inelastic scattering processes will not be investigated in this book.

Six coordinates are required to unambiguously describe the photon at time t. These are the three coordinates of the position vector $\mathbf{r}$, the magnitude of momentum $p(\nu)$ and two angles characterizing the direction of flight Ω. At a certain point in space a local system of Cartesian coordinates $(\tilde{x}, \tilde{y}, \tilde{z})$ is introduced. At the origin of this system we define a spherical coordinate system $\tilde{r}, \vartheta, \varphi$, where $\tilde{r}$ is the radial distance from the origin located at $\mathbf{r}$ and (ϑ, φ) are the zenith and azimuthal angle, respectively, see Figure 1.13. In the latter system the direction Ω may be described

Table 1.1 *Definition of special radiance fields*

Radiance field	List of variables
Stationary	$I_\nu = I_\nu(\mathbf{r}, \mathbf{\Omega})$
Isotropic	$I_\nu = I_\nu(\mathbf{r}, t)$
Homogeneous	$I_\nu = I_\nu(\mathbf{\Omega}, t)$
Homogeneous and isotropic	$I_\nu = I_\nu(t)$

by the triple set of coordinates ($\tilde{r} = 1, \vartheta, \varphi$). The differential *solid angle element* $d\Omega$ is defined by

$$d\Omega = \frac{dA}{\tilde{r}^2} \tag{1.18}$$

Here, $dA = \tilde{r}^2 \sin\vartheta \, d\vartheta \, d\varphi$ is the differential area element on a sphere with radius $\tilde{r}$, see Figure 1.13. Thus we obtain

$$\boxed{d\Omega = \sin\vartheta \, d\vartheta \, d\varphi} \tag{1.19}$$

Integration over the unit sphere yields

$$\int_{4\pi} d\Omega = \int_0^{2\pi} \int_0^{\pi} \sin\vartheta \, d\vartheta \, d\varphi = \int_0^{2\pi} \int_{-1}^{1} d\mu \, d\varphi = 4\pi \tag{1.20}$$

where the abbreviation $\mu = \cos\vartheta$ has been introduced.

The *distribution function of photons* $f(\nu, \mathbf{r}, \mathbf{\Omega}, t) = f_\nu(\mathbf{r}, \mathbf{\Omega}, t)$ is defined by[2]

$$N_\nu(\mathbf{r}, \mathbf{\Omega}, t)d\nu = f_\nu(\mathbf{r}, \mathbf{\Omega}, t) \, dV \, d\Omega \, d\nu \tag{1.21}$$

where $N_\nu d\nu$ represents the number of photons at time t contained within the volume element dV centered at $\mathbf{r}$, within the solid angle element $d\Omega$ about the flight direction $\mathbf{\Omega}$, and within the frequency interval $(\nu, \nu + d\nu)$. Therefore, f_ν has units of ($\text{m}^{-3} \, \text{sr}^{-1} \, \text{Hz}^{-1}$). In place of the photon distribution function f_ν, in radiative transfer theory it is customary to use the *radiance* $I_\nu(\mathbf{r}, \mathbf{\Omega}, t)$ as defined by

$$I_\nu(\mathbf{r}, \mathbf{\Omega}, t) = ch\nu f_\nu(\mathbf{r}, \mathbf{\Omega}, t) \tag{1.22}$$

From this equation it is seen that the monochromatic radiance is expressed in units of ($\text{W} \, \text{m}^{-2} \, \text{sr}^{-1} \, \text{Hz}^{-1}$). In the most general case the radiance field is time dependent, it varies in space, direction, and frequency. Table 1.1 briefly lists some special cases of I_ν.

---

[2] The dependence of radiative quantities on frequency is commonly denoted by the subscript ν.

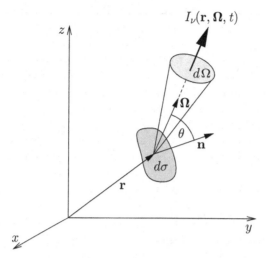

Fig. 1.14 Radiative energy streaming through the infinitesimal surface element $d\sigma$ with surface normal $\mathbf{n}$ into the solid angle element $d\Omega$ around the flight direction Ω of the photons.

The physical meaning of the radiance can be illustrated with the help of the energy relation

$$u_\nu(\mathbf{r}, \Omega, t)d\nu = I_\nu(\mathbf{r}, \Omega, t)\cos\theta \, d\Omega \, d\sigma \, dt \, d\nu \qquad (1.23)$$

Thus $u_\nu d\nu$ is the radiative energy contained within the frequency interval $(\nu, \nu + d\nu)$ streaming during dt at $\mathbf{r}$ through the surface element $d\sigma$ with unit surface normal $\mathbf{n}$ into the solid angle element $d\Omega$ along Ω. The angle between Ω and the surface normal of $d\sigma$ is denoted by θ, see Figure 1.14. Therefore, $I_\nu d\nu$ is expressed in $(\mathrm{W\,m^{-2}\,sr^{-1}})$.

The *energy density* $\hat{u}(\mathbf{r}, t)$ of the radiation field, expressed in units of $(\mathrm{J\,m^{-3}})$, is obtained by integrating the term $h\nu f_\nu$ over all directions and frequencies

$$\boxed{\hat{u}(\mathbf{r}, t) = \int_0^\infty \int_{4\pi} h\nu f_\nu(\mathbf{r}, \Omega, t)\, d\Omega \, d\nu = \frac{1}{c}\int_0^\infty \int_{4\pi} I_\nu(\mathbf{r}, \Omega, t)\, d\Omega \, d\nu} \qquad (1.24)$$

Let us now consider the important case that the radiance is described by the *Planck function* B_ν $(\mathrm{W\,m^{-2}\,sr^{-1}\,Hz^{-1}})$, which is also known as the *spectral black body radiance*. This special radiation field which is stationary, isotropic and homogeneous coexists with matter in *perfect thermodynamic equilibrium* at temperature T. The expression

$$B_\nu d\nu = \frac{2h\nu^3}{c^2}(e^{h\nu/kT} - 1)^{-1}d\nu \qquad (1.25)$$

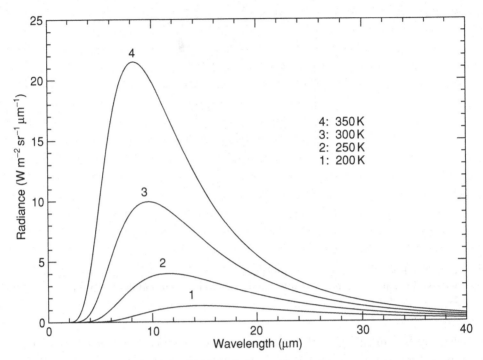

Fig. 1.15 Planckian black body curves for various temperatures.

represents the energy (unpolarized radiation) emitted by a black unit surface area per unit time interval within a cone of solid angle $\Omega_0 = 1$ sr vertical to the emitting surface in the frequency range between ν and $\nu + d\nu$.

Figure 1.15 depicts four Planck curves as function of the wavelength for the temperatures 200, 250, 300 and 350 K. It is clearly seen that with decreasing temperature the maxima of the curves are shifted towards larger wavelengths. This phenomenon is also known as *Wien's displacement law*. The Planck curve of a black body with temperature 6000 K (the Sun) has its maximum around 0.5 μm while for a black body with $T = 300$ K (the Earth) the maximum is found at 10 μm. For a further discussion of Wien's displacement law see also Problem 1.1.

The constant $k = 1.380662 \times 10^{-23}$ J K^{-1} appearing in (1.25) is known as the *Boltzmann constant*. The corresponding energy density follows from

$$\hat{u} = \frac{1}{c} \int_0^\infty \int_{4\pi} B_\nu d\Omega \, d\nu = \frac{4\pi}{c} \int_0^\infty \frac{2h\nu^3}{c^2} (e^{h\nu/kT} - 1)^{-1} d\nu \qquad (1.26)$$

The integral over frequency can be evaluated by substituting the new variable $x = h\nu/kT$ and developing the exponential term $(e^x - 1)^{-1}$ into a power series,

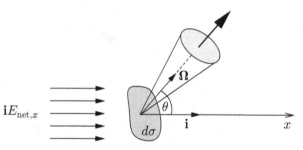

Fig. 1.16 Radiative energy streaming through the infinitesimal surface element $d\sigma$ in x-direction.

yielding

$$\hat{u} = \frac{8\pi k^4 T^4}{(hc)^3} \int_0^\infty x^3 (e^x - 1)^{-1} dx = \frac{48\pi (kT)^4}{(hc)^3} \sum_{n=1}^\infty \frac{1}{n^4} \tag{1.27}$$

Since

$$\sum_{n=1}^\infty \frac{1}{n^4} = \frac{\pi^4}{90} \tag{1.28}$$

the final result is

$$\boxed{\hat{u} = \frac{4}{c}\sigma T^4, \qquad \sigma = \frac{2\pi^5 k^4}{15 h^3 c^2} = 5.67032 \times 10^{-8}\ \text{W m}^{-2}\,\text{K}^{-4}} \tag{1.29}$$

where σ is the *Stefan–Boltzmann constant*. Equation (1.29) can also be derived from purely thermodynamic arguments as shown, for example, in THD (2004).

1.5 The net radiative flux density vector

Consider the special case that Ω is located in the (x, z)-plane and that the normal unit vector $\mathbf{n}$ of the surface element $d\sigma$ points in the x-direction of a Cartesian coordinate system, that is $\mathbf{n} = \mathbf{i}$. According to (1.23) for the spectral differential radiative energy crossing $d\sigma$ during dt we find the expression

$$E_{\text{net},x,\nu}(\mathbf{r}, \Omega, t)d\nu = \frac{u_\nu(\mathbf{r}, \Omega, t)d\nu}{d\sigma\, dt} = I_\nu(\mathbf{r}, \Omega, t) \cos\theta\, d\Omega\, d\nu$$
$$= I_\nu(\mathbf{r}, \Omega, t)\Omega_x\, d\Omega\, d\nu \tag{1.30}$$

where $\Omega_x = \Omega \cdot \mathbf{i} = \cos\theta = \sin\vartheta$ is the projection of Ω onto the x-axis, see Figures 1.16 and 1.17. Integrating this relation over the solid angle and over all

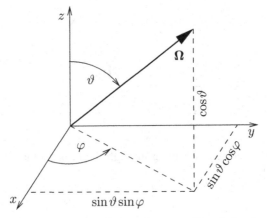

Fig. 1.17 Cartesian components of the unit vector $\mathbf{\Omega}$ pointing in the direction of photon travel.

frequencies yields the radiative energy streaming within unit time through the surface element in the x-direction

$$E_{\text{net},x}(\mathbf{r}, t) = \int_0^\infty \int_{4\pi} \Omega_x I_\nu(\mathbf{r}, \mathbf{\Omega}, t) d\Omega \, d\nu = \int_0^\infty \int_{4\pi} \Omega_x ch\nu f_\nu(\mathbf{r}, \mathbf{\Omega}, t) d\Omega \, d\nu$$
(1.31)

In the general case Ω_x will be a more complicated expression. If $(E_{\text{net},x}, E_{\text{net},y}, E_{\text{net},z})$ are the three components of the *net radiative flux density vector*, then $\mathbf{E}_{\text{net}}$ is given by

$$\mathbf{E}_{\text{net}}(\mathbf{r}, t) = \int_0^\infty \int_{4\pi} \mathbf{\Omega} I_\nu(\mathbf{r}, \mathbf{\Omega}, t) d\Omega \, d\nu = \mathbf{i} E_{\text{net},x}(\mathbf{r}, t) + \mathbf{j} E_{\text{net},y}(\mathbf{r}, t) + \mathbf{k} E_{\text{net},z}(\mathbf{r}, t)$$

(1.32)

where

$$E_{\text{net},y}(\mathbf{r}, t) = \int_0^\infty \int_{4\pi} \Omega_y I_\nu(\mathbf{r}, \mathbf{\Omega}, t) d\Omega \, d\nu$$

$$E_{\text{net},z}(\mathbf{r}, t) = \int_0^\infty \int_{4\pi} \Omega_z I_\nu(\mathbf{r}, \mathbf{\Omega}, t) d\Omega \, d\nu$$
(1.33)

We will now derive an explicit form of $\mathbf{E}_{\text{net}}$ in Cartesian coordinates. From (1.32) follows the definition of the *spectral net radiative flux density vector*

$$\mathbf{E}_{\text{net},\nu}(\mathbf{r}, t) = \int_{4\pi} \mathbf{\Omega} I_\nu(\mathbf{r}, \mathbf{\Omega}, t) d\Omega$$
(1.34)

Thus the component of $\mathbf{E}_{net,\nu}(\mathbf{r}, t)$ in the arbitrary direction $\mathbf{n}$ is given by

$$E_{net,n,\nu}(\mathbf{r}, t) = \mathbf{E}_{net,\nu}(\mathbf{r}, t) \cdot \mathbf{n} = \int_{4\pi} \boldsymbol{\Omega} \cdot \mathbf{n} I_\nu(\mathbf{r}, \boldsymbol{\Omega}, t) d\Omega = \int_{4\pi} \cos(\boldsymbol{\Omega}, \mathbf{n}) I_\nu(\mathbf{r}, \boldsymbol{\Omega}, t) d\Omega$$

$$(1.35)$$

To find the Cartesian components $(\Omega_x, \Omega_y, \Omega_z)$, of the unit vector $\boldsymbol{\Omega} = (1, \vartheta, \varphi)$ we perform the scalar multiplication with the Cartesian unit vectors $\mathbf{i}, \mathbf{j}$ and $\mathbf{k}$. From Figure 1.17 we find immediately

$$\begin{aligned}
\boldsymbol{\Omega} \cdot \mathbf{i} &= \Omega_x = \cos(\boldsymbol{\Omega}, \mathbf{i}) = \sin\vartheta \cos\varphi \\
\boldsymbol{\Omega} \cdot \mathbf{j} &= \Omega_y = \cos(\boldsymbol{\Omega}, \mathbf{j}) = \sin\vartheta \sin\varphi \\
\boldsymbol{\Omega} \cdot \mathbf{k} &= \Omega_z = \cos(\boldsymbol{\Omega}, \mathbf{k}) = \cos\vartheta = \mu
\end{aligned} \qquad (1.36)$$

Thus, from (1.36) the Cartesian components of $\mathbf{E}_{net,\nu}(\mathbf{r}, t)$ are finally given as

$$\begin{aligned}
&\text{(a)} \ E_{net,x,\nu}(\mathbf{r}, t) = \int_0^{2\pi} \int_{-1}^{1} I_\nu(\mathbf{r}, \mu, \varphi, t) \cos\varphi (1 - \mu^2)^{1/2} d\mu \, d\varphi \\
&\text{(b)} \ E_{net,y,\nu}(\mathbf{r}, t) = \int_0^{2\pi} \int_{-1}^{1} I_\nu(\mathbf{r}, \mu, \varphi, t) \sin\varphi (1 - \mu^2)^{1/2} d\mu \, d\varphi \\
&\text{(c)} \ E_{net,z,\nu}(\mathbf{r}, t) = \int_0^{2\pi} \int_{-1}^{1} I_\nu(\mathbf{r}, \mu, \varphi, t) \mu \, d\mu \, d\varphi
\end{aligned} \qquad (1.37)$$

with $(1 - \mu^2)^{1/2} = \sin\vartheta$ and $d\Omega = -d\mu d\varphi$.

It is straightforward to show that for an isotropic radiation field $\mathbf{E}_{net,\nu} = 0$. For example, evaluating in (1.37a) for $I_\nu = const$ the integral of the x-component yields

$$\int_0^{2\pi} \cos\varphi \int_{-1}^{1} (1 - \mu^2)^{1/2} d\mu \, d\varphi = \int_{4\pi} \Omega_x d\Omega = 0 \qquad (1.38)$$

Similarly we obtain for the integrals of the y- and z-component

$$\int_{4\pi} \Omega_y d\Omega = 0, \quad \int_{4\pi} \Omega_z d\Omega = 0 \qquad (1.39)$$

1.6 The interaction of radiation with matter

1.6.1 Absorption

If a photon travels through space filled with matter, a certain *absorption probability* can be defined. For the mathematical description of this process the *absorption coefficient* $k_{abs,\nu}(\mathbf{r}, t)$ with units (m^{-1}) is introduced. The dimensionless differential

$$\boxed{d\tau_{abs}(\mathbf{r}, t) = k_{abs,\nu}(\mathbf{r}, t) \, ds} \qquad (1.40)$$

is the so-called *differential optical depth for absorption* where ds is the geometrical distance travelled by the photon. Thus the differential $d\tau_{abs}(\mathbf{r}, t)$ is a measure for the probability that the photon is absorbed along ds so that the photon disappears. It is important to realize that (1.40) is valid only for *isotropic media*. In general, for *anisotropic media*, the absorption coefficient not only depends on position, frequency and time but also on the direction $\mathbf{\Omega}$. For all practical purposes, the atmosphere can be considered an isotropic medium.

Sometimes it is preferable to use the *mass absorption coefficient* $\kappa_{abs,\nu}(\mathbf{r}, t)$ which is defined by the relation

$$k_{abs,\nu}(\mathbf{r}, t) = \rho_{abs}(\mathbf{r}, t)\kappa_{abs,\nu}(\mathbf{r}, t) \qquad (1.41)$$

where $\rho_{abs}(\mathbf{r}, t)$ is the density of the absorbing medium, and $\kappa_{abs}(\mathbf{r}, t)$ has units of $(m^2\,kg^{-1})$.

1.6.2 Scattering

In a similar manner the photon may suffer an elastic scattering process after having travelled a certain distance ds. The occurrence of a scattering process does not mean that the photon disappears at the location of scattering, instead of that it changes its flight direction from $\mathbf{\Omega}'$ to $\mathbf{\Omega}$. Let us denote the *differential scattering coefficient* by $k_{sca,\nu}(\mathbf{r}, \mathbf{\Omega}' \to \mathbf{\Omega}, t)$. In analogy to (1.40), the *differential optical depth for scattering* is defined as

$$d\tau_{sca}(\mathbf{r}, \mathbf{\Omega}' \to \mathbf{\Omega}, t) = k_{sca,\nu}(\mathbf{r}, \mathbf{\Omega}' \to \mathbf{\Omega}, t)\,d\Omega\,ds \qquad (1.42)$$

This expression is a measure of the probability that a photon of frequency ν with initial direction $\mathbf{\Omega}'$, in traveling the distance ds, is scattered into $d\Omega$ having the new direction $\mathbf{\Omega}$. From (1.42) it is clear that $k_{sca,\nu}(\mathbf{r}, \mathbf{\Omega}' \to \mathbf{\Omega}, t)$ is expressed in units of $(m^{-1}\,sr^{-1})$. It is noteworthy that the differential scattering coefficient agrees with the *scattering function* $\tilde{P}(\cos\Theta)$ which will be introduced in a later chapter. The notation $\mathbf{\Omega}' \to \mathbf{\Omega}$ has been chosen since in general the differential scattering probability depends explicitly on both directions $\mathbf{\Omega}'$ and $\mathbf{\Omega}$. However, for applications involving homogeneous spherical particles (e.g. cloud droplets) it is obvious that the scattering process depends only on the cosine of the *scattering angle*

$$\cos\Theta = \mathbf{\Omega}' \cdot \mathbf{\Omega} \qquad (1.43)$$

This means that scattering is rotationally symmetric about the direction of incidence, see Figure 1.18. In case of randomly oriented inhomogeneous or non-spherical

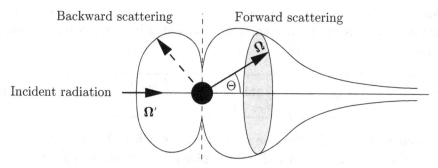

Backward scattering   Forward scattering

Incident radiation

Fig. 1.18 Illustration of the rotationally symmetric scattering phase function. $\mathcal{P}(\mathbf{r}, \mathbf{\Omega}' \to \mathbf{\Omega}) = const$ on the circle defined by all points with $\Theta = const$.

particles (e.g. ice particles or aerosol particles) it is often assumed that the scattering angle may be defined analogously.

If we integrate the differential scattering coefficient over all possible directions $\mathbf{\Omega}$, we obtain the ordinary scattering coefficient $k_{\text{sca},\nu}(\mathbf{r}, t)$

$$k_{\text{sca},\nu}(\mathbf{r}, t) = \int_{4\pi} k_{\text{sca},\nu}(\mathbf{r}, \mathbf{\Omega}' \to \mathbf{\Omega}, t) d\Omega \qquad (1.44)$$

This relation together with (1.42) can be used to define the *scattering phase function* or simply *phase function* $\mathcal{P}_\nu(\mathbf{r}, \mathbf{\Omega}' \to \mathbf{\Omega}, t)$

$$\boxed{k_{\text{sca},\nu}(\mathbf{r}, \mathbf{\Omega}' \to \mathbf{\Omega}, t) = \frac{1}{4\pi} k_{\text{sca},\nu}(\mathbf{r}, t) \mathcal{P}_\nu(\mathbf{r}, \mathbf{\Omega}' \to \mathbf{\Omega}, t)} \qquad (1.45)$$

The scattering phase function is a measure of the probability density distribution for a scattering process from the incident direction $\mathbf{\Omega}'$ into the direction $\mathbf{\Omega}$. The normalization of $\mathcal{P}$ is guaranteed since integrating (1.45) over the unit sphere, utilizing (1.44), yields

$$\frac{1}{4\pi} \int_{4\pi} \mathcal{P}_\nu(\mathbf{r}, \mathbf{\Omega}' \to \mathbf{\Omega}, t) d\Omega = 1 \qquad (1.46)$$

It is instructive to discuss a particularly simple form of scattering, namely an *isotropic scattering process*. In this case the phase function is simply given by

$$\mathcal{P}_\nu(\mathbf{r}, \mathbf{\Omega}' \to \mathbf{\Omega}, t) = 1 \qquad (1.47)$$

i.e. for each direction there is equal probability of scattering.

The sum of absorption and scattering is called extinction. The *extinction coefficient* is defined by

$$\boxed{k_{\text{ext},\nu}(\mathbf{r}, t) = k_{\text{abs},\nu}(\mathbf{r}, t) + k_{\text{sca},\nu}(\mathbf{r}, t)} \qquad (1.48)$$

Another important optical parameter is the *single scattering albedo* $\omega_{0,v}(\mathbf{r}, t)$ which is defined as the relative amount of scattering involved in the extinction process

$$\omega_{0,v}(\mathbf{r}, t) = \frac{k_{\text{sca},v}(\mathbf{r}, t)}{k_{\text{ext},v}(\mathbf{r}, t)} = 1 - \frac{k_{\text{abs},v}(\mathbf{r}, t)}{k_{\text{ext},v}(\mathbf{r}, t)} \qquad (1.49)$$

Of particular importance is the case $\omega_{0,v}(\mathbf{r}, t) = 1$ for which the scattering process is *conservative*. In a medium with conservative scattering no absorption of radiation occurs. Later it will be shown that a conservative plane-parallel medium is characterized by a vertically constant net radiative flux density.

1.6.3 Emission

Emission is a process that generates photons within the medium. In the long-wave spectral region photons are emitted and absorbed by atmospheric trace gases such as water vapor, carbon dioxide, ozone, by cloud and aerosol particles, and by the Earth's surface. As already mentioned previously, in case of local thermodynamic equilibrium these emission processes can be described by the Planckian function.

For the mathematical formulation of emission processes we introduce the so-called *emission coefficient* $j_v(\mathbf{r}, t)$ for isotropic radiation sources. This coefficient defines the number of photons emitted per unit time and unit volume within the frequency interval $(v, v + dv)$. The photons are contained in the solid angle element $d\Omega = \mathbf{\Omega} \, d\Omega$

$$\left. \frac{\partial}{\partial t} N_v(\mathbf{r}, t) \right|_{\text{em}} = j_v(\mathbf{r}, t) \, dV \, d\Omega \, dv \qquad (1.50)$$

The emission coefficient is expressed in units of $(\text{m}^{-3} \, \text{s}^{-1} \, \text{sr}^{-1} \, \text{Hz}^{-1})$.

1.7 Problems

1.1: With increasing temperature the maximum of the Planckian black body curve is shifted to shorter wavelengths. Observing that $dv = -cd\lambda/\lambda^2$, express (1.25) in terms of wavelength.

(a) Differentiate Planck's law with respect to wavelength and estimate the wavelength λ_{max} of maximum emission for a fixed temperature T. The resulting formula is known as Wien's displacement law.

(b) Find λ_{max} for the solar temperature $T = 6000$ K and for the terrestrial temperature $T = 300$ K.

1.2: Calculate for the two asymptotic situations

    (a) $v \ll 1$:   *Rayleigh–Jeans distribution*
    (b) $v \gg 1$:   *Wien distribution*

the resulting simplified radiation laws of Planck.

1.3: Integrate Planck's formula (1.25) over all frequencies and directions to find the hemispheric flux density $E_b = \sigma T^4$. This is known as the *Stefan–Boltzmann law*.

1.4: A black horizontal receiving element (radiometer) of unit area is located directly below the center of a circular cloud at height z having the temperature T_c. The cloud radius is R. Find an expression for the flux density E incident on the receiving element in terms of the Stefan–Boltzmann law, z and R. Assume that the cloud is a black body radiator whose radiance is $\sigma T^4/\pi$. Ignore any interactions of the radiation with the atmosphere.

    (a) Start your analysis using Lambert's law of photometry.
    (b) Rework the problem using equation (1.37c).

1.5: An idealized valley may be considered as the interior part of a spherical surface of radius a. The valley surface is assumed to radiate as a black body of temperature T.

    (a) Find an expression for the radiation received by a radiometer which is located at a distance $z > a$ above the lowest part of the valley. Ignore any interaction of the radiation with the atmosphere.
    (b) Repeat the calculation with the radiometer located below the center of curvature, that is $z < a$.

Hint: Use Lambert's law of photometry, see Problem 1.4.

1.6: A spherical emitter of radius a emits isotropically radiation into empty space.

    (a) Find the flux density $\mathbf{E}_r = E_r(\mathbf{r})\mathbf{e}_r$ at a distance $r \geq a$ from the center of the sphere. $\mathbf{e}_r$ is a unit vector along the radius.
    (b) From $\mathbf{E}_r$ obtain the power ϕ emitted by the sphere.
    (c) Find the energy density $\hat{u}(\mathbf{r})$.

1.7: For a monochromatic homogeneous plane parallel radiation field (solar radiation $S_{0,v}$) find the energy density $\hat{u}_v$ and the net flux density $\mathbf{E}_{net,v}$. Ignore any interaction of the radiation with the atmosphere.

2

The radiative transfer equation

2.1 Eulerian derivation of the radiative transfer equation

In the following section we will derive a *budget equation for photons* in a medium in which scattering, absorption and emission processes take place. The photon budget equation finally results in the so-called *radiative transfer equation* (RTE) which is a linear integro-differential equation for the radiance $I_\nu(\mathbf{r}, \mathbf{\Omega}, t)$. Let us consider a six-dimensional (6-D) volume element in $(x, y, z, \vartheta, \varphi, \nu)$-space with side lengths $(\Delta x, \Delta y, \Delta z, \Delta \vartheta, \Delta \varphi, \Delta \nu)$. This volume element is assumed to be fixed in time t. According to (1.21), at point $\mathbf{r}$ the total number of photons N_ν is given by

$$N_\nu(\mathbf{r}, \mathbf{\Omega}, t) = f_\nu(\mathbf{r}, \mathbf{\Omega}, t) \Delta V \Delta \Omega \Delta \nu \qquad (2.1)$$

where $\Delta V = \Delta x \Delta y \Delta z$ is the ordinary volume element in space. In order to simplify the notation, the dependence of different variables on $(\mathbf{r}, \mathbf{\Omega}, t)$ will henceforth be omitted except where confusion is likely to occur.

The derivation of the photon budget equation requires the knowledge of the local time rate of change of the number of photons leaving and entering the 6-D volume element

$$\frac{\partial N_\nu}{\partial t} = \frac{\partial f_\nu}{\partial t} \Delta V \Delta \Omega \Delta \nu \qquad (2.2)$$

This change consists of the following processes.

$\left.\dfrac{\partial N_\nu}{\partial t}\right|_{\text{exch}}$:   Exchange of photons of the considered volume element with the exterior surrounding.

$\left.\dfrac{\partial N_\nu}{\partial t}\right|_{\text{abs}}$:   Absorption of photons with frequency ν and direction $\mathbf{\Omega}$.

$\left.\dfrac{\partial N_\nu}{\partial t}\right|_{\text{outsc}}$:   Scattering of photons with frequency ν and direction $\mathbf{\Omega}$ into all other directions $\mathbf{\Omega}'$ (*outscattering*).

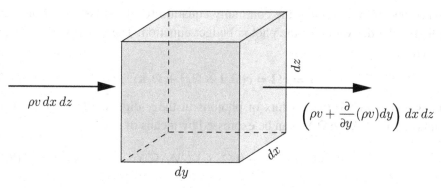

$$\rho v\, dx\, dz \qquad\qquad \left(\rho v + \frac{\partial}{\partial y}(\rho v)dy\right) dx\, dz$$

Fig. 2.1 Mass flux entering the left face and leaving the right face of a cube with volume $dx\, dy\, dz$.

$$\left.\frac{\partial N_\nu}{\partial t}\right|_{\text{insc}} :$$ Scattering of photons with frequency ν and arbitrary direction Ω' into the desired direction Ω (*inscattering*).

$$\left.\frac{\partial N_\nu}{\partial t}\right|_{\text{em}} :$$ Emission of photons with frequency ν in direction Ω.

The individual contributions of these processes to the photon budget equation will now be discussed in detail.

2.1.1 The exchange of photons

The exchange of photons can be treated in analogy to the *continuity equation for the mass* in fluid mechanics. We will consider a cube with side lengths (dx, dy, dz). The velocity of a fluid particle is given by $\mathbf{v} = u\mathbf{i} + v\mathbf{j} + w\mathbf{k}$, and ρ is the mass density of the medium. The volume dV is assumed to be fixed in space. The local time rate of change of the mass $dM = \rho\, dx\, dy\, dz$ of the cube is given by adding all mass fluxes through its surface. Figure 2.1 depicts the mass fluxes entering and leaving the infinitesimal volume element $dx\, dy\, dz$ through the vertical sides with area $dx\, dz$. Thus the net flux in y-direction is given by

$$F_{\rho,y} = -\left(\rho v + \frac{\partial}{\partial y}(\rho v)dy\right) dx\, dz + \rho v\, dx\, dz = -\frac{\partial}{\partial y}(\rho v)dx\, dy\, dz \quad (2.3)$$

In a similar manner we obtain the net fluxes $F_{\rho,x}$ and $F_{\rho,z}$ in the x- and z-direction. Adding up all three contributions and dividing by $dx\, dy\, dz$ yields the well-known continuity equation

$$\boxed{\frac{\partial \rho}{\partial t} = -\nabla\cdot(\rho\mathbf{v})} \quad (2.4)$$

The concept of obtaining the continuity equation for the mass M will now be applied to the derivation of the photon budget equation. The velocity of a photon is given by

$$\mathbf{v} = c\mathbf{\Omega} = c(\Omega_x\mathbf{i} + \Omega_y\mathbf{j} + \Omega_z\mathbf{k}) \tag{2.5}$$

Analogously to (2.3) the net flux of photons in the y-direction through the 6-D volume element $\Delta V \Delta \Omega \Delta \nu$ can be expressed by means of

$$F_{y,\nu} = -\frac{\partial}{\partial y}\left(c\Omega_y f_\nu\right)\Delta V \Delta \Omega \Delta \nu \tag{2.6}$$

Note that $F_{y,\nu}$ is expressed in units of (s^{-1}). Adding up the contributions of the three directions yields the time rate of change for the number of photons due to the exchange with the surroundings

$$\left.\frac{\partial N_\nu}{\partial t}\right|_{\text{exch}} = -\left(\frac{\partial}{\partial x}(\Omega_x f_\nu) + \frac{\partial}{\partial y}(\Omega_y f_\nu) + \frac{\partial}{\partial z}(\Omega_z f_\nu)\right)c\Delta V \Delta \Omega \Delta \nu \tag{2.7}$$

The exchange term has been derived under the assumption that the medium's *index of refraction* is constant in space and time. This assumption is sufficient for most atmospheric applications. Otherwise, the photon path will be subject to refraction leading to photon trajectories which are curved in space.

2.1.2 The absorption of photons

The absorption rate of photons within the 6-D volume element is given by the product of the photon number N_ν and the probability that a photon will be absorbed during the time interval $(t, t + dt)$. Dividing (1.40) by dt yields

$$\frac{d\tau_{\text{abs}}}{dt} = k_{\text{abs},\nu}\frac{ds}{dt} = k_{\text{abs},\nu}c \tag{2.8}$$

with $c = ds/dt$. Hence, for the total absorption rate of photons we obtain the expression

$$\left.\frac{\partial N_\nu}{\partial t}\right|_{\text{abs}} = N_\nu\frac{d\tau_{\text{abs}}}{dt} = f_\nu k_{\text{abs},\nu}c\Delta V \Delta \Omega \Delta \nu \tag{2.9}$$

2.1.3 The scattering of photons

Figure 2.2 illustrates the inscattering and outscattering process of photons. For inscattering the direction of the photons is indicated by solid arrows, outscattering is denoted by dashed arrows. It will be noticed that for the direction $\mathbf{\Omega}$ inscattering represents a gain of photons whereas outscattering results in a reduction of the number of photons in this particular direction.

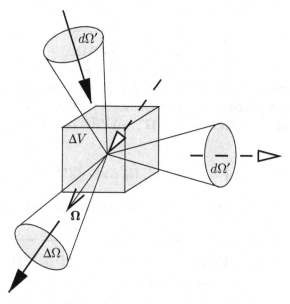

Fig. 2.2 Schematic view of the inscattering (solid arrows) and the outscattering (dashed arrows) processes.

Analogously to (1.42) we may express the probability that at time t a photon is scattered from $\Omega \to \Omega'$ by means of

$$d\tau_{\text{sca}}(\Omega \to \Omega') = k_{\text{sca},\nu}(\Omega \to \Omega')d\Omega'ds \qquad (2.10)$$

Dividing this equation by dt and multiplying by the number of photons N_ν yields the time rate of change of photons resulting from the outscattering process $\Omega \to \Omega'$

$$N_\nu \frac{d}{dt}\left[\tau_{\text{sca}}(\Omega \to \Omega')\right] = N_\nu k_{\text{sca},\nu}(\Omega \to \Omega')d\Omega'c \qquad (2.11)$$

The total loss of photons due to outscattering is obtained by integrating (2.11) over all directions Ω'

$$\frac{\partial N_\nu}{\partial t}\bigg|_{\text{outsc}} = N_\nu c \int_{4\pi} k_{\text{sca},\nu}(\Omega \to \Omega')d\Omega'$$

$$= f_\nu \Delta V \Delta \Omega \Delta \nu \frac{c}{4\pi} \int_{4\pi} k_{\text{sca},\nu} \mathcal{P}_\nu(\Omega \to \Omega')d\Omega' \qquad (2.12)$$

Here, use was made of (1.45) and (2.1). Utilizing the normalization condition for the *phase function*

$$\frac{1}{4\pi} \int_{4\pi} \mathcal{P}_\nu(\Omega \to \Omega')d\Omega' = 1 \qquad (2.13)$$

finally gives

$$\frac{\partial N_\nu}{\partial t}\bigg|_{\text{outsc}} = f_\nu k_{\text{sca},\nu} c \Delta V \Delta \Omega \Delta \nu \qquad (2.14)$$

In a similar manner we may find the gain of photons for the direction Ω due to inscattering from all directions Ω'. The number of photons moving in direction Ω', before inscattering takes place, is $f_\nu(\mathbf{r}, \Omega', t)\Delta V d\Omega' \Delta \nu$. In analogy to (2.11) we have

$$N_\nu \frac{d}{dt}\left[\tau_{\text{sca}}(\Omega' \to \Omega)\right] = f_\nu(\Omega')\,\Delta V d\Omega' \Delta \nu k_{\text{sca},\nu}(\Omega' \to \Omega)\Delta \Omega \, c \qquad (2.15)$$

Integrating over all directions Ω', using equation (1.45), we find for the inscattering rate

$$\frac{\partial N_\nu}{\partial t}\bigg|_{\text{insc}} = k_{\text{sca},\nu}\Delta V \Delta \Omega \Delta \nu \frac{c}{4\pi}\int_{4\pi} \mathcal{P}_\nu(\Omega' \to \Omega) f_\nu(\Omega')d\Omega' \qquad (2.16)$$

2.1.4 The emission rate

Finally, according to (1.50) the time rate of change of photons due to *emission* is given by

$$\frac{\partial N_\nu}{\partial t}\bigg|_{\text{em}} = j_\nu \, \Delta V \, \Delta \Omega \, \Delta \nu \qquad (2.17)$$

Now we have derived mathematical expressions for the five contributions for the photon budget equation as listed at the beginning of this section.

2.1.5 The budget equation of the photon distribution function

The budget equation for the photon distribution function f_ν is obtained by adding up the individual contributions. Considering absorption and outscattering processes as negative contributions, we may write

$$\frac{\partial N_\nu}{\partial t} = \frac{\partial N_\nu}{\partial t}\bigg|_{\text{exch}} - \frac{\partial N_\nu}{\partial t}\bigg|_{\text{abs}} - \frac{\partial N_\nu}{\partial t}\bigg|_{\text{outsc}} + \frac{\partial N_\nu}{\partial t}\bigg|_{\text{insc}} + \frac{\partial N_\nu}{\partial t}\bigg|_{\text{em}} \qquad (2.18)$$

Combination of (2.7), (2.9), (2.14), (2.16), and (2.17) gives

$$\frac{\partial f_\nu}{\partial t} = -c\left(\frac{\partial}{\partial x}(\Omega_x f_\nu) + \frac{\partial}{\partial y}(\Omega_y f_\nu) + \frac{\partial}{\partial z}(\Omega_z f_\nu)\right) - c f_\nu k_{\text{abs},\nu} - c f_\nu k_{\text{sca},\nu}$$

$$+ \frac{c}{4\pi}k_{\text{sca},\nu}\int_{4\pi} \mathcal{P}_\nu(\Omega' \to \Omega) f_\nu(\Omega')d\Omega' + j_\nu \qquad (2.19)$$

where the common factor $\Delta V \, \Delta \Omega \, \Delta \nu$ has been cancelled out.

Obviously, the unit vector $\mathbf{\Omega}$ is divergence-free, that is

$$\nabla \cdot \mathbf{\Omega} = \frac{\partial \Omega_x}{\partial x} + \frac{\partial \Omega_y}{\partial y} + \frac{\partial \Omega_z}{\partial z} = 0 \qquad (2.20)$$

Therefore, the streaming term on the right-hand side of (2.19) may further be simplified yielding the final form of the photon budget equation for a nonstationary situation

$$\frac{\partial f_\nu}{\partial t} = -c\mathbf{\Omega} \cdot \nabla f_\nu - cf_\nu k_{\text{ext},\nu} + \frac{c}{4\pi} k_{\text{sca},\nu} \int_{4\pi} \mathcal{P}_\nu(\mathbf{\Omega}' \to \mathbf{\Omega}) f_\nu(\mathbf{\Omega}') d\Omega' + j_\nu \qquad (2.21)$$

Here, the extinction coefficient $k_{\text{ext},\nu}$ as defined in (1.48) has been introduced.

Since we consider the spatial change of the photon distribution function along ds in $\mathbf{\Omega}$-direction, ∇f_ν may be expressed in terms of

$$\nabla f_\nu = \mathbf{\Omega} \frac{df_\nu}{ds} \qquad (2.22)$$

so that

$$\mathbf{\Omega} \cdot \nabla f_\nu = \frac{df_\nu}{ds} = \Omega_x \frac{\partial f_\nu}{\partial x} + \Omega_y \frac{\partial f_\nu}{\partial y} + \Omega_z \frac{\partial f_\nu}{\partial z} \qquad (2.23)$$

According to (1.36) the Cartesian components of $\mathbf{\Omega}$ are given by

$$\Omega_x = \mathbf{\Omega} \cdot \mathbf{i} = \sin \vartheta \cos \varphi, \qquad \Omega_y = \mathbf{\Omega} \cdot \mathbf{j} = \sin \vartheta \sin \varphi, \qquad \Omega_z = \mathbf{\Omega} \cdot \mathbf{k} = \cos \vartheta \qquad (2.24)$$

By introducing into (2.21) the radiance I_ν as defined in (1.22), one obtains the general *nonstationary form of the radiative transfer equation*

$$\boxed{\frac{1}{c} \frac{\partial I_\nu}{\partial t} + \mathbf{\Omega} \cdot \nabla I_\nu = -k_{\text{ext},\nu} I_\nu + \frac{k_{\text{sca},\nu}}{4\pi} \int_{4\pi} \mathcal{P}_\nu(\mathbf{\Omega}' \to \mathbf{\Omega}) I_\nu(\mathbf{\Omega}') d\Omega' + J_\nu^e} \qquad (2.25)$$

Here, the *source function for true emission*

$$\boxed{J_\nu^e(\mathbf{r}, t) = h\nu j_\nu(\mathbf{r}, t)} \qquad (2.26)$$

has been introduced. This function has units of $(\text{W m}^{-3}\,\text{sr}^{-1}\,\text{Hz}^{-1})$. Its relation to the *Planck function* will be described later.

For most atmospheric applications the term $1/c(\partial I/\partial t)$ in the RTE can be neglected in comparison to the remaining terms since the propagation speed c is very high. Thus (2.25) simplifies to

$$\mathbf{\Omega} \cdot \nabla I_\nu = -k_{\text{ext},\nu} I_\nu + \frac{k_{\text{sca},\nu}}{4\pi} \int_{4\pi} \mathcal{P}_\nu(\mathbf{\Omega}' \to \mathbf{\Omega}) I_\nu(\mathbf{\Omega}') d\Omega' + J_\nu^e \qquad (2.27)$$

In the following we will assume that scattering takes place on spherical particles. Since according to (1.43) in this case the scattering process $\Omega' \to \Omega$ depends on the cosine of the scattering angle $\cos \Theta = \Omega' \cdot \Omega$ only, henceforth the term $\Omega' \to \Omega$ will be replaced by $\Omega' \cdot \Omega$ or by $\cos \Theta$. Utilizing in (2.27) the definition of the *single scattering albedo* as given in (1.49), we obtain the standard form of the RTE for a three-dimensional medium

$$-\frac{1}{k_{\text{ext},\nu}} \Omega \cdot \nabla I_\nu = I_\nu - \frac{\omega_{0,\nu}}{4\pi} \int_{4\pi} \mathcal{P}_\nu(\Omega' \cdot \Omega) I_\nu(\Omega') d\Omega' - \frac{1}{k_{\text{ext},\nu}} J_\nu^e \qquad (2.28)$$

The derivation of the RTE is based on arguments of radiation hydrodynamics as presented by Pomraning (1973) where many additional and interesting details may be found. The RTE can also be derived on the basis of geometric reasoning in the manner described by Chandrasekhar (1960).

In passing we would like to remark that the RTE is part of the atmospheric predictive system. With changing composition of the atmospheric constituents the radiation parameters are also changing so that radiance continues to be a function of time.

2.2 The direct–diffuse splitting of the radiance field

The *total solar radiation field* is defined as the sum of the *direct solar beam* and the *diffuse solar radiation*. Usually one writes the RTE (2.28) in a different form by splitting $I_\nu(\mathbf{r}, \Omega)$ into the unscattered direct light $S_\nu(\mathbf{r})$ and the diffuse light $I_{d,\nu}(\mathbf{r}, \Omega)$

$$I_\nu = I_{d,\nu} + S_\nu \delta(\Omega - \Omega_0) \qquad (2.29)$$

where δ is the Dirac δ-function and Ω_0 is the direction of the solar radiation. While $I_{d,\nu}$ is expressed in $(\text{W m}^{-2} \text{ sr}^{-1} \text{ Hz}^{-1})$, the units of the parallel solar radiation are $(\text{W m}^{-2} \text{ Hz}^{-1})$. In order to have a consistent set of units, the Dirac δ-function $\delta(\Omega - \Omega_0)$ must refer to the unit solid angle, i.e. (sr^{-1}).

According to (2.23) we may write

$$\Omega \cdot \nabla I_\nu = \frac{dI_\nu}{ds} = \frac{dI_{d,\nu}}{ds} + \frac{dS_\nu}{ds} \delta(\Omega - \Omega_0) \qquad (2.30)$$

The attenuation of the direct Sun beam S_ν along its way from the top of the atmosphere ($s = 0$) to the location s at $\mathbf{r}$ follows from *Beer's law*

$$\frac{dS_\nu}{ds} = -k_{\text{ext},\nu}(s) S_\nu, \qquad \Omega = \Omega_0 \qquad (2.31)$$

Using the boundary condition $S_\nu(s = 0) = S_{0,\nu}$, where $S_{0,\nu}$ is the solar radiation at the top of the atmosphere, the solution of (2.31) is

$$S_\nu(s) = S_{0,\nu} \exp\left(-\int_0^s k_{\text{ext},\nu}(s')ds'\right), \qquad \Omega = \Omega_0 \qquad (2.32)$$

$S_{0,\nu}$ is also called the *solar constant* or the *extraterrestrial solar flux*. The exponential function occurring in (2.32) is also known as the *transmission function* since it determines the fraction of radiation which is transmitted through the path s. Obviously, the transmission function is bounded by 0 (no transmission) and 1 (total transmission).

The integral expression on the right-hand side of (2.28) contains the contributions due to multiple scattering processes. Introducing into this term the definition (2.29) yields

$$\frac{k_{\text{sca},\nu}}{4\pi} \int_{4\pi} \mathcal{P}_\nu(\Omega' \cdot \Omega) I_\nu(\Omega') d\Omega' = \frac{k_{\text{sca},\nu}}{4\pi} \int_{4\pi} \mathcal{P}_\nu(\Omega' \cdot \Omega) I_{d,\nu}(\Omega') d\Omega'$$

$$+ \frac{k_{\text{sca},\nu}}{4\pi} \int_{4\pi} \mathcal{P}_\nu(\Omega' \cdot \Omega) S_\nu \delta(\Omega' - \Omega_0) d\Omega' \qquad (2.33)$$

The last expression in this equation represents the direct solar radiation scattered from Ω_0 into the direction Ω. Hence, we may also write

$$\int_{4\pi} \mathcal{P}_\nu(\Omega' \cdot \Omega) S_\nu \delta(\Omega' - \Omega_0) d\Omega' = \mathcal{P}_\nu(\Omega_0 \cdot \Omega) S_\nu \qquad (2.34)$$

We are now ready to obtain the RTE for the diffuse radiation. Substituting (2.29), (2.31), and (2.33) together with (2.34) into (2.28) gives

$$\Omega \cdot \nabla I_{d,\nu} - k_{\text{ext},\nu} S_\nu \delta(\Omega - \Omega_0) = J_\nu^e - k_{\text{ext},\nu} I_{d,\nu} - k_{\text{ext},\nu} S_\nu \delta(\Omega - \Omega_0)$$

$$+ \frac{k_{\text{sca},\nu}}{4\pi} \int_{4\pi} \mathcal{P}_\nu(\Omega' \cdot \Omega) I_{d,\nu}(\Omega') d\Omega'$$

$$+ \frac{k_{\text{sca},\nu}}{4\pi} \mathcal{P}_\nu(\Omega_0 \cdot \Omega) S_\nu \qquad (2.35)$$

It can been seen that the two terms involving the direct solar light cancel. Thus, after introducing the direct–diffuse splitting (2.29), the final version of the RTE for the stationary diffuse radiation field of a three-dimensional medium reads

$$\boxed{\Omega \cdot \nabla I_{d,\nu} = -k_{\text{ext},\nu} I_{d,\nu} + \frac{k_{\text{sca},\nu}}{4\pi} \int_{4\pi} \mathcal{P}_\nu(\Omega' \cdot \Omega) I_{d,\nu}(\Omega') d\Omega' + \frac{k_{\text{sca},\nu}}{4\pi} \mathcal{P}_\nu(\Omega_0 \cdot \Omega) S_\nu + J_\nu^e}$$

$$(2.36)$$

The third term on the right-hand side describes the generation of diffuse light from the unscattered part of the direct sunlight.

2.3 The radiatively induced temperature change

According to (1.34) the net flux density vector is obtained by taking the first moment of the radiance with respect to $\mathbf{\Omega}$

$$\mathbf{E}_{\text{net},\nu} = \int_{4\pi} \mathbf{\Omega} I_\nu d\Omega \tag{2.37}$$

The total net flux density vector follows from an integration of this equation over the entire electromagnetic spectrum

$$\mathbf{E}_{\text{net}} = \int_0^\infty \mathbf{E}_{\text{net},\nu} d\nu = \int_0^\infty \int_{4\pi} \mathbf{\Omega} I_\nu d\Omega \, d\nu \tag{2.38}$$

Since $\nabla \cdot \mathbf{\Omega} = 0$ the RTE for the total (diffuse plus direct) radiation field (2.28) can be written as

$$\nabla \cdot (\mathbf{\Omega} I_\nu) + k_{\text{ext},\nu} I_\nu = \frac{k_{\text{sca},\nu}}{4\pi} \int_{4\pi} \mathcal{P}_\nu(\mathbf{\Omega}' \cdot \mathbf{\Omega}) I_\nu(\mathbf{\Omega}') d\Omega' + J_\nu^{\text{e}} \tag{2.39}$$

Employing the integral operator $\int_{4\pi} d\Omega \dots$ to the above equation we find in view of (2.37)

$$\nabla \cdot \mathbf{E}_{\text{net},\nu} + k_{\text{ext},\nu} \int_{4\pi} I_\nu d\Omega = \frac{k_{\text{sca},\nu}}{4\pi} \int_{4\pi} \int_{4\pi} \mathcal{P}_\nu(\mathbf{\Omega}' \cdot \mathbf{\Omega}) I_\nu(\mathbf{\Omega}') d\Omega \, d\Omega' + 4\pi \, J_\nu^{\text{e}} \tag{2.40}$$

since the source function J_ν was assumed to be isotropic. Applying the normalization condition for the *scattering phase function*

$$\frac{1}{4\pi} \int_{4\pi} \mathcal{P}_\nu(\mathbf{\Omega}' \cdot \mathbf{\Omega}) d\Omega = 1 \tag{2.41}$$

to the integral on the right-hand side, we obtain for the divergence of the net flux density vector

$$\begin{aligned}
\nabla \cdot \mathbf{E}_{\text{net},\nu} &= k_{\text{sca},\nu} \int_{4\pi} I_\nu d\Omega - k_{\text{ext},\nu} \int_{4\pi} I_\nu d\Omega + 4\pi \, J_\nu^{\text{e}} \\
&= -k_{\text{abs},\nu} \int_{4\pi} I_\nu d\Omega + 4\pi \, J_\nu^{\text{e}}
\end{aligned} \tag{2.42}$$

If the medium does not contain any true interior sources ($J_\nu^{\text{e}} = 0$), then the vector of the net flux density is given by

$$\nabla \cdot \mathbf{E}_{\text{net},\nu} = -k_{\text{abs},\nu} \int_{4\pi} I_\nu d\Omega < 0 \tag{2.43}$$

Whenever $\nabla \cdot \mathbf{E}_{\text{net},\nu} = 0$ the medium is said to be in *radiative equilibrium* for the frequency ν. The concept of radiative equilibrium will be discussed in more detail in a later chapter.

In the absence of frictional effects the *first law of thermodynamics* can be written as

$$\frac{de}{dt} + p\frac{dv}{dt} = \frac{dh}{dt} - \frac{1}{\rho}\frac{dp}{dt} = -\frac{1}{\rho}\nabla \cdot (\mathbf{J}^q + \mathbf{E}_{net}) \tag{2.44}$$

see e.g. Chapter 3 of THD (2004). Here, $v = 1/\rho$ is the specific volume of the air with density ρ. The quantities e, h, p and $\mathbf{J}^q$ stand for the *specific internal energy*, the *specific enthalpy*, the total pressure, and the *heat flux*, respectively. Here we are interested only in the contribution of radiative processes to the atmospheric temperature change. For an isobaric process $(dp = 0)$ enthalpy and temperature changes are related by

$$\frac{dh}{dt} = c_p\frac{dT}{dt} = -\frac{1}{\rho}\nabla \cdot \mathbf{E}_{net} \tag{2.45}$$

where c_p is the *specific heat at constant pressure*. The local time rate of change of the temperature caused by radiative processes alone is then given by

$$\boxed{\left.\frac{\partial T}{\partial t}\right|_{rad,\nu} = -\frac{1}{\rho c_p}\nabla \cdot \mathbf{E}_{net,\nu} = -\frac{1}{\rho c_p}\left(-k_{abs,\nu}\int_{4\pi} I_\nu d\Omega + 4\pi J_\nu^e\right)} \tag{2.46}$$

From this equation the following conclusions are drawn.

(i) The first term on the right-hand side describes the absorption of photons. Since this term is never negative it causes local warming.

(ii) The second term on the right-hand side describing the emission of photons is never positive, thus resulting in local cooling.

(iii) In the absence of absorption and emission no radiatively induced temperature changes take place.

2.4 The radiative transfer equation for a horizontally homogeneous atmosphere

The simplest geometry for a scattering and absorbing medium is the so-called *plane–parallel approximation*, where in the horizontal direction the medium stretches to infinity. In such a homogeneous plane–parallel slab all optical properties are independent of the horizontal position. Moreover, the incident radiation, including the parallel solar beam, is assumed to be independent of the horizontal coordinates along the upper and lower boundaries of the atmosphere. In many cases the plane–parallel assumption represents a good approximation to a planetary atmosphere. It is important to note that the plane–parallel approximation to the RTE is best whenever the vertical variations of all radiative quantities dominate over the horizontal variability which is often the case. Two specific examples for which the plane–parallel theory is inadequate are: (i) radiative transfer in finite clouds located

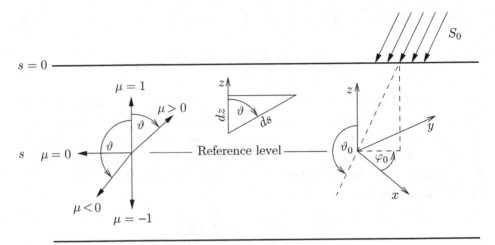

Fig. 2.3 Illustration of the plane–parallel geometry. Upwelling radiation: $\mu > 0$. Downwelling radiation: $\mu < 0$. The zenith ($\mu = 1$), nadir ($\mu = -1$), and horizontal ($\mu = 0$) directions are indicated. Solar radiation is incident from (ϑ_0, φ_0).

over a heterogeneously reflecting ground; and (ii) radiative transfer in a spherical atmosphere for solar positions near or below the horizon.

In the following we derive the RTE for a plane–parallel medium. For ease of notation the subscript ν will henceforth be omitted from all radiative quantities. We start with the RTE for the diffuse radiation in the form (2.36) (omitting for simplicity the subscript d) and using $\mathbf{\Omega} \cdot \nabla I = dI/ds$,

$$\frac{1}{k_{ext}} \frac{dI(s)}{ds} = -I(s) + \frac{\omega_0}{4\pi} \int_{4\pi} \mathcal{P}(\mathbf{\Omega}' \cdot \mathbf{\Omega}) I(s, \mathbf{\Omega}') d\Omega'$$

$$+ \frac{\omega_0}{4\pi} \mathcal{P}(\mathbf{\Omega} \cdot \mathbf{\Omega}_0) S(s) + \frac{1}{k_{ext}} J^e(s) \tag{2.47}$$

For a *black body* the *emission source function* is given by *Kirchhoff's law*

$$J^e(s) = k_{abs} B(s) \quad \text{with} \quad \frac{k_{abs}}{k_{ext}} = 1 - \omega_0 \tag{2.48}$$

where $B(s) = B(T(s))$ is the Planck function which depends on frequency and local temperature. The form of (2.48) will be motivated in the Appendix to this chapter.

In a plane–parallel medium the only spatial variable is the altitude z. The direction of the radiation is defined by the angle ϑ with respect to the z-axis and by the azimuth angle φ counted from an arbitrary origin. As illustrated in Figure 2.3, the path element ds is related to dz by

$$ds = \frac{dz}{\mu} \quad \text{with} \quad \mu = \cos \vartheta \tag{2.49}$$

From the figure it is also seen that for upward directed radiance $\mu > 0$ while for downward directed radiance $\mu < 0$. At an arbitrary reference level the horizontal direction is characterized by $\mu = 0$. Radiation propagating in the upward and downward vertical direction is specified by $\mu = 1$ and $\mu = -1$, respectively. However, in order to simplify the notation, for the downward directed radiance with $\vartheta > 90°$ the cosine of the zenith angle will henceforth be denoted by $-\mu$ so that for arbitrary directions of the radiance μ is positive definite.

The direct solar radiation is expressed by

$$S(s) = S_0 \exp\left(-\int_0^s k_{ext}(s')ds'\right) = S_0 \exp\left(-\frac{1}{\mu_0}\int_z^\infty k_{ext}(z')dz'\right) \qquad (2.50)$$

where according to (2.49) and due to the fact that $\theta_0 > 90°$, the path increment ds' has been replaced by $-dz'/\mu_0$. The integration of the direct beam starts at the upper boundary $s = 0$ of the medium and proceeds along a straight path of length s.

The *optical depth* τ is defined as the integral of the extinction coefficient over height along a path perpendicular to the horizontal plane, i.e.

$$\tau = \int_z^\infty k_{ext}(z')dz' \quad \text{or} \quad d\tau = -k_{ext}dz \qquad (2.51)$$

Since according to (2.49) and (2.51) $k_{ext}ds = k_{ext}dz/\mu = -d\tau/\mu$ we obtain for the solar radiation

$$S(\tau) = S_0 \exp\left(-\frac{\tau}{\mu_0}\right) \quad \text{with} \quad \mu_0 > 0 \qquad (2.52)$$

With these results one finds for the plane–parallel approximation of the RTE the integro-differential equation

$$\mu\frac{d}{d\tau}I(\tau, \mu, \varphi) = I(\tau, \mu, \varphi) - \frac{\omega_0}{4\pi}\int_0^{2\pi}\int_{-1}^1 \mathcal{P}(\cos\Theta)I(\tau, \mu', \varphi')d\mu'd\varphi'$$
$$- \frac{\omega_0}{4\pi}\mathcal{P}(\cos\Theta_0)S_0\exp\left(-\frac{\tau}{\mu_0}\right) - (1-\omega_0)B(\tau) \qquad (2.53)$$

We repeat that the cosine of the scattering angle is given by $\cos\Theta = \Omega'\cdot\Omega$, which is a function of the four angles ϑ', φ', ϑ, φ. An expression for $\cos\Theta$ will be derived later.

The scattering phase function $P(\cos\Theta)$ is assumed to be rotationally symmetric along the direction of the incident light, an assumption that is exact for spherical scattering particles. The scattering of light by homogeneous spheres is rigorously described by the so-called *Mie theory* which will be thoroughly discussed in a later chapter. The scattering properties of a single particle are solely a function of two physical parameters; the *complex index of refraction* $N = n + i\kappa$,[1] and the so-called *Mie size parameter*

$$x = \frac{2\pi r}{\lambda} \tag{2.54}$$

where r is the radius of the scattering sphere and λ is the wavelength of the incident electromagnetic wave. The schematic illustration of the rotationally symmetric form of $P(\cos\Theta)$ shown in Figure 1.18 indicates a strong forward peak of the scattering phase function. This is typical of cloud droplets and spherical aerosol particles when they are illuminated by solar radiation. Having specified the size parameter and the index of refraction, the Mie theory provides a table of numbers of P versus Θ.

In order to give a mathematical description of P, the scattering phase function will be expanded as an infinite series of *Legendre polynomials* $P_l(\cos\Theta)$

$$P(\cos\Theta) = \sum_{l=0}^{\infty} p_l P_l(\cos\Theta) \tag{2.55}$$

For practical purposes the phase function will be truncated after a finite number of terms.

The Legendre polynomials for argument x, with $-1 \leq x \leq 1$, are defined by

$$P_l(x) = \frac{1}{2^l l!} \frac{d^l}{dx^l} \left(x^2 - 1\right)^l \tag{2.56}$$

It will be readily seen that the first three polynomials are given by

$$P_0(x) = 1, \qquad P_1(x) = x, \qquad P_2(x) = \frac{3}{2}x^2 - \frac{1}{2} \tag{2.57}$$

The Legendre polynomials are a special case of the more general *associated Legendre polynomials* $P_l^m(x)$ as defined by

$$P_l^m(x) = \frac{(1 - x^2)^{m/2}}{2^l l!} \frac{d^{l+m}}{dx^{l+m}} (x^2 - 1)^l = (1 - x^2)^{m/2} \frac{d^m}{dx^m} P_l(x) \tag{2.58}$$

---

[1] n and κ are, respectively, the real and the imaginary part of the index of refraction.

They have the following properties

(a) $P_l^m(x) = 0, \; m > l,$ $P_l^{m=0}(x) = P_l(x),$ $P_l^m(-x) = (-1)^{l+m} P_l^m(x)$

(b) $\int_{-1}^1 P_n^m(x) P_l^m(x) dx = \dfrac{2}{2l+1} \dfrac{(l+m)!}{(l-m)!} \delta_{nl}$ (2.59)

Equation (2.59b) expresses the *orthogonality relation of the associated Legendre polynomials* on the interval $[-1, 1]$.

In order to obtain an analytical expression for the phase function $\mathcal{P}$ we need to evaluate the expansion coefficients p_l occurring in (2.55). This is done by multiplying both sides of (2.55) by $P_n(\cos \Theta)$ and then integrating over the unit sphere

$$\frac{1}{4\pi} \int_{4\pi} \mathcal{P}(\cos \Theta) P_n(\cos \Theta) d\Omega = \frac{1}{4\pi} \int_{4\pi} \sum_{l=0}^{\infty} p_l P_l(\cos \Theta) P_n(\cos \Theta) d\Omega \quad (2.60)$$

In this equation the integration $\int_{4\pi} \ldots d\Omega$ may be replaced by $\int_0^{2\pi} d\varphi \int_{-1}^1 \ldots d\cos \Theta$. Application of the orthogonality relations for P_l, which are given by (2.59b) if we set $m = 0$, finally yields

$$\boxed{p_l = \frac{2l+1}{2} \int_{-1}^1 P_l(\cos \Theta) \mathcal{P}(\cos \Theta) d\cos \Theta} \quad (2.61)$$

In addition to specifying $\mathcal{P}$ with the help of the Mie theory, mathematical models or experimentally determined values of the phase function may be employed. Due to the normalization of $\mathcal{P}$, i.e.

$$\frac{1}{4\pi} \int_{4\pi} \mathcal{P}(\cos \Theta) d\Omega = \frac{1}{4\pi} \int_0^{2\pi} \int_{-1}^1 \mathcal{P}(\cos \Theta) d\cos \Theta \, d\varphi$$

$$= \frac{1}{2} \int_{-1}^1 \mathcal{P}(\cos \Theta) d\cos \Theta = 1$$
(2.62)

the first expansion coefficient p_0 is always given by

$$p_0 = \frac{1}{2} \int_{-1}^1 \mathcal{P}(\cos \Theta) d\cos \Theta = 1 \quad (2.63)$$

Thus the coefficient p_0 expresses conservation of energy in case of pure scattering.

In the following we require an explicit expression for the cosine of the scattering angle Θ. From Figure 1.16 and equation (1.36) one easily obtains for $\mathbf{\Omega'} \cdot \mathbf{\Omega} = \cos \Theta$ the expression

$$\cos \Theta = (\mathbf{i} \cos \varphi' \sin \vartheta' + \mathbf{j} \sin \varphi' \sin \vartheta' + \mathbf{k} \cos \vartheta')$$

$$\cdot (\mathbf{i} \cos \varphi \sin \vartheta + \mathbf{j} \sin \varphi \sin \vartheta + \mathbf{k} \cos \vartheta)$$
(2.64)

With $\mu = \cos \vartheta$, $\mu' = \cos \vartheta'$ we find

$$\boxed{\cos \Theta = \mu\mu' + (1 - \mu^2)^{1/2}(1 - \mu'^2)^{1/2} \cos(\varphi - \varphi')} \qquad (2.65)$$

This relation allows us to express the scattering angle with respect to the angles ϑ', φ' and ϑ, φ. From (2.53) it may be seen that the angles ϑ, φ are fixed while ϑ', φ' vary over the unit sphere.

To continue the mathematical development we also need to state the *addition theorem for the associated Legendre polynomials* which is given by

$$\boxed{P_n(\cos \Theta) = P_n(\mu)P_n(\mu') + 2 \sum_{m=1}^{n} \frac{(n-m)!}{(n+m)!} P_n^m(\mu)P_n^m(\mu') \cos m(\varphi - \varphi')} \qquad (2.66)$$

(cf. Jackson, 1975). This theorem will now be used to separate the angles of the scattering phase function. Substituting (2.66) into (2.55) gives

$$P(\cos \Theta) = \sum_{l=0}^{\infty} p_l \left(P_l(\mu)P_l(\mu') + 2 \sum_{m=1}^{l} \frac{(l-m)!}{(l+m)!} P_l^m(\mu)P_l^m(\mu') \cos m(\varphi - \varphi') \right)$$

$$= \sum_{l=0}^{\infty} \sum_{m=0}^{l} (2 - \delta_{0m}) p_l^m P_l^m(\mu)P_l^m(\mu') \cos m(\varphi - \varphi') \qquad (2.67)$$

$$\text{with} \quad p_l^m = p_l \frac{(l-m)!}{(l+m)!}$$

Since $P_l^m(x) = 0$ for $m > l$, $P(\cos \Theta)$ can be reformulated as

$$\boxed{\begin{aligned} P(\cos \Theta) &= \sum_{l=0}^{\infty} \sum_{m=0}^{\infty} (2 - \delta_{0m}) p_l^m P_l^m(\mu)P_l^m(\mu') \cos m(\varphi - \varphi') \\ &= \sum_{m=0}^{\infty} (2 - \delta_{0m}) \sum_{l=m}^{\infty} p_l^m P_l^m(\mu)P_l^m(\mu') \cos m(\varphi - \varphi') \\ &= P(\mu, \varphi, \mu', \varphi') \end{aligned}} \qquad (2.68)$$

It can now be seen that in the scattering phase function the (μ, μ')-dependence has been completely separated from the (φ, φ')-dependence.

In order to remove the azimuthal dependence of the radiance field we assume that the radiance can be expressed by means of a Fourier expansion for an even function of the azimuth angle φ

$$I(\tau, \mu, \varphi) = \sum_{m=0}^{\infty} (2 - \delta_{0m}) I^m(\tau, \mu) \cos m(\varphi - \varphi_0) \qquad (2.69)$$

where φ_0 is the azimuth angle of the direct solar beam. It is customary to orient the Cartesian (x, y, z)-coordinate system in such a way that $\varphi_0 = 0$, see Figure 2.3, so that

$$I(\tau, \mu, \varphi) = \sum_{m=0}^{\infty}(2 - \delta_{0m})I^m(\tau, \mu)\cos m\varphi \qquad (2.70)$$

The expansion (2.70) may be used whenever the phase function is expressed as a series of Legendre polynomials.

Substituting (2.68) and (2.70) into the RTE (2.53) yields

$$\mu \sum_{m=0}^{\infty}(2 - \delta_{0m})\frac{d}{d\tau}I^m(\tau, \mu)\cos m\varphi$$

$$= \sum_{m=0}^{\infty}(2 - \delta_{0m})I^m(\tau, \mu)\cos m\varphi$$

$$-\frac{\omega_0}{4\pi}S_0\exp\left(-\frac{\tau}{\mu_0}\right)\sum_{m=0}^{\infty}(2 - \delta_{0m})\sum_{l=m}^{\infty}p_l^m P_l^m(\mu)P_l^m(-\mu_0)\cos m\varphi$$

$$-\frac{\omega_0}{4\pi}\int_0^{2\pi}\int_{-1}^{1}\left(\sum_{m=0}^{\infty}(2 - \delta_{0m})\sum_{l=m}^{\infty}p_l^m P_l^m(\mu)P_l^m(\mu')\cos m(\varphi - \varphi')\right.$$

$$\left.\times\sum_{i=0}^{\infty}(2 - \delta_{0i})I^i(\tau, \mu')\cos i\varphi'\right)d\mu'd\varphi' - (1 - \omega_0)\sum_{m=0}^{\infty}\delta_{0m}B(\tau)$$

$$(2.71)$$

The above equation can be simplified by employing the *orthogonality relations of the trigonometric functions*

$$\int_0^{2\pi}\cos m(\varphi - \varphi')\cos l\varphi\, d\varphi = (1 + \delta_{0l})\delta_{lm}\pi\cos l\varphi' \qquad (2.72)$$

An expression for $I^m(\tau, \mu)$ can be isolated from (2.71) by carrying out the following two steps: (i) Multiply (2.71) by $\cos k\varphi$; (ii) Carry out the operation $\int_0^{2\pi}\ldots d\varphi$.

As a consequence, the following relations are needed

$$\int_0^{2\pi}\cos m\varphi\cos k\varphi\, d\varphi = (1 + \delta_{0k})\delta_{mk}\pi$$

$$\int_0^{2\pi}\cos k\varphi\, d\varphi = (1 + \delta_{0k})\delta_{0k}\pi \qquad (2.73)$$

$$\int_0^{2\pi}\cos m(\varphi - \varphi')\cos k\varphi\, d\varphi = (1 + \delta_{0k})\delta_{mk}\pi\cos k\varphi'$$

From (2.71) it is seen that for the multiple scattering term an integration over φ' remains to be done for which we obtain the result

$$\int_0^{2\pi} \cos i\varphi' \cos k\varphi' \, d\varphi' = (1 + \delta_{0i})\delta_{ik}\pi \tag{2.74}$$

Using the above intermediate steps we find from (2.71)

$$\mu \sum_{m=0}^{\infty}(2 - \delta_{0m})\frac{d}{d\tau}I^m(\tau, \mu)(1 + \delta_{0k})\delta_{mk}\pi$$

$$= \sum_{m=0}^{\infty}(2 - \delta_{0m})(1 + \delta_{0k})\delta_{mk}\pi I^m(\tau, \mu)$$

$$- \frac{\omega_0}{4\pi}S_0 \exp\left(-\frac{\tau}{\mu_0}\right)\sum_{m=0}^{\infty}(2 - \delta_{0m})\sum_{l=m}^{\infty}p_l^m P_l^m(\mu)P_l^m(-\mu_0)(1 + \delta_{0k})\delta_{mk}\pi$$

$$- \frac{\omega_0}{4\pi}\int_{-1}^{1}\left(\sum_{m=0}^{\infty}(2 - \delta_{0m})\sum_{l=m}^{\infty}p_l^m P_l^m(\mu)P_l^m(\mu')(1 + \delta_{0k})\delta_{mk}\pi\right.$$

$$\left.\times \sum_{i=0}^{\infty}(2 - \delta_{0i})I^i(\tau, \mu')(1 + \delta_{0i})\delta_{ik}\pi\right)d\mu' - (1 - \omega_0)\sum_{m=0}^{\infty}\delta_{0m}B(\tau)(1 + \delta_{0k})\delta_{0k}\pi \tag{2.75}$$

We recognize immediately that, with the exception of $m = k$, all terms vanish. Therefore, after separation of the azimuthal dependence, for the m-th Fourier mode of the radiance the RTE is given by

$$\boxed{\begin{aligned}
&\text{(a) } \mu\frac{d}{d\tau}I^m(\tau, \mu) = I^m(\tau, \mu) - \frac{\omega_0}{2}\int_{-1}^{1}\sum_{l=m}^{\infty}p_l^m P_l^m(\mu)P_l^m(\mu')I^m(\tau, \mu')d\mu'\\
&\qquad\qquad\qquad\qquad - \frac{\omega_0}{4\pi}S_0 \exp\left(-\frac{\tau}{\mu_0}\right)\sum_{l=m}^{\infty}p_l^m P_l^m(\mu)P_l^m(-\mu_0) - (1 - \omega_0)B(\tau)\delta_{0m}\\
&\text{(b) } \mu\frac{d}{d\tau}I^m(\tau, \mu) = I^m(\tau, \mu) - J(\tau, \mu)
\end{aligned}} \tag{2.76}$$

(2.76b) is the standard form of the RTE for plane–parallel atmospheres. The source function $J(\tau, \mu)$ summarizes the terms describing multiple scattering, primary scattering of solar radiation and thermal emissions.

For the special case $m = 0$ the same result is obtained by performing an azimuthal average of (2.53). Consider the averaging process for each individual term separately.

(1) Averaging the intensity $I(\tau, \mu, \varphi)$ as defined by (2.70) yields

$$\frac{1}{2\pi} \int_0^{2\pi} I(\tau, \mu, \varphi)d\varphi = \frac{1}{2\pi} \sum_{m=0}^{\infty} (2 - \delta_{0m}) \int_0^{2\pi} I^m(\tau, \mu) \cos m\varphi \, d\varphi$$

$$= I^{m=0}(\tau, \mu) = I(\tau, \mu) \tag{2.77}$$

Only $m = 0$ contributes to the azimuthally averaged radiation field since the integral of the cosine function over a complete cycle vanishes.

(2) Since the Planck function is isotropic the *thermal emission term* retains its form.

(3) Analogously to (1) the primary scattering term can be averaged according to the expansion (2.68)

$$\frac{1}{2\pi} \int_0^{2\pi} \mathcal{P}(\cos \Theta_0)d\varphi = \sum_{l=0}^{\infty} p_l P_l(\mu) P_l(-\mu_0) = \mathcal{P}(\mu, -\mu_0) \tag{2.78}$$

(4) For the multiple scattering term we proceed as in (2.77) and (2.78) obtaining

$$\frac{\omega_0}{8\pi^2} \int_0^{2\pi} \int_{-1}^{1} I(\tau, \mu', \varphi') \int_0^{2\pi} \mathcal{P}(\cos \Theta)d\varphi \, d\mu' \, d\varphi'$$

$$= \frac{\omega_0}{4\pi} \int_0^{2\pi} \int_{-1}^{1} I(\tau, \mu', \varphi') \mathcal{P}(\mu, \mu')d\mu' \, d\varphi' \tag{2.79}$$

$$= \frac{\omega_0}{2} \int_{-1}^{1} I(\tau, \mu') \mathcal{P}(\mu, \mu')d\mu'$$

The azimuthally integrated form of the phase function $\mathcal{P}(\mu, \mu')$ is given by

$$\mathcal{P}(\mu, \mu') = \frac{1}{2\pi} \int_0^{2\pi} \mathcal{P}(\cos \Theta)d\varphi$$

$$= \frac{1}{2\pi} \int_0^{2\pi} \mathcal{P}(\mu, \varphi, \mu', \varphi')d\varphi = \sum_{l=0}^{\infty} p_l P_l(\mu) P_l(\mu') \tag{2.80}$$

Application of steps (1)–(4) results in the *azimuthally integrated form* of the RTE

$$\mu \frac{d}{d\tau} I(\tau, \mu) = I(\tau, \mu) - \frac{\omega_0}{2} \int_{-1}^{1} \mathcal{P}(\mu, \mu')I(\tau, \mu')d\mu'$$

$$- \frac{\omega_0}{4\pi} S_0 \exp\left(-\frac{\tau}{\mu_0}\right) \mathcal{P}(\mu, -\mu_0) - (1 - \omega_0)B(\tau) \tag{2.81}$$

Comparison of (2.81) with (2.76) reveals that for $m = 0$ both equations are identical.

The separation of the φ-dependence of the radiance field results in an infinite system of first-order ordinary differential equations for the Fourier modes $I^m(\tau, \mu)$.

It is convenient to introduce into (2.76) the following abbreviations

(a) $\quad R^m(\mu, \mu') = \sum\limits_{l=m}^{\infty} p_l^m P_l^m(\mu) P_l^m(\mu')$

(b) $\quad J^m(\tau, \mu) = \dfrac{\omega_0}{2} \displaystyle\int_{-1}^{1} R^m(\mu, \mu') I^m(\tau, \mu') d\mu' \qquad\qquad\qquad$ (2.82)

$\qquad\qquad + \dfrac{\omega_0}{4\pi} S_0 \exp\left(-\dfrac{\tau}{\mu_0}\right) R^m(\mu, -\mu_0) + (1 - \omega_0) B(\tau) \delta_{0m}$

With these definitions the RTE reads in short-hand notation

$$\mu \frac{d}{d\tau} I^m(\tau, \mu) = I^m(\tau, \mu) - J^m(\tau, \mu), \qquad m = 0, 1, \ldots, \Lambda \qquad (2.83)$$

For practical applications the infinite series (2.70) must be truncated after a finite number of terms, Λ, as indicated above.

2.5 Splitting of the radiance field into upwelling and downwelling radiation

It is common practice to distinguish between the upwelling ($\mu > 0$) and downwelling ($\mu < 0$) radiation, see Figure 2.3. For the radiance field I^m and the source term J^m the following quantities are introduced[2]

$$\begin{aligned} I_+^m(\tau, \mu) &= I^m(\tau, \mu > 0), & I_-^m(\tau, \mu) &= I^m(\tau, \mu < 0) \\ J_+^m(\tau, \mu) &= J^m(\tau, \mu > 0), & J_-^m(\tau, \mu) &= J^m(\tau, \mu < 0) \end{aligned} \qquad (2.84)$$

Now the integral $\int_{-1}^{1} \ldots d\mu'$ in (2.82b) will be split into the two parts $\int_{-1}^{0} \ldots d\mu'$ and $\int_{0}^{1} \ldots d\mu'$

$$\int_{-1}^{1} R^m(\mu, \mu') I^m(\tau, \mu') d\mu' = \int_{-1}^{0} R^m(\mu, \mu') I^m(\tau, \mu') d\mu' + \int_{0}^{1} R^m(\mu, \mu') I^m(\tau, \mu') d\mu'$$

$$= \int_{0}^{1} R^m(\mu, -\mu') I^m(\tau, -\mu') d\mu' + \int_{0}^{1} R^m(\mu, \mu') I^m(\tau, \mu') d\mu'$$

$$= \int_{0}^{1} \left[R^m(\mu, -\mu') I_-^m(\tau, \mu') + R^m(\mu, \mu') I_+^m(\tau, \mu') \right] d\mu'$$

$$(2.85)$$

With the above convention μ' and μ are always positive quantities. As a consequence of this, the system (2.83) can be treated as two separate differential equations, one governing the upwelling part and the other one the downwelling part of the radiation

---

[2] Some authors define the zenith angle ϑ in the opposite way, that is $\mu > 0$ for downwelling radiation and $\mu < 0$ for upwelling radiation.

field

$$\mu \frac{d}{d\tau} I_+^m(\tau, \mu) = I_+^m(\tau, \mu) - J_+^m(\tau, \mu)$$

$$\mu \frac{d}{d\tau} I_-^m(\tau, \mu) = -I_-^m(\tau, \mu) + J_-^m(\tau, \mu) \qquad (2.86)$$

$$m = 0, 1, \ldots, \Lambda, \qquad \mu > 0$$

In a similar manner the source terms of (2.82b) can be split to give

(a) $$J_+^m(\tau, \mu) = \frac{\omega_0}{2} \int_0^1 \left[I_+^m(\tau, \mu') R^m(\mu, \mu') + I_-^m(\tau, \mu') R^m(\mu, -\mu') \right] d\mu'$$

$$+ \frac{\omega_0}{4\pi} S_0 \exp\left(-\frac{\tau}{\mu_0}\right) R^m(\mu, -\mu_0) + (1 - \omega_0) B(\tau) \delta_{0m}$$

(b) $$J_-^m(\tau, \mu) = \frac{\omega_0}{2} \int_0^1 \left[I_+^m(\tau, \mu') R^m(-\mu, \mu') + I_-^m(\tau, \mu') R^m(-\mu, -\mu') \right] d\mu'$$

$$+ \frac{\omega_0}{4\pi} S_0 \exp\left(-\frac{\tau}{\mu_0}\right) R^m(-\mu, -\mu_0) + (1 - \omega_0) B(\tau) \delta_{0m}$$

$$(2.87)$$

with $\mu > 0$ and $\mu' > 0$.

The above expressions demonstrate that (2.86) is a system of $2(\Lambda + 1)$ coupled integro-differential equations. The source terms J_+^m and J_-^m contain integrals over μ' which may be approximated by means of quadrature formulas. Best suited for the interval $[-1, 1]$ are the *Gauss–Legendre quadrature formulas*, which are obtained by neglecting in the expression

$$\int_{-1}^1 f(\mu) d\mu = \sum_{i=1}^r w_i f(\mu_i) + R_r$$

$$(2.88)$$

$$\text{with} \quad R_r = \frac{2^{2r+1}(r!)^4}{(2r+1)\left[(2r)!\right]^3} f^{(2r)}(\xi), \qquad -1 < \xi < 1$$

the error term R_r (see, e.g. Abramowitz and Stegun, 1972). The w_i are the weights and the μ_i are the zeros or roots of the Legendre polynomials $P_r(\mu)$. The weights are given by

$$w_i = \frac{2}{(1 - \mu_i^2)\left[P_r'(\mu_i)\right]^2} > 0, \qquad \sum_{i=1}^r w_i = 2 \qquad (2.89)$$

where P_r' is the derivative of P_r. The last expression follows immediately from (2.88) by setting there $f(\mu) = 1$.

It is convenient to divide the interval $[-1, 1]$ according to the zeros of even-order Legendre polynomials, which is achieved by choosing $r = 2s$. For such even divisions we have $2s$ different weights and zeros. By denoting them as $(w_k, \mu_k,$

$k = -s, \ldots, -1, 1, \ldots, s)$ the following relations hold

$$w_i = w_{-i}, \qquad \mu_i = -\mu_{-i}, \qquad i = 1, \ldots, s \qquad (2.90)$$

Now we split the integral of (2.88) into two parts yielding

$$\int_{-1}^{1} f(\mu)d\mu = \int_{-1}^{0} f(\mu)d\mu + \int_{0}^{1} f(\mu)d\mu = \int_{0}^{1} f(-\mu)d\mu + \int_{0}^{1} f(\mu)d\mu$$

$$(2.91)$$

For the sum on the right-hand side of (2.88) we may write

$$\sum_{i=-s}^{s}{}' w_i f(\mu_i) = \sum_{i=-s}^{-1} w_i f(\mu_i) + \sum_{i=1}^{s} w_i f(\mu_i)$$

$$= \sum_{i=1}^{s} w_i f(-\mu_i) + \sum_{i=1}^{s} w_i f(\mu_i) \qquad (2.92)$$

where use was made of (2.90). Hence, the notation $\sum'$ is equivalent to the summation from $i = -s$ to $i = s$ thereby omitting the term $i = 0$. By comparing (2.91) with (2.92) we obtain the Gauss–Legendre quadrature formulas of order $2s$ in the form

$$\boxed{\int_{0}^{1} f(-\mu)d\mu \approx \sum_{i=1}^{s} w_i f(-\mu_i), \qquad \int_{0}^{1} f(\mu)d\mu \approx \sum_{i=1}^{s} w_i f(\mu_i)} \qquad (2.93)$$

For ease of notation (2.93) will henceforth be called the Gaussian quadrature formulas. It is important to note that the $\mu_i, i = 1, \ldots, s$ are the positive zeros of the Legendre polynomials $P_{2s}(\mu)$. Furthermore, from (2.89) and (2.90) we see that the weights fulfill the requirement

$$\sum_{i=1}^{s} w_i = 1 \qquad (2.94)$$

We conclude that a particular Gaussian quadrature formula is defined by the set of quantities $(w_i, \mu_i, \ i = 1, \ldots, s)$. The weights and the nodes may be found in tabulated form (see e.g. Abramowitz and Stegun, 1972) or may be found using efficient numerical algorithms (Press *et al.*, 1992). Thus the nodes define the directions of the radiances for which the system of equations (2.86) will be discretized.

In order to express the directional dependence of the radiance it is convenient to introduce a compact matrix–vector notation. One defines two vectors for the upwelling and downwelling radiation field by means of

$$\mathbf{I}_+^m(\tau) = \begin{pmatrix} I_+^m(\tau, \mu_1) \\ \vdots \\ I_+^m(\tau, \mu_s) \end{pmatrix}, \qquad \mathbf{I}_-^m(\tau) = \begin{pmatrix} I_-^m(\tau, \mu_1) \\ \vdots \\ I_-^m(\tau, \mu_s) \end{pmatrix} \qquad (2.95)$$

In a similar manner vectors for the sources of the up- and downwelling primary scattered solar radiation and the thermal emission will be defined

$$\mathbf{J}^m_{+,1}(\tau) = \frac{\omega_0}{4\pi} S_0 \exp\left(-\frac{\tau}{\mu_0}\right) \begin{pmatrix} R^m(\mu_1, -\mu_0) \\ \vdots \\ R^m(\mu_s, -\mu_0) \end{pmatrix}, \qquad \mathbf{J}^m_{+,2}(\tau) = (1 - \omega_0)B(\tau)\delta_{0m} \begin{pmatrix} 1 \\ \vdots \\ 1 \end{pmatrix}$$

$$\mathbf{J}^m_{-,1}(\tau) = \frac{\omega_0}{4\pi} S_0 \exp\left(-\frac{\tau}{\mu_0}\right) \begin{pmatrix} R^m(-\mu_1, -\mu_0) \\ \vdots \\ R^m(-\mu_s, -\mu_0) \end{pmatrix}, \qquad \mathbf{J}^m_{-,2}(\tau) = \mathbf{J}^m_{+,2}(\tau) \qquad (2.96)$$

where all vectors contain s rows.

For the evaluation of the multiple scattering integral the phase function term $R^m(\mu, \mu')$ needs to be discretized. To give an example, let us first evaluate the multiple scattering term in $\mathbf{J}^m_+$ for a particular direction μ_k by employing the quadrature formula (2.93)

$$\frac{\omega_0}{2} \int_0^1 \left[I^m_+(\tau, \mu')R^m(\mu_k, \mu') + I^m_-(\tau, \mu')R^m(\mu_k, -\mu')\right] d\mu'$$

$$\approx \frac{\omega_0}{2} \sum_{i=1}^s \left[w_i I^m_+(\tau, \mu_i)R^m(\mu_k, \mu_i) + w_i I^m_-(\tau, \mu_i)R^m(\mu_k, -\mu_i)\right]$$

$$\qquad (2.97)$$

$$= \frac{\omega_0}{2} \sum_{i=1}^s \left(\sum_{r=1}^s p^m_{kr,++} w_r \delta_{ri} I^m_+(\tau, \mu_i) + \sum_{r=1}^s p^m_{kr,+-} w_r \delta_{ri} I^m_-(\tau, \mu_i)\right)$$

$$= \frac{\omega_0}{2} \left[\mathbb{P}^m_{++} \mathbb{W} \mathbf{I}^m_+(\tau) + \mathbb{P}^m_{+-} \mathbb{W} \mathbf{I}^m_-(\tau)\right]$$

Here the following matrices have been introduced

$$\mathbb{P}^m_{++} = (p^m_{kr,++}) = \begin{pmatrix} R^m(\mu_1, \mu_1) & \cdots & R^m(\mu_1, \mu_s) \\ \vdots & & \vdots \\ R^m(\mu_s, \mu_1) & \cdots & R^m(\mu_s, \mu_s) \end{pmatrix}$$

$$\mathbb{P}^m_{+-} = (p^m_{kr,+-}) = \begin{pmatrix} R^m(\mu_1, -\mu_1) & \cdots & R^m(\mu_1, -\mu_s) \\ \vdots & & \vdots \\ R^m(\mu_s, -\mu_1) & \cdots & R^m(\mu_s, -\mu_s) \end{pmatrix} \qquad (2.98)$$

$$\mathbb{W} = (w_i \delta_{ij}) = \begin{pmatrix} w_1 & 0 & \cdots & 0 \\ 0 & w_2 & \cdots & 0 \\ \vdots & \vdots & \ddots & \vdots \\ 0 & 0 & \cdots & w_s \end{pmatrix}$$

Since

$$P_l^m(-\mu) = (-1)^{l+m} P_l^m(\mu)$$
$$P_l^m(-\mu)P_l^m(-\mu') = (-1)^{2(l+m)} P_l^m(\mu)P_l^m(\mu') = P_l^m(\mu)P_l^m(\mu') \tag{2.99}$$

with the help of (2.82a) we obtain the equalities

$$\mathbb{P}_{++}^m = \mathbb{P}_{--}^m, \qquad \mathbb{P}_{+-}^m = \mathbb{P}_{-+}^m \tag{2.100}$$

With the above definitions the RTE can now be written in the form

$$\mathbb{M}\frac{d}{d\tau}\mathbf{I}_+^m(\tau) = \mathbf{I}_+^m(\tau) - \mathbf{J}_+^m(\tau)$$
$$\mathbb{M}\frac{d}{d\tau}\mathbf{I}_-^m(\tau) = -\mathbf{I}_-^m(\tau) + \mathbf{J}_-^m(\tau), \qquad m = 0, 1, \ldots, \Lambda \tag{2.101}$$

where the matrix $\mathbb{M}$ is defined by

$$\mathbb{M} = (\mu_i \delta_{ij}) = \begin{pmatrix} \mu_1 & 0 & \cdots & 0 \\ 0 & \mu_2 & \cdots & 0 \\ \vdots & \vdots & \ddots & \vdots \\ 0 & 0 & \cdots & \mu_s \end{pmatrix} \tag{2.102}$$

and the source vectors $\mathbf{J}_+^m$, $\mathbf{J}_-^m$ are given in matrix notation as

$$\mathbf{J}_+^m(\tau) = \mathbf{J}_{+,1}^m(\tau) + \mathbf{J}_{+,2}^m(\tau) + \frac{\omega_0}{2}\left[\mathbb{P}_{++}^m\mathbb{W}\mathbf{I}_+^m(\tau) + \mathbb{P}_{+-}^m\mathbb{W}\,\mathbf{I}_-^m(\tau)\right]$$
$$\mathbf{J}_-^m(\tau) = \mathbf{J}_{-,1}^m(\tau) + \mathbf{J}_{-,2}^m(\tau) + \frac{\omega_0}{2}\left[\mathbb{P}_{+-}^m\mathbb{W}\mathbf{I}_+^m(\tau) + \mathbb{P}_{++}^m\mathbb{W}\,\mathbf{I}_-^m(\tau)\right] \tag{2.103}$$

Multiplying (2.101) from the left by $\mathbb{M}^{-1}$ we find the *matrix–vector form of the RTE* as

$$\boxed{\frac{d}{d\tau}\begin{pmatrix}\mathbf{I}_+^m(\tau) \\ \mathbf{I}_-^m(\tau)\end{pmatrix} = \begin{pmatrix}\Gamma_{++}^m & -\Gamma_{+-}^m \\ \Gamma_{+-}^m & -\Gamma_{++}^m\end{pmatrix}\begin{pmatrix}\mathbf{I}_+^m(\tau) \\ \mathbf{I}_-^m(\tau)\end{pmatrix} + \begin{pmatrix}-\Sigma_+^m(\tau) \\ \Sigma_-^m(\tau)\end{pmatrix}} \tag{2.104}$$

where the following abbreviations have been introduced

$$\Gamma_{++}^m = \mathbb{M}^{-1}\left(\mathbb{E} - \frac{\omega_0}{2}\mathbb{P}_{++}^m\mathbb{W}\right)$$
$$\Gamma_{+-}^m = \mathbb{M}^{-1}\frac{\omega_0}{2}\mathbb{P}_{+-}^m\mathbb{W}$$
$$\Sigma_+^m(\tau) = \mathbb{M}^{-1}\left[\mathbf{J}_{+,1}^m(\tau) + \mathbf{J}_{+,2}^m(\tau)\right]$$
$$\Sigma_-^m(\tau) = \mathbb{M}^{-1}\left[\mathbf{J}_{-,1}^m(\tau) + \mathbf{J}_{-,2}^m(\tau)\right] \tag{2.105}$$

$\mathbb{E}$ is the $s \times s$ unit matrix. From the first two expressions of (2.105) it may be easily seen that due to (2.100) $\Gamma_{++}^m = \Gamma_{--}^m$ and $\Gamma_{+-}^m = \Gamma_{-+}^m$.

The system (2.104) represents the discretized form of the RTE, discretized in μ-space. It has been derived under the assumptions that

(i) the *phase function* $\mathcal{P}$ can be developed as an infinite series of Legendre polynomials; and

(ii) the φ-dependence of both the radiance as well as the phase function can be separated in product form from the dependency of the other variables.

If the solution of the Fourier modes I_+^m, I_-^m of the radiance is known, the full dependence of the radiance function follows from (2.70)[3]

$$
\mathbf{I}_+(\tau, \varphi) = \sum_{m=0}^{\infty}(2 - \delta_{0m})\mathbf{I}_+^m(\tau)\cos m\varphi
$$

$$
\mathbf{I}_-(\tau, \varphi) = \sum_{m=0}^{\infty}(2 - \delta_{0m})\mathbf{I}_-^m(\tau)\cos m\varphi
$$

(2.106)

The μ_i-dependency of the radiance vectors $\mathbf{I}_+$, $\mathbf{I}_-$ is expressed by (2.95).

2.6 The solution of the radiative transfer equation for a horizontally homogeneous atmosphere

The energy budget of the atmosphere is determined by the solar insolation and the emission of infrared radiation by the Earth and the atmosphere. Thus scattering, absorption and emission of radiation influence the radiative exchange between the individual atmospheric layers. These processes depend on both the wavelength of the radiation as well as on the different radiatively active atmospheric species. As already mentioned in Chapter 1, the solar radiation spectrum ranges from roughly 0.2–3.5 μm while the thermal radiation spectrum of the Earth and the atmosphere covers the region of about 3.5–100 μm. Hence, for all practical purposes both spectral regions may be treated separately. As a consequence of this, in the solar spectral region one is justified to neglect the Planck function $B(\tau)$, while in the infrared spectral region S_0 may be omitted. This is most easily achieved by setting in the source vectors $\mathbf{J}_\pm^m$ for the short-wave region $\mathbf{J}_{\pm,2}^m(\tau) = 0$ and for the long-wave region $\mathbf{J}_{\pm,1}^m(\tau) = 0$, see (2.96) and (2.103).

In later chapters we will show that in the infrared spectral region scattering processes are important only in the so-called *atmospheric window region* extending from about 8 to 12.5 μm. In this spectral region the clear atmosphere is almost transparent to infrared radiation while scattering and absorption by aerosol and cloud particles must be accounted for. The atmospheric window is of particular

---

[3] Recall that the coordinate system has been rotated so that $\varphi_0 = 0$.

importance for the energy budget of the Earth since in this region the largest fraction of thermal radiation is emitted to space. Outside the window region scattering,
in general, plays almost no role in the troposphere. Unless in the upper troposphere
layers are extremely dry, outside the CO_2 absorption spectrum it is sufficient to
include H_2O absorption in the radiative transfer equation. Since even in the atmospheric window scattering is relatively weak, very often scattering processes are
completely neglected in the entire infrared spectral region. In this case the source
function $\mathbf{J}_\pm$ of the RTE reduces to $\mathbf{J}^m_{\pm,2}$ with $\omega_0 = 0$.

Neglecting all scattering processes yields a relatively simple form of the RTE
which has a particularly simple solution. This motivates us to present the solution
of the RTE for two different cases. In the first case, scattering and absorption
processes are included while in the second case only absorption is accounted for.
According to the above discussion, the first case applies to the short-wave region
and to the atmospheric window, whereas the second case may be applied to the
infrared spectral region excepting the atmospheric window.

2.6.1 The scattering atmosphere

Inspection of the system (2.101) shows that the RTE is an inhomogeneous first-order
differential equation of the form

$$\frac{dy}{dx} + P(x)y(x) + Q(x) = 0, \qquad y(x = x_0) = y_0 \qquad (2.107)$$

with the general solution

$$y(x) = y_0 \exp\left(-\int_{x_0}^{x} P(t)dt\right) - \int_{x_0}^{x} Q(x') \exp\left(-\int_{x'}^{x} P(t)dt\right) dx' \quad (2.108)$$

Let τ_g denote the total optical depth of the medium. By comparing (2.101) with
(2.107) we may substitute:

(a) Upwelling radiation

$$x = \tau, \qquad x_0 = \tau_g, \qquad x' = \tau'$$
$$y(x) = \mathbf{I}^m_+(\tau), \qquad P(x) = -\mathbf{M}^{-1}, \qquad Q(x) = \mathbf{M}^{-1}\mathbf{J}^m_+(\tau)$$

(b) Downwelling radiation   (2.109)

$$x = \tau, \qquad x_0 = 0, \qquad x' = \tau'$$
$$y(x) = \mathbf{I}^m_-(\tau), \qquad P(x) = \mathbf{M}^{-1}, \qquad Q(x) = -\mathbf{M}^{-1}\mathbf{J}^m_-(\tau)$$

According to (2.108) the formal solution for the upwelling and downwelling radiation is then given by

$$\mathbf{I}_+^m(\tau) = \exp\left[-\mathbb{M}^{-1}(\tau_g - \tau)\right]\mathbf{I}_+^m(\tau_g) + \int_\tau^{\tau_g} \exp\left[-\mathbb{M}^{-1}(\tau' - \tau)\right]\mathbb{M}^{-1}\mathbf{J}_+^m(\tau')d\tau'$$

$$\mathbf{I}_-^m(\tau) = \exp\left(-\mathbb{M}^{-1}\tau\right)\mathbf{I}_-^m(0) + \int_0^\tau \exp\left[-\mathbb{M}^{-1}(\tau - \tau')\right]\mathbb{M}^{-1}\mathbf{J}_-^m(\tau')d\tau'$$

$$(2.110)$$

These solutions are called *formal solutions of the RTE* because they do not represent explicit expressions for $\mathbf{I}_+^m$ and $\mathbf{I}_-^m$. This is due to the fact that the source terms $\mathbf{J}_+^m$ and $\mathbf{J}_-^m$ implicitly contain the radiance field $\mathbf{I}_+^m$ and $\mathbf{I}_-^m$ as follows from (2.103).

Some remarks concerning the so-called *exponential matrices* appearing in (2.110) may be helpful. A few important rules for these matrices are stated below.

$$\exp(\mathbb{A}) = \sum_{k=0}^\infty \frac{1}{k!}\mathbb{A}^k, \qquad\qquad \exp(\mathbb{O}) = \mathbb{E} \qquad (2.111)$$

$$\exp(\mathbb{A})\exp(\mathbb{B}) \neq \exp(\mathbb{A}+\mathbb{B}), \qquad \exp(\mathbb{A})\exp(\mathbb{B}) \neq \exp(\mathbb{B})\exp(\mathbb{A})$$

Thus exponential matrices do not commute. If $\mathbb{A}$ is a diagonal $(s \times s)$ matrix, i.e. $\mathbb{A} = (\lambda_i\delta_{ij})$, then

$$\exp(\mathbb{A}) = \left(\exp(\lambda_i\delta_{ij})\right) = \begin{pmatrix} \exp(\lambda_1) & 0 & \cdots & 0 \\ 0 & \exp(\lambda_2) & \cdots & 0 \\ \vdots & \vdots & \ddots & \vdots \\ 0 & 0 & \cdots & \exp(\lambda_s) \end{pmatrix} \qquad (2.112)$$

2.6.2 The nonscattering atmosphere

Now we will solve the radiative transfer equation in an atmosphere where only absorption takes place. In this case the scattering coefficient and, thus, the single scattering albedo ω_0 vanishes so that (2.53) reduces to

$$\mu\frac{d}{d\tau}I(\tau,\mu) = I(\tau,\mu) - B(\tau) \qquad (2.113)$$

For a given frequency interval the Planck function B depends on temperature only. The functional notation $B(\tau)$ simply implies that B has to be evaluated using the temperature existing at τ. Due to the fact that the interior long-wave sources as well as the radiation field at the boundaries of the medium are isotropic, the long-wave radiance field is axially symmetric with respect to the z-axis. Therefore, the radiative transfer equation does not depend on the azimuth angle φ so that, in contrast to the solar spectral region, a Fourier cosine expansion of the radiance is not necessary.

Splitting the radiation field into upwelling and downwelling radiation yields

$$
\mu \frac{d}{d\tau} I_+(\tau, \mu) = I_+(\tau, \mu) - B(\tau)
$$
$$
\mu \frac{d}{d\tau} I_-(\tau, \mu) = -I_-(\tau, \mu) + B(\tau)
$$

(2.114)

From (2.114) it is seen that, in contrast to (2.86), the system of ordinary differential equations for I_+ and I_- is decoupled. The reason for this is that the multiple scattering term, which couples the upwelling and downwelling radiation field, has been neglected. Hence it is possible to obtain an analytical solution of (2.114). This solution requires proper boundary conditions at the top and the bottom of the atmosphere.

By comparing (2.114) with (2.107) we obtain the following correspondences

(a) Upwelling radiation

$$
x = \tau, \qquad\qquad x_0 = \tau_g, \qquad\qquad x' = \tau'
$$
$$
y(x) = I_+(\tau), \quad P(x) = -\frac{1}{\mu}, \quad Q(x) = \frac{1}{\mu} B(\tau)
$$

(a) Downwelling radiation   (2.115)

$$
x = \tau, \qquad\qquad x_0 = 0, \qquad\qquad x' = \tau'
$$
$$
y(x) = I_-(\tau), \quad P(x) = \frac{1}{\mu}, \quad Q(x) = -\frac{1}{\mu} B(\tau)
$$

The boundary conditions at the lower and upper boundary of the atmosphere are given by

$$
I_+(\tau_g, \mu) = B_g, \quad I_-(0, \mu) = \begin{cases} B(0) & \text{if a black body is present at } \tau = 0 \\ 0 & \text{else} \end{cases}
$$

(2.116)

B_g is the *Planckian black body radiation* of the ground. Substituting (2.115) into the general solution (2.108) we obtain for the up- and downwelling radiances at the reference level τ:

$$
\boxed{
\begin{aligned}
I_+(\tau, \mu) &= B_g \exp\left(-\frac{\tau_g - \tau}{\mu}\right) + \frac{1}{\mu} \int_\tau^{\tau_g} B(\tau') \exp\left(-\frac{\tau' - \tau}{\mu}\right) d\tau' \\
I_-(\tau, \mu) &= I_-(0, \mu) \exp\left(-\frac{\tau}{\mu}\right) + \frac{1}{\mu} \int_0^\tau B(\tau') \exp\left(-\frac{\tau - \tau'}{\mu}\right) d\tau'
\end{aligned}
}
$$

(2.117)

These equations can be interpreted as follows: the first term on the right-hand side of each equation describes that part of the radiation which is emitted at the boundary

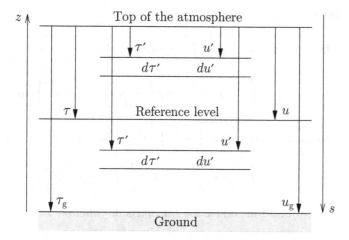

Fig. 2.4 Correspondences between the optical depth τ and the absorbing mass u.

of the medium and transmitted to level τ. The second term on the right-hand side of each equation expresses the total contribution of all elementary layers of optical thickness $d\tau'$. The black body radiations $B(\tau')$ are multiplied by the transmittance from level τ' to the level of observation τ.

According to (1.40) and (1.41) for an absorbing medium the differential of the optical depth $d\tau$ and the mass absorption coefficient κ_{abs} are related by

$$d\tau = k_{abs}ds = \kappa_{abs}du' \implies \tau = \int_0^u \kappa_{abs}du' \tag{2.118}$$

with $\quad du' = \rho_{abs}ds = -\rho_{abs}dz$

Here, u refers to the absorbing mass between the top of the atmosphere and the reference level, see Figure 2.4. This figure also depicts some other correspondences between the optical depth and the absorbing mass which will be used below. Substituting (2.118) into (2.117) yields

$$I_+(u, \mu) = B_g \exp\left(-\frac{1}{\mu}\int_u^{u_g} \kappa_{abs}(u')du'\right)$$
$$+ \frac{1}{\mu}\int_u^{u_g} B(u')\exp\left(-\frac{1}{\mu}\int_u^{u'} \kappa_{abs}(t)dt\right)\kappa_{abs}(u')du'$$
$$I_-(u, \mu) = I_-(0, \mu)\exp\left(-\frac{1}{\mu}\int_0^u \kappa_{abs}(u')du'\right) \tag{2.119}$$
$$+ \frac{1}{\mu}\int_0^u B(u')\exp\left(-\frac{1}{\mu}\int_{u'}^u \kappa_{abs}(t)dt\right)\kappa_{abs}(u')du'$$

These equations can also be reformulated as

$$I_+(u, \mu) = B_g \exp\left(-\frac{1}{\mu}\int_u^{u_g} \kappa_{abs}(u')du'\right)$$

$$- \int_u^{u_g} B(u')\frac{\partial}{\partial u'}\exp\left(-\frac{1}{\mu}\int_u^{u'}\kappa_{abs}(t)dt\right)du' \tag{2.120}$$

$$I_-(u, \mu) = I_-(0, \mu)\exp\left(-\frac{1}{\mu}\int_0^u \kappa_{abs}(u')du'\right)$$

$$+ \int_0^u B(u')\frac{\partial}{\partial u'}\exp\left(-\frac{1}{\mu}\int_{u'}^u\kappa_{abs}(t)dt\right)du'$$

For ease of notation we introduce the *transmission function* T via

$$\boxed{T(u, u', \mu) = \exp\left(-\frac{1}{\mu}\int_u^{u'}\kappa_{abs}(t)dt\right)} \tag{2.121}$$

Now (2.120) assumes the form

$$I_+(u, \mu) = B_g T(u, u_g, \mu) - \int_u^{u_g} B(u')\frac{\partial}{\partial u'}T(u, u', \mu)du'$$

$$I_-(u, \mu) = I_-(0, \mu)T(0, u, \mu) + \int_0^u B(u')\frac{\partial}{\partial u'}T(u', u, \mu)du' \tag{2.122}$$

Partial integration of these equations with respect to u' yields the final result

$$\boxed{\begin{aligned}I_+(u, \mu) &= [B_g - B(u_g)]T(u, u_g, \mu) + B(u) + \int_u^{u_g}\frac{dB}{du'}T(u, u', \mu)du' \\ I_-(u, \mu) &= [I_-(0, \mu) - B(0)]T(0, u, \mu) + B(u) - \int_0^u\frac{dB}{du'}T(u', u, \mu)du'\end{aligned}}$$

$$\tag{2.123}$$

Here, $B(u_g)$ describes the Planck radiation of the air temperature existing directly above the ground. This means that the Earth's surface itself may have a temperature T_g which is different from the air temperature $T(u_g)$. A similar situation applies to the upper boundary of the atmosphere where the black body radiation B must be evaluated at the temperature existing at $u = 0$. A radiation source from outside the atmosphere may radiate with the intensity $I_-(0, \mu)$.

An interesting interpretation can be found for the integral terms in (2.123). Depending on the sign of dB/du', positive or negative contributions are registered at the reference level u. Consider, for example, the situation of a temperature inversion at the reference level as depicted in Figure 2.5. In this case both integral terms will make a positive contribution to the radiances at the level u.

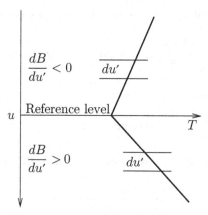

Fig. 2.5 Contributions of the integral terms in (2.123) to the upward and downward radiances.

In contrast to the formal solution of the RTE for the scattering atmosphere, see (2.110), equation (2.123) describes the analytical solution for a nonscattering atmosphere, provided the vertical distributions of the temperature and the absorption coefficient are known. This is certainly a great advantage of (2.123) compared to (2.110). However, as will be seen in later chapters, the solution of the RTE for the purely absorbing atmosphere is still rather complicated. This is due to the fact that the absorption coefficient strongly depends on the wavelength so that the integration of (2.123) over a certain wavelength interval causes some problems.

2.7 Radiative flux densities and heating rates

For the computation of the radiative energy balance and the calculation of radiative heating rates it is sufficient to determine the net radiative flux densities, see Sections 1.2 and 2.3.1. In a horizontally homogeneous atmosphere the radiance depends only on the variables (z, μ, φ) or alternatively on (τ, μ, φ) but not on x and y. Hence, only the vertical component of the net flux density needs to be considered. Since for a purely absorbing atmosphere we have the analytical solution of the RTE, see (2.123), it is interesting to determine the radiative flux densities and heating rates for this particular case. However, before doing so we will again start the discussion with the general situation of a scattering atmosphere.

2.7.1 The scattering atmosphere

It is convenient to rotate the Cartesian coordinate system in such a way that the solar azimuth vanishes, that is $\varphi_0 = 0$, see Figure 2.3. According to (1.37c) and

(2.70) the z-component of the radiative net flux density is then given by

$$E_{\text{net},z}(\tau) = \int_0^{2\pi} \int_{-1}^{1} \sum_{m=0}^{\infty} (2 - \delta_{0m}) I^m(\tau, \mu) \mu \cos m\varphi \, d\mu \, d\varphi \qquad (2.124)$$

Since

$$\int_0^{2\pi} \cos m\varphi \, d\varphi = 2\pi \delta_{0m} \qquad (2.125)$$

we find that only the zeroth Fourier mode $m = 0$ of the radiance is required for evaluating $E_{\text{net},z}$, i.e.

$$E_{\text{net},z}(\tau) = 2\pi \int_{-1}^{1} I^0(\tau, \mu) \mu \, d\mu \qquad (2.126)$$

This is a very important result since it shows that for heating rate computations knowledge of the full directional dependence of the radiation field is not necessary.

Analogously to the radiance field, the term $E_{\text{net},z}$ will be split into an upwelling and a downwelling component by defining

$$
\begin{aligned}
E_{\text{net},z} &= 2\pi \int_0^{1} I^0(\tau, \mu) \mu \, d\mu - 2\pi \int_0^{-1} I^0(\tau, \mu) \mu \, d\mu \\
&= 2\pi \int_0^{1} I_+^0(\tau, \mu) \mu \, d\mu - 2\pi \int_0^{1} I_-^0(\tau, \mu) \mu \, d\mu = E_{+,z} - E_{-,z}
\end{aligned}
\qquad (2.127)
$$

Obviously, the upward and downward directed flux densities $E_{\pm,z}$ are positive definite.

In a similar manner the net radiative flux densities for the x- and y-direction can be computed. Utilizing (1.37a) one finds for $E_{\text{net},x}$

$$E_{\text{net},x}(\tau) = \int_{-1}^{1} \sum_{m=0}^{\infty} (2 - \delta_{0m}) I^m(\tau, \mu)(1 - \mu^2)^{1/2} d\mu \int_0^{2\pi} \cos \varphi \cos m\varphi \, d\varphi \qquad (2.128)$$

Since

$$\int_0^{2\pi} \cos \varphi \cos m\varphi \, d\varphi = (1 + \delta_{01})\delta_{1m}\pi = \pi \delta_{1m} \qquad (2.129)$$

we obtain

$$E_{\text{net},x}(\tau) = 2\pi \int_{-1}^{1} (1 - \mu^2)^{1/2} I^1(\tau, \mu) d\mu \qquad (2.130)$$

Owing to the particular choice of the coordinate system with $\varphi_0 = 0$ and due to the orthogonality of the trigonometric functions the y-component of the net radiative flux density vanishes

$$E_{\text{net},y}(\tau) = \int_{-1}^{1} \sum_{m=0}^{\infty} (2 - \delta_{0m}) I^m(\tau, \mu)(1 - \mu^2)^{1/2} d\mu \int_0^{2\pi} \sin \varphi \cos m\varphi \, d\varphi = 0$$

(2.131)

where now use was made of (1.37b). If the coordinate system is rotated about the z-axis so that $\varphi_0 \neq 0$, then we obtain a nonvanishing value for $E_{\text{net},y}$. However, in this case $E_{\text{net},x}$ will assume a different value as well.

Obviously, for a horizontally homogeneous medium the net flux density does not depend on the horizontal variables x and y. Hence we may write

$$\nabla \cdot \mathbf{E}_{\text{net}} = \frac{\partial E_{\text{net},x}}{\partial x} + \frac{\partial E_{\text{net},y}}{\partial y} + \frac{\partial E_{\text{net},z}}{\partial z} = \frac{\partial E_{\text{net},z}}{\partial z}$$

(2.132)

According to (2.46) for the horizontally homogeneous medium the *radiatively induced temperature change* due to the diffuse radiation at height z is then given by

$$\frac{\partial T}{\partial t}\bigg|_{\text{rad,dif}} = -\frac{1}{c_p \rho(z)} \frac{\partial E_{\text{net},z}}{\partial z} = \frac{k_{\text{ext}}}{c_p \rho(z)} \frac{\partial E_{\text{net},z}}{\partial \tau} = \frac{k_{\text{ext}}}{c_p \rho(z)} \left(\frac{\partial E_{+,z}}{\partial \tau} - \frac{\partial E_{-,z}}{\partial \tau} \right)$$

(2.133)

where k_{ext} is the volume extinction coefficient as defined in (1.48) and ρ is the density of the air. Recall, however, that owing to the splitting of the radiance field into diffuse and direct radiation, see (2.29), (2.133) represents only the contribution of the diffuse radiation. The total temperature change by radiative effects is obtained by adding to $\partial T / \partial t|_{\text{rad,dif}}$ the contribution by the direct solar radiation. Substituting (2.29) together with (2.52) into (1.37c) yields the direct solar net flux density

$$E_{\text{net},z,\text{dir}} = \int_0^{2\pi} \int_{-1}^{1} S(\tau) \delta(\mathbf{\Omega} - \mathbf{\Omega}_0) \mu \, d\mu \, d\varphi = -\mu_0 S_0 \exp\left(-\frac{\tau}{\mu_0}\right)$$

(2.134)

where $\delta(\mathbf{\Omega} - \mathbf{\Omega}_0) = \delta(\mu - \mu_0)\delta(\varphi - \varphi_0)$. Hence, analogously to (2.133), the temperature change due to the extinction of solar radiation is given by

$$\frac{\partial T}{\partial t}\bigg|_{\text{rad,dir}} = \frac{k_{\text{ext}}}{c_p \rho(z)} \frac{\partial E_{\text{net},z,\text{dir}}}{\partial \tau} = \frac{k_{\text{ext}}}{c_p \rho(z)} S_0 \exp\left(-\frac{\tau}{\mu_0}\right)$$

(2.135)

Combining (2.133) and (2.135) we finally obtain

$$\begin{aligned} \frac{\partial T}{\partial t}\bigg|_{\text{rad},\nu} &= \frac{\partial T}{\partial t}\bigg|_{\text{rad,dif},\nu} + \frac{\partial T}{\partial t}\bigg|_{\text{rad,dir},\nu} \\ &= \frac{k_{\text{ext},\nu}}{c_p \rho(z)} \left[\frac{\partial E_{+,z,\nu}}{\partial \tau} - \frac{\partial E_{-,z,\nu}}{\partial \tau} + S_{0,\nu} \exp\left(-\frac{\tau}{\mu_0}\right) \right] \end{aligned}$$

(2.136)

Here, we have again included the subscript ν in order to remind the reader that this is a spectral equation. Hence, the total radiative heating rate is obtained by integrating (2.136) over the entire spectral region. If the vertical net flux density of the total radiation including the parallel solar radiation is constant with height then the radiative temperature change vanishes.

2.7.2 The nonscattering atmosphere

For the nonscattering atmosphere the upward and downward directed flux densities are obtained by integrating (2.123) over the upper and lower hemisphere. This gives

$$
\begin{aligned}
E_+(u) &= \int_0^{2\pi}\int_0^1 I_+(u,\mu)\mu d\mu\, d\varphi \\
&= 2\pi[B_g - B(u_g)]\int_0^1 T(u,u_g,\mu)\mu\, d\mu \\
&\quad + \pi B(u) + 2\pi\int_u^{u_g}\frac{dB}{du'}\int_0^1 T(u,u',\mu)\mu d\mu du' \\
E_-(u) &= \int_0^{2\pi}\int_0^1 I_-(u,\mu)\mu d\mu d\varphi \\
&= 2\pi[I_-(0) - B(0)]\int_0^1 T(0,u,\mu)\mu d\mu \\
&\quad + \pi B(u) - 2\pi\int_0^u\frac{dB}{du'}\int_0^1 T(u',u,\mu)\mu d\mu\, du'
\end{aligned}
\tag{2.137}
$$

where for simplicity we have assumed that the radiation field incident at the top of the atmosphere is isotropic. It is convenient to define the so-called *flux-transmission function* T_f via

$$
T_f(u,u') = 2\int_0^1 T(u,u',\mu)\mu d\mu
\tag{2.138}
$$

Substituting (2.121) into this equation yields

$$
T_f(u,u') = 2\int_0^1 \exp\left(-\frac{1}{\mu}\int_u^{u'}\kappa_{abs}(t)dt\right)\mu d\mu = 2\int_1^\infty \xi^{-3}\exp(-\xi x)d\xi
\tag{2.139}
$$

where the following substitutions have been utilized

$$
\xi = \frac{1}{\mu}, \qquad d\xi = -\frac{1}{\mu^2}d\mu, \qquad x = \int_u^{u'}\kappa_{abs}(t)dt
\tag{2.140}
$$

The last integral expression in (2.139) allows for the introduction of the *exponential integral* of third order

$$E_3(x) = \int_1^\infty \xi^{-3} \exp(-\xi x) d\xi \qquad (2.141)$$

Thus we obtain

$$T_f(u, u') = 2E_3 \left(\int_u^{u'} \kappa_{abs}(t) dt \right) \qquad (2.142)$$

Substitution of the flux-transmission function into (2.137); gives finally

$$E_+(u) = \pi [B_g - B(u_g)] T_f(u, u_g) + \pi B(u) + \pi \int_u^{u_g} \frac{dB}{du'} T_f(u, u') du'$$

$$E_-(u) = \pi [I_-(0) - B(0)] T_f(0, u) + \pi B(u) - \pi \int_0^u \frac{dB}{du'} T_f(u', u) du' \qquad (2.143)$$

These expressions may be used in (2.133) to obtain the infrared contributions of the radiative heating rates whereby the substitution $du = -\rho_{abs} dz$ has to be applied.

2.8 Appendix

2.8.1 Local thermodynamic equilibrium

Assuming conditions of *local thermodynamic equilibrium*, the source function J_ν^e for infrared emission should be proportional to the Planck black body function B_ν. We will now attempt to determine the proportionality factor between the source function and the Planck function for a simple physical situation. The result, however, is more general. From (2.36) we obtain

$$\Omega \cdot \nabla I_\nu = \frac{d}{ds} I_\nu(\mathbf{r}, \Omega) = -k_{ext,\nu} I_\nu(\mathbf{r}, \Omega) + \frac{k_{sca,\nu}}{4\pi} \int_{4\pi} P_\nu(\Omega' \cdot \Omega) I_\nu(\mathbf{r}, \Omega') d\Omega' + J_\nu^e(\mathbf{r})$$

$$(2.144)$$

Here, we have omitted the primary scattering term of the solar radiation since in the range of infrared emission the solar radiation is practically zero for atmospheric conditions in the troposphere and lower stratosphere. Integration of (2.144) over the unit sphere, assuming isotropic and homogeneous radiation, gives

$$\int_{4\pi} \frac{dI_\nu(\mathbf{r})}{ds} d\Omega = -k_{ext,\nu} \int_{4\pi} I_\nu(\mathbf{r}) d\Omega + \frac{k_{sca,\nu}}{4\pi} \int_{4\pi} I_\nu(\mathbf{r}) \int_{4\pi} P_\nu d\Omega' \, d\Omega + \int_{4\pi} J_\nu^e(\mathbf{r}) d\Omega$$

$$= -k_{abs,\nu} \int_{4\pi} I_\nu(\mathbf{r}) d\Omega + \int_{4\pi} J_\nu^e(\mathbf{r}) d\Omega = 0 \qquad (2.145)$$

where the integral of the phase function over unit sphere has been evaluated according to (1.46). From (2.145) we obtain for the source function

$$\boxed{J_\nu^e(\mathbf{r}) = k_{\text{abs},\nu} I_\nu(\mathbf{r}) = k_{\text{abs},\nu} B_\nu(\mathbf{r})} \tag{2.146}$$

Since $J_\nu^e(\mathbf{r})$ is the source function for true emission of heat radiation we set $I_\nu = B_\nu$. (2.146) is known as *Kirchhoff's law*. In case that the conditions of local thermodynamic equilibrium are not met, the emission of radiative energy becomes a function of the energy states in the gas. Equation (2.146) is not an exact derivation but a plausible discussion.

2.9 Problems

2.1: Find the normalization constant C for the phase function $\mathcal{P}(\cos\Theta) = C(1 + \cos^2\Theta)$. Plot your result in polar coordinates and Θ versus $\mathcal{P}(\cos\Theta)$.

2.2: Obtain the solution to the RTE in the form (2.76) for radiative transfer within a plane–parallel cloud of total optical thickness τ_c. Assume that no diffuse radiation is incident on the cloud either from above or from below. Ignore the multiple scattering term and the Planckian emission.

2.3: Consider the following special cases for the solution you obtained from Problem 2.2

(a) Find $I_-^m(\tau, \mu = \mu_0)$.
(b) Find $I_+^m(\tau, 0)$ and $I_-(\tau, 0)$ for $\tau \neq 0$ and $\tau \neq \tau_c$.
(c) Find $I_+^m(\tau, \mu)$ and $I_-^m(\tau, \mu)$ for $\tau = 0$ and $\mu = 0$.

2.4: In the radiative transfer theory we have to deal with integrals of the type

$$\int_\tau^{\tau_c} \exp\left(-\frac{\tau' - \tau}{\mu}\right) A(\tau')\frac{d\tau'}{\mu} \quad \text{and} \quad \int_0^\tau \exp\left(-\frac{\tau - \tau'}{\mu}\right) A(\tau')\frac{d\tau'}{\mu}$$

Evaluate these integrals for $\mu = 0$ where $A(\tau')$ is an arbitrary function.

2.5: Assume isothermal conditions for the system Earth–atmosphere.

(a) From (2.117) find $I_+(\tau, \mu)$.
(b) Which additional condition must apply to obtain the same result for $I_-(\tau, \mu)$?
(c) Show that the identical results follow from (2.123).

2.6: The *exponential integral* of order n is defined by

$$E_n(x) = \int_1^\infty \frac{\exp(-\xi x)}{\xi^n} d\xi$$

(a) Find $E_n(0)$ for $n = 1, 2, \ldots$
(b) Find dE_n/dx.

(c) Verify

$$n E_{n+1}(x) = \exp(-x) - x E_n(x) = \exp(-x) + x \frac{d E_{n+1}(x)}{dx}, \qquad n = 1, 2, \ldots$$

(d) Verify $E_n(x) = \int_0^1 \exp\left(-\frac{x}{\mu}\right) \mu^{n-2} d\mu.$

2.7: According to (2.28) for a plane–parallel atmosphere the RTE can be written in the form

$$\mu \frac{d}{d\tau} I(\tau, \mu, \varphi) = I(\tau, \mu, \varphi) - \frac{1}{4\pi} \int_0^{2\pi} \int_{-1}^{1} P(\mu, \varphi, \mu', \varphi') d\mu' d\varphi'$$

if $\omega_0 = 1$ and $J^e = 0$. In this case the net flux E_{net} is a constant. Show that the solution to this equation can be written as

$$K(\tau) = \frac{E_{net}}{4\pi} \left[\left(1 - \frac{p_1}{3}\right) \tau + C \right]$$

where C is a constant and K is given by

$$K(\tau) = \frac{1}{4\pi} \int_0^{2\pi} \int_{-1}^{1} I(\tau, \mu, \varphi) \mu^2 d\mu d\varphi$$

Hint: Expand the phase function according to (2.55) and make use of the addition theorem (2.66).

2.8: Consider the RTE in the form stated in Problem 2.7. Assume that the phase function is given by $P = 3/4(1 + \cos^2 \Theta)$. Find the source function for the azimuthally averaged RTE.

2.9: Consider the RTE as given in Problem 2.7. For a semi-infinite plane–parallel atmosphere and the boundary condition $I(0, -\mu) = 0$.

(a) Find the solution to the azimuthally averaged RTE in case of isotropic scattering.
(b) Introduce the solution found under (a) into the integral $I_n(\tau) = \int_{-1}^{1} I(\tau, \mu) \mu^n d\mu$. Finally, introduce the exponential integral defined in Problem 2.6 into the latter integral.
(c) Find the net flux density in terms of the exponential integral.

3

Principles of invariance

The *principle of invariance* in the original form was stated by Ambartsumian (1942) expressing the invariance of the diffusely reflected radiation emerging from a semi-infinite atmosphere to the addition or subtraction of an infinitely thin atmospheric layer. Chandrasekhar (1960) advanced the original form and stated four general principles of invariance which apply to finite atmospheric layers. These principles are not based on the radiative transfer equation, but they are of equal physical validity. We accept Goody's (1964a) statement that the principles of invariance may be viewed as a series of common-sense relations between the scattering and transmission functions with the radiances emerging from the upper and lower boundaries of an atmospheric layer and at some intermediate variable level.

3.1 Definitions of the scattering and transmission functions

Let us consider a plane–parallel atmospheric layer of vertical optical thickness τ_1 bounded on both sides by a vacuum, see Figure 3.1. The upper boundary of this layer is illuminated by a beam of parallel downward directed radiation S_0, while at $\tau = \tau_1$ no radiation is incident in the upward direction. For simplicity, only short-wave radiation will be considered. However, inclusion of infrared radiation causes no particular difficulties. We call this situation the *restricted* or *standard problem*. We define the *scattering function* $S(\tau_1, \mu, \varphi, \mu_0, \varphi_0)$ and the *transmission function* $T(\tau_1, \mu, \varphi, \mu_0, \varphi_0)$ by formulating the reflected and transmitted radiances $I(0, \mu, \varphi)$ and $I(\tau_1, -\mu, \varphi)$ as

$$
\begin{aligned}
I(0, \mu, \varphi) &= \frac{S_0}{4\pi\mu} S(\tau_1, \mu, \varphi, \mu_0, \varphi_0) \\
I(\tau_1, -\mu, \varphi) &= \frac{S_0}{4\pi\mu} T(\tau_1, \mu, \varphi, \mu_0, \varphi_0)
\end{aligned}
\tag{3.1}
$$

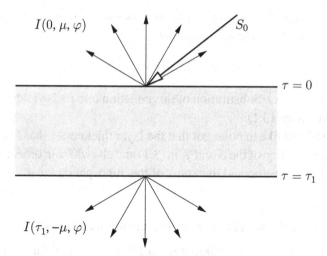

Fig. 3.1 Schematic diagram of reflected and transmitted radiances at $\tau = 0$ and $\tau = \tau_1$ of a plane–parallel atmospheric layer.

where here and in the following $0 < \mu \leq 1$. Thus, $I(0, \mu, \varphi)$ denotes upward directed radiation at the top of the atmospheric layer while $I(\tau_1, -\mu, \varphi)$ describes downward directed radiation at $\tau = \tau_1$. The reason that the factor $1/\mu$ has been introduced in the definitions (3.1) is to obtain symmetry of the S and T functions in the variables (μ, φ) and (μ_0, φ_0) as required by *Helmholtz's principle of reciprocity* to be discussed soon.

It is noteworthy that the reflected and transmitted intensities include only the diffuse radiation which has suffered one or more scattering processes, i.e. the directly transmitted radiation $S_0 \exp(-\tau_1/\mu_0)$ is not included in (3.1). The radiances appearing in (3.1) are solutions of the standard problem. The extended solution which includes ground reflection will be taken up in a later section.

So far we have assumed that the atmospheric layer is illuminated by an incident parallel beam. The solution for an arbitrary incident radiation field $I_{\text{in}}(0, -\mu', \varphi')$ with the same angular distribution at every point on the surface $\tau = 0$ can also be expressed in terms of the functions S and T. In this situation we would have to integrate the incoming radiance over all directions of the incoming light yielding

$$I(0, \mu, \varphi) = \frac{1}{4\pi\mu} \int_0^{2\pi} \int_0^1 \mathcal{S}(\tau_1, \mu, \varphi, \mu', \varphi') I_{\text{in}}(0, -\mu', \varphi') d\mu' d\varphi'$$

$$I(\tau_1, -\mu, \varphi) = \frac{1}{4\pi\mu} \int_0^{2\pi} \int_0^1 \mathcal{T}(\tau_1, \mu, \varphi, \mu', \varphi') I_{\text{in}}(0, -\mu', \varphi') d\mu' d\varphi'$$

(3.2)

By assuming that the incoming radiance is the solar radiation, I_{in} may be expressed in terms of the Dirac δ-functions as

$$I_{in}(0, -\mu, \varphi) = S_0 \delta(\mu - \mu_0)\delta(\varphi - \varphi_0) \tag{3.3}$$

Note that $\delta(x) = \delta(-x)$. Substitution of this equation into (3.2) yields, as it should, the original equations (3.1).

Finally, we would like to point out that the layer thickness τ_1 has been explicitly included in the definition of the S and T in (3.1) and elsewhere in order to emphasize their dependence on the optical thickness of the atmospheric layer.

3.2 Diffuse reflection in a semi-infinite atmosphere

Consider a semi-infinite plane–parallel atmosphere being illuminated by a parallel beam of radiation. In this case only the law of reflection will be of interest. Now the emerging diffuse radiation at the top of the layer is written as

$$I(0, \mu, \varphi) = \frac{S_0}{4\pi\mu} S(\mu, \varphi, \mu_0, \varphi_0) \tag{3.4}$$

omitting specific reference to the infinite optical thickness of the atmosphere.

The incident solar flux density S_0 at the upper boundary of the layer is reduced according to *Beer's law*, see (2.52). Thus at the optical depth τ we obtain the so-called *reduced parallel solar flux density* $S_0 \exp(-\tau/\mu_0)$. At this level in the downward direction we not only have the reduced parallel solar radiation but also a diffuse radiation field $I(\tau, -\mu, \varphi)$. Both parts of the downward radiation will be reflected by the atmospheric layer below τ so that the upward radiation at τ is given by

$$\begin{aligned}
I(\tau, \mu, \varphi) &= \frac{1}{4\pi\mu} \int_0^{2\pi} \int_0^1 S(\mu, \varphi, \mu', \varphi')I(\tau, -\mu', \varphi')d\mu'd\varphi' \\
&+ \frac{S_0}{4\pi\mu} \exp\left(-\frac{\tau}{\mu_0}\right) S(\mu, \varphi, \mu_0, \varphi_0)
\end{aligned} \tag{3.5}$$

Owing to the boundary condition

$$I(0, -\mu, \varphi) = 0 \tag{3.6}$$

at the top of the layer where $\tau = 0$, (3.5) reduces to the first equation of (3.1) in case of a finite optical thickness. Furthermore, by observing that for a semi-infinite optical thickness the atmosphere below the level τ is still infinitely large, at level τ the total downward radiation is reflected according to the same law of diffuse reflection as (3.4). Thus equation (3.5) can be considered as a *statement* of the invariance of $S(\mu, \varphi, \mu_0, \varphi_0)$ to the addition and subtraction of layers in a semi-infinite atmosphere.

In order to obtain the required integral equation for S, we differentiate (3.5) with respect to τ, set $\tau = 0$ and find

$$\left[\frac{d}{d\tau} I(\tau, \mu, \varphi)\right]_{\tau=0} = \frac{1}{4\pi\mu} \int_0^{2\pi} \int_0^1 S(\mu, \varphi, \mu', \varphi') \left[\frac{d}{d\tau} I(\tau, -\mu', \varphi')\right]_{\tau=0} d\mu' d\varphi'$$
$$- \frac{S_0}{4\pi\mu\mu_0} S(\mu, \varphi, \mu_0, \varphi_0) \tag{3.7}$$

The derivatives occurring in this equation can be evaluated with the help of the RTE in the form

$$\mu \frac{d}{d\tau} I(\tau, \mu, \varphi) = I(\tau, \mu, \varphi) - J(\tau, \mu, \varphi) \tag{3.8}$$

where J is the source function which, according to equation (2.53), may be written as

$$J(\tau, \mu, \varphi) = \frac{\omega_0}{4\pi} \int_0^{2\pi} \int_{-1}^1 P(\mu, \varphi, \mu', \varphi') I(\tau, \mu', \varphi') d\mu' d\varphi'$$
$$+ \frac{\omega_0}{4\pi} S_0 \exp\left(-\frac{\tau}{\mu_0}\right) P(\mu, \varphi, -\mu_0, \varphi_0) \tag{3.9}$$

The two differential expressions appearing in (3.7) follow from (3.8) and are given by

$$\left[\frac{d}{d\tau} I(\tau, \mu, \varphi)\right]_{\tau=0} = \frac{1}{\mu} [I(0, \mu, \varphi) - J(0, \mu, \varphi)]$$
$$\left[\frac{d}{d\tau} I(\tau, -\mu, \varphi)\right]_{\tau=0} = \frac{1}{\mu} J(0, -\mu, \varphi) \tag{3.10}$$

where use was made of the boundary condition (3.6). Substituting (3.10) into (3.7) yields

$$I(0, \mu, \varphi) - J(0, \mu, \varphi) = \frac{1}{4\pi} \int_0^{2\pi} \int_0^1 S(\mu, \varphi, \mu', \varphi') J(0, -\mu', \varphi') \frac{d\mu'}{\mu'} d\varphi'$$
$$- \frac{S_0}{4\pi\mu_0} S(\mu, \varphi, \mu_0, \varphi_0) \tag{3.11}$$

With the help of (3.4), equation (3.11) can be rewritten as

$$\frac{S_0}{4\pi} \left(\frac{1}{\mu} + \frac{1}{\mu_0}\right) S(\mu, \varphi, \mu_0, \varphi_0) = \frac{1}{4\pi} \int_0^{2\pi} \int_0^1 S(\mu, \varphi, \mu', \varphi') J(0, -\mu', \varphi') \frac{d\mu'}{\mu'} d\varphi'$$
$$+ J(0, \mu, \varphi) \tag{3.12}$$

In order to replace $J(0, \mu, \varphi)$ on the right-hand side of (3.12), we again use (3.9). To make the mathematical operation more obvious, we first rewrite the integral term as

$$\int_{-1}^{1} P(\mu, \varphi, \mu', \varphi')I(\tau, \mu', \varphi')d\mu' = \int_{0}^{1} P(\mu, \varphi, \mu', \varphi')I(\tau, \mu', \varphi')d\mu'$$

$$+ \int_{0}^{1} P(\mu, \varphi, -\mu', \varphi')I(\tau, -\mu', \varphi')d\mu'$$

(3.13)

At $\tau = 0$, due to the upper boundary condition, the second integral in (3.13) vanishes so that

$$J(0, \mu, \varphi) = \frac{\omega_0 S_0}{(4\pi)^2} \int_{0}^{2\pi} \int_{0}^{1} P(\mu, \varphi, \mu', \varphi')S(\mu', \varphi', \mu_0, \varphi_0)\frac{d\mu'}{\mu'}d\varphi'$$

$$+ \frac{\omega_0}{4\pi} S_0 P(\mu, \varphi, -\mu_0, \varphi_0)$$

(3.14)

Introducing this expression into (3.12) finally yields

$$\left(\frac{1}{\mu} + \frac{1}{\mu_0}\right)S(\mu, \varphi, \mu_0, \varphi_0)$$

$$= \frac{\omega_0}{(4\pi)^2} \int_{0}^{2\pi} \int_{0}^{1} \int_{0}^{2\pi} \int_{0}^{1} P(-\mu', \varphi', \mu'', \varphi'')S(\mu'', \varphi'', \mu_0, \varphi_0)S(\mu, \varphi, \mu', \varphi')$$

$$\times \frac{d\mu''}{\mu''}d\varphi'' \frac{d\mu'}{\mu'}d\varphi' + \frac{\omega_0}{4\pi} \int_{0}^{2\pi} \int_{0}^{1} P(-\mu', \varphi', -\mu_0, \varphi_0)S(\mu, \varphi, \mu', \varphi')\frac{d\mu'}{\mu'}d\varphi'$$

$$+ \frac{\omega_0}{4\pi} \int_{0}^{2\pi} \int_{0}^{1} P(\mu, \varphi, \mu', \varphi')S(\mu', \varphi', \mu_0, \varphi_0)\frac{d\mu'}{\mu'}d\varphi' + \omega_0 P(\mu, \varphi, -\mu_0, \varphi_0)$$

(3.15)

which is the desired integral equation for S. This integral equation is nonlinear, but for simple phase functions there exist methods to evaluate it without excessive numerical difficulties.

The scattering function S and the transmission function T have the important property of being symmetric in the pair of variables (μ, φ) and (μ_0, φ_0). In order to recognize this, we first refer to the definition of the *phase function* in the form (2.68). Recalling the symmetry relations of the associated Legendre polynomials as stated in (2.59) we immediately recognize the validity of the relations

$$P(\mu, \varphi, \mu', \varphi') = P(\mu', \varphi', \mu, \varphi)$$
$$P(-\mu, \varphi, -\mu', \varphi') = P(\mu, \varphi, \mu', \varphi')$$
$$P(\mu, \varphi, -\mu', \varphi') = P(-\mu, \varphi, \mu', \varphi') = P(\mu', \varphi', -\mu, \varphi)$$

(3.16)

It is convenient to introduce the function $\tilde{f}(\mu, \varphi, \mu', \varphi')$ which is obtained from $f(\mu, \varphi, \mu', \varphi')$ by transposing the variables (μ, φ) and (μ', φ')

$$\tilde{f}(\mu, \varphi, \mu', \varphi') = f(\mu', \varphi', \mu, \varphi) \tag{3.17}$$

Using this notation, the symmetry properties of $\mathcal{P}(\mu, \varphi, \mu', \varphi')$ may be expressed by means of

$$\tilde{\mathcal{P}}(\mu, \varphi, \mu', \varphi') = \mathcal{P}(\mu', \varphi', \mu, \varphi), \quad \tilde{\mathcal{P}}(\mu, \varphi, -\mu', \varphi') = \mathcal{P}(\mu', \varphi', -\mu, \varphi) \tag{3.18}$$

Transposing the variables (μ, φ) and (μ_0, φ_0) in (3.15) applying the relations (3.16), we observe that $\tilde{S}$ satisfies the same equation as S. Thus if S is a solution of (3.15) so is $\tilde{S}$. However, it does not follow that S is necessarily symmetrical in the variables (μ, φ) and (μ_0, φ_0). We omit the complete proof of the symmetry relations but simply assume that the scattering and the transmission functions are symmetric, i.e.

$$\boxed{\tilde{S}(\mu, \varphi, \mu_0, \varphi_0) = S(\mu, \varphi, \mu_0, \varphi_0), \quad \tilde{T}(\mu, \varphi, \mu_0, \varphi_0) = T(\mu, \varphi, \mu_0, \varphi_0)}$$

$$\tag{3.19}$$

Verbally, the scattering and the transmission functions are unaltered when the directions of incidence and emergence are interchanged. This is the *principle of invariance* as applied to the functions S and T. A rigorous proof of (3.19) may be found in Chandrasekhar (1960).

We will now briefly consider the simple case of isotropic radiation which, according to (2.13), is given by $\mathcal{P} = 1$. Owing to the axial symmetry of the radiation field, (3.15) reduces to

$$\left(\frac{1}{\mu} + \frac{1}{\mu_0}\right) S(\mu, \mu_0) = \frac{\omega_0}{4} \int_0^1 \int_0^1 S(\mu'', \mu_0) S(\mu, \mu') \frac{d\mu'' \, d\mu'}{\mu'' \, \mu'}$$
$$+ \frac{\omega_0}{2} \int_0^1 S(\mu, \mu') \frac{d\mu'}{\mu'} + \frac{\omega_0}{2} \int_0^1 S(\mu', \mu_0) \frac{d\mu'}{\mu'} + \omega_0 \tag{3.20}$$

which can be also be written in the form

$$\left(\frac{1}{\mu} + \frac{1}{\mu_0}\right) S(\mu, \mu_0) = \omega_0 \left(1 + \frac{1}{2} \int_0^1 S(\mu, \mu') \frac{d\mu'}{\mu'}\right) \left(1 + \frac{1}{2} \int_0^1 S(\mu', \mu_0) \frac{d\mu'}{\mu'}\right) \tag{3.21}$$

Since $S(\mu, \mu')$ is symmetric in the variables (μ, μ'), on the right-hand side of this equation the two terms in parentheses must be the values of μ and μ_0 of the same function. It is customary to denote this function by the symbol $H(\mu)$, as defined by

$$
H(\mu) = 1 + \frac{1}{2} \int_0^1 S(\mu, \mu') \frac{d\mu'}{\mu'} = 1 + \frac{1}{2} \int_0^1 S(\mu', \mu) \frac{d\mu'}{\mu'} \qquad (3.22)
$$

For the conservative case we obtain

$$
\int_0^1 H(\mu) d\mu = 2, \qquad \int_0^1 \mu H(\mu) d\mu = \frac{2}{\sqrt{3}} \qquad (3.23)
$$

With (3.22) equation (3.21) assumes the more simple form

$$
\left(\frac{1}{\mu} + \frac{1}{\mu_0} \right) S(\mu, \mu_0) = \omega_0 H(\mu) H(\mu_0) \qquad (3.24)
$$

By substituting S back into equation (3.22), we find the following expression for the so-called *H-function* which is a nonlinear integral equation

$$
H(\mu) = 1 + \frac{\omega_0 \mu}{2} H(\mu) \int_0^1 \frac{H(\mu')}{\mu + \mu'} d\mu' \qquad (3.25)
$$

The H-function plays an important role in some parts of radiative tansfer theory. A table of the H-function can be found, for example, in Chandrasekhar (1960).

3.3 Chandrasekhar's four statements of the principles of invariance

Let us consider a plane–parallel atmospheric layer of total optical thickness τ_1 so that $0 \le \tau \le \tau_1$. We will now state the following four principles.

(1) The radiance $I(\tau, \mu, \varphi)$ in the upward direction at any level τ results from the reflection of the reduced incident flux density $S_0 \exp(-\tau/\mu_0)$ and the diffuse downward radiation $I(\tau, -\mu', \varphi')$ incident on the surface τ, reflected by the atmosphere below τ having the optical thickness $(\tau_1 - \tau)$. Hence we may write

$$
\begin{aligned}
I(\tau, \mu, \varphi) = &\frac{1}{4\pi\mu} \int_0^{2\pi} \int_0^1 S(\tau_1 - \tau, \mu, \varphi, \mu', \varphi') I(\tau, -\mu', \varphi') d\mu' d\varphi' \\
&+ \frac{S_0}{4\pi\mu} \exp\left(-\frac{\tau}{\mu_0} \right) S(\tau_1 - \tau, \mu, \varphi, \mu_0, \varphi_0)
\end{aligned} \qquad (3.26)
$$

(2) The radiance $I(\tau, -\mu, \varphi)$ in the downward direction at any level τ results from the transmission of the incident flux density by the atmosphere of optical thickness τ, above the surface, and the reflection by this same surface[1] of the diffuse radiation

---

[1] The formulation 'reflection by this same surface' sounds as if some sort of a mirror in horizontal position is placed in the atmosphere reflecting the radiation. Chandrasekhar simply implies that the atmospheric sublayer above the surface acts as if it were an imperfect mirror.

$I(\tau, \mu, \varphi)$ incident on it from below, that is

$$
\begin{aligned}
I(\tau, -\mu, \varphi) = {} & \frac{1}{4\pi\mu} \int_0^{2\pi} \int_0^1 S(\tau, \mu, \varphi, \mu', \varphi') I(\tau, \mu', \varphi') d\mu' d\varphi' \\
& + \frac{S_0}{4\pi\mu} T(\tau, \mu, \varphi, \mu_0, \varphi_0)
\end{aligned}
\tag{3.27}
$$

(3) The diffuse reflection of the incident light by the entire atmospheric layer of optical thickness τ_1 is equivalent to the reflection by the part of the atmosphere above the level τ and the transmission by the same part of the atmosphere of the diffuse radiation $I(\tau, \mu', \varphi')$ incident on the surface τ from below. The corresponding equation reads

$$
\begin{aligned}
\frac{S_0}{4\pi\mu} S(\tau_1, \mu, \varphi, \mu_0, \varphi_0) = {} & \frac{1}{4\pi\mu} \int_0^{2\pi} \int_0^1 T(\tau, \mu, \varphi, \mu', \varphi') I(\tau, \mu', \varphi') d\mu' d\varphi' \\
& + \frac{S_0}{4\pi\mu} S(\tau, \mu, \varphi, \mu_0, \varphi_0) + I(\tau, \mu, \varphi) \exp\left(-\frac{\tau}{\mu}\right)
\end{aligned}
\tag{3.28}
$$

The last term on the right-hand side refers to the direct transmission of the diffuse radiance $I(\tau, \mu, \varphi)$ which is already in the direction (μ, φ).

(4) The diffuse transmission of the incident light by the entire atmospheric layer of optical thickness τ_1 is equivalent to the transmission of the reduced incident flux density $S_0 \exp(-\tau/\mu_0)$ and the diffuse radiation $I(\tau, -\mu', \varphi')$ incident on the surface τ and transmitted by the atmosphere of optical thickness $(\tau_1 - \tau)$ below τ. This principle is expressed by

$$
\begin{aligned}
\frac{S_0}{4\pi\mu} T(\tau_1, \mu, \varphi, \mu_0, \varphi_0) = {} & \frac{1}{4\pi\mu} \int_0^{2\pi} \int_0^1 T(\tau_1 - \tau, \mu, \varphi, \mu', \varphi') I(\tau, -\mu', \varphi') d\mu' d\varphi' \\
& + \frac{S_0}{4\pi\mu} \exp\left(-\frac{\tau}{\mu_0}\right) T(\tau_1 - \tau, \mu, \varphi, \mu_0, \varphi_0) \\
& + I(\tau, -\mu, \varphi) \exp\left(-\frac{\tau_1 - \tau}{\mu}\right)
\end{aligned}
\tag{3.29}
$$

The last term on the right-hand side refers to the direct transmission of the diffuse radiance $I(\tau, -\mu, \varphi)$, already in the direction $(-\mu, \varphi)$.

It is important to note that equations (3.26)–(3.29) are sufficient to uniquely determine the radiation field in terms of the scattering and the transmission functions S and T for plane–parallel atmospheres of finite optical thickness.

The four principles of invariance listed in the previous section will now be used to derive a set of four integral equations for the scattering and the transmission functions. These will be obtained by differentiating equations (3.26)–(3.29) with respect to τ and then passing (3.26) and (3.29) to the limit $\tau = 0$ and (3.27) and (3.28) to the limit $\tau = \tau_1$. Furthermore, we will make use of the boundary conditions

$$I(0, -\mu, \varphi) = 0, \qquad I(\tau_1, \mu, \varphi) = 0 \tag{3.30}$$

which apply to the standard problem. The resulting equations are

$$\left[\frac{d}{d\tau} I(\tau, \mu, \varphi) \right]_{\tau=0} = \frac{1}{4\pi\mu} \int_0^{2\pi} \int_0^1 S(\tau_1, \mu, \varphi, \mu', \varphi') \left[\frac{d}{d\tau} I(\tau, -\mu', \varphi') \right]_{\tau=0} d\mu' d\varphi'$$
$$+ \frac{S_0}{4\pi\mu} \left(-\frac{1}{\mu_0} S(\tau_1, \mu, \varphi, \mu_0, \varphi_0) \right.$$
$$\left. + \left[\frac{d}{d\tau} S(\tau_1 - \tau, \mu, \varphi, \mu_0, \varphi_0) \right]_{\tau=0} \right)$$

$$\left[\frac{d}{d\tau} I(\tau, -\mu, \varphi) \right]_{\tau=\tau_1} = \frac{1}{4\pi\mu} \int_0^{2\pi} \int_0^1 S(\tau_1, \mu, \varphi, \mu', \varphi') \left[\frac{d}{d\tau} I(\tau, \mu', \varphi') \right]_{\tau=\tau_1} d\mu' d\varphi'$$
$$+ \frac{S_0}{4\pi\mu} \left[\frac{d}{d\tau} T(\tau, \mu, \varphi, \mu_0, \varphi_0) \right]_{\tau=\tau_1}$$

$$0 = \frac{1}{4\pi\mu} \int_0^{2\pi} \int_0^1 T(\tau_1, \mu, \varphi, \mu', \varphi') \left[\frac{d}{d\tau} I(\tau, \mu', \varphi') \right]_{\tau=\tau_1} d\mu' d\varphi'$$
$$+ \frac{S_0}{4\pi\mu} \left[\frac{d}{d\tau} S(\tau, \mu, \varphi, \mu_0, \varphi_0) \right]_{\tau=\tau_1} + \left[\frac{d}{d\tau} I(\tau, \mu, \varphi) \right]_{\tau=\tau_1} \exp\left(-\frac{\tau_1}{\mu} \right)$$

$$0 = \frac{1}{4\pi\mu} \int_0^{2\pi} \int_0^1 T(\tau_1, \mu, \varphi, \mu', \varphi') \left[\frac{d}{d\tau} I(\tau, -\mu', \varphi') \right]_{\tau=0} d\mu' d\varphi'$$
$$+ \frac{S_0}{4\pi\mu} \left(-\frac{1}{\mu_0} T(\tau_1, \mu, \varphi, \mu_0, \varphi_0) + \left[\frac{d}{d\tau} T(\tau_1 - \tau, \mu, \varphi, \mu_0, \varphi_0) \right]_{\tau=0} \right)$$
$$+ \left[\frac{d}{d\tau} I(\tau, -\mu, \varphi) \right]_{\tau=0} \exp\left(-\frac{\tau_1}{\mu} \right) \tag{3.31}$$

The derivatives $dI/d\tau$ appearing in these equations can be replaced with the help of the RTE in the form (3.8) by employing the basic definitions (3.1) and the boundary conditions (3.30). Thus we obtain

$$\left[\frac{d}{d\tau} I(\tau, \mu, \varphi) \right]_{\tau=0} = \frac{1}{\mu} \left(\frac{S_0}{4\pi\mu} S(\tau_1, \mu, \varphi, \mu_0, \varphi_0) - J(0, \mu, \varphi) \right)$$

$$\left[\frac{d}{d\tau} I(\tau, -\mu, \varphi) \right]_{\tau=0} = \frac{1}{\mu} J(0, -\mu, \varphi)$$

$$\left[\frac{d}{d\tau} I(\tau, \mu, \varphi) \right]_{\tau=\tau_1} = -\frac{1}{\mu} J(\tau_1, \mu, \varphi) \tag{3.32}$$

$$\left[\frac{d}{d\tau} I(\tau, -\mu, \varphi) \right]_{\tau=\tau_1} = -\frac{1}{\mu} \left(\frac{S_0}{4\pi\mu} T(\tau_1, \mu, \varphi, \mu_0, \varphi_0) - J(\tau_1, -\mu, \varphi) \right)$$

Substituting (3.32) into (3.31) results in the four integral equations

(a) $\dfrac{S_0}{4\pi}\left(\left(\dfrac{1}{\mu}+\dfrac{1}{\mu_0}\right)S(\tau_1,\mu,\varphi,\mu_0,\varphi_0)-\left[\dfrac{d}{d\tau}S(\tau_1-\tau,\mu,\varphi,\mu_0,\varphi_0)\right]_{\tau=0}\right)$

$=\dfrac{1}{4\pi}\displaystyle\int_0^{2\pi}\int_0^1 S(\tau_1,\mu,\varphi,\mu',\varphi')J(0,-\mu',\varphi')\dfrac{d\mu'}{\mu'}d\varphi'+J(0,\mu,\varphi)$

(b) $\dfrac{S_0}{4\pi}\left(\dfrac{1}{\mu}T(\tau_1,\mu,\varphi,\mu_0,\varphi_0)+\left[\dfrac{d}{d\tau}T(\tau,\mu,\varphi,\mu_0,\varphi_0)\right]_{\tau=\tau_1}\right)$

$=\dfrac{1}{4\pi}\displaystyle\int_0^{2\pi}\int_0^1 S(\tau_1,\mu,\varphi,\mu',\varphi')J(\tau_1,\mu',\varphi')\dfrac{d\mu'}{\mu'}d\varphi'+J(\tau_1,-\mu,\varphi)$

(c) $\dfrac{S_0}{4\pi}\left[\dfrac{d}{d\tau}S(\tau,\mu,\varphi,\mu_0,\varphi_0)\right]_{\tau=\tau_1}$ (3.33)

$=\dfrac{1}{4\pi}\displaystyle\int_0^{2\pi}\int_0^1 T(\tau_1,\mu,\varphi,\mu',\varphi')J(\tau_1,\mu',\varphi')\dfrac{d\mu'}{\mu'}d\varphi'+J(\tau_1,\mu,\varphi)\exp\left(-\dfrac{\tau_1}{\mu}\right)$

(d) $\dfrac{S_0}{4\pi}\left(\dfrac{1}{\mu_0}T(\tau_1,\mu,\varphi,\mu_0,\varphi_0)-\left[\dfrac{d}{d\tau}T(\tau_1-\tau,\mu,\varphi,\mu_0,\varphi_0)\right]_{\tau=0}\right)$

$=\dfrac{1}{4\pi}\displaystyle\int_0^{2\pi}\int_0^1 T(\tau_1,\mu,\varphi,\mu',\varphi')J(0,-\mu',\varphi')\dfrac{d\mu'}{\mu'}d\varphi'+J(0,-\mu,\varphi)\exp\left(-\dfrac{\tau_1}{\mu}\right)$

The validity of

(a) $\left[\dfrac{d}{d\tau}S(\tau,\mu,\varphi,\mu_0,\varphi_0)\right]_{\tau=\tau_1}=-\left[\dfrac{d}{d\tau}S(\tau_1-\tau,\mu,\varphi,\mu_0,\varphi_0)\right]_{\tau=0}$

(3.34)

(b) $\left[\dfrac{d}{d\tau}T(\tau,\mu,\varphi,\mu_0,\varphi_0)\right]_{\tau=\tau_1}=-\left[\dfrac{d}{d\tau}T(\tau_1-\tau,\mu,\varphi,\mu_0,\varphi_0)\right]_{\tau=0}$

may be easily verified. Multiplying (3.33b) by $-1/\mu_0$ and (3.33d) by $1/\mu$ and adding the results, with the help of (3.34b) we obtain an expression for the derivative of the transmission function with respect to τ analogously to (3.33c)

$$\dfrac{S_0}{4\pi}\left(\dfrac{1}{\mu}-\dfrac{1}{\mu_0}\right)\left[\dfrac{d}{d\tau}T(\tau,\mu,\varphi,\mu_0,\varphi_0)\right]_{\tau=\tau_1}$$

$$=\dfrac{1}{4\pi\mu}\int_0^{2\pi}\int_0^1 T(\tau_1,\mu,\varphi,\mu',\varphi')J(0,-\mu',\varphi')\dfrac{d\mu'}{\mu'}d\varphi'$$

$$-\dfrac{1}{4\pi\mu_0}\int_0^{2\pi}\int_0^1 S(\tau_1,\mu,\varphi,\mu',\varphi')J(\tau_1,\mu',\varphi')\dfrac{d\mu'}{\mu'}d\varphi'$$

$$-\dfrac{1}{\mu_0}J(\tau_1,-\mu,\varphi)+\dfrac{1}{\mu}J(0,-\mu,\varphi)\exp\left(-\dfrac{\tau_1}{\mu}\right)$$ (3.35)

Finally, in (3.33a,b) the derivatives of $\mathcal{S}$ and $\mathcal{T}$ with respect to τ may be replaced by means of (3.33c,d) yielding

$$
\frac{S_0}{4\pi}\left(\frac{1}{\mu}+\frac{1}{\mu_0}\right)\mathcal{S}(\tau_1,\mu,\varphi,\mu_0,\varphi_0)=\frac{1}{4\pi}\int_0^{2\pi}\int_0^1 \mathcal{S}(\tau_1,\mu,\varphi,\mu',\varphi')J(0,-\mu',\varphi')\frac{d\mu'}{\mu'}d\varphi'
$$
$$
-\frac{1}{4\pi}\int_0^{2\pi}\int_0^1 \mathcal{T}(\tau_1,\mu,\varphi,\mu',\varphi')J(\tau_1,\mu',\varphi')\frac{d\mu'}{\mu'}d\varphi'
$$
$$
+J(0,\mu,\varphi)-J(\tau_1,\mu,\varphi)\exp\left(-\frac{\tau_1}{\mu}\right)
$$

$$
\frac{S_0}{4\pi}\left(\frac{1}{\mu}-\frac{1}{\mu_0}\right)\mathcal{T}(\tau_1,\mu,\varphi,\mu_0,\varphi_0)=\frac{1}{4\pi}\int_0^{2\pi}\int_0^1 \mathcal{S}(\tau_1,\mu,\varphi,\mu',\varphi')J(\tau_1,\mu',\varphi')\frac{d\mu'}{\mu'}d\varphi'
$$
$$
-\frac{1}{4\pi}\int_0^{2\pi}\int_0^1 \mathcal{T}(\tau_1,\mu,\varphi,\mu',\varphi')J(0,-\mu',\varphi')\frac{d\mu'}{\mu'}d\varphi'
$$
$$
+J(\tau_1,-\mu,\varphi)-J(0,-\mu,\varphi)\exp\left(-\frac{\tau_1}{\mu}\right)
$$

$$(3.36)$$

Inspection shows that these equations still contain the source functions J at the boundaries $\tau=0$ and $\tau=\tau_1$. These may be eliminated with the help of equations (3.1), (3.9) and the boundary conditions (3.30). Without difficulties we obtain the required expressions

$$
J(0,\mu,\varphi)=\frac{\omega_0 S_0}{(4\pi)^2}\int_0^{2\pi}\int_0^1 \mathcal{S}(\tau_1,\mu',\varphi',\mu_0,\varphi_0)\,\mathcal{P}(\mu,\varphi,\mu',\varphi')\frac{d\mu'}{\mu'}d\varphi'
$$
$$
+\frac{\omega_0}{4\pi}S_0\mathcal{P}(\mu,\varphi,-\mu_0,\varphi_0)
$$

$$
J(\tau_1,\mu,\varphi)=\frac{\omega_0 S_0}{(4\pi)^2}\int_0^{2\pi}\int_0^1 \mathcal{T}(\tau_1,\mu',\varphi',\mu_0,\varphi_0)\mathcal{P}(\mu,\varphi,-\mu',\varphi')\frac{d\mu'}{\mu'}d\varphi'
$$
$$
+\frac{\omega_0}{4\pi}S_0\exp\left(-\frac{\tau_1}{\mu_0}\right)\mathcal{P}(\mu,\varphi,-\mu_0,\varphi_0)
$$

$$(3.37)$$

Equations (3.36) together with (3.37) are the desired integral relations for $\mathcal{S}$ and $\mathcal{T}$ in analogy to equation (3.15) for $\mathcal{S}$ in case of a semi-infinite atmosphere. These integral relations describe the problem of diffuse reflection and transmission of plane–parallel atmospheres of finite optical thickness. As Chandrasekhar points out, (3.36) may be regarded as the expression of the invariance of the laws of diffuse reflection and transmission to the addition (or removal) of layers of arbitrary optical thickness to (or from) the atmosphere at the top and the simultaneous removal (or addition) of layers of equal optical thickness from (or to) the atmosphere at the base of the layer.

It is possible to extend the analysis to include the effects of polarization. Without going into details at this point, the scattering and transmission functions S and T will have to be replaced by scattering and transmission matrices. This makes the analysis quite complicated. In a later chapter we are going to show how to include polarization in the radiative transfer equation.

We will conclude this section with a few additional remarks of how to evaluate the preceding equations by expanding the scattering and transmission functions in Fourier type series. In Section 2.4 we have shown that if the phase function can be expressed in the form (2.68), then it is possible to expand the radiance $I(\tau, \mu, \varphi)$ by means of (2.69). Similarly, by expanding the phase function in the form (2.68), the scattering and the transmission functions may be written as

$$S(\tau_1, \mu, \varphi, \mu_0, \varphi_0) = \sum_{m=0}^{N} S^m(\tau_1, \mu, \mu_0) \cos m(\varphi - \varphi_0)$$

$$T(\tau_1, \mu, \varphi, \mu_0, \varphi_0) = \sum_{m=0}^{N} T^m(\tau_1, \mu, \mu_0) \cos m(\varphi - \varphi_0)$$

(3.38)

By substituting these expansions into (3.36), using the proper orthogonality conditions, we can eliminate the azimuthal dependence. We will not carry out the analysis but refer the interested reader to Chandrasekhar (1960).

3.4 The inclusion of surface reflection

The boundary conditions (3.30) of the standard problem did not include the effects of surface reflection which may be of great importance in a number of realistic situations. Whenever the surface reflection is included in the transfer analysis, we speak of the *planetary problem*. To simplify the analysis of the planetary problem, we will assume that the light is reflected according to *Lambert's law*, that is the reflected light is uniform and independent of the angular distribution of the incoming light. In this case we also speak of a *Lambertian surface*. For convenience, as before, we set the azimuthal angle of the Sun $\varphi_0 = 0$. Furthermore, we ignore the effects of polarization.

The two azimuthal independent terms $S^0(\tau_1, \mu, \mu_0)$ and $T^0(\tau_1, \mu, \mu_0)$ in the expansions (3.38) are given by

$$S^0(\tau_1, \mu, \mu_0) = \frac{1}{2\pi} \int_0^{2\pi} S(\tau_1, \mu, \varphi, \mu_0, 0) d\varphi$$

$$T^0(\tau_1, \mu, \mu_0) = \frac{1}{2\pi} \int_0^{2\pi} T(\tau_1, \mu, \varphi, \mu_0, 0) d\varphi$$

(3.39)

To obtain a simple mathematical structure of the equations to be derived, we introduce the abbreviations

$$
\text{(a)} \quad s(\tau_1, \mu) = \frac{1}{2} \int_0^1 S^0(\tau_1, \mu, \mu') d\mu'
$$

$$
\text{(b)} \quad t(\tau_1, \mu) = \frac{1}{2} \int_0^1 T^0(\tau_1, \mu, \mu') d\mu' \qquad (3.40)
$$

$$
\text{(c)} \quad \bar{s}(\tau_1) = 2 \int_0^1 s(\tau_1, \mu) d\mu
$$

Furthermore, in order to distinguish the solution of the planetary problem from that of the standard problem we place an asterisk to all quantities referring to the planetary problem. Thus I^* refers to the radiance in the presence of ground reflection.

Since we have assumed that the ground reflection is Lambertian, the reflected intensity I_g at τ_1 will be the same in all upward directions. At the top of the atmospheric layer the emergent intensity $I^*(0, \mu, \varphi)$ is expressed as the sum of three terms, that is

$$
I^*(0, \mu, \varphi) = I(0, \mu, \varphi) + \frac{1}{4\pi\mu} \int_0^{2\pi} \int_0^1 T(\tau_1, \mu, \varphi, \mu', \varphi') I_g d\mu' d\varphi'
$$

$$
+ I_g \exp\left(-\frac{\tau_1}{\mu}\right) \qquad (3.41)
$$

The first term on the right-hand side represents the diffusely reflected radiance of the standard problem, i.e. in the absence of ground reflection. The second term stands for the radiance I_g which (under the conditions of the standard problem) is transmitted into the upward hemisphere. The third term represents the transmission of the radiance I_g, already in the direction (μ, φ) which is not scattered out of the beam.

Substituting (3.40b) into (3.41) gives

$$
I^*(0, \mu, \varphi) = I(0, \mu, \varphi) + I_g \left[\frac{t(\tau_1, \mu)}{\mu} + \exp\left(-\frac{\tau_1}{\mu}\right)\right] \qquad (3.42)
$$

$$
= I(0, \mu, \varphi) + I_g \gamma(\tau_1, \mu)
$$

where the abbreviation

$$
\gamma(\tau_1, \mu) = \frac{t(\tau_1, \mu)}{\mu} + \exp\left(-\frac{\tau_1}{\mu}\right) \qquad (3.43)
$$

has been introduced.

The isotropic radiance I_g incident on the surface $\tau = \tau_1$ will also be scattered in the downward direction by the atmosphere. This amount of radiation is given by

$$
I_{g,sca}(-\mu) = \frac{1}{4\pi\mu} \int_0^{2\pi} \int_0^1 S(\tau_1, \mu, \varphi, \mu', \varphi') I_g d\mu' d\varphi' = I_g \frac{s(\tau_1, \mu)}{\mu} \qquad (3.44)
$$

where use was made of (3.40a). Adding $I_{g,\text{sca}}$ to the diffusely transmitted downward radiation we obtain

$$I^*(\tau_1, -\mu, \varphi) = \frac{S_0}{4\pi\mu}T(\tau_1, \mu, \varphi, \mu_0, 0) + I_g\frac{s(\tau_1, \mu)}{\mu} \tag{3.45}$$

Next, we need to find an explicit mathematical expression for I_g. We recognize that the downward radiative flux density arriving at the level $\tau = \tau_1$ consists of three parts:

(1) the flux density of the diffusely transmitted radiance;
(2) the downward scattered radiative flux density; and
(3) the reduced incident flux density.

Adding all three contributions yields

$$
\begin{aligned}
&\int_0^{2\pi}\int_0^1 \frac{S_0}{4\pi\mu}T(\tau_1, \mu, \varphi, \mu_0, 0)\mu\, d\mu\, d\varphi + \int_0^{2\pi}\int_0^1 I_{g,\text{sca}}(-\mu)\mu\, d\mu\, d\varphi + \mu_0 S_0 \exp\left(-\frac{\tau_1}{\mu_0}\right) \\
&= \frac{S_0}{2}\int_0^1 T^0(\tau_1, \mu, \mu_0)d\mu + 2\pi I_g\int_0^1 s(\tau_1, \mu)d\mu + \mu_0 S_0 \exp\left(-\frac{\tau_1}{\mu_0}\right) \\
&= \mu_0 S_0\left[\frac{t(\tau_1, \mu_0)}{\mu_0} + \exp\left(-\frac{\tau_1}{\mu_0}\right)\right] + \pi I_g\bar{s} = \mu_0 S_0\gamma(\tau_1, \mu_0) + \pi I_g\bar{s} \tag{3.46}
\end{aligned}
$$

where use was made of (3.40) and (3.43). The total downward radiation given by (3.46) will be reflected at the ground. By introducing the *albedo of the ground* A_g, the reflected flux density is

$$\pi I_g = A_g[\mu_0 S_0\gamma(\tau_1, \mu_0) + \pi I_g\bar{s}] \tag{3.47}$$

so that we finally obtain

$$\boxed{I_g = \frac{A_g\mu_0 S_0\gamma(\tau_1, \mu_0)}{\pi(1 - A_g\bar{s})}} \tag{3.48}$$

Substituting this expression into (3.42) and (3.45), we obtain the desired expressions for the upward radiance emerging at $\tau = 0$ and the downward radiance emerging at $\tau = \tau_1$

$$
\boxed{
\begin{aligned}
I^*(0, \mu, \varphi) &= \frac{S_0}{4\pi\mu}\left[S(\tau_1, \mu, \varphi, \mu_0, 0) + \frac{4A_g}{1 - A_g\bar{s}}\mu\mu_0\gamma(\tau_1, \mu_0)\gamma(\tau_1, \mu)\right] \\
I^*(\tau_1, -\mu, \varphi) &= \frac{S_0}{4\pi\mu}\left[T(\tau_1, \mu, \varphi, \mu_0, 0) + \frac{4A_g}{1 - A_g\bar{s}}\mu_0\gamma(\tau_1, \mu_0)s(\tau_1, \mu)\right]
\end{aligned}
}
$$

$$\tag{3.49}$$

3.5 Diffuse reflection and transmission for isotropic scattering

We will conclude this chapter by presenting a relatively simple example showing in which way the scattering and the transmission functions may be employed. Again we return to the standard problem. To arrive at the equations of isotropic scattering, we introduce average radiances, scattering and transmission functions according to

$$I(0, \mu) = \frac{1}{2\pi} \int_0^{2\pi} I(0, \mu, \varphi) d\varphi, \qquad I(\tau_1, -\mu) = \frac{1}{2\pi} \int_0^{2\pi} I(\tau_1, -\mu, \varphi) d\varphi$$

$$S(\tau_1, \mu, \mu_0) = \frac{1}{2\pi} \int_0^{2\pi} S(\tau_1, \mu, \varphi, \mu_0, 0) d\varphi, \qquad T(\tau_1, \mu, \mu_0) = \frac{1}{2\pi} \int_0^{2\pi} T(\tau_1, \mu, \varphi, \mu_0, 0) d\varphi \tag{3.50}$$

With these quantities the averaged form of (3.1) is given as

$$I(0, \mu) = \frac{S_0}{4\pi\mu} S(\tau_1, \mu, \mu_0), \qquad I(\tau_1, -\mu) = \frac{S_0}{4\pi\mu} T(\tau_1, \mu, \mu_0) \tag{3.51}$$

Now we proceed to find explicit formulas for $S(\tau_1, \mu, \mu_0)$ and $T(\tau_1, \mu, \mu_0)$. Isotropic scattering is expressed by the phase function $\mathcal{P} = 1$, see (1.47). Utilizing (3.50) and averaging the source functions, in the case of isotropic scattering (3.37) reduces to

$$J(0) = \frac{\omega_0}{4\pi} S_0 \left(1 + \frac{1}{2} \int_0^1 S(\tau_1, \mu', \mu_0) \frac{d\mu'}{\mu'} \right)$$

$$J(\tau_1) = \frac{\omega_0}{4\pi} S_0 \left[\exp\left(-\frac{\tau_1}{\mu_0}\right) + \frac{1}{2} \int_0^1 T(\tau_1, \mu', \mu_0) \frac{d\mu'}{\mu'} \right] \tag{3.52}$$

In order to have a compact notation, we introduce the functions

$$\boxed{\begin{aligned} X(\mu) &= 1 + \frac{1}{2} \int_0^1 S(\tau_1, \mu', \mu) \frac{d\mu'}{\mu'} \\ Y(\mu) &= \exp\left(-\frac{\tau_1}{\mu}\right) + \frac{1}{2} \int_0^1 T(\tau_1, \mu', \mu) \frac{d\mu'}{\mu'} \end{aligned}} \tag{3.53}$$

Substituting (3.53) into (3.52) results in

$$J(0) = \frac{\omega_0}{4\pi} S_0 X(\mu_0), \qquad J(\tau_1) = \frac{\omega_0}{4\pi} S_0 Y(\mu_0) \tag{3.54}$$

From (3.53) it may be easily seen that in case of a semi-infinite atmosphere, i.e. $\tau_1 \to \infty$, $Y(\mu)$ approaches zero while $X(\mu)$ is equivalent to $H(\mu)$ as defined in (3.22).

Now we rewrite equations (3.36) for the condition of isotropic scattering. Utilizing (3.54) yields the integral relations for S and T in the case of isotropic scattering

$$\left(\frac{1}{\mu}+\frac{1}{\mu_0}\right)S(\tau_1,\mu,\mu_0)=\frac{\omega_0}{2}\int_0^1 S(\tau_1,\mu,\mu')X(\mu_0)\frac{d\mu'}{\mu'}-\frac{\omega_0}{2}\int_0^1 T(\tau_1,\mu,\mu')Y(\mu_0)\frac{d\mu'}{\mu'}$$

$$+\omega_0 X(\mu_0)-\omega_0 Y(\mu_0)\exp\left(-\frac{\tau_1}{\mu}\right)$$

$$\left(\frac{1}{\mu}-\frac{1}{\mu_0}\right)T(\tau_1,\mu,\mu_0)=\frac{\omega_0}{2}\int_0^1 S(\tau_1,\mu,\mu')Y(\mu_0)\frac{d\mu'}{\mu'}-\frac{\omega_0}{2}\int_0^1 T(\tau_1,\mu,\mu')X(\mu_0)\frac{d\mu'}{\mu'}$$

$$+\omega_0 Y(\mu_0)-\omega_0 X(\mu_0)\exp\left(-\frac{\tau_1}{\mu}\right) \tag{3.55}$$

Substituting (3.53) into these expressions results in

$$\boxed{\begin{aligned}\left(\frac{1}{\mu}+\frac{1}{\mu_0}\right)S(\tau_1,\mu,\mu_0)&=\omega_0\left[X(\mu)X(\mu_0)-Y(\mu)Y(\mu_0)\right]\\ \left(\frac{1}{\mu}-\frac{1}{\mu_0}\right)T(\tau_1,\mu,\mu_0)&=\omega_0\left[X(\mu)Y(\mu_0)-Y(\mu)X(\mu_0)\right]\end{aligned}} \tag{3.56}$$

For the derivatives of S and T with respect to τ we obtain from (3.33c) and (3.35) for the case of isotropic scattering

$$\left[\frac{d}{d\tau}S(\tau,\mu,\mu_0)\right]_{\tau=\tau_1}=\omega_0 Y(\mu)Y(\mu_0)$$

$$\left(\frac{1}{\mu}-\frac{1}{\mu_0}\right)\left[\frac{d}{d\tau}T(\tau,\mu,\mu_0)\right]_{\tau=\tau_1}=\omega_0\left[\frac{1}{\mu}Y(\mu)X(\mu_0)-\frac{1}{\mu_0}X(\mu)Y(\mu_0)\right] \tag{3.57}$$

Details leading to (3.55) and (3.57) will be worked out in the exercises to this chapter.

Substitution of (3.56) into (3.53) gives integral relations for $X(\mu)$ and $Y(\mu)$, analogously to (3.25) for the H-function

$$X(\mu)=1+\frac{\omega_0}{2}\int_0^1[X(\mu)X(\mu')-Y(\mu)Y(\mu')]\frac{\mu}{\mu+\mu'}d\mu'$$

$$Y(\mu)=\exp\left(-\frac{\tau_1}{\mu}\right)+\frac{\omega_0}{2}\int_0^1[Y(\mu)X(\mu')-X(\mu)Y(\mu')]\frac{\mu}{\mu-\mu'}d\mu' \tag{3.58}$$

The functions $X(\mu)$ and $Y(\mu)$ are special cases of the general forms

$$\boxed{\begin{aligned}X(\mu)&=1+\int_0^1[X(\mu)X(\mu')-Y(\mu)Y(\mu')]\frac{\mu}{\mu+\mu'}\psi(\mu')d\mu'\\ Y(\mu)&=\exp\left(-\frac{\tau_1}{\mu}\right)+\int_0^1[Y(\mu)X(\mu')-X(\mu)Y(\mu')]\frac{\mu}{\mu-\mu'}\psi(\mu')d\mu'\end{aligned}} \tag{3.59}$$

These are known as *Chandrasekhar's X- and Y- functions*. The quantity $\psi(\mu)$ is called the *characteristic function* which differs from problem to problem. The numerical evaluation of the nonlinear integral equations (3.59) may be accomplished by an iteration procedure.

Obviously, for the case of isotropic scattering we obtain $\psi(\mu) = \omega_0/2$. For *Rayleigh scattering* the characteristic function $\psi(\mu)$ still has a fairly simple algebraic form. As pointed out by Liou (2002), for the more complicated forms of the Mie-type phase functions, the characteristic functions are rather complicated and are not available for practical applications.

Finally, we wish to point out that in case of conservative perfect scattering, i.e. $\omega_0 = 1$, the integral equations (3.59) are not sufficient to characterize the physical situation uniquely. In the simple situation of isotropic scattering it is not particularly difficult to resolve the ambiguity. We will refrain from further discussing this topic and refer to Chapter IX of Chandrasekhar (1960) where a full treatment is given.

As stated above, for highly peaked Mie type phase functions it becomes increasingly difficult to apply the principles of invariance to find exact solutions to transfer problems. Even if we succeeded in obtaining such solutions, we are still faced with the specification of realistic input data for atmospheric problems. In practice, we are usually compelled to apply model data which may not always be sufficient to simulate real physical situations. Thus the application of model atmospheric data to an exact solution of a transfer problem at best results in an approximation to the solution of a real physical problem. Usually the numerical evaluation of the exact solution is difficult and time consuming, particularly if the calculations have to be carried out at many wavelengths. Instead of evaluating exact or quasi-exact solutions, for many practical purposes it might be sufficient to use approximate methods. Usually these offer the advantage that they can be quickly evaluated which is important in case of climate modeling and weather prediction.

In the following chapters we will discuss various quasi-exact as well as some approximate solution methods for the RTE at various levels of sophistication. Of course, whenever possible the more exact solutions are used in order to test the validity of the approximate methods.

3.6 Problems

3.1: Verify equation (3.34).
3.2: Verify equation (3.37).
3.3: Carry out all steps in detail in Section 4.4 to obtain (3.48).
3.4: Reduce (3.36) to the isotropic form (3.56).

3.5: Use (3.33c) and (3.35) to obtain (3.57) which refers to isotropic scattering.

3.6: The *law of darkening* describes the angular distribution $I(0, \mu)$, $0 \le \mu \le 1$ of the emergent radiation in case of a plane–parallel semi-infinite atmosphere with constant net flux. Show that $I(0, \mu) = J(0)H(\mu)$ where $J(0) = 1/2 \int_0^1 I(0, \mu)d\mu$. Assume the validity of

$$I(0, \mu) = \frac{1}{2} \int_0^1 I(0, \mu')d\mu' + \frac{1}{4} \int_0^1 \int_0^1 S(\mu, \mu')I(0, \mu'')\frac{d\mu'}{\mu'}d\mu''$$

3.7: (a) The so-called *Hopf–Bronstein relation* for conservative isotropic scattering (in a plane–parallel semi-infinite atmosphere) is given by $I(0, \mu) = \sqrt{3}/(4\pi)S_0 H(\mu)$. If the emergent radiation $I(0, \mu)$ is assumed to be given by

$$I(0, \mu) = \frac{3}{4\pi} S_0 \left(\mu + \frac{1}{2\mu} \int_0^1 S(\mu, \mu')\mu'd\mu' \right)$$

find $J(0) = 1/2 \int_0^1 I(0, \mu)d\mu$ and $\Psi = 1 + \int_0^1 \int_0^1 S(\mu, \mu')(\mu'/\mu)d\mu'd\mu$.

(b) Use the information of part (a) to show that $I(0, \mu)$ can also be written as

$$I(0, \mu) = \frac{1}{2} \int_0^1 I(0, \mu')d\mu' + \frac{1}{4} \int_0^1 \int_0^1 S(\mu, \mu')I(0, \mu'')\frac{d\mu'}{\mu'}d\mu''$$

4

Quasi-exact solution methods for the radiative transfer equation

In this chapter we will discuss various techniques that can be used to solve the radiative transfer equation. Although these techniques employ quite different mathematical models, they all produce very accurate solutions. Therefore, we call them *quasi-exact solution methods*. A disadvantage of the quasi-exact procedures is their mathematical complexity which causes them to be very expensive computationally. However, the quasi-exact methods can be used to produce benchmark computations to test the quality of the computationally very efficient approximate solution methods which for practical reasons are employed in weather prediction and climate models. The most common and efficient solution schemes are the *two-stream methods* which will be discussed in detail in Chapter 6.

4.1 The matrix operator method

The *matrix operator method* (MOM) is one of the most accurate techniques for solving the radiative transfer equation in a planetary atmosphere. It is based on the fact that the Fourier components $I^m(\tau, \mu)$ of the radiation field as introduced in (2.70) can be represented in discretized form as vectors $\mathbf{I}^m_\pm$. In order to apply the method we need to determine the transmission and reflection properties of each individual layer of the medium. In the following subsections it will be shown how to compute the transmitted and reflected radiation for the entire medium and at its interior levels.

4.1.1 Derivation of the addition theorems

Let us start with the discretization of the radiances $\mathbf{I}^m_\pm(\tau)$, see (2.95). We will consider two different but homogeneous sublayers $(0, 1) = \tau_1 - \tau_0$ and

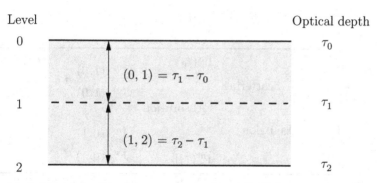

Fig. 4.1 Subdivision of an inhomogeneous layer $(0, 2)$ into two homogeneous sublayers $(0, 1)$ and $(1, 2)$ having different optical properties.

$(1, 2) = \tau_2 - \tau_1$. In general, however, the combined layer $(0, 2) = \tau_2 - \tau_0$ is inhomogeneous, see Figure 4.1.

For sublayer $(0, 1)$ we will define the following optical quantities for the m-th Fourier mode of the radiance which fully describe the transmission and reflection properties.

$\mathbf{t}^m(0, 1)$ discretized *transmissivity* (directions $i = 1, \ldots, s$) of sublayer $(0, 1)$ applied to the radiance $\mathbf{I}_-^m(\tau_0)$ incident at τ_0,

$\mathbf{t}^m(1, 0)$ discretized transmissivity for sublayer $(1, 0)$ applied to the radiance $\mathbf{I}_+^m(\tau_1)$ incident at τ_1,

$\mathbf{r}^m(0, 1)$ discretized *reflectivity* (directions $i = 1, \ldots, s$) of sublayer $(0, 1)$ for the radiance $\mathbf{I}_-^m(\tau_0)$ incident at τ_0,

$\mathbf{r}^m(1, 0)$ discretized reflectivity for sublayer $(1, 0)$ but for the radiance $\mathbf{I}_+^m(\tau_1)$,

$\mathbf{J}_{-,1}^m(0, 1)$ discretized source function for the downward directed primary scattered sunlight generated in sublayer $(0, 1)$,

$\mathbf{J}_{+,1}^m(1, 0)$ discretized source function for the upward directed primary scattered sunlight generated in sublayer $(1, 0)$,

$\mathbf{J}_{-,2}^m(0, 1)$ discretized source function for the downward thermal emission generated in sublayer $(0, 1)$,

$\mathbf{J}_{+,2}^m(1, 0)$ discretized source function for the upward directed thermal emission generated in sublayer $(1, 0)$.

Similar definitions apply to the corresponding matrix and vector quantities for sublayer $(1, 2)$ and the combined layer $(0, 2)$.

Let us consider Figure 4.2 describing the basic properties of the radiation model. The radiation transmitted and reflected by an arbitrary layer depends linearly on the incident radiation from above and below. This is the so-called *linear interaction*

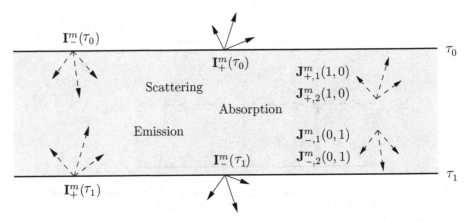

Fig. 4.2 The linear interaction principle for the emanating radiation $\mathbf{I}_+^m(\tau_0)$, $\mathbf{I}_-^m(\tau_1)$ from sublayer $(0, 1)$ expressed in terms of the incident radiation $\mathbf{I}_+^m(\tau_1)$, $\mathbf{I}_-^m(\tau_0)$ and the interior sources for primary scattering and thermal emission $\mathbf{J}_{+,1}^m(1, 0)$, $\mathbf{J}_{+,2}^m(1, 0)$, $\mathbf{J}_{-,1}^m(0, 1)$, $\mathbf{J}_{-,2}^m(0, 1)$. In general, scattering, absorption and emission take place.

principle. Therefore, the radiance emanating at τ_0 and τ_1, expressed in terms of the radiation $\mathbf{I}_-^m(\tau_0)$, $\mathbf{I}_+^m(\tau_1)$ incident at the boundaries of the sublayer $(0, 1)$ and the interior sources for primary scattering and thermal emission $\mathbf{J}_{+,1}^m(1, 0)$, $\mathbf{J}_{-,1}^m(0, 1)$, $\mathbf{J}_{+,2}^m(1, 0)$, $\mathbf{J}_{-,2}^m(0, 1)$ is given by

(a) $\mathbf{I}_+^m(\tau_0) = \mathfrak{t}^m(1, 0)\mathbf{I}_+^m(\tau_1) + \mathfrak{r}^m(0, 1)\mathbf{I}_-^m(\tau_0) + \mathbf{J}_{+,1}^m(1, 0) + \mathbf{J}_{+,2}^m(1, 0)$

(b) $\mathbf{I}_-^m(\tau_1) = \mathfrak{r}^m(1, 0)\mathbf{I}_+^m(\tau_1) + \mathfrak{t}^m(0, 1)\mathbf{I}_-^m(\tau_0) + \mathbf{J}_{-,1}^m(0, 1) + \mathbf{J}_{-,2}^m(0, 1)$ (4.1)

In a similar manner we find for sublayer $(1, 2)$

(a) $\mathbf{I}_+^m(\tau_1) = \mathfrak{t}^m(2, 1)\mathbf{I}_+^m(\tau_2) + \mathfrak{r}^m(1, 2)\mathbf{I}_-^m(\tau_1) + \mathbf{J}_{+,1}^m(2, 1) + \mathbf{J}_{+,2}^m(2, 1)$

(b) $\mathbf{I}_-^m(\tau_2) = \mathfrak{r}^m(2, 1)\mathbf{I}_+^m(\tau_2) + \mathfrak{t}^m(1, 2)\mathbf{I}_-^m(\tau_1) + \mathbf{J}_{-,1}^m(1, 2) + \mathbf{J}_{-,2}^m(1, 2)$ (4.2)

If we replace in (4.1) the sublayer index 1 by index 2, we obtain for the combined layer $(0, 2)$ the result

(a) $\mathbf{I}_+^m(\tau_0) = \mathfrak{t}^m(2, 0)\mathbf{I}_+^m(\tau_2) + \mathfrak{r}^m(0, 2)\mathbf{I}_-^m(\tau_0) + \mathbf{J}_{+,1}^m(2, 0) + \mathbf{J}_{+,2}^m(2, 0)$

(b) $\mathbf{I}_-^m(\tau_2) = \mathfrak{r}^m(2, 0)\mathbf{I}_+^m(\tau_2) + \mathfrak{t}^m(0, 2)\mathbf{I}_-^m(\tau_0) + \mathbf{J}_{-,1}^m(0, 2) + \mathbf{J}_{-,2}^m(0, 2)$ (4.3)

It should be noted that in the last two equations the matrices $\mathfrak{t}^m(0, 2)$, $\mathfrak{t}^m(2, 0)$, $\mathfrak{r}^m(0, 2)$, $\mathfrak{r}^m(2, 0)$ as well as the source vectors $\mathbf{J}_{-,1}^m(0, 2)$, $\mathbf{J}_{-,2}^m(0, 2)$, $\mathbf{J}_{+,1}^m(2, 0)$, $\mathbf{J}_{+,2}^m(2, 0)$ are still unknown. We will show, however, that for the combined layer these unknown quantities can be determined as functions of the radiative quantities of the individual sublayers $(0, 1)$ and $(1, 2)$.

In a first step we will express the radiances $\mathbf{I}_\pm^m(\tau_1)$ at level τ_1 in terms of the radiances $\mathbf{I}_-^m(\tau_0)$, $\mathbf{I}_+^m(\tau_2)$ illuminating the combined layer $(0, 2)$. Replacing $\mathbf{I}_-^m(\tau_1)$ in (4.2a) by (4.1b) and in (4.1b) the term $\mathbf{I}_+^m(\tau_1)$ by (4.2a), we obtain two rather complicated expressions

(a) $\mathbf{I}_+^m(\tau_1) = \mathfrak{r}^m(1, 2)[\mathfrak{r}^m(1, 0)\mathbf{I}_+^m(\tau_1) + \mathfrak{t}^m(0, 1)\mathbf{I}_-^m(\tau_0) + \mathbf{J}_{-,1}^m(0, 1) + \mathbf{J}_{-,2}^m(0, 1)]$
$\qquad + \mathfrak{t}^m(2, 1)\mathbf{I}_+^m(\tau_2) + \mathbf{J}_{+,1}^m(2, 1) + \mathbf{J}_{+,2}^m(2, 1)$

(b) $\mathbf{I}_-^m(\tau_1) = \mathfrak{r}^m(1, 0)\left[\mathfrak{t}^m(2, 1)\mathbf{I}_+^m(\tau_2) + \mathfrak{r}^m(1, 2)\mathbf{I}_-^m(\tau_1) + \mathbf{J}_{+,1}^m(2, 1) + \mathbf{J}_{+,2}^m(2, 1)\right]$
$\qquad + \mathfrak{t}^m(0, 1)\mathbf{I}_-^m(\tau_0) + \mathbf{J}_{-,1}^m(0, 1) + \mathbf{J}_{-,2}^m(0, 1)$ (4.4)

Solving for $\mathbf{I}_\pm^m(\tau_1)$ yields

(a) $\mathbf{I}_+^m(\tau_1) = [\mathbb{E} - \mathfrak{r}^m(1, 2)\mathfrak{r}^m(1, 0)]^{-1}[\mathfrak{t}^m(2, 1)\mathbf{I}_+^m(\tau_2) + \mathfrak{r}^m(1, 2)\mathfrak{t}^m(0, 1)\mathbf{I}_-^m(\tau_0)$
$\qquad + \mathfrak{r}^m(1, 2)[\mathbf{J}_{-,1}^m(0, 1) + \mathbf{J}_{-,2}^m(0, 1)] + \mathbf{J}_{+,1}^m(2, 1) + \mathbf{J}_{+,2}^m(2, 1)]$

(b) $\mathbf{I}_-^m(\tau_1) = [\mathbb{E} - \mathfrak{r}^m(1, 0)\mathfrak{r}^m(1, 2)]^{-1}[\mathfrak{r}^m(1, 0)\mathfrak{t}^m(2, 1)\mathbf{I}_+^m(\tau_2) + \mathfrak{t}^m(0, 1)\mathbf{I}_-^m(\tau_0)$
$\qquad + \mathfrak{r}^m(1, 0)[\mathbf{J}_{+,1}^m(2, 1) + \mathbf{J}_{+,2}^m(2, 1)] + \mathbf{J}_{-,1}^m(0, 1) + \mathbf{J}_{-,2}^m(0, 1)]$

$$(4.5)$$

where $\mathbb{E}$ is the unit matrix.

It is important to realize that (4.5a,b) contain the optical properties of both sublayers $(0, 1)$ and $(1, 2)$ only, but none of layer $(0, 2)$. This fact can be used to derive the so-called *addition theorems* for the optical properties of the layer $(0, 2)$. Substituting $\mathbf{I}_+^m(\tau_1)$ from (4.5a) into (4.1a) and $\mathbf{I}_-^m(\tau_1)$ from (4.5b) into (4.2b) yields

(a) $\mathbf{I}_+^m(\tau_0) = \mathfrak{t}^m(1, 0)[\mathbb{E} - \mathfrak{r}^m(1, 2)\mathfrak{r}^m(1, 0)]^{-1}\mathfrak{t}^m(2, 1)\mathbf{I}_+^m(\tau_2)$
$\qquad + \mathfrak{t}^m(1, 0)[\mathbb{E} - \mathfrak{r}^m(1, 2)\mathfrak{r}^m(1, 0)]^{-1}\mathfrak{r}^m(1, 2)\mathfrak{t}^m(0, 1)\mathbf{I}_-^m(\tau_0)$
$\qquad + \mathfrak{t}^m(1, 0)[\mathbb{E} - \mathfrak{r}^m(1, 2)\mathfrak{r}^m(1, 0)]^{-1}\mathfrak{r}^m(1, 2)[\mathbf{J}_{-,1}^m(0, 1) + \mathbf{J}_{-,2}^m(0, 1)]$
$\qquad + \mathfrak{t}^m(1, 0)[\mathbb{E} - \mathfrak{r}^m(1, 2)\mathfrak{r}^m(1, 0)]^{-1}[\mathbf{J}_{+,1}^m(2, 1) + \mathbf{J}_{+,2}^m(2, 1)]$
$\qquad + \mathfrak{r}^m(0, 1)\mathbf{I}_-^m(\tau_0) + \mathbf{J}_{+,1}^m(1, 0) + \mathbf{J}_{+,2}^m(1, 0)$

$$(4.6)$$

(b) $\mathbf{I}_-^m(\tau_2) = \mathfrak{t}^m(1, 2)[\mathbb{E} - \mathfrak{r}^m(1, 0)\mathfrak{r}^m(1, 2)]^{-1}\mathfrak{r}^m(1, 0)\mathfrak{t}^m(2, 1)\mathbf{I}_+^m(\tau_2)$
$\qquad + \mathfrak{t}^m(1, 2)[\mathbb{E} - \mathfrak{r}^m(1, 0)\mathfrak{r}^m(1, 2)]^{-1}\mathfrak{t}^m(0, 1)\mathbf{I}_-^m(\tau_0)$
$\qquad + \mathfrak{t}^m(1, 2)[\mathbb{E} - \mathfrak{r}^m(1, 0)\mathfrak{r}^m(1, 2)]^{-1}\mathfrak{r}^m(1, 0)[\mathbf{J}_{+,1}^m(2, 1) + \mathbf{J}_{+,2}^m(2, 1)]$
$\qquad + \mathfrak{t}^m(1, 2)[\mathbb{E} - \mathfrak{r}^m(1, 0)\mathfrak{r}^m(1, 2)]^{-1}[\mathbf{J}_{-,1}^m(0, 1) + \mathbf{J}_{-,2}^m(0, 1)]$
$\qquad + \mathfrak{r}^m(2, 1)\mathbf{I}_+^m(\tau_2) + \mathbf{J}_{-,1}^m(1, 2) + \mathbf{J}_{-,2}^m(1, 2)$

Comparison of these expressions with (4.3a,b) gives the required optical properties for the combined layer (0, 2)

$$
\begin{aligned}
&\mathfrak{t}^m(0,2) = \mathfrak{t}^m(1,2)[\mathbb{E} - \mathfrak{r}^m(1,0)\mathfrak{r}^m(1,2)]^{-1}\mathfrak{t}^m(0,1) \\
&\mathfrak{t}^m(2,0) = \mathfrak{t}^m(1,0)[\mathbb{E} - \mathfrak{r}^m(1,2)\mathfrak{r}^m(1,0)]^{-1}\mathfrak{t}^m(2,1) \\
&\mathfrak{r}^m(0,2) = \mathfrak{r}^m(0,1) + \mathfrak{t}^m(1,0)[\mathbb{E} - \mathfrak{r}^m(1,2)\mathfrak{r}^m(1,0)]^{-1}\mathfrak{r}^m(1,2)\mathfrak{t}^m(0,1) \\
&\mathfrak{r}^m(2,0) = \mathfrak{r}^m(2,1) + \mathfrak{t}^m(1,2)[\mathbb{E} - \mathfrak{r}^m(1,0)\mathfrak{r}^m(1,2)]^{-1}\mathfrak{r}^m(1,0)\mathfrak{t}^m(2,1)
\end{aligned}
\tag{4.7}
$$

and

$$
\begin{aligned}
\mathbf{J}^m_{-,1}(0,2) &= \mathbf{J}^m_{-,1}(1,2) + \mathfrak{t}^m(1,2)[\mathbb{E} - \mathfrak{r}^m(1,0)\mathfrak{r}^m(1,2)]^{-1} \\
&\quad \times \left[\mathbf{J}^m_{-,1}(0,1) + \mathfrak{r}^m(1,0)\mathbf{J}^m_{+,1}(2,1)\right] \\
\mathbf{J}^m_{-,2}(0,2) &= \mathbf{J}^m_{-,2}(1,2) + \mathfrak{t}^m(1,2)[\mathbb{E} - \mathfrak{r}^m(1,0)\mathfrak{r}^m(1,2)]^{-1} \\
&\quad \times \left[\mathbf{J}^m_{-,2}(0,1) + \mathfrak{r}^m(1,0)\mathbf{J}^m_{+,2}(2,1)\right] \\
\mathbf{J}^m_{+,1}(2,0) &= \mathbf{J}^m_{+,1}(1,0) + \mathfrak{t}^m(1,0)[\mathbb{E} - \mathfrak{r}^m(1,2)\mathfrak{r}^m(1,0)]^{-1} \\
&\quad \times \left[\mathbf{J}^m_{+,1}(2,1) + \mathfrak{r}^m(1,2)\mathbf{J}^m_{-,1}(0,1)\right] \\
\mathbf{J}^m_{+,2}(2,0) &= \mathbf{J}^m_{+,2}(1,0) + \mathfrak{t}^m(1,0)[\mathbb{E} - \mathfrak{r}^m(1,2)\mathfrak{r}^m(1,0)]^{-1} \\
&\quad \times \left[\mathbf{J}^m_{+,2}(2,1) + \mathfrak{r}^m(1,2)\mathbf{J}^m_{-,2}(0,1)\right]
\end{aligned}
\tag{4.8}
$$

It is clear that the above formulas can be generalized to state the addition formulas of two arbitrary adjacent layers $(i, i+1)$, $(i+1, i+2)$ by the substitutions $0 \to i$, $1 \to i+1, 2 \to i+2$.[1]

The physical interpretation of the inverse matrix $\mathbb{A} = [\mathbb{E} - \mathfrak{r}^m(1,2)\mathfrak{r}^m(1,0)]^{-1}$ occurring in (4.6a) will be demonstrated by means of Figure 4.3. It should be observed that the inverse matrix can be developed analogously to the scalar expression

$$
\frac{1}{1-x} = 1 + x + x^2 + x^3 + \dots, \qquad |x| < 1
\tag{4.9}
$$

i.e. $\mathbb{A}$ can formally be developed as the infinite series

$$
\mathbb{A} = \sum_{k=0}^{\infty} [\mathfrak{r}^m(1,2)\mathfrak{r}^m(1,0)]^k
\tag{4.10}
$$

The series of matrix products converges if the eigenvalues λ_k of $\mathbb{A}$ fulfill the condition $|\lambda_k| < 1$, for more details see Gantmacher (1986). Summing up the individual contributions of the multiple reflections between the two sublayers in Figure 4.3

---

[1] Note that $\mathfrak{t}^m(i, i+1)$, $\mathfrak{r}^m(i, i+1)$ represent $(s \times s)$ matrices for the transmission and reflection of downward radiation incident at level i.

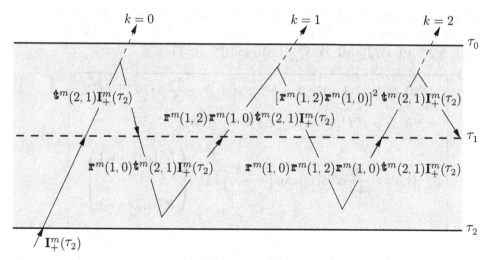

Fig. 4.3 Physical interpretation of the matrix $\mathbb{A} = [\mathbb{E} - \mathbb{r}^m(1,2)\mathbb{r}^m(1,0)]^{-1}$ as an infinite sum of successive reflections between layer $(0,1)$ and layer $(1,2)$.

is equivalent to the expansion terms given in (4.10) pre-multiplying the expression $\mathbb{t}^m(2,1)\mathbf{I}^m_+(\tau_2)$ in (4.6a).

4.1.2 The optical properties of a homogeneous elementary layer

So far we have not determined the optical properties of a homogeneous elementary layer. This layer is required to be sufficiently thin so that the photons entering this layer will experience primary scattering only. In admitting only primary scattering, the optical transmission, reflection and emission properties of the layer can be directly extracted from the RTE (2.104). For $\Delta\tau = \tau_1 - \tau_0 \ll 1$, the left-hand side of (2.104) can be written in finite difference form as

$$\frac{d}{d\tau}\begin{pmatrix} \mathbf{I}^m_+(\tau) \\ \mathbf{I}^m_-(\tau) \end{pmatrix} \approx \frac{1}{\Delta\tau}\begin{pmatrix} \mathbf{I}^m_+(\tau_1) - \mathbf{I}^m_+(\tau_0) \\ \mathbf{I}^m_-(\tau_1) - \mathbf{I}^m_-(\tau_0) \end{pmatrix} \tag{4.11}$$

with $\tau_0 < \tau_1$ and $\tau = (\tau_0 + \tau_1)/2$.

Equations (4.1a,b) can be used to express the difference of the upwelling and downwelling radiances at levels τ_1 and τ_0 so that we obtain

$$\frac{1}{\Delta\tau}\begin{pmatrix} \mathbf{I}^m_+(\tau_1) - \mathbf{I}^m_+(\tau_0) \\ \mathbf{I}^m_-(\tau_1) - \mathbf{I}^m_-(\tau_0) \end{pmatrix} = \frac{1}{\Delta\tau}\begin{pmatrix} \mathbb{E} - \mathbb{t}^m(1,0) & -\mathbb{r}^m(0,1) \\ \mathbb{r}^m(1,0) & \mathbb{t}^m(0,1) - \mathbb{E} \end{pmatrix}\begin{pmatrix} \mathbf{I}^m_+(\tau_1) \\ \mathbf{I}^m_-(\tau_0) \end{pmatrix}$$
$$+ \frac{1}{\Delta\tau}\begin{pmatrix} -\mathbf{J}^m_{+,1}(1,0) - \mathbf{J}^m_{+,2}(1,0) \\ \mathbf{J}^m_{-,1}(0,1) + \mathbf{J}^m_{-,2}(0,1) \end{pmatrix} \tag{4.12}$$

By comparing (4.11) and (4.12) with the RTE (2.104) we find explicit expressions for the optical matrices $\mathfrak{t}^m(0, 1)$, $\mathfrak{t}^m(1, 0)$, $\mathfrak{r}^m(1, 0)$, $\mathfrak{r}^m(0, 1)$, and for the source vectors $\mathbf{J}^m_{+,1}(0, 1)$, $\mathbf{J}^m_{-,1}(0, 1)$, $\mathbf{J}^m_{+,2}(1, 0)$, and $\mathbf{J}^m_{-,2}(0, 1)$

$$
\mathfrak{t}^m(0, 1) = \mathfrak{t}^m(1, 0) = \mathbb{E} - \mathbb{\Gamma}^m_{++}\Delta\tau = \mathbb{F} + \frac{\omega_0}{2}\Delta\tau \mathrm{M}^{-1}\mathbb{P}^m_{++}\mathrm{W}
$$

$$
\mathfrak{r}^m(0, 1) = \mathfrak{r}^m(1, 0) = \mathbb{\Gamma}^m_{+-}\Delta\tau = \frac{\omega_0}{2}\Delta\tau \mathrm{M}^{-1}\mathbb{P}^m_{+-}\mathrm{W}
$$

$$
\mathbf{J}^m_{+,1}(1, 0) = \frac{\omega_0}{4\pi}\Delta\tau S(\tau_0)\exp\left(-\frac{\Delta\tau}{\mu_0}\right)\mathrm{M}^{-1}\begin{pmatrix} R^m(\mu_1, -\mu_0) \\ \vdots \\ R^m(\mu_s, -\mu_0) \end{pmatrix}
$$

(4.13)

$$
\mathbf{J}^m_{-,1}(0, 1) = \frac{\omega_0}{4\pi}\Delta\tau S(\tau_0)\exp\left(-\frac{\Delta\tau}{\mu_0}\right)\mathrm{M}^{-1}\begin{pmatrix} R^m(-\mu_1, -\mu_0) \\ \vdots \\ R^m(-\mu_s, -\mu_0) \end{pmatrix}
$$

$$
\mathbf{J}^m_{+,2}(1, 0) = \mathbf{J}^m_{-,2}(0, 1) = \Delta\tau(1 - \omega_0)B(\tau)\delta_{0m}\mathrm{M}^{-1}\begin{pmatrix} 1 \\ \vdots \\ 1 \end{pmatrix}
$$

where we have used the definitions (2.96) and (2.105). Furthermore, the abbreviation

$$
\mathbb{F} = \mathbb{E} - \mathrm{M}^{-1}\Delta\tau = \left(\left(1 - \frac{\Delta\tau}{\mu_i}\right)\delta_{ij}\right) \approx \left(\exp\left[-\frac{\Delta\tau}{\mu_i}\right]\delta_{ij}\right) \tag{4.14}
$$

has been introduced. It is important to recall that in the homogeneous elementary layer with optical depth $\Delta\tau$ only single scattering is assumed to take place. Numerical experimentation shows that a choice of $\Delta\tau \lesssim 2^{-15}$ is sufficiently small so that the single scattering approximation holds.

4.1.3 The doubling algorithm

The addition theorems for the optical properties (4.7) and (4.8) can be used to build up the optical properties of an arbitrarily thick homogeneous layer. This is done in the following way.

We start with a homogeneous elementary layer $(0, 1)$. Assuming that this layer is small enough so that only single scattering takes place, the optical properties of this layer are given by (4.13). Now we construct the layer $(0, 2)$ by adding to $(0, 1)$ the layer $(1, 2)$ having the same optical properties as $(0, 1)$. Since the layers are homogeneous and their optical properties are identical, the addition theorems (4.7) and (4.8) reduce to

$$\mathfrak{t}^m(0, 2) = \mathfrak{t}^m(2, 0) = \mathfrak{t}^m(0, 1)[\mathbb{E} - \mathfrak{r}^m(0, 1)\mathfrak{r}^m(0, 1)]^{-1}\mathfrak{t}^m(0, 1)$$

$$\mathfrak{r}^m(0, 2) = \mathfrak{r}^m(2, 0) = \mathfrak{r}^m(0, 1) + \mathfrak{t}^m(0, 1)[\mathbb{E} - \mathfrak{r}^m(0, 1)\mathfrak{r}^m(0, 1)]^{-1}\mathfrak{r}^m(0, 1)\mathfrak{t}^m(0, 1)$$

with (4.15)

$$\mathfrak{t}^m(0, 1) = \mathfrak{t}^m(1, 0) = \mathfrak{t}^m(1, 2) = \mathfrak{t}^m(2, 1)$$

$$\mathfrak{r}^m(0, 1) = \mathfrak{r}^m(1, 0) = \mathfrak{r}^m(1, 2) = \mathfrak{r}^m(2, 1)$$

and

$$\mathbf{J}^m_{-,1}(0, 2) = \mathbf{J}^m_{-,1}(1, 2) + \mathfrak{t}^m(0, 1)[\mathbb{E} - \mathfrak{r}^m(0, 1)\mathfrak{r}^m(0, 1)]^{-1}$$
$$\times \left[\mathbf{J}^m_{-,1}(0, 1) + \mathfrak{r}^m(0, 1)\mathbf{J}^m_{+,1}(2, 1)\right]$$

$$\mathbf{J}^m_{-,2}(0, 2) = \mathbf{J}^m_{-,2}(0, 1) + \mathfrak{t}^m(0, 1)[\mathbb{E} - \mathfrak{r}^m(0, 1)\mathfrak{r}^m(0, 1)]^{-1}$$
$$\times \left[\mathbf{J}^m_{-,2}(0, 1) + \mathfrak{r}^m(0, 1)\mathbf{J}^m_{+,2}(1, 0)\right]$$

$$\mathbf{J}^m_{+,1}(2, 0) = \mathbf{J}^m_{+,1}(1, 0) + \mathfrak{t}^m(0, 1)[\mathbb{E} - \mathfrak{r}^m(0, 1)\mathfrak{r}^m(0, 1)]^{-1}$$
$$\times \left[\mathbf{J}^m_{+,1}(2, 1) + \mathfrak{r}^m(0, 1)\mathbf{J}^m_{-,1}(0, 1)\right]$$

$$\mathbf{J}^m_{+,2}(2, 0) = \mathbf{J}^m_{+,2}(1, 0) + \mathfrak{t}^m(0, 1)[\mathbb{E} - \mathfrak{r}^m(0, 1)\mathfrak{r}^m(0, 1)]^{-1}$$
$$\times \left[\mathbf{J}^m_{+,2}(1, 0) + \mathfrak{r}^m(0, 1)\mathbf{J}^m_{-,2}(0, 1)\right] \quad (4.16)$$

with

$$\mathbf{J}^m_{+,1}(2, 1) = \mathbf{J}^m_{+,1}(1, 0) \exp\left(-\frac{\tau_1 - \tau_0}{\mu_0}\right)$$

$$\mathbf{J}^m_{-,1}(1, 2) = \mathbf{J}^m_{-,1}(0, 1) \exp\left(-\frac{\tau_1 - \tau_0}{\mu_0}\right)$$

$$\mathbf{J}^m_{+,2}(2, 1) = \mathbf{J}^m_{+,2}(1, 0)$$
$$\mathbf{J}^m_{-,2}(1, 2) = \mathbf{J}^m_{-,2}(0, 1)$$

Sometimes these addition theorems are called the *star-product algorithm*. No distinction has been made between the upward and downward *transmission* and *reflection matrices* as well as for the *thermal source vectors*. Only the expressions $\mathbf{J}^m_{+,1}(2, 1)$ and $\mathbf{J}^m_{-,1}(1, 2)$ involving primary scattering require a special distinction.

The optical properties of layer $(0, 4)$ are obtained by adding to $(0, 2)$ the layer $(2, 4)$ having the same optical properties as layer $(0, 2)$. Layer $(0, 8)$ is obtained by adding layers $(0, 4)$ and $(4, 8)$ where again layer $(4, 8)$ has the same optical properties as $(0, 4)$. This process is continued until the final thickness of the homogeneous layer is reached.

Figure 4.4 illustrates the adding and doubling procedure. One may easily see that after l doubling steps one arrives at a homogeneous layer with total optical depth $2^l(0, 1)$, that is after 15 doubling steps a layer with optical depth 1 is generated if the thickness of the starting layer is $(0, 1) = \tau_1 - \tau_0 = 2^{-15}$. In the case of a

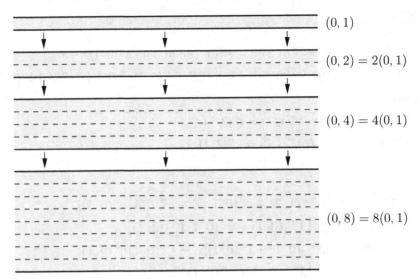

Fig. 4.4 The doubling algorithm for homogeneous layers.

homogeneous cloud with a typical optical thickness of 32, only five additional
doubling steps are required. Since the final layer is required to be homogeneous it
must also be isothermal in order to specify the thermal emission.

4.1.4 Inhomogeneous atmospheres

Each vertically inhomogeneous atmosphere can be well approximated by N sublayers of varying optical properties and thermal structure. Each of these, however, is
homogeneous and isothermal. Let τ_N denote the total optical depth of such an atmosphere. Reflection of radiation at the ground can be treated by means of an additional
fictitious layer $(N, N + 1)$ to which the following special properties are assigned

$$\mathfrak{t}^m(N, N + 1) = \mathfrak{t}^m(N + 1, N) = 0$$

$$\mathfrak{r}^m(N, N + 1) = \mathfrak{r}^m_g$$

$$\mathfrak{r}^m(N + 1, N) = 0$$

$$\mathbf{J}^m_{-,1}(N, N + 1) = \mathbf{J}^m_{-,2}(N, N + 1) = 0$$

$$\mathbf{J}^m_{+,1}(N + 1, N) = \frac{A_g \mu_0}{\pi} S_0 \exp\left(-\frac{\tau_N}{\mu_0}\right) \delta_{0m} \begin{pmatrix} 1 \\ \vdots \\ 1 \end{pmatrix} \tag{4.17}$$

$$\mathbf{J}^m_{+,2}(N + 1, N) = (1 - A_g) B_g \delta_{0m} \begin{pmatrix} 1 \\ \vdots \\ 1 \end{pmatrix}$$

Here B_g and A_g are the *thermal emission of the ground* and the albedo of the isotropically reflecting ground, respectively. The assumption of isotropic ground reflection leads to the following form of the $(s \times s)$ reflection matrix $\mathbf{r}_g^m$

$$\mathbf{r}_g^m = 2A_g \begin{pmatrix} \mu_1 w_1 & \cdots & \mu_s w_s \\ \vdots & \ddots & \vdots \\ \mu_1 w_1 & \cdots & \mu_s w_s \end{pmatrix} \delta_{0m} \tag{4.18}$$

Note that for isotropy the reflection matrix $\mathbf{r}_g^m$ is independent of the azimuthal expansion index m. The derivation of this particular form for $\mathbf{r}_g^m$ will be given in the Appendix to this chapter.

In addition to the reflection at the ground we need to specify the boundary conditions for the diffuse radiation at the top of the atmosphere and at the ground. Usually, external diffuse illumination will be ignored, that is

$$\mathbf{I}_-^m(\tau = 0) = 0, \quad \mathbf{I}_+^m(\tau_{N+1}) = 0, \quad \mathbf{I}_-^m(\tau_{N+1}) = 0 \tag{4.19}$$

In the following we will summarize the main steps of the MOM. The algorithm proceeds in the following way.

(1) Calculation of the optical properties.
   (i) First calculate the optical properties of the elementary layers by means of (4.13).
   (ii) Utilizing the adding and doubling formulas (4.15) and (4.16), the optical properties of all homogeneous sublayers $(i, i+1)$, $i = 0, \ldots, N$ are calculated.
   (iii) The optical properties of the combined layer $(0, 2)$ are calculated by means of (4.7) and (4.8).
   (iv) The combined layer $(0, 3)$ is obtained by again applying the addition theorems (4.7) and (4.8) to the sublayers $(0, 2)$ and $(2, 3)$.
   (v) This procedure is continued until the optical properties of all required combinations of sublayers (i, j), $i = 0, \ldots, N$, $j = 0, \ldots, N$ are determined.
   (vi) Utilizing the special properties of the fictitious layer $(N, N+1)$ as listed in (4.17), the optical properties of the total layer $(0, N+1)$ are obtained by replacing in (4.7) and (4.8) $1 \to N$ and $2 \to N+1$. This yields

$$\mathfrak{t}^m(N+1, 0) = 0, \qquad \mathfrak{t}^m(0, N+1) = 0$$

$$\mathbf{r}^m(0, N+1) = \mathbf{r}^m(0, N) + \mathfrak{t}^m(N, 0)\left[\mathbb{E} - \mathbf{r}_g^m \mathbf{r}^m(N, 0)\right]^{-1} \mathbf{r}_g^m \mathfrak{t}^m(0, N)$$

$$\mathbf{r}^m(N+1, 0) = 0$$

$$\mathbf{J}_{-,1}^m(0, N+1) = 0, \qquad \mathbf{J}_{-,2}^m(0, N+1) = 0$$

$$\mathbf{J}_{+,1}^m(N+1, 0) = \mathbf{J}_{+,1}^m(N, 0) + \mathfrak{t}^m(N, 0)\left[\mathbb{E} - \mathbf{r}_g^m \mathbf{r}^m(N, 0)\right]^{-1} \tag{4.20}$$
$$\times \left[\mathbf{J}_{+,1}^m(N+1, N) + \mathbf{r}_g^m \mathbf{J}_{-,1}^m(0, N)\right]$$

$$\mathbf{J}_{+,2}^m(N+1, 0) = \mathbf{J}_{+,2}^m(N, 0) + \mathfrak{t}^m(N, 0)\left[\mathbb{E} - \mathbf{r}_g^m \mathbf{r}^m(N, 0)\right]^{-1}$$
$$\times \left[\mathbf{J}_{+,2}^m(N+1, N) + \mathbf{r}_g^m \mathbf{J}_{-,2}^m(0, N)\right]$$

(2) Calculation of the radiances.

Replacing in (4.1b) and (4.2a) $1 \rightarrow i$ we may formally write

$$
\begin{aligned}
\mathbf{I}_+^m(\tau_i) &= \alpha_+(\tau_i)\mathbf{I}_+^m(\tau_{i+1}) + \beta_+(\tau_i)\mathbf{I}_-^m(\tau_i) + \gamma_+(\tau_i) \\
\mathbf{I}_-^m(\tau_i) &= \alpha_-(\tau_i)\mathbf{I}_+^m(\tau_i) + \beta_-(\tau_i)\mathbf{I}_-^m(\tau_{i-1}) + \gamma_-(\tau_i)
\end{aligned}
\tag{4.21}
$$

The terms $\alpha_\pm(\tau_i)$, $\beta_\pm(\tau_i)$, $\gamma_\pm(\tau_i)$ are functions of the optical properties of the medium which have been determined in part (1) of the procedure, i.e. they are known quantities. Writing (4.21) for $i = 0, \ldots, N+1$ we obtain a system of $2(N+2)$ equations for the radiances $\mathbf{I}_\pm(\tau_i)$, $i = 0, \ldots, N+1$. Utilizing the boundary conditions (4.19) this system may be solved by means of standard methods of linear algebra.

Finally, we list various properties of the MOM.

(1) MOM can be applied to layers with arbitrary large optical thickness.
(2) MOM offers high accuracy so that it can be used to compute benchmark results for testing simpler techniques.
(3) MOM requires a significant amount of computer time due to the numerous inversions and matrix-vector multiplications. This is particularly true if the radiance field needs to be calculated at all interior levels.
(4) MOM can be extended by known methods to account for polarization effects.

The mathematical development as described above in large parts follows the work of Plass *et al.* (1973). Numerous other authors have also contributed to the development of the matrix operator method, for details see Lenoble (1985).

4.2 The successive order of scattering method

As already mentioned at the beginning of this chapter, the solutions (2.110) of the radiative transfer equation are only formal because the source functions $\mathbf{J}_\pm$ themselves are functions of the radiances $I_\pm$. However, this fact might be a motivation to construct an iterative solution of the RTE. For this iterative approach we first define the following vectors

(a) $\quad \mathbf{Y}_+^m(\tau) = \exp\left[-\mathbf{M}^{-1}(\tau_N - \tau)\right]\mathbf{I}_+^m(\tau_N)$

$$
+ \int_\tau^{\tau_N} \exp[-\mathbf{M}^{-1}(\tau' - \tau)]\mathbf{M}^{-1}\left[\mathbf{J}_{+,1}^m(\tau') + \mathbf{J}_{+,2}^m(\tau')\right]d\tau'
\tag{4.22}
$$

(b) $\quad \mathbf{Y}_-^m(\tau) = \exp(-\mathbf{M}^{-1}\tau)\mathbf{I}_-^m(0)$

$$
+ \int_0^\tau \exp[-\mathbf{M}^{-1}(\tau - \tau')]\mathbf{M}^{-1}\left[\mathbf{J}_{-,1}^m(\tau') + \mathbf{J}_{-,2}^m(\tau')\right]d\tau'
$$

The source vectors $\mathbf{J}_{\pm,1}, \mathbf{J}_{\pm,2}$ are given by (2.96). From these expressions it is seen that $\mathbf{Y}_+^m$ and $\mathbf{Y}_-^m$ contain only those quantities that are fixed with respect to the iteration process, namely the boundary conditions, the primary solar scattering term and the thermal emission.

For the remaining contributions to the formal solution (2.110) we define

(a) $\quad Q_+^m[\tau, I_+^m(\tau), I_-^m(\tau)] = \dfrac{1}{2} \displaystyle\int_\tau^{\tau_N} \omega_0 \exp[-M^{-1}(\tau' - \tau)]M^{-1}$

$\qquad\qquad\qquad\qquad \times [\mathbb{P}_{++}^m(\tau')W\, I_+^m(\tau') + \mathbb{P}_{+-}^m(\tau')W\, I_-^m(\tau')]d\tau'$

(b) $\quad Q_-^m[\tau, I_+^m(\tau), I_-^m(\tau)] = \dfrac{1}{2} \displaystyle\int_0^\tau \omega_0 \exp[-M^{-1}(\tau - \tau')]M^{-1}$

$\qquad\qquad\qquad\qquad \times [\mathbb{P}_{+-}^m(\tau')W\, I_+^m(\tau') + \mathbb{P}_{++}^m(\tau')W\, I_-^m(\tau')]d\tau'$

$$(4.23)$$

where use was made of (2.103). Utilizing these equations the formal solution of the RTE can now be written as

$$I_+^m(\tau) = Y_+^m(\tau) + Q_+^m\left[\tau, I_+^m(\tau), I_-^m(\tau)\right]$$
$$I_-^m(\tau) = Y_-^m(\tau) + Q_-^m\left[\tau, I_+^m(\tau), I_-^m(\tau)\right]$$

$$(4.24)$$

For these equations we can set up an iteration process by defining

$$\boxed{\begin{aligned} I_+^m(\tau)_n &= I_+^m(\tau)_{n=0} + Q_+^m\left(\tau, I_+^m(\tau)_{n-1}, I_-^m(\tau)_{n-1}\right) \\ I_-^m(\tau)_n &= I_-^m(\tau)_{n=0} + Q_-^m\left(\tau, I_+^m(\tau)_{n-1}, I_-^m(\tau)_{n-1}\right), \quad n \geq 1 \end{aligned}}$$

$$(4.25)$$

where n is the iteration step. For the starting value $n = 0$ we have

$$\boxed{I_+^m(\tau)_{n=0} = Y_+^m(\tau), \quad I_-^m(\tau)_{n=0} = Y_-^m(\tau)}$$

$$(4.26)$$

Equations (4.25) and (4.26) are known as the *successive order of scattering* (SOS) method of radiative transfer.

Equation (4.25) is based on the following method of solution. Consider the integral equation

$$y(x) = g(x) + \int_{x_0}^x f\left[t, y(t)\right] dt$$

$$(4.27)$$

A solution of this integral equation can be found by successive iteration. Let n represent the iteration step then

$$y_{(n)} = y_{(n=0)}(x) + \int_{x_0}^x f[t, y_{(n-1)}(t)]\, dt, \quad n \geq 1$$

$$(4.28)$$

with $\quad y_{(n=0)}(x) = g(x)$

The boundary conditions in (4.25) can be treated in the following manner.

(1) At the upper boundary of the medium we assume that no diffuse radiation enters

$$\mathbf{I}_-^m(\tau = 0) = 0 \tag{4.29}$$

(2) At the lower boundary we again adopt an isotropically reflecting ground with albedo A_g. The diffusely reflected radiation can then be expressed as

$$\mathbf{I}_+^m(\tau = \tau_N) = \mathbf{r}_g^m \mathbf{I}_-^m(\tau_N) + \frac{A_g}{\pi} \delta_{0m} \mu_0 S_0 \exp\left(-\frac{\tau}{\mu_0}\right) \begin{pmatrix} 1 \\ \vdots \\ 1 \end{pmatrix} + B_g(1 - A_g)\delta_{0m} \begin{pmatrix} 1 \\ \vdots \\ 1 \end{pmatrix} \tag{4.30}$$

where $\mathbf{r}_g^m$ is the reflection matrix of the ground, see (4.18).

The individual contributions in (4.30) can be interpreted in the following way.

(1) The first term on the right represents the isotropic reflection of the diffuse light $\mathbf{I}_-^m(\tau_N)$ which is incident at the ground.
(2) The second term stands for the isotropic reflection of the direct sunlight.
(3) The third term accounts for the black body emission B_g of the ground with temperature T_g.

The azimuthally dependent radiance field can be reconstructed with the help of (2.106). For the flux densities the expansion term $m = 0$ is required only.

In the successive order of scattering (SOS) method the first and the n-th iteration step can be interpreted as follows.

(1) For $n = 1$ in (4.25) the radiance vectors $\mathbf{I}_\pm^m(\tau)_{n=0}$ include:
    (i) the influence of the boundary conditions;
    (ii) primary scattering of the direct sunlight;
    (iii) the isotropic thermal emission of each layer.
(2) The source vectors $\mathbf{Q}_\pm^m(\tau, \mathbf{I}_\pm^m(\tau)_1)$ include:
    (i) secondary scattered sunlight;
    (ii) primary scattered thermal radiation.
(3) In going from iteration step n to step $n + 1$ one more scattering process is simulated.

The iteration is continued until convergence is achieved, i.e. the changes in the radiance vectors remain below a certain tolerance.

The convergence properties of the SOS method depend primarily on:

(1) the total optical depth of the medium;
(2) the total number of terms considered in the expansion of the phase function;
(3) the inhomogeneity of the optical parameters within the medium;
(4) the total number s of discrete directions for representing the vectors $\mathbf{I}_\pm^m$.

The SOS method has the following two advantages over the MOM.

(1) Each iteration step has a physical significance, i.e. each additional iteration means that one additional scattering process is simulated.
(2) The inhomogeneity of the optical parameters can be easily handled, that is – in contrast to the MOM – an *a priori* subdivision of the atmosphere into different individual homogeneous sublayers is not necessary.

However, the SOS method has the disadvantage of requiring a large number of iteration steps thus converging very slowly. This is particularly true if the medium is practically conservative ($\omega_0 \to 1$) or if it has a very large optical thickness. In these situations, however, techniques for speeding up the convergence process may partly eliminate the problem.

The SOS method in the form described above follows an unpublished lecture given by Z. Sekera. A detailed description is given in Korb and Zdunkowski (1970). A fairly complete list of references for SOS may be found in Lenoble (1985).

4.3 The discrete ordinate method

The discrete ordinate method (DOM) is another very elegant approach for solving the RTE in a plane–parallel atmosphere. It also belongs to the most accurate techniques and may be used for calculating benchmark solutions to certain problems. The formulation of the DOM dates back to Chandrasekhar (1960). Starting point for the DOM is the discretization of the *m*-th Fourier mode of the radiance field, see (2.69).

In the following we discuss the DOM for the azimuthally averaged radiation field, i.e. for $m = 0$. Only the case $m = 0$ is needed to calculate important quantities such as radiative flux densities, *actinic fluxes*, and heating rates. Actinic fluxes are important for photochemistry and result from integrating the radiance over the unit sphere. The case $m \neq 0$ is needed to account for the directional dependence of the radiation field as required, for example, in remote sensing. In the sequel, for $I^{m=0}(\tau, \mu)$ we will simply write $I(\tau, \mu)$. From (2.76) it may be seen that for the case $m \neq 0$ the same procedure applies.

Let us consider a total of $2s$ directions for discretizing the radiation streams, that is $-1 \leq \mu_i \leq 1$, $i = -s, \ldots, -1, 1, \ldots, s$, as illustrated in Figure 4.5. In the following it will be shown how to solve analytically the resulting coupled system of linear differential equations for homogeneous sublayers.

Evaluating (2.76) for $m = 0$ at the discrete direction μ_i and approximating the multiple scattering integral with the help of the Gaussian quadrature (2.88) leads to

$$\mu_i \frac{dI(\tau, \mu_i)}{d\tau} = I(\tau, \mu_i) - \frac{\omega_0}{2} \sum_{j=-s}^{s}{}' w_j I(\tau, \mu_j) \mathcal{P}(\mu_i, \mu_j)$$

$$- \frac{\omega_0}{4\pi} S_0 \exp\left(-\frac{\tau}{\mu_0}\right) \mathcal{P}(\mu_i, -\mu_0) - (1 - \omega_0) B(\tau) \quad (4.31)$$

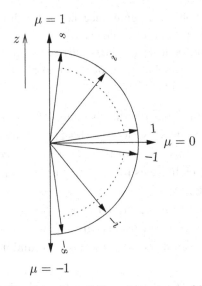

Fig. 4.5 Discretization of $I(\tau, \mu)$ in a total of $2s$ streams.

where the azimuthally averaged *phase function* $\mathcal{P}(\mu, \mu')$ has been used as defined in (2.80). For practical reasons the infinite series in (2.80) must be truncated after a sufficient number of terms. The special case that the direction μ_0 of the direct solar beam coincides with one of the quadrature directions must be avoided, otherwise singularities would occur.

From (2.88) it may be easily seen that the Gaussian quadrature of order r is exact on the interval $[-1, 1]$ for Legendre polynomials $P_m(\mu)$, $m = 0, 1, \ldots, 2r - 1$, that is

$$\int_{-1}^{1} P_m(\mu) \, d\mu = \sum_{i=1}^{r} w_i \, P_m(\mu_i) = 2\delta_{0m}, \quad m = 0, 1, \ldots, 2r - 1 \qquad (4.32)$$

An important consequence of this is that the normalization condition of the phase function (2.62) is also satisfied in its discretized version when developing this function into a finite series of Legendre polynomials which is truncated after the term $l = 2r - 1$. We will now demonstrate this. Starting with (2.68) we find the following expression

$$\mathcal{P}(\cos \Theta) = \sum_{l=0}^{2r-1} p_l P_l(\cos \Theta)$$

$$= \sum_{l=0}^{2r-1} p_l P_l(\mu) P_l(\mu') + 2 \sum_{m=1}^{2r-1} \sum_{l=m}^{2r-1} p_l^m P_l^m(\mu) P_l^m(\mu') \cos m(\varphi - \varphi')$$

$$= \mathcal{P}(\mu, \mu') + 2 \sum_{m=1}^{2r-1} \sum_{l=m}^{2r-1} p_l^m P_l^m(\mu) P_l^m(\mu') \cos m(\varphi - \varphi') \qquad (4.33)$$

To see that there is no need to renormalize the phase function in DOM, consider the quadrature form of the normalization condition as applied to $\mathcal{P}(\mu, \mu')$

$$
\begin{aligned}
\frac{1}{2} \int_{-1}^{1} \mathcal{P}(\mu, \mu') \, d\mu' &= \frac{1}{2} \sum_{i=1}^{r} w_i \mathcal{P}(\mu, \mu_i) \\
&= \frac{1}{2} \sum_{i=1}^{r} w_i \sum_{m=0}^{2r-1} p_m P_m(\mu) P_m(\mu_i) \\
&= \frac{1}{2} \sum_{m=0}^{2r-1} p_m P_m(\mu) \sum_{i=1}^{r} w_i P_m(\mu_i) \qquad (4.34) \\
&= \frac{1}{2} \sum_{m=0}^{2r-1} p_m P_m(\mu) \int_{-1}^{1} P_m(\mu) \, d\mu \\
&= \sum_{m=0}^{2r-1} p_m P_m(\mu) \delta_{0m} = 1
\end{aligned}
$$

As demonstrated by Wiscombe (1977), the above property also holds for the so-called *δ-scaled phase function* $\mathcal{P}^*(\mu, \mu')$ as defined by

$$
\boxed{\mathcal{P}^*(\mu, \mu') = 2f\delta(\mu - \mu') + (1 - f) \sum_{m=0}^{2r-1} p_m^* P_m(\mu) P_m(\mu')} \qquad (4.35)
$$

if it is truncated after the term $l = 2r - 1$. The quantity f is the fraction of radiation scattered into the forward peak of the phase function. We refrain from proving this equation since the proof is carried out analogously to the development leading to (4.34). Note also that $p_m \neq p_m^*$. The δ-scaled phase function will be discussed in more detail in a later chapter.

If the expansion of the phase function is continued beyond $2r - 1$ then the phase function is no longer correctly normalized so that artificial absorption may occur as pointed out by Wiscombe (1977). It is important to realize that the normalization of the phase function is a basic requirement for the numerical algorithm to be energy conserving.

An additional comment is due regarding an improved discretization of the term $I(\tau, \mu) = \int_{-1}^{1} I(\tau, \mu') \mathcal{P}(\mu, \mu') \, d\mu'$. The ordinary Gaussian quadrature, used before, reads

$$
\int_{-1}^{1} I(\tau, \mu') \mathcal{P}(\mu, \mu') \, d\mu' \approx \sum_{i=-n}^{n} {}' w_i I(\tau, \mu_i) \mathcal{P}(\mu, \mu_i) \qquad (4.36)
$$

For increasing quadrature order n the nodes μ_i still cluster near $\mu = 1$ and $\mu = -1$, but only a few nodes are located near the horizon $\mu = 0$. Therefore, for strongly anisotropic phase functions the accuracy of the radiance field does not improve significantly by simply increasing the number of nodes. In order to improve this situation one can use a simple trick by splitting the integration over $[-1, 1]$ into the subintervals $[-1, 0]$ and $[0, 1]$. By using the transformations $\tilde{\mu} = 2\mu + 1$ for the interval $[-1, 0]$ and $\tilde{\mu} = 2\mu - 1$ for the interval $[0, 1]$ and applying a s-point Gaussian quadrature, we can write for integrals of the type

$$\int_{-1}^{1} f(\mu) \, d\mu = \frac{1}{2} \int_{-1}^{1} f\left(\frac{\tilde{\mu} - 1}{2}\right) d\tilde{\mu} + \frac{1}{2} \int_{-1}^{1} f\left(\frac{\tilde{\mu} + 1}{2}\right) d\tilde{\mu}$$

$$\approx \frac{1}{2} \sum_{i=1}^{s} w_i \left[f\left(\frac{\tilde{\mu}_i - 1}{2}\right) + f\left(\frac{\tilde{\mu}_i + 1}{2}\right) \right]$$

(4.37)

Figure 4.6 illustrates how the ordinary s-point Gaussian quadrature for the variable $\tilde{\mu}$ is mapped onto the intervals $[-1, 0]$ and $[0, 1]$ for the variable μ. The latter nodes are now symmetric with respect to the locations $\mu = -0.5$ and $\mu = 0.5$ and cluster near the end points of each interval. It can also be seen that the nodes μ_i are antisymmetric with respect to $\mu = 0$.

The zeros and weights of the Legendre polynomials occurring in (4.37) have the following properties

$$\tilde{\mu}_i = -\tilde{\mu}_{s+1-i}, \quad w_i = w_{s+1-i}, \quad i = 1, \ldots, s$$

(4.38)

Utilizing these expressions together with the substitution $\mu_i' = (\tilde{\mu}_i + 1)/2$, the right-hand side of (4.37) may be written as

$$\frac{1}{2} \sum_{i=1}^{s} w_i \left[f\left(\frac{\tilde{\mu}_i - 1}{2}\right) + f\left(\frac{\tilde{\mu}_i + 1}{2}\right) \right]$$

$$= \frac{1}{2} \sum_{i=1}^{s} w_i \left[f\left(-\frac{\tilde{\mu}_{s+1-i} + 1}{2}\right) + f\left(\frac{\tilde{\mu}_i + 1}{2}\right) \right]$$

$$= \frac{1}{2} \sum_{i=1}^{s} w_{s+1-i} f\left(-\mu_{s+1-i}'\right) + \frac{1}{2} \sum_{i=1}^{s} w_i f\left(\mu_i'\right)$$

$$= \frac{1}{2} \sum_{i=1}^{s} w_i \left[f\left(-\mu_i'\right) + f\left(\mu_i'\right) \right]$$

(4.39)

Introducing the nodes μ_i' and weights w_i' according to

$$\boxed{\begin{aligned} w_i' &= \frac{1}{2} w_i, \quad w_{-i}' = w_i', \\ \mu_i' &= \frac{\tilde{\mu}_i + 1}{2}, \quad \mu_{-i}' = -\mu_i', \end{aligned} \quad i = 1, \ldots, s}$$

(4.40)

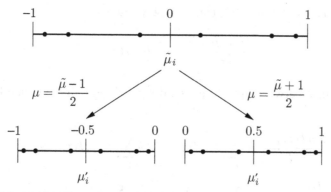

Fig. 4.6 The double-Gaussian quadrature rule. The dots mark the locations of the quadrature nodes for $i = 1, \ldots, s$.

results in the so-called *double-Gaussian quadrature* of order $2s$

$$\boxed{\int_{-1}^{1} f(\mu)\, d\mu \approx \sum_{i=-s}^{s}{'} w_i' f(\mu_i')}$$

(4.41)

Recall that the notation $\sum'$ means that in the summation the term $i = 0$ is omitted. Hence for the $2s$-point double-Gaussian quadrature formula only the nodes and zeros (w_i, μ_i) of the original s-point Gaussian quadrature formula are needed. The advantage of the double-Gaussian quadrature formulas is that the nodes are not only clustered in both the upward and downward directions but also near the horizon.

Let us now find the solution to the RTE for the special case that no thermal emission exists, $B(\tau) = 0$. Inclusion of this term, however, poses no particular difficulty. The treatment of thermal emission will be taken care of in a later chapter. For a layer with constant optical properties the inhomogeneous system of linear differential equations (4.31) can be solved exactly. The solution is composed of the general solution of the homogeneous system plus a particular solution of the inhomogeneous part. First let us define the set of coefficients

$$b_{i,j} = \begin{cases} -\dfrac{1}{\mu_i} \dfrac{\omega_0}{2} w_j \mathcal{P}(\mu_i, \mu_j) & i \neq j \\[2mm] \dfrac{1}{\mu_i}\left[1 - \dfrac{\omega_0}{2} w_j \mathcal{P}(\mu_j, \mu_j) \right] & i = j \end{cases}$$

(4.42)

These coefficients satisfy the asymmetry relations

$$b_{i,j} = -b_{-i,-j}, \quad b_{i,-j} = -b_{-i,j}$$

(4.43)

The homogeneous part of (4.31) can then be written as

$$\frac{dI(\tau, \mu_i)}{d\tau} = \sum_{j=-s}^{s}{}' b_{i,j} I(\tau, \mu_j) \tag{4.44}$$

If we distinguish between upwelling and downwelling radiation, then from (4.44) we obtain

$$\frac{dI(\tau, \mu_i)}{d\tau} = \sum_{j=1}^{s} b_{i,j} I(\tau, \mu_j) + \sum_{j=1}^{s} b_{i,-j} I(\tau, -\mu_j)$$

$$\frac{dI(\tau, -\mu_i)}{d\tau} = \sum_{j=1}^{s} b_{-i,j} I(\tau, \mu_j) + \sum_{j=1}^{s} b_{-i,-j} I(\tau, -\mu_j) \tag{4.45}$$

which can be written in vector–matrix form as

$$\frac{d}{d\tau} \begin{pmatrix} \mathbf{I}_+ \\ \mathbf{I}_- \end{pmatrix} = \begin{pmatrix} \mathbb{B}_+ & \mathbb{B}_- \\ -\mathbb{B}_- & -\mathbb{B}_+ \end{pmatrix} \begin{pmatrix} \mathbf{I}_+ \\ \mathbf{I}_- \end{pmatrix} \tag{4.46}$$

Here the vectors $\mathbf{I}_\pm$ and the matrices $\mathbb{B}_\pm$ are defined as

$$\mathbf{I}_\pm = \begin{pmatrix} I(\tau, \pm\mu_1) \\ \vdots \\ I(\tau, \pm\mu_s) \end{pmatrix}, \qquad \begin{aligned} \mathbb{B}_+ &= (b_{i,j}) & i &= 1, \ldots, s \\ \mathbb{B}_- &= (b_{i,-j}) \, , & j &= 1, \ldots, s \end{aligned} \tag{4.47}$$

With the exponential trial solution

$$\mathbf{I}_\pm = \mathbf{F}_\pm \exp(-k\tau) \tag{4.48}$$

involving the vectors $\mathbf{F}_\pm = (F_{\pm,i})$, $i = 1, \ldots, s$ which are independent of τ, the homogeneous system (4.46) may be transformed into an eigenvalue problem of order $2s$

$$\begin{pmatrix} \mathbb{B}_+ & \mathbb{B}_- \\ -\mathbb{B}_- & -\mathbb{B}_+ \end{pmatrix} \begin{pmatrix} \mathbf{F}_+ \\ \mathbf{F}_- \end{pmatrix} = -k \begin{pmatrix} \mathbf{F}_+ \\ \mathbf{F}_- \end{pmatrix} \tag{4.49}$$

Without going into details we need to mention that all eigenvalues occur in pairs $\pm k$ (see Chandrasekhar, 1960) and that they are all real (Kuščer and Vidav, 1969).

Due to the particular form of the nonsymmetrical matrices $\mathbb{B}_-$ and $\mathbb{B}_+$ the above eigenvalue problem of order $2s$ can be reduced to a corresponding problem of order s as will be shown next. Adding and subtracting the first and second equation in (4.49) leads to

$$\begin{aligned} \text{(a)} \quad & (\mathbb{B}_+ - \mathbb{B}_-)(\mathbf{F}_+ - \mathbf{F}_-) = -k\,(\mathbf{F}_+ + \mathbf{F}_-) \\ \text{(b)} \quad & (\mathbb{B}_+ + \mathbb{B}_-)(\mathbf{F}_+ + \mathbf{F}_-) = -k\,(\mathbf{F}_+ - \mathbf{F}_-) \end{aligned} \tag{4.50}$$

If we now multiply (4.50b) by $(\mathbb{B}_+ - \mathbb{B}_-)$ and insert on the right-hand side of the resulting expression (4.50a), we obtain

$$(\mathbb{B}_+ - \mathbb{B}_-)(\mathbb{B}_+ + \mathbb{B}_-)(\mathbf{F}_+ + \mathbf{F}_-) = k^2(\mathbf{F}_+ + \mathbf{F}_-) \tag{4.51}$$

Thus we obtain an eigenvalue problem of order s involving the matrix $(\mathbb{B}_+ - \mathbb{B}_-)(\mathbb{B}_+ + \mathbb{B}_-)$ with eigenvalues k^2 and eigenvectors $\mathbf{F}_+ + \mathbf{F}_-$. We may then use (4.50b) to determine $\mathbf{F}_+ - \mathbf{F}_-$. The eigenvectors $\mathbf{F}_+$, $\mathbf{F}_-$ of the original equation (4.49) can be found from

$$\begin{aligned} \mathbf{X}_+ &= \mathbf{F}_+ + \mathbf{F}_-, & \mathbf{X}_- &= \mathbf{F}_+ - \mathbf{F}_- \\ \mathbf{F}_+ &= \frac{1}{2}(\mathbf{X}_+ + \mathbf{X}_-), & \mathbf{F}_- &= \frac{1}{2}(\mathbf{X}_+ - \mathbf{X}_-) \end{aligned} \tag{4.52}$$

As mentioned above, the $2s$ eigenvalues of (4.49) are all distinct and occur in pairs $\pm k_j$, $(j = 1, \ldots, s)$. These eigenvalues and the corresponding i-th component of the j-th eigenvector, $F_j(\mu_i)$, for eigenvalue k_j can be efficiently computed with numerical standard algorithms (see, e.g. Press *et al.*, 1992).

The general solution I_h of the homogeneous system is then given by

$$I_h(\tau, \mu_i) = \sum_{j=-s}^{s}{}' D_j F_j(\mu_i) \exp(-k_j \tau), \quad i = -s, \ldots, -1, 1, \ldots, s \tag{4.53}$$

The constants D_j, $j = -s, \ldots, -1, 1, \ldots, s$ follow from the boundary conditions at the upper and the lower boundary of the homogeneous layer.

A particular solution I_p of the inhomogeneous system can be obtained from a trial solution resembling the functional form of the right-hand side of (4.31)

$$I_p(\tau, \mu_i) = Z(\mu_i) \exp\left(-\frac{\tau}{\mu_0}\right), \quad i = -s, \ldots, -1, 1, \ldots, s \tag{4.54}$$

where the $2s$ unknown coefficients $Z(\mu_i)$, $i = -s, \ldots, -1, 1, \ldots, s$ can be found by inserting (4.54) into (4.31). In summary, for a homogeneous layer the complete solution of the azimuthally averaged RTE can now be written down by adding the homogeneous solution (4.53) and the inhomogeneous solution (4.54) yielding

$$\boxed{I(\tau, \mu_i) = \sum_{j=-s}^{s}{}' D_j F_j(\mu_i) \exp(-k_j \tau) + Z(\mu_i) \exp\left(-\frac{\tau}{\mu_0}\right)} \tag{4.55}$$

So far we have considered the particular case of a single homogeneous layer. The inhomogeneous atmosphere will now be subdivided into Q different homogeneous sublayers, see Figure 4.7. The complete solution for each homogeneous sublayer

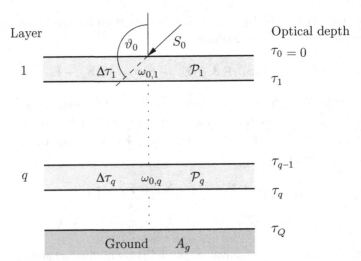

Fig. 4.7 Subdivision of the atmosphere in a total number of Q different homogeneous sublayers with optical thickness $\Delta\tau_q$, single scattering albedo $\omega_{0,q}$, and phase function $\mathcal{P}_q$, $q = 1, \ldots, Q$. The albedo of the diffusely reflecting ground is A_g.

is already known from (4.55). For layer q we now write

$$I_q(\tau, \mu_i) = \sum_{j=-s}^{s}{}' D_{q,j} F_{q,j}(\mu_i) \exp[-k_{q,j}(\tau - \tau_{q-1})] + Z_q(\mu_i) \exp\left(-\frac{\tau}{\mu_0}\right)$$

(4.56)

with $1 \leq q \leq Q$. Note that in the homogeneous part of the solution the integration constants $D_{q,j}$, the eigenvalues $k_{q,j}$, the components of the eigenvectors $F_{q,j}$, and the constants Z_q of the particular solution refer to layer number q. It is also clear that the solution (4.56) is only valid for the optical depth $\tau_{q-1} \leq \tau \leq \tau_q$.

In total $2sQ$ equations are required to determine the integration constants $D_{q,j}$. Two of these equations are given by the specification of the boundary conditions at $\tau = 0$ and $\tau = \tau_Q$. Assuming isotropic reflection at the ground with albedo A_g we obtain

$$I_1(\tau = 0, -\mu_i) = 0$$
$$I_Q(\tau_Q, \mu_i) = \frac{A_g}{\pi}\left[E_{-,z}(\tau_Q) + \mu_0 S_0 \exp\left(-\frac{\tau_Q}{\mu_0}\right)\right]$$
$$= \frac{A_g}{\pi}\left[2\pi \sum_{j=1}^{s} w_j I_Q(\tau_Q, -\mu_j)\mu_j + \mu_0 S_0 \exp\left(-\frac{\tau_Q}{\mu_0}\right)\right], \quad i = 1, \ldots, s$$

(4.57)

The upwelling radiance at the ground is assumed to be isotropic and, therefore, is identical for all directions μ_i as expressed by the last equation of (4.57). The remaining $2(s-1)Q$ equations are determined by the requirement that the radiance for each μ_i must be continuous at each level interface $\tau_q, (q = 1, \ldots, Q - 1)$, i.e.

$$I_q(\tau_q, \mu_i) = I_{q+1}(\tau_q, \mu_i), \quad q = 1, \ldots, Q - 1, \quad i = -s, \ldots, -1, 1, \ldots, s \tag{4.58}$$

Inserting the complete solution as stated by (4.56) into relations (4.57) and (4.58) then leads to a linear system for the coefficients $D_{q,i}$

$$\sum_{j=-s}^{s}{}' D_{1,j} F_{1,j}(-\mu_i) = -Z_1(-\mu_i), \quad i = 1, \ldots, s$$

$$\sum_{j=-s}^{s}{}' [D_{q,j} \gamma_{q,j}(\mu_i) - D_{q+1,j} F_{q+1,j}(\mu_i)] = \eta_q(\mu_i), \quad \begin{aligned} i &= 1, \ldots, s \\ q &= 1, \ldots, Q - 1 \end{aligned}$$

$$\sum_{j=-s}^{s}{}' D_{Q,j} \beta_j(\mu_i) = -\varepsilon(\mu_i), \quad i = 1, \ldots, s \tag{4.59}$$

where the following abbreviations have been introduced

$$\gamma_{q,j}(\mu_i) = F_{q,j}(\mu_i) \exp[-k_{q,j}(\tau_q - \tau_{q-1})]$$

$$\eta_q(\mu_i) = [Z_{q+1}(\mu_i) - Z_q(\mu_i)] \exp\left(-\frac{\tau_q}{\mu_0}\right)$$

$$\beta_j(\mu_i) = \left(F_{Q,j}(\mu_i) - 2A_g \sum_{l=1}^{s} w_l F_{Q,l}(-\mu_l)\mu_l\right) \exp[-k_{Q,j}(\tau_Q - \tau_{Q-1})] \tag{4.60}$$

$$\varepsilon(\mu_i) = \left(Z_Q(\mu_i) - 2A_g \sum_{j=1}^{s} w_j Z_Q(-\mu_j)\mu_j - \frac{A_g}{\pi}\mu_0 S_0\right) \exp\left(-\frac{\tau_Q}{\mu_0}\right)$$

The linear equation system (4.59) may also be written in matrix form as

$$\mathbb{K}\mathbf{D} = \mathbf{X} \tag{4.61}$$

$\mathbb{K}$ is a square matrix of dimension $(2s\,Q \times 2s\,Q)$, and the vectors $\mathbf{D}$ and $\mathbf{X}$ are $2sQ$-dimensional. From this system of linear equations for $2s\,Q$ unknown coefficients we may obtain the required constants of integration $D_{q,j}$ by numerical inversion.

The major advantages and disadvantages of the DOM are listed here.

(1) The solution of the RTE can be derived in a completely explicit form.
(2) The computational effort for each individual layer is independent of its optical depth.
(3) The accuracy of the method compares well with the MOM and, therefore, DOM can also be used to perform benchmark calculations.
(4) Unless one uses a δ-approximation to the phase function, see (4.35), sharp phase functions may produce unrealistic oscillating radiance patterns.
(5) DOM is computationally too expensive for routine calculations in climate models.

The DOM in the form described above mainly follows the derivation given in Stamnes *et al.* (1988). These authors also supplied the very reliable programme package *DISORT* to the scientific community, a package which is widely used for various applications. The discretization of the radiance field into streams traveling along the directions of the Gaussian quadrature nodes is fully described in Chandrasekhar (1960). Important developments regarding both the algorithmic formulation of the DOM for an inhomogeneous atmosphere as well as the correct evaluation of the numerical eigenvalue-eigenvector problem have been contributed by Liou (1973) and Asano (1975).

4.4 The spherical harmonics method

In this section we will discuss the principles of the spherical harmonics method (SHM) and illustrate the solution of the radiance field for a homogeneous atmospheric layer. The SHM employs the transfer equation (2.76) which is based on the cosine Fourier expansion of the radiance

$$I(\tau, \mu, \varphi) = \sum_{m=0}^{\Lambda} (2 - \delta_{0m}) I^m(\tau, \mu) \cos m\varphi \tag{4.62}$$

There are several equivalent ways of separating the μ and τ dependencies. Here we choose the following expansion

$$I^m(\tau, \mu) = \sum_{l=m}^{M} \frac{2l+1}{2} I_l^m(\tau) P_l^m(\mu) \tag{4.63}$$

with $M = 2p - 1 + m$, where p is chosen as the smallest integer number that fulfills the condition $2p - 1 + m \geq \Lambda$. Recall also that, according to (2.59a), $P_l^m(\mu) = 0$ for $m > l$. Inserting (4.63) into (2.76) and employing the integral operation

$$\int_{-1}^{1} \ldots P_n^m(\mu) \, d\mu \tag{4.64}$$

leads to a system of $\Lambda + 1$ ordinary inhomogeneous linear differential equations. Each such system contains $2p = M + 1 - m$ differential equations for the determination of the unknown functions $I_l^m(\tau)$ and has the form

$$(l + m + 1)\frac{dI_{l+1}^m(\tau)}{d\tau} + (l - m)\frac{dI_{l-1}^m(\tau)}{d\tau} + [\omega_0 p_l - (2l+1)] I_l^m(\tau)$$

$$= -\frac{\omega_0}{2\pi} S_0 \exp\left(-\frac{\tau}{\mu_0}\right) p_l^m P_l^m(-\mu_0) - 2(1 - \omega_0) B(\tau) \delta_{0m} \delta_{0l} \tag{4.65}$$

with $m = 0, 1, \ldots, \Lambda$ and $l = m, m + 1, \ldots, M$. In order to have as many unknowns I_l^m as equations we require

$$I_{m-1}^m(\tau) = 0, \quad I_{M+1}^m(\tau) = 0 \tag{4.66}$$

These two terms appear in (4.65) for $l = m$ and $l = M$, respectively. Otherwise the system would be under-determined.

We return to the discussion why there should be an even number $(2p)$ of equations for the unknown functions $I_l^m(\tau)$. From (4.63) we see that the expansion coefficients $I_l^m(\tau)$ fully determine the m-th azimuthal Fourier expansion coefficient $I^m(\tau, \mu)$ of the radiance function. Considering an inhomogeneous atmosphere consisting of individual homogeneous sublayers, for each m the radiance function $I^m(\tau, \mu)$ is required to be a continuous function of μ in the range $(-1, 1)$. Since at the top of a particular layer we have the same number of determining equations for the incoming downwelling radiance as for the incoming upwelling radiance at the bottom of that layer, in total this leads to an even number of equations over the full range of μ. For further details see also Dave (1975).

It is convenient to introduce the following vector and matrix symbols

$$\mathbf{I}^m(\tau) = \begin{pmatrix} I_m^m(\tau) \\ I_{m+1}^m(\tau) \\ \vdots \\ I_M^m(\tau) \end{pmatrix} \tag{4.67}$$

for $m = 0, 1, \ldots, \Lambda$. Note that this column vector has exactly $2p$ rows. For the primary scattered sunlight and the thermal emission term, i.e. the terms on the right-hand side of (4.65), we introduce the $2p$-dimensional vector $\mathbf{f}^m(\tau)$. Furthermore, we need to define two $2p \times 2p$ coefficient matrices $\mathbb{A}^m$, $\mathbb{B}^m$. The first of these contains the factors in (4.65) multiplying the derivatives with respect to τ, while the second matrix incorporates the factors premultiplying I_l^m. In this manner we obtain from (4.65) for each Fourier mode m the matrix differential equation

$$\mathbb{B}^m \frac{d\mathbf{I}^m(\tau)}{d\tau} - \mathbb{A}^m \mathbf{I}^m(\tau) = \mathbf{f}^m(\tau) \tag{4.68}$$

The matrix elements of $\mathbb{A}^m$ and $\mathbb{B}^m$ can be determined as follows: let us consider a fixed but arbitrary value of m. The first two and the last two equations of the system

(4.65) are given by

$$(2m + 1)\frac{d I_{m+1}^m(\tau)}{d\tau} + [\omega_0\, p_m - (2m + 1)]\, I_m^m(\tau)$$

$$= -\frac{\omega_0}{2\pi} S_0 \exp\left(-\frac{\tau}{\mu_0}\right) p_m^m P_m^m(-\mu_0) - 2(1 - \omega_0)B(\tau)\delta_{0m}\delta_{0m}$$

$$\times (2m + 2)\frac{d I_{m+2}^m(\tau)}{d\tau} + \frac{d I_m^m(\tau)}{d\tau} + [\omega_0\, p_{m+1} - (2m + 3)]\, I_{m+1}^m(\tau)$$

$$= -\frac{\omega_0}{2\pi} S_0 \exp\left(-\frac{\tau}{\mu_0}\right) p_{m+1}^m P_{m+1}^m(-\mu_0) - 2(1 - \omega_0)B(\tau)\delta_{0m}\delta_{0,m+1}$$

$$\vdots$$

$$(M + m)\frac{d I_M^m(\tau)}{d\tau} + (M - 1 - m)\frac{d I_{M-2}^m(\tau)}{d\tau} + [\omega_0\, p_{M-1} - (2M - 1)]\, I_{M-1}^m(\tau)$$

$$= -\frac{\omega_0}{2\pi} S_0 \exp\left(-\frac{\tau}{\mu_0}\right) p_{M-1}^m P_{M-1}^m(-\mu_0) - 2(1 - \omega_0)B(\tau)\delta_{0m}\delta_{0,M-1}$$

$$\times (M - m)\frac{d I_{M-1}^m(\tau)}{d\tau} + [\omega_0\, p_M - (2M + 1)]\, I_M^m(\tau)$$

$$= -\frac{\omega_0}{2\pi} S_0 \exp\left(-\frac{\tau}{\mu_0}\right) p_M^m P_M^m(-\mu_0) - 2(1 - \omega_0)B(\tau)\delta_{0m}\delta_{0,M} \qquad (4.69)$$

It can be readily seen that the matrix $\mathbb{B}^m$ is tridiagonal with zero entries on the main diagonal. The matrix elements $B_{i,k}^m$ are given by

$$\begin{aligned}
B_{i,i-1}^m &= i - 1, & i &= 2, \ldots, 2p \\
B_{i,i+1}^m &= 2m + i, & i &= 1, \ldots, 2p - 1 \\
B_{i,k}^m &= 0 & i &= 1, \ldots, 2p \\
& & k &\neq i - 1, i + 1
\end{aligned} \qquad (4.70)$$

Similarly we find for the matrix $\mathbb{A}^m$ nonzero entries on the main diagonal

$$A_{i,k}^m = 0, \qquad\qquad\qquad i \neq k, \quad \begin{aligned} i &= 1, \ldots, 2p \\ k &= 1, \ldots, 2p \end{aligned} \qquad (4.71)$$

$$A_{i,i}^m = [2(m + i) - 1] - \omega_0\, p_{m+i-1}, \qquad\qquad i = 1, \ldots, 2p$$

For the components of the vector $\mathbf{f}^m(\tau)$ one obtains

$$f_i^m(\tau) = -\frac{\omega_0}{2\pi} S_0 \exp\left(-\frac{\tau}{\mu_0}\right) p_{i+m-1}^m P_{i+m-1}^m(-\mu_0) - 2(1 - \omega_0)B(\tau)\delta_{0m}\delta_{0,i+m-1}$$

$$i = 1, \ldots, 2p \qquad\qquad\qquad\qquad\qquad\qquad\qquad\qquad\qquad\qquad\qquad (4.72)$$

Multiplication of (4.68) by $(\mathbb{B}^m)^{-1}$ and using the abbreviations

$$\mathbb{G}^m = (\mathbb{B}^m)^{-1}\mathbb{A}^m, \qquad \mathbf{D}^m(\tau) = (\mathbb{B}^m)^{-1}\mathbf{f}^m(\tau) \tag{4.73}$$

yields

$$\frac{d\mathbf{I}^m(\tau)}{d\tau} - \mathbb{G}^m\mathbf{I}^m(\tau) = \mathbf{D}^m(\tau) \tag{4.74}$$

The solution of this ordinary differential equation is

$$\boxed{\mathbf{I}^m(\tau) = \exp\left(\mathbb{G}^m\tau\right)\mathbf{C}^m + \int_0^\tau \exp[\mathbb{G}^m(\tau - \tau')]\mathbf{D}^m(\tau')d\tau'} \tag{4.75}$$

where $\mathbf{C}^m$ is the vector of integration constants

$$\mathbf{C}^m = \begin{pmatrix} C_1^m \\ C_2^m \\ \vdots \\ C_{2p}^m \end{pmatrix} \tag{4.76}$$

It should be noted that the apparently complicated solution (4.75) contains parts of the boundary conditions. For a homogeneous layer extending between $\tau = 0$ and $\tau = \tau_1$ we directly obtain at the upper boundary

$$\mathbf{I}^m(\tau = 0) = \mathbf{C}^m \tag{4.77}$$

which determines the constants of integration after specification of the radiation incident at $\tau = 0$. For more details the reader may consult the work of Flatau and Stephens (1988).

There are many ways to evaluate the exponential matrix appearing in (4.75), see Moler and van Loan (1978). One way is to determine the exponential matrix with the help of the *Jordan matrix* $\mathbb{J}^m$ in normal form which is defined as

$$\mathbb{J}^m = \begin{pmatrix} \lambda_1^m & \cdots & 0 \\ \vdots & \ddots & \vdots \\ 0 & \cdots & \lambda_{2p}^m \end{pmatrix} \tag{4.78}$$

The exponential matrix can be found from

$$\exp(\mathbb{G}^m\tau) = \mathbb{P}^m \exp(\mathbb{J}^m\tau)(\mathbb{P}^m)^{-1} \tag{4.79}$$

where $\mathbb{P}^m$ is the so-called *modal matrix* containing the eigenvectors of $\mathbb{G}^m$. The individual eigenvectors result from the distinct eigenvalues $\lambda_1^m, \ldots, \lambda_{2p}^m$. A simple discussion on the subject may be found in Bronson (1972).

However, for systems not too large the Putzer method (Putzer, 1966) provides an entirely analytical procedure which will be outlined next. The exponential matrix occurring in (4.75) may be evaluated using the following algorithm

$$\mathbb{N}^m(\tau) = \exp(\mathbb{G}^m\tau) = \sum_{j=0}^{M-m} \eta_{j+1}^m(\tau)\mathbb{X}_j^m \tag{4.80}$$

In the previous formula we have used the definitions

$$\mathbb{X}_0^m = \mathbb{E}, \quad \mathbb{X}_j^m = \prod_{i=1}^{j}\left(\mathbb{G}^m - \lambda_i^m\mathbb{E}\right), \quad j = 1, 2, \ldots, M+1-m \tag{4.81}$$

where $\mathbb{E}$ is the $2p \times 2p$ unit matrix. The λ_i^m are the eigenvalues of the matrix $\mathbb{G}^m$ which can be determined using standard numerical routines. The scalar coefficients $\eta_1^m(\tau), \ldots, \eta_{M+1-m}^m(\tau)$ occurring in (4.80) can be recursively determined via the solution of the following linear system of ordinary differential equations

$$\frac{d\eta_1^m}{d\tau} = \lambda_1^m\eta_1^m$$

$$\frac{d\eta_{j+1}^m}{d\tau} = \lambda_{j+1}^m\eta_{j+1}^m + \eta_j^m, \quad j = 1, \ldots, M-m \tag{4.82}$$

The initial conditions for this system are given by

$$\eta_1^m(0) = 1, \quad \eta_{j+1}^m(0) = 0 \tag{4.83}$$

Although this method requires solving a system of linear differential equations, it has a triangular structure. Thus the solutions can be determined in succession.

Using the definition (4.80) we may formally write for the integral term in (4.75)

$$\mathbf{H}^m(\tau) = \int_0^\tau \exp\left[\mathbb{G}^m(\tau - \tau')\right]\mathbf{D}^m(\tau')d\tau' = \int_0^\tau \mathbb{N}^m(\tau - \tau')\mathbf{D}^m(\tau')d\tau' \tag{4.84}$$

Utilizing (4.80) and (4.84) the solution (4.75) of the m-th differential equation system may be written as

$$\mathbf{I}^m(\tau) = \mathbb{N}^m(\tau)\mathbf{C}^m + \mathbf{H}^m(\tau) \tag{4.85}$$

Alternatively one may write this equation in component form as

$$\begin{pmatrix} I_m^m(\tau) \\ \vdots \\ I_M^m(\tau) \end{pmatrix} = \begin{pmatrix} N_{1,1}^m(\tau) & \cdots & N_{1,2p}^m(\tau) \\ \vdots & \ddots & \vdots \\ N_{2p,1}^m(\tau) & \cdots & N_{2p,2p}^m(\tau) \end{pmatrix} \begin{pmatrix} C_1^m \\ \vdots \\ C_{2p}^m \end{pmatrix} + \begin{pmatrix} H_1^m(\tau) \\ \vdots \\ H_{2p}^m(\tau) \end{pmatrix} \tag{4.86}$$

From this equation we immediately see that for $m = 0, 1, \ldots, \Lambda$ we have a total of $\Lambda + 1$ different solutions. For each particular value of m there are in total $2p$

different rows in (4.86). It is noteworthy that for an increasing Fourier mode m the system of differential equations decreases in size up to the point where for $m = M$ only a single scalar differential equation has to be solved.

We now return to (4.75) to involve the boundary conditions. The integration constants C_i^m are determined by first combining the solutions $I_l^m(\tau)$ as contained in (4.63) to obtain the m-th Fourier mode of the radiance. There is no prescribed way to specify the boundary conditions. A method which has found wide acceptance is the so-called *Marshak boundary condition* which will be used in the following. We will assume that no downwelling diffuse radiation enters at the top of the atmosphere. For the downwelling and the upwelling radiance fields $I^m(\tau = 0, -\mu)$ and $I(\tau_Q, \mu)$ the following relations must hold

(a) $\displaystyle\int_{-1}^{0} I^m(\tau = 0, \mu) P_{m+2j-1}^m(\mu)\, d\mu = 0$

(b) $\displaystyle\int_{0}^{1} I^m(\tau_Q, \mu) P_{m+2j-1}^m(\mu)\, d\mu$

$$= \int_0^1 \left[2A_g \int_{-1}^0 I^m(\tau_Q, \mu')\mu'd\mu' + \left(A_g \frac{\mu_0 S_0}{\pi} \exp\left(-\frac{\tau_Q}{\mu_0} \right) \right. \right.$$

$$\left. \left. + (1 - A_g)B_g \right) \delta_{0m} \right] P_{m+2j-1}^m(\mu)d\mu \qquad (4.87)$$

where $j = 1, 2, \ldots, p$ and $m = 0, 1, \ldots, \Lambda$. Note that the scalar version of the boundary condition (4.30) has already been employed in (4.87b). The integration constants may be determined using standard numerical algorithms for systems of linear equations. A vertically inhomogeneous atmosphere is handled as in DOM by requiring continuity conditions for the radiation field at the interior boundaries for each layer.

The main advantages and disadvantages of the SHM are listed here.

(1) The solution of the radiance field, to a large degree, can be derived in an analytic manner.
(2) In contrast to DOM, a discretization of the μ-dependence is not required. Therefore, the computation time does not increase when the radiance is needed for a large number of directions μ.
(3) The SHM circumvents the problems involved in integrating the highly oscillatory P_l^m functions for large l by carrying out these integrations analytically.
(4) The radiances at all depths inside as well as the reflected and transmitted radiation field may be obtained simultaneously.
(5) An increasing number of directions does not notably change the total computation time. The same fact applies to the total optical depth of the medium.
(6) The SHM needs less computation time as MOM and SOS if the same amount of information is required (internal radiation field).

(7) The SHM is closely related to DOM. Benchmark computations of both methods essentially yield identical results.

(8) As is the case for any exact treatment of the RTE, the computation time drastically increases with the number of expansion terms (Λ) in the phase function.

The SHM as formulated above follows the work of Deuze et al. (1973), Zdunkowski and Korb (1974), and Zdunkowski and Korb (1985). It should also be emphasized that for a total of four streams and for the azimuthally independent radiation field ($m = 0$) very efficient and accurate solutions may be obtained for fluxes and heating rates, since both the eigenvalue problem as well as the vector $\mathbf{H}^{m=0}(\tau)$ can be solved analytically. For more details interested readers are directed to the work of Li and Ramaswamy (1996) and Zdunkowski et al. (1998).

4.5 The finite difference method

The finite difference method (FDM) is based on the integro-differential form of the RTE (2.76), using a phase function truncation after Λ terms, which for the m-th Fourier mode is repeated below

$$
\mu \frac{d}{d\tau} I^m(\tau, \mu) = I^m(\tau, \mu) - \frac{\omega_0}{2} \int_{-1}^{1} \sum_{l=m}^{\Lambda} p_l^m P_l^m(\mu) P_l^m(\mu') I^m(\tau, \mu') d\mu'
$$

$$
- \frac{\omega_0}{4\pi} S_0 \exp\left(-\frac{\tau}{\mu_0}\right) \sum_{l=m}^{\Lambda} p_l^m P_l^m(\mu) P_l^m(-\mu_0) - (1 - \omega_0) B(\tau) \delta_{0m}
$$

$$
(4.88)
$$

In the following it is more convenient to formulate the RTE in z-space rather than the usual τ-space, i.e.

$$
\mu \frac{d}{dz} I^m(z, \mu) = -k_{\text{ext}}(z) I^m(z, \mu) + \frac{k_{\text{sca}}(z)}{2} \int_{-1}^{1} \sum_{l=m}^{\Lambda} p_l^m P_l^m(\mu) P_l^m(\mu') I^m(z, \mu') d\mu'
$$

$$
+ \frac{k_{\text{sca}}(z)}{4\pi} S_0 \exp\left(-\frac{\tau(z)}{\mu_0}\right) \sum_{l=m}^{\Lambda} p_l^m P_l^m(\mu) P_l^m(-\mu_0) + k_{\text{abs}}(z) B(z) \delta_{0m}
$$

$$
(4.89)
$$

Let us discretize μ by introducing a discrete set of $2s$ Gaussian quadrature points (w_i, μ_i) with the usual properties

$$
w_{-i} = w_i, \quad \mu_{-i} = -\mu_i, \quad i = -s, \ldots, -1, 1, \ldots, s \qquad (4.90)
$$

Furthermore, symmetric and antisymmetric sums of the radiation field in direction μ_i are introduced via

$$I_i^{m,+}(z) = \frac{1}{2}\left[I_i^m(z) + I_{-i}^m(z)\right], \quad I_i^{m,-}(z) = \frac{1}{2}\left[I_i^m(z) - I_{-i}^m(z)\right] \qquad (4.91)$$

where $I_i^m(z) = I^m(z, \mu_i)$ and $I_{-i}^m(z) = I^m(z, -\mu_i)$. Based on these definitions, and after some tedious but simple steps, we obtain a set of coupled, ordinary first-order differential equations approximating the original transfer equation

$$(a) \quad \mu_i \frac{dI_i^{m,+}}{dz} + k_{ext} I_i^{m,-} - k_{sca} \sum_{j=1}^{s} w_j P_{ij}^{m,-} I_j^{m,-} = k_{sca} S_i^{m,-}$$

$$i = 1, \ldots, s \quad (4.92)$$

$$(b) \quad \mu_i \frac{dI_i^{m,-}}{dz} + k_{ext} I_i^{m,+} - k_{sca} \sum_{j=1}^{s} w_j P_{ij}^{m,+} I_j^{m,+} = k_{sca} S_i^{m,+} + k_{abs} B \delta_{0m}$$

Here, $S_i^{m,+}$, $S_i^{m,-}$ symbolize the primary scattered light. Note that for the symmetric sum in (4.92a) no Planckian emission term occurs. For the symmetric and the antisymmetric sum of the phase function in directions $\pm\mu_j$ we have defined

$$P_{ij}^{m,+} = p^m(\mu_i, \mu_j) + p^m(\mu_i, -\mu_j), \quad P_{ij}^{m,-} = p^m(\mu_i, \mu_j) - p^m(\mu_i, -\mu_j)$$
$$(4.93)$$

where the coefficients p^m are given by

$$p^m(\mu_i, \mu_j) = \frac{1}{2}\sum_{l=m}^{\Lambda} p_l^m P_l^m(\mu_i) P_l^m(\mu_j) \qquad (4.94)$$

The definitions (4.93) make use of the *symmetry properties of the phase function*

$$\boxed{\mathcal{P}(-\mu_i, -\mu_j) = \mathcal{P}(\mu_i, \mu_j), \quad \mathcal{P}(-\mu_i, \mu_j) = \mathcal{P}(\mu_i, -\mu_j)} \qquad (4.95)$$

Finally, for the primary scattered sunlight the following expressions have been introduced in (4.92)

$$S_i^{m,+} = \frac{S_0}{4\pi} \exp\left(-\frac{\tau(z)}{\mu_0}\right) \frac{1}{2}\sum_{l=m}^{\Lambda} p_l^m \left[P_l^m(\mu_i) + P_l^m(-\mu_i)\right] P_l^m(-\mu_0)$$

$$(4.96)$$

$$S_i^{m,-} = \frac{S_0}{4\pi} \exp\left(-\frac{\tau(z)}{\mu_0}\right) \frac{1}{2}\sum_{l=m}^{\Lambda} p_l^m \left[P_l^m(\mu_i) - P_l^m(-\mu_i)\right] P_l^m(-\mu_0)$$

The special case $m = 0$ is of particular interest if fluxes and heating rates are required. This is the situation treated by Barkstrom (1976) which will be discussed next.[2] Since $P_l(-\mu) = (-1)^l P_l(\mu)$, see (2.59a), the Legendre expansion of the

---

[2] The case $m \neq 0$ has also been investigated, see Rozanov *et al.* (1997).

phase function allows us to write $P_{ij}^+ = P_{ij}^{m=0,+}$ and $P_{ij}^- = P_{ij}^{m=0,-}$ as fully symmetric expressions. We recognize this from the following expression.

$$
\begin{aligned}
P_{ij}^+ &= \frac{1}{2} \sum_{l=0}^{\Lambda} p_l P_l(\mu_i) P_l(\mu_j) + \frac{1}{2} \sum_{l=0}^{\Lambda} p_l P_l(\mu_i) P_l(-\mu_j) \\
&= \frac{1}{2} \sum_{l=0}^{\Lambda/2} p_{2l} P_{2l}(\mu_i) P_{2l}(\mu_j) + \frac{1}{2} \sum_{l=0}^{\Lambda'} p_{2l+1} P_{2l+1}(\mu_i) P_{2l+1}(\mu_j) \\
&\quad + \frac{1}{2} \sum_{l=0}^{\Lambda/2} p_{2l} P_{2l}(\mu_i) P_{2l}(-\mu_j) + \frac{1}{2} \sum_{l=0}^{\Lambda'} p_{2l+1} P_{2l+1}(\mu_i) P_{2l+1}(-\mu_j) \\
&= \sum_{l=0}^{\Lambda/2} p_{2l} P_{2l}(\mu_i) P_{2l}(\mu_j)
\end{aligned}
\tag{4.97}
$$

where Λ' is the next smallest integer value to $(\Lambda - 1)/2$ and $p_l = p_l^{m=0}$. Analogously we obtain for P_{ij}^- the expression

$$
P_{ij}^- = \sum_{l=0}^{\Lambda/2} p_{2l+1} P_{2l+1}(\mu_i) P_{2l+1}(\mu_j)
\tag{4.98}
$$

In a similar manner we can define for the primary scattered sunlight the expressions

$$
P_{i0}^+ = \sum_{l=0}^{\Lambda/2} p_{2l} P_{2l}(\mu_i) P_{2l}(\mu_0), \quad P_{i0}^- = \sum_{l=0}^{\Lambda/2} p_{2l+1} P_{2l+1}(\mu_i) P_{2l+1}(\mu_0)
\tag{4.99}
$$

so that the terms involving the primary scattered direct sunlight are given by

$$
S_i^+ = \frac{S_0}{4\pi} \exp\left(-\frac{\tau(z)}{\mu_0}\right) P_{i0}^+, \quad S_i^- = \frac{S_0}{4\pi} \exp\left(-\frac{\tau(z)}{\mu_0}\right) P_{i0}^-
\tag{4.100}
$$

4.5.1 Vertical discretization in the finite difference method

In order to solve the coupled system of differential equations (4.92) for $m = 0$, we introduce an odd number of discrete values z_k, $k = 1, 2, \ldots, 2K + 1$. The derivatives occurring in (4.92) will be approximated by means of centered differences

$$
\left.\frac{df}{dz}\right|_{z_k} \approx \frac{f(z_{k+1}) - f(z_{k-1})}{\Delta z_k} \quad \text{with} \quad \Delta z_k = z_{k+1} - z_{k-1}
\tag{4.101}
$$

For numerical convenience Barkstrom (1976) recommends to evaluate $I_i^- = I_i^{m=0,-}$ at all even-numbered grid points and at the boundaries z_1, z_{2K+1}, and the function $I_i^+ = I_i^{m=0,+}$ at all odd-numbered grid points. Experience shows that these approximations produce rather accurate results for the case of diffuse incident radiation.

In case of an incident direct solar beam the primary scattered light varies very rapidly with optical depth so that the above procedure is not sufficiently accurate. However, this deficiency can be eliminated by analytically integrating the primary scattered light as will be shown below.

Integrating (4.92) between the interior points z_{k-1} and z_{k+1} and approximating all integrals (excepting the primary scattering term) by means of

$$\int_{z_{k-1}}^{z_{k+1}} f(z)dz \approx \Delta z_k f(z_k) \tag{4.102}$$

gives

$$I_i^+(z_{k+1}) - I_i^+(z_{k-1}) + \frac{\Delta z_k}{\mu_i}\left[k_{\text{ext}}(z_k)I_i^-(z_k) - k_{\text{sca}}(z_k)\sum_{j=1}^{s} w_j P_{ij}^-(z_k)I_j^-(z_k)\right]$$

$$= \int_{z_{k-1}}^{z_{k+1}} \frac{k_{\text{sca}}(z)}{\mu_i} S_i^-(z)dz$$

$$I_i^-(z_{k+1}) - I_i^-(z_{k-1}) + \frac{\Delta z_k}{\mu_i}\left[k_{\text{ext}}(z_k)I_i^+(z_k) - k_{\text{sca}}(z_k)\sum_{j=1}^{s} w_j P_{ij}^+(z_k)I_j^+(z_k)\right]$$

$$= \int_{z_{k-1}}^{z_{k+1}} \frac{k_{\text{sca}}(z)}{\mu_i} S_i^+(z)dz + \frac{\Delta z_k}{\mu_i} k_{\text{abs}}(z_k)B(z_k) \tag{4.103}$$

Due to the identity

$$\exp\left(-\frac{\tau(z)}{\mu_0}\right) = \frac{\mu_0}{k_{\text{ext}}(z)}\frac{d}{dz}\exp\left(-\frac{\tau(z)}{\mu_0}\right) \tag{4.104}$$

we obtain

$$\int_{z_{k-1}}^{z_{k+1}} \frac{k_{\text{sca}}(z)}{\mu_i} S_i^{\pm}(z)dz \approx \frac{k_{\text{sca}}(z_k)}{k_{\text{ext}}(z_k)}\frac{\mu_0}{\mu_i}\frac{S_0}{4\pi} P_{i0}^{\pm}(z_k)$$

$$\times \left[\exp\left(-\frac{\tau(z_{k+1})}{\mu_0}\right) - \exp\left(-\frac{\tau(z_{k-1})}{\mu_0}\right)\right] \tag{4.105}$$

As already mentioned above, (4.105) provides a very accurate approximation for the integration of the solar term.

Substituting (4.105) into (4.103) yields the discretized form of the RTE

$$
\begin{aligned}
&I_i^+(z_{k+1}) - I_i^+(z_{k-1}) \\
&\quad + \frac{\Delta z_k}{\mu_i}\left[k_{\text{ext}}(z_k)I_i^-(z_k) - k_{\text{sca}}(z_k)\sum_{j=1}^{s} w_j P_{ij}^-(z_k)I_j^-(z_k)\right] \\
&= \frac{k_{\text{sca}}(z_k)}{k_{\text{ext}}(z_k)}\frac{\mu_0}{\mu_i}\frac{S_0}{4\pi}P_{i0}^-(z_k)\left[\exp\left(-\frac{\tau(z_{k+1})}{\mu_0}\right) - \exp\left(-\frac{\tau(z_{k-1})}{\mu_0}\right)\right] \\
&I_i^-(z_{k+1}) - I_i^-(z_{k-1}) \\
&\quad + \frac{\Delta z_k}{\mu_i}\left[k_{\text{ext}}(z_k)I_i^+(z_k) - k_{\text{sca}}(z_k)\sum_{j=1}^{s} w_j P_{ij}^+(z_k)I_j^+(z_k)\right] \\
&= \frac{k_{\text{sca}}(z_k)}{k_{\text{ext}}(z_k)}\frac{\mu_0}{\mu_i}\frac{S_0}{4\pi}P_{i0}^+(z_k)\left[\exp\left(-\frac{\tau(z_{k+1})}{\mu_0}\right) - \exp\left(-\frac{\tau(z_{k-1})}{\mu_0}\right)\right] \\
&\quad + \frac{\Delta z_k}{\mu_i}k_{\text{abs}}(z_k)B(z_k)
\end{aligned}
$$

$$(4.106)$$

4.5.2 Treatment of the boundary conditions

Next we will consider the specification of the boundary conditions. We will assume that no diffuse radiation is incident at the top of the atmosphere, i.e. at $z = z_t = z_{2K+1}$

$$I(z_t, \mu < 0) = 0 \qquad (4.107)$$

At the lower boundary $z = z_g = 0$ we assume an isotropically emitting ground with temperature T_g. However, in accordance with observations the ground emits only a fraction ε_g of the black body radiation $B(T_g)$. The term ε_g is called the *emissivity of the ground*. This concept will be discussed in more detail in a later chapter. Consequently, we must allow for isotropic reflection of diffuse radiation with albedo A_g. The boundary condition then is given by

$$I(0, \mu > 0) = 2A_g \int_0^1 I(0, -\mu')\mu'd\mu' + \frac{A_g}{\pi}\mu_0 S_0 \exp\left(-\frac{\tau(0)}{\mu_0}\right) + \varepsilon_g B(T_g)$$

$$(4.108)$$

According to (4.91) the upwelling and downwelling diffuse radiation can be recovered from the I^+- and I^--functions via

$$I_i(z) = I_i^+(z) + I_i^-(z), \quad I_{-i}(z) = I_i^+(z) - I_i^-(z), \quad i = 1, \dots, s \qquad (4.109)$$

In terms of I^+ and I^- the boundary conditions (4.106) and (4.107) can be expressed as

(a) $I_i^+(0) + I_i^-(0) = 2A_g \sum\limits_{j=1}^{s} w_j \mu_j \left[I_j^+(0) - I_j^-(0) \right]$

$$+ \frac{A_g}{\pi} \mu_0 S_0 \exp\left(-\frac{\tau(0)}{\mu_0} \right) + \varepsilon_g B(T_g), \qquad i = 1, \ldots, s$$

(b) $I_i^+(z_t) - I_i^-(z_t) = 0$ (4.110)

It is convenient to define s-dimensional vectors

$$\mathbf{I}^+(z) = \begin{pmatrix} I_1^+(z) \\ \vdots \\ I_s^+(z) \end{pmatrix}, \quad \mathbf{I}^-(z) = \begin{pmatrix} I_1^-(z) \\ \vdots \\ I_s^-(z) \end{pmatrix} \qquad (4.111)$$

Furthermore, boundary matrices $\mathbb{A}_t$, $\mathbb{B}_t$ and $\mathbb{A}_g$, $\mathbb{B}_g$ will be used for the upper and lower boundary, respectively. Using these definitions equations (4.110) may be reformulated as

$$\mathbb{A}_t \mathbf{I}^+(z_t) + \mathbb{B}_t \mathbf{I}^-(z_t) = 0$$

$$\mathbb{A}_g \mathbf{I}^+(0) + \mathbb{B}_g \mathbf{I}^-(0) = \frac{A_g}{\pi} \mu_0 S_0 \exp\left(-\frac{\tau(0)}{\mu_0} \right) + \varepsilon_g B(T_g) \begin{pmatrix} 1 \\ \vdots \\ 1 \end{pmatrix} \qquad (4.112)$$

By writing (4.110b) as

$$\sum_{j=1}^{s} \left[\delta_{ij} I_j^+(z_t) - \delta_{ij} I_j^-(z_t) \right] = 0, \quad i = 1, \ldots, s \qquad (4.113)$$

it may be easily seen that the elements of the boundary matrices $\mathbb{A}_t$ and $\mathbb{B}_t$ are given by

$$A_{t,ij} = \delta_{ij}, \quad B_{t,ij} = -\delta_{ij}, \quad i = 1, \ldots, s, \quad j = 1, \ldots, s \qquad (4.114)$$

Analogously we find for the lower boundary condition

$$A_{g,ij} = \delta_{ij} - 2A_g w_j \mu_j, \quad B_{g,ij} = \delta_{ij} + 2A_g w_j \mu_j, \quad i = 1, \ldots, s, \quad j = 1, \ldots, s \qquad (4.115)$$

The system of matrix equations with $(2K + 3)s$ equations to be solved in the FDM can then be formulated in block tridiagonal form as

$$
\begin{pmatrix}
\mathbb{B}_g & \mathbb{A}_g \\
-\mathbb{E} & \mathbb{A}_1 & \mathbb{E} \\
& -\mathbb{E} & \mathbb{B}_2 & \mathbb{E} \\
& & -\mathbb{E} & \mathbb{A}_3 & \mathbb{E} \\
& & & -\mathbb{E} & \mathbb{B}_4 & \mathbb{E} \\
& & & & & \ddots \\
& & & & & -\mathbb{E} & \mathbb{A}_{2K-1} & \mathbb{E} \\
& & & & & & -\mathbb{E} & \mathbb{B}_{2K} & \mathbb{E} \\
& & & & & & & -\mathbb{E} & \mathbb{A}_{2K+1} & \mathbb{E} \\
& & & & & & & & \mathbb{A}_t & \mathbb{B}_t
\end{pmatrix}
\begin{pmatrix}
\mathbf{I}_1^- \\
\mathbf{I}_1^+ \\
\mathbf{I}_2^- \\
\mathbf{I}_3^+ \\
\mathbf{I}_4^- \\
\vdots \\
\mathbf{I}_{2K-1}^+ \\
\mathbf{I}_{2K}^- \\
\mathbf{I}_{2K+1}^+ \\
\mathbf{I}_{2K+1}^-
\end{pmatrix}
=
\begin{pmatrix}
\mathbf{X}_g \\
\mathbf{X}_1 \\
\mathbf{X}_2 \\
\mathbf{X}_3 \\
\mathbf{X}_4 \\
\vdots \\
\mathbf{X}_{2K-1} \\
\mathbf{X}_{2K} \\
\mathbf{X}_{2K+1} \\
\mathbf{X}_t
\end{pmatrix}
$$

$$(4.116)$$

In view of (4.106) the $(s \times s)$ block matrix elements $A_{k,ij}$ (k odd) and $B_{k,ij}$ (k even) in index form are given by

$$
A_{k,ij} = \frac{z_\beta - z_\alpha}{\mu_i} \left[k_{\text{ext}}(z_k)\delta_{ij} - k_{\text{sca}}(z_k)w_j P_{ij}^+(z_k) \right]
$$
$$
k = 1, 3, 5, \ldots, 2K + 1, \quad i = 1, \ldots, s, \quad j = 1, \ldots, s
$$
$$
B_{k,ij} = \frac{\Delta z_k}{\mu_i} \left[k_{\text{ext}}(z_k)\delta_{ij} - k_{\text{sca}}(z_k)w_j P_{ij}^-(z_k) \right]
$$
$$
k = 2, 4, 6, \ldots, 2K, \quad i = 1, \ldots, s, \quad j = 1, \ldots, s
$$

$$(4.117)$$

For odd $k \geq 3$ we have to take $z_\beta = z_{k+1}$, $z_\alpha = z_{k-1}$. For $k = 1$ and $k = 2K + 1$ we require $z_\beta = z_2$, $z_\alpha = z_1$ and $z_\beta = z_{2K+1}$, $z_\alpha = z_{2K}$, respectively. The components of the vectors $\mathbf{X}_k$ are

$$
X_{k,i} = \frac{k_{\text{sca}}(z_k)}{k_{\text{ext}}(z_k)} \frac{\mu_0}{\mu_i} \frac{S_0}{4\pi} P_{i0}^+(z_k) \left[\exp\left(-\frac{\tau(z_{k+1})}{\mu_0} \right) - \exp\left(-\frac{\tau(z_{k-1})}{\mu_0} \right) \right]
$$
$$
+ \frac{z_\beta - z_\alpha}{\mu_i} k_{\text{abs}}(z_k)B(z_k)
$$
$$
k = 1, 3, 5, \ldots, 2K + 1, \quad i = 1, \ldots, s
$$
$$
X_{k,i} = \frac{k_{\text{sca}}(z_k)}{k_{\text{ext}}(z_k)} \frac{\mu_0}{\mu_i} \frac{S_0}{4\pi} P_{i0}^-(z_k) \left[\exp\left(-\frac{\tau(z_{k+1})}{\mu_0} \right) - \exp\left(-\frac{\tau(z_{k-1})}{\mu_0} \right) \right]
$$
$$
k = 2, 4, 6, \ldots, 2K, \quad i = 1, \ldots, s
$$

$$(4.118)$$

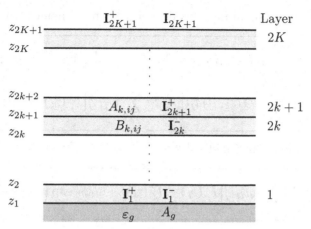

Fig. 4.8 Model atmosphere and assignment of the vectors $\mathbf{I}^+$, $\mathbf{I}^-$, and the block matrix elements $A_{k,ij}$ and $B_{k,ij}$.

The vectors $\mathbf{X}_g$, $\mathbf{X}_t$ are abbreviations for the right-hand sides of the boundary conditions (4.112). Figure 4.8 depicts the arrangement of the various vector and matrix quantities for the FDM.

4.5.3 Computation of mean radiances and flux densities

Once we have found the solutions for the vectors $\mathbf{I}^+(z_k)$, $\mathbf{I}^-(z_k)$, $k = 1, \ldots, 2K + 1$ we may easily compute the internal diffuse radiation field at all levels z_k. The mean radiance is defined as

$$\bar{I}(z_k) = \frac{1}{4\pi} \int_0^{2\pi} \int_{-1}^{1} I(z_k, \mu) d\mu \, d\varphi \tag{4.119}$$

and can be directly obtained from $\mathbf{I}^+(z_k)$

$$\boxed{\bar{I}(z_k) = \sum_{j=1}^{s} w_j I_j^+(z_k)} \tag{4.120}$$

Here we have used

$$\frac{1}{2} \int_{-1}^{1} I(z, \mu) \, d\mu = \frac{1}{2} \int_{-1}^{0} I(z, \mu) d\mu + \frac{1}{2} \int_{0}^{1} I(z, \mu) d\mu = \sum_{j=1}^{s} w_j I_j^+(z) \tag{4.121}$$

The total mean radiance is given by the sum of the mean diffuse radiance and the direct beam

$$\boxed{\bar{I}_{\text{tot}}(z_k) = \bar{I}(z_k) + S_0 \exp\left(-\frac{\tau(z_k)}{\mu_0}\right)} \tag{4.122}$$

The up- and downwelling diffuse flux densities can be computed from

$$
\begin{aligned}
E_+(z_k) &= 2\pi \int_0^1 \mu I(z_k, \mu) d\mu = 2\pi \sum_{j=1}^s w_j \mu_j \big[I_j^+(z_k) + I_j^-(z_k)\big] \\
E_-(z_k) &= 2\pi \left| \int_{-1}^0 \mu I(z_k, \mu) d\mu \right| = 2\pi \sum_{j=1}^s w_j \mu_j \big[I_j^+(z_k) - I_j^-(z_k)\big]
\end{aligned}
\tag{4.123}
$$

and the total downwelling flux density is

$$
E_{-,\text{tot}}(z_k) = 2\pi \sum_{j=1}^s w_j \mu_j \big[I_j^+(z_k) - I_j^-(z_k)\big] + \mu_0 S_0 \exp\left(-\frac{\tau(z_k)}{\mu_0}\right)
\tag{4.124}
$$

Utilizing (4.123) the diffuse net flux density is given by

$$
E_{\text{net}}(z_k) = 2\pi \int_{-1}^1 \mu I(z_k, \mu) d\mu = E_+(z_k) - E_-(z_k) = 4\pi \sum_{j=1}^s w_j \mu_j I_j^-(z_k)
\tag{4.125}
$$

To obtain the total net flux density we have to add the direct radiation, cf. (2.134), yielding

$$
E_{\text{net,tot}}(z_k) = E_{\text{net}}(z_k) - \mu_0 S_0 \exp\left(-\frac{\tau(z_k)}{\mu_0}\right)
\tag{4.126}
$$

4.6 The Monte Carlo method

The Monte Carlo method (MCM) is based on the direct physical simulation of the scattering process for the transfer of solar or thermal radiation in the atmosphere. In the MCM a very large number of model photons enters the medium in consideration (entire atmosphere or cloud elements). We envision a model photon to represent a package of real photons. Neglecting the effect of atmospheric refraction, the straight line paths of these photons between particles interacting with the radiation is changed by scattering processes. The method requires the calculation of a large number of photons propagating in a particular direction as they are passing a certain test surface. From these counts we obtain the radiance as a function of position. This permits us to compute physically relevant quantities such as flux densities for arbitrarily oriented test surfaces, the mean radiance field, and heating rates due to the (partial) absorption of model photons within the medium. It is very important to realize that one has to use a sufficiently high number of such model photons so that for the particular radiative quantity of interest a reliable statistics is achieved. If the statistics does not satify this requirement one can simply continue the simulation by increasing the total number of photons modeled until some accuracy criterion is met.

The MCM has the chief advantage that one can treat arbitrarily complex problems. For example, it is relatively easy to determine the radiative transfer through a three-dimensional spatial volume partially filled with cloud elements. In contrast to analytical methods based on the numerical solution of the differential or integral form of the RTE, the MCM has no difficulty at all in accounting for the horizontal and/or vertical inhomogeneity of the optical parameters. Therefore, radiances or flux densities can be computed at each location within a specified medium. The only real but decisive disadvantage limiting the applicability of the MCM is related to the statistical nature of the simulation process which, in certain situations, requires an excessively large amount of computer time.

To give an example, the accuracy with which a certain quantity can be determined increases only with the square root of total number of photons processed. Thus it is very difficult to reach with MCM an accuracy limit below, say, 0.1%. In addition, one has to make sure that a reliable random number generator is employed. If this is not the case the computed radiance for a specific direction, for example, cannot be determined very accurately. A good random number generator provides a large number of significant decimal places for any random number between 0 and 1 and also is able to generate a long random sequence before repetition occurs. For some strategic choices to select a good random number generator the reader is referred to Press *et al.* (1992). The MCM has been proven to be a very valuable research tool for many applications which presently cannot be treated by other methods.

4.6.1 Determination of photon paths

For simplicity we will only discuss the determination of photon paths for a homogeneous plane–parallel medium of horizontally infinite extent. Let us assume that the upper boundary of the medium is uniformly illuminated by parallel solar radiation. For simplicity thermal radiation will not be treated in the discussion that follows. At the lower boundary of the atmosphere we will assume isotropic reflection of the ground with albedo A_g. Let us consider a model photon reaching the ground. The energy fraction $1 - A_g$ of the model photon will be absorbed by the ground while the remaining part is reflected.

As stated above, a model photon is assumed to represent a package of real photons. The initial energy of the model photon is found by dividing the solar energy in a certain spectral interval per unit area and unit time by the total number of model photons used in the simulation. If an interaction with an absorbing gas molecule or aerosol particle takes place, as expressed by the single *scattering albedo*

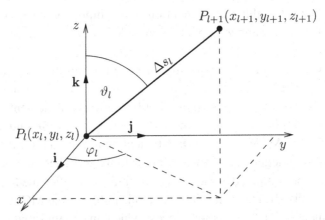

Fig. 4.9 Definition of the coordinates of two arbitrary interaction points P_l and P_{l+1}.

ω_0, the fraction $1 - \omega_0$ is lost by the model photon. Scattering changes the flight direction only, but it conserves energy.

An arbitrary photon trajectory can be defined as follows: let P_0 be the point of entrance of the photon as it enters the atmosphere through the upper boundary. The distance from P_0 to the first interaction point P_1 will be called Δs_0, and the distance between two successive points P_l and P_{l+1} is Δs_l. The direction of flight of the test photon along the distance Δs_l will be expressed by the zenith and azimuth angles ϑ_l and φ_l. By introducing a fixed Cartesian coordinate system whose z-axis is pointing towards the zenith, see Figure 4.9, we find the coordinates $(x_{l+1}, y_{l+1}, z_{l+1})$ of the point P_{l+1} as

$$x_{l+1} = x_l + \Delta s_l \sin \vartheta_l \cos \varphi_l, \quad y_{l+1} = y_l + \Delta s_l \sin \vartheta_l \sin \varphi_l, \quad z_{l+1} = z_l + \Delta s_l \cos \vartheta_l$$

$$(4.127)$$

To begin with, we will ignore the effect of gaseous absorption, but we will admit particle absorption. Let us assume that the test photon is located at point P_l with coordinates (x_l, y_l, z_l) where scattering occurs. Now the photon will travel the distance Δs_l to point P_{l+1} where the next scattering interaction with the medium takes place. Due to the scattering event at P_l, the new direction of the photon path may be expressed by selecting the local zenith and azimuth scattering angles ϑ_l and φ_l as shown in Figure 4.9. We will not yet specify the type of scattering, but simply follow the zig-zag path of the model photon through the atmosphere.

The entire atmosphere is discretized by a set of reference levels z_j, ($j = 0, \ldots, J$) where $z_0 = z_g$ and $z_J = z_t$ denote the ground and the top of the atmosphere, respectively. On its path through the atmosphere the test photon will intersect the level z_j as shown in Figure 4.10. We will label the intersection points with the symbol $D_{i,j}$ whereby the index i denotes the number of the departure point P_i after

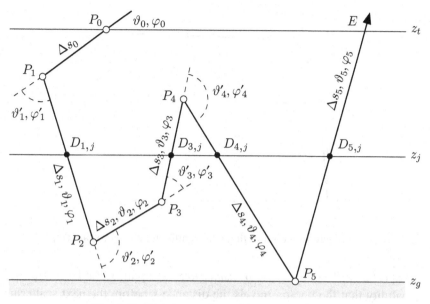

Fig. 4.10 Trajectory of an arbitrary test photon.

the previous scattering process. As a computational step we store the direction of flight for all test photons as they pass z_j. In Section 4.6.2 we will briefly describe how this information can be used to obtain radiative fluxes and radiances at all reference levels. If a photon will escape to space we will use the symbol 'E' to designate this particular event.

Figure 4.10 gives an example trajectory of a test photon through the atmosphere. The photon enters at the top of the atmosphere at point P_0. The angle of incidence is (ϑ_0, φ_0). At points P_1 through P_4 the photon is scattered from the incident direction $(\vartheta_{l-1}, \varphi_{l-1})$ to the new directions (ϑ_l, φ_l). The angles $(\vartheta_l', \varphi_l')$ measure the local zenith and azimuthal angles of scattering with respect to the direction of incidence at point P_l. At point P_5 a particular event occurs, that is isotropic reflection and partial absorption at the ground. The pair of angles $(\vartheta_r, \varphi_r) = (\vartheta_5, \varphi_5)$ is used to describe this reflection process. The figure also illustrates the intersection points $D_{i,j}$ which are used for photon counting at the reference level z_j.

The flight distances Δs_l can be determined from *Beer's law*, cf. (2.31). According to (2.32) for a homogeneous medium with extinction coefficient k_{ext} the transmission of photons traveling the distance s is given by

$$T(s) = \exp(-k_{ext}s) \tag{4.128}$$

Recall that k_{ext} is wavelength dependent, i.e. $k_{ext} = k_{ext,\nu}$, so that Monte Carlo simulations have to be carried out for several wavelengths separately to derive wavelength-integrated radiative quantities.

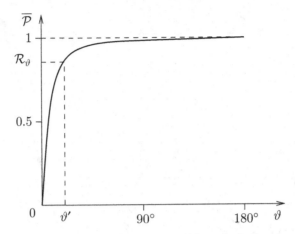

Fig. 4.11 Determination of the local zenith angle ϑ_1 for scattering.

In Chapter 2 we found that the *transmission* for a path s can be interpreted as the probability that the photon travels the distance s before the next scattering or absorption process occurs. Therefore, by choosing in (4.128) for the transmission $\mathcal{T}(s)$ a random number $\mathcal{R}_{t,0}$ between 0 and 1 we may determine the path length $s = \Delta s_0$. In this manner we find the coordinates (x_1, y_1, z_1) of the first interaction point P_1.

So far the test photon flew along the direction (ϑ_0, φ_0) of the direct solar beam. Due to scattering at P_1 the test photon will now travel in a new direction as given by the local scattering angles $(\vartheta_1', \varphi_1')$, see Figure 4.10. The local zenith angle for scattering ϑ_1' can be determined from the specified phase function $\mathcal{P}(\cos \vartheta)$ in the following way. The probability that a test photon is scattered into the interval $(0, \vartheta')$ is given by the probability distribution function

$$\bar{\mathcal{P}}(\vartheta') = \frac{\displaystyle\int_0^{\vartheta'} \mathcal{P}(\cos \vartheta) \sin \vartheta \, d\vartheta}{\displaystyle\int_0^{\pi} \mathcal{P}(\cos \vartheta) \sin \vartheta \, d\vartheta} \qquad (4.129)$$

Figure 4.11 illustrates schematically in which way the probability distribution function $\bar{\mathcal{P}}(\vartheta')$ depends on ϑ'. As soon as a particular phase function $\mathcal{P}(\cos \vartheta)$ is chosen the function $\bar{\mathcal{P}}(\vartheta')$ can be determined by numerical integration. Now we choose a random number $\mathcal{R}_{\vartheta,1}$ between 0 and 1 which picks a certain value for $\bar{\mathcal{P}}(\vartheta_1')$. Numerical inversion of the graph, depicted in Figure 4.11, leads to the scattering angle ϑ_1'.

The local azimuth angle for scattering can be found in a similar way. Owing to the rotational symmetry of $\mathcal{P}(\cos \vartheta_1')$, see Figure 1.18, for a constant value of ϑ_1' the phase function is independent of the azimuth angle φ' measured in a

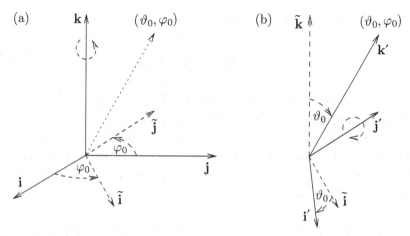

Fig. 4.12 Rotation of the $(\mathbf{i}, \mathbf{j}, \mathbf{k})$ system about the angles φ_0 (a) and ϑ_0 (b) yielding the $(\mathbf{i}', \mathbf{j}', \mathbf{k}')$ system.

plane perpendicular to the direction of incidence. Thus we obtain φ_1' by selecting a random number $\mathcal{R}_{\varphi,1}$ from the interval $[0, 2\pi)$. Now we have found the local scattering angles $(\vartheta_1', \varphi_1')$ at point P_1 which are defined with respect to the direction of incidence (ϑ_0, φ_0) of the test photon at this point.

In the next step we need to determine the new flight direction (ϑ_1, φ_1) with respect to the fixed Cartesian coordinate system. Let us introduce a new Cartesian system at point P_1 defined by the unit vectors $(\mathbf{i}', \mathbf{j}', \mathbf{k}')$. This primed coordinate system is oriented so that $\mathbf{k}'$ points along the direction (ϑ_0, φ_0). This can be achieved by first rotating the system $(\mathbf{i}, \mathbf{j}, \mathbf{k})$ by an angle φ_0 about the z-axis yielding an intermediate $(\tilde{\mathbf{i}}, \tilde{\mathbf{j}}, \tilde{\mathbf{k}})$ system. This intermediate coordinate system will then be rotated about the $\tilde{\mathbf{j}}$-axis by an angle ϑ_0 giving the final $(\mathbf{i}', \mathbf{j}', \mathbf{k}')$ system. Analytically the two rotations are given by

$$
\text{(a)} \quad
\begin{aligned}
\tilde{\mathbf{i}} &= \mathbf{i} \cos \varphi_0 + \mathbf{j} \sin \varphi_0 \\
\tilde{\mathbf{j}} &= -\mathbf{i} \sin \varphi_0 + \mathbf{j} \cos \varphi_0 \\
\tilde{\mathbf{k}} &= \mathbf{k}
\end{aligned}
\qquad
\text{(b)} \quad
\begin{aligned}
\mathbf{i}' &= \tilde{\mathbf{i}} \cos \vartheta_0 - \tilde{\mathbf{k}} \sin \vartheta_0 \\
\mathbf{j}' &= \tilde{\mathbf{j}} \\
\mathbf{k}' &= \tilde{\mathbf{i}} \sin \vartheta_0 + \tilde{\mathbf{k}} \cos \vartheta_0
\end{aligned}
\qquad (4.130)
$$

Figure 4.12 depicts the two rotations about φ_0 and ϑ_0. Substituting (4.130a) into (4.130b), using matrix notation, yields the transformation formula for $(\mathbf{i}', \mathbf{j}', \mathbf{k}')$ as function of the original $(\mathbf{i}, \mathbf{j}, \mathbf{k})$

$$
\begin{pmatrix} \mathbf{i}' \\ \mathbf{j}' \\ \mathbf{k}' \end{pmatrix}
=
\begin{pmatrix}
\cos \vartheta_0 \cos \varphi_0 & \cos \vartheta_0 \sin \varphi_0 & -\sin \vartheta_0 \\
-\sin \varphi_0 & \cos \varphi_0 & 0 \\
\sin \vartheta_0 \cos \varphi_0 & \sin \vartheta_0 \sin \varphi_0 & \cos \vartheta_0
\end{pmatrix}
\begin{pmatrix} \mathbf{i} \\ \mathbf{j} \\ \mathbf{k} \end{pmatrix}
\qquad (4.131)
$$

Denoting in (4.131) the matrix elements by $A'_{i,j}$ we may write

$$\mathbf{i}'_k = \sum_{n=1}^{3} A'_{kn} \mathbf{i}_n, \quad k = 1, 2, 3 \tag{4.132}$$

where here and in the following we identify $\mathbf{i} = \mathbf{i}_1, \mathbf{j} = \mathbf{i}_2$ and $\mathbf{k} = \mathbf{i}_3$.

At P_1 we will define a third coordinate system with double-primed unit vectors $(\mathbf{i}'', \mathbf{j}'', \mathbf{k}'')$ where the unit vector $\mathbf{k}''$ points into the flight direction (ϑ_1, φ_1) of the test photon after the scattering event. Two successive rotations by the angles φ'_1 and ϑ'_1 transform the primed coordinate system into the double-primed system. In analogy to (4.131) we find

$$\begin{pmatrix} \mathbf{i}'' \\ \mathbf{j}'' \\ \mathbf{k}'' \end{pmatrix} = \begin{pmatrix} \cos\vartheta'_1 \cos\varphi'_1 & \cos\vartheta'_1 \sin\varphi'_1 & -\sin\vartheta'_1 \\ -\sin\varphi'_1 & \cos\varphi'_1 & 0 \\ \sin\vartheta'_1 \cos\varphi'_1 & \sin\vartheta'_1 \sin\varphi'_1 & \cos\vartheta'_1 \end{pmatrix} \begin{pmatrix} \mathbf{i}' \\ \mathbf{j}' \\ \mathbf{k}' \end{pmatrix} \tag{4.133}$$

or

$$\mathbf{i}''_k = \sum_{n=1}^{3} A''_{kn} \mathbf{i}'_n, \quad k = 1, 2, 3 \tag{4.134}$$

Combination of (4.132) with (4.134) leads to

$$\mathbf{i}''_k = \sum_{n=1}^{3} \sum_{m=1}^{3} A''_{kn} A'_{nm} \mathbf{i}_m, \quad k = 1, 2, 3 \tag{4.135}$$

Finally, we need a fourth coordinate system $(\mathbf{i}^*, \mathbf{j}^*, \mathbf{k}^*)$ resulting from the rotation of the fixed coordinate system $(\mathbf{i}, \mathbf{j}, \mathbf{k})$ in such a way that $\mathbf{k}^*$ points in the photon's new direction (ϑ_1, φ_1) at P_1. From (4.131) we obtain again

$$\begin{pmatrix} \mathbf{i}^* \\ \mathbf{j}^* \\ \mathbf{k}^* \end{pmatrix} = \begin{pmatrix} \cos\vartheta_1 \cos\varphi_1 & \cos\vartheta_1 \sin\varphi_1 & -\sin\vartheta_1 \\ -\sin\varphi_1 & \cos\varphi_1 & 0 \\ \sin\vartheta_1 \cos\varphi_1 & \sin\vartheta_1 \sin\varphi_1 & \cos\vartheta_1 \end{pmatrix} \begin{pmatrix} \mathbf{i} \\ \mathbf{j} \\ \mathbf{k} \end{pmatrix} \tag{4.136}$$

Obviously, both unit vectors $\mathbf{k}^*$ and $\mathbf{k}''$ are identical. Before the scattering event occurs at point P_1 the photon travels in direction $\mathbf{k}'$, after the scattering process its new direction is $\mathbf{k}''$. Therefore, by comparing (4.135) for $k = 3$ with the last row of (4.136) we find the identities

$$\sin\vartheta_1 \cos\varphi_1 = \sum_{n=1}^{3} A''_{3n} A'_{n1}, \quad \sin\vartheta_1 \sin\varphi_1 = \sum_{n=1}^{3} A''_{3n} A'_{n2}, \quad \cos\vartheta_1 = \sum_{n=1}^{3} A''_{3n} A'_{n3}$$

$$\tag{4.137}$$

A new random number $\mathcal{R}_{t,1}$ will determine the path length Δs_1 between P_1 and the new point P_2. Substituting this path length together with (4.137) into (4.127) yields the new coordinates (x_2, y_2, z_2) of the point P_2. At this point new local scattering angles are chosen as random numbers and the computational process is repeated until the test photon either is absorbed within the medium or at the ground or it leaves the atmosphere into space, (event E in Figure 4.10). Since reflection at the ground is assumed to be isotropic, two uniformly distributed random numbers from the intervals $[0, \pi/2)$ and $[0, 2\pi)$ can be used to find the new direction of the test photon after ground reflection.

The above discussion was based on the assumption that the medium is homogeneous. In case of a vertically and horizontally inhomogeneous medium the transmission calculation must fully account for the (x, y, z)-dependence when determining the optical depth, i.e. (4.128) has to be generalized to

$$T(\Delta s) = \exp\left(-\int_{\Delta s} k_{ext}(s)\, ds\right) \tag{4.138}$$

In addition to the already defined levels z_j for the vertically inhomogeneous medium, we have to introduce similar grids for discretizing the (x, y)-space. Therefore, for a general three-dimensional medium the space is discretized into small volume elements ΔV. The photon paths are then traced through these individual volume elements. Radiative fluxes or actinic fluxes may be determined at the midpoints of the six faces of each volume element by weighting them with the corresponding projection factor as the photons intersect a particular reference area. In case of actinic fluxes this projection factor is always 1 since the weight of each test photon is independent of its flight direction. For radiative flux densities through area elements $x = const$ the weighting factor is equal to the cosine of the angle α subtended by the outward normal $\mathbf{n}_x$ of the area element and the unit vector $\mathbf{\Omega}$ specifying the flight direction of the photon, see Figure 4.13. The energy is counted positive if the photon travels into the interior of ΔV, otherwise it is negative.

4.6.2 Treatment of absorption

The initial energy of the model photons is reduced by absorption due to gases and atmospheric particles. Let N represent the total number of model photons, e.g. some 10 000, entering the top of the atmosphere. On entry the initial energy carried by such a photon is given by $E_0 = \mu_0 S_0/N$ where S_0 refers to the solar constant within a small spectral interval. Let $s_{i,j}$ specify the total photon path between the starting point P_0 and an arbitrary intersection point $D_{i,j}$ as defined in Figure 4.10.

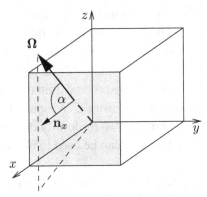

Fig. 4.13 Definition of the angle α between the photon's flight direction Ω and the outward normal $\mathbf{n}_x$ of the area $x = const$ of the volume element ΔV.

The length of this path is then given by

$$s_{i,j} = \sum_{k=0}^{i-1} \Delta s_k + \Delta s_{i,j} \qquad (4.139)$$

where $\Delta s_{i,j}$ is the distance between point P_i and $D_{i,j}$. If $\omega_{0,i}$ is the single scattering albedo of the particulate material located at point P_i, then the energy of the model photon after traversing the path s_i, may be expressed as

$$E(s_{i,j}) = E_0[1 - \mathcal{A}(s_{i,j})] \prod_{k=1}^{i} \omega_{0,k} \qquad (4.140)$$

where $\mathcal{A}(s_{i,j})$ is the so-called *absorption function* of the atmospheric absorber gas. This function will be discussed in detail in a later chapter. In the special case of reflection at the ground the value of the corresponding $\omega_{0,k}$ has to be replaced by the ground albedo A_g. From (4.140) it is seen that the effect of gaseous and particulate absorption can be determined after the photon trajectories have been computed in a purely scattering atmosphere. Some authors treat the absorption process differently, namely by reducing the photon's energy at each interaction point during the simulation of each photon trajectory.

Finally, we have to decide what kind of interaction takes place at a particular interaction point P_i. In general, three different processes are possible, that is molecular scattering, extinction by aerosol particles or extinction by cloud droplets. Gaseous absorption is already treated in terms of the absorption function $\mathcal{A}(s_{i,j})$ in (4.140). In order to determine the kind of interaction at P_i additional random numbers are drawn. Usually it is assumed that only one of the possible interactions takes place. However, some authors also allow for combined extinction processes of aerosols and water drops. In these cases average values of the extinction ccefficients and phase functions of the different particles have to be calculated. Monte Carlo algorithms designed in the above manner have been employed, for example, by O'Hirok and Gautier (1998a,b) and by Trautmann *et al.* (1999), to name a few. Radiative characteristics of finite cloud fields have also been calculated using MCM, see e.g. Welch and Zdunkowski (1981). For more details the interested reader is referred to the original literature.

In the previous subsections we described how the random paths of test photons can be determined. In order to obtain a statistically sound result a huge number N of such photon trajectories has to be simulated. All we need to do for the computation of the radiance field and the flux density is to count the number and energy of the test photons passing all intersection points $D_j = \sum_i D_{i,j}$ at height z_j. For simplicity we assume a horizontally homogeneous atmosphere so that all photon paths can be related to a single vertical atmospheric reference column having unit area. This treatment can be justified by the fact that for uniform external illumination in each such reference column, on average, the same number of photons with a certain energy and flight direction can be found.

Let us assume that for D_j a total of $J(z_j)$ test photons are registered. Let us further introduce a total of $2s$ solid angle elements $\Delta\Omega_i$ which cover the 4π unit sphere, i.e.

$$
\begin{array}{lll}
\text{lower } 2\pi \text{ hemisphere:} & \Delta\Omega_i, & i = 1, \ldots, s \\
\text{upper } 2\pi \text{ hemisphere:} & \Delta\Omega_i, & i = s+1, \ldots, 2s
\end{array}
\tag{4.141}
$$

The number of test photons confined in $\Delta\Omega_i$ is designated by $J_i(z_j)$ so that

$$
J(z_j) = \sum_{i=1}^{2s} J_i(z_j)
\tag{4.142}
$$

Each of the $J_i(z_j)$ photons may carry a particular energy $E_l(\Delta\Omega_i, z_j)$ which is the energy of model photon number l within $\Delta\Omega_i$ penetrating the reference surface z_j. Adding the energy contributions of all photons yields the radiance at level z_j within the solid angle element $\Delta\Omega_i$

$$
I_i(z_j) = \frac{1}{\Delta\Omega_i \, \Delta t \, \Delta F} \sum_{l=1}^{J_i(z_j)} E_l(\Delta\Omega_i, z_j)
\tag{4.143}
$$

Here Δt and ΔF are the time interval and the reference area referring to photons entering the top of the atmosphere at point P_0. From (4.143) we may compute the upward and downward flux densities as

$$E_+(z_j) = \sum_{i=s+1}^{2s} I_i(z_j)\Delta\Omega_i, \quad E_-(z_j) = \sum_{i=1}^{s} I_i(z_j)\Delta\Omega_i \qquad (4.144)$$

In a three-dimensional atmosphere similar expressions can be defined to determine the flux densities crossing area elements $x = const$ and $y = const$. We will not discuss this topic any further.

The Monte Carlo method described in the previous subsections is not limited to a horizontally homogeneous situation and for media infinitely extending in the lateral directions. Obeying the general principles introduced above, one can construct a Monte Carlo model for a three-dimensional situation. This can be achieved by introducing in analogy to the area $z_j = const$ vertically oriented areas $x = const$ and $y = const$ with corresponding arrays to register the intersections of the photons through these area elements. Moreover, the optical properties for the extinction coefficient, the single scattering albedo and the details of the scattering phase function must be supplied to represent the physical characteristics of the finite medium. In addition, one has to realize that solar photons not only illuminate the upper boundary of the model domain but also enter it through the lateral sunlit surfaces.

A comprehensive description of the MCM may be found in Davis *et al.* (1979). A full discussion of the rotation matrices determining the flight direction of a model photon after a scattering event is given in Zdunkowski and Korb (1985). For various applications of the MCM to two- and three-dimensional spatially inhomogeneous media we refer, for example, to the work of Barker and Davies (1992), Cahalan *et al.* (1994), Los *et al.* (1997), and more recently to Trautmann *et al.* (1999).

4.7 Appendix

4.7.1 The reflection matrix at the ground

We will now derive the reflection matrix $\mathbf{r}_g^m$ for $m = 0$ under the assumption of isotropic ground reflection for both the incident diffuse light and the direct beam. In addition, we include possible contributions due to the thermal radiation of the ground.

The upward directed flux density at the ground ($\tau = \tau_N$) is given by

$$E_{+,z}(\tau_N) = A_g\left[E_{-,z}(\tau_N) + \mu_0 S_0 \exp\left(-\frac{\tau_N}{\mu_0}\right)\right] + (1 - A_g)\pi B_g \qquad (4.145)$$

where A_g is the albedo of the isotropically reflecting ground, $E_{\pm,z}(\tau_N)$ are the upward and downward traveling flux densities at ground level, $\mu_0 S_0 \exp(-\tau_N/\mu_0)$

the direct solar flux density reaching the ground, and B_g the black body emission of the ground with temperature T_g.

Owing to the isotropy of the reflected radiation the flux densities in (4.145) can be replaced by corresponding isotropic radiances, i.e.

$$2\pi \int_0^1 \mu I_+^{m=0}(\tau_N)d\mu = A_g 2\pi \int_0^1 \mu I_-^{m=0}(\tau_N, \mu)d\mu + A_g \mu_0 S_0 \exp\left(-\frac{\tau_N}{\mu_0}\right)\delta_{0m}$$
$$+ (1 - A_g)\pi B_g \delta_{0m} \qquad (4.146)$$

The integral on the left-hand side of this equation may be evaluated yielding

$$2\pi \int_0^1 \mu I_+^{m=0}(\tau_N)d\mu = 2\pi I_+^{m=0}(\tau_N)\left[\frac{\mu^2}{2}\right]_{\mu=0}^{\mu=1} = \pi I_+^{m=0}(\tau_N) \quad (4.147)$$

Expressing in (4.146) the integral containing the downwelling radiation by means of the Gaussian quadrature gives together with (4.147)

$$I_+^{m=0}(\tau_N) = 2A_g \sum_{i=1}^s w_i \mu_i I_-^{m=0}(\tau_N, \mu_i) + \frac{A_g}{\pi}\mu_0 S_0 \exp\left(-\frac{\tau_N}{\mu_0}\right)\delta_{0m}$$
$$+ (1 - A_g)B_g\delta_{0m} \qquad (4.148)$$

Since the reflected radiation is isotropic it follows that $I_+^{m=0}(\tau_N)$ has the same value for each direction μ_i $(i = 1, \ldots, s)$. Therefore, the upwelling radiation can also be written in vector–matrix form as

$$
\begin{pmatrix} I_+^{m=0}(\tau_N) \\ \vdots \\ I_+^{m=0}(\tau_N) \end{pmatrix} = 2A_g \begin{pmatrix} \mu_1 w_1 & \cdots & \mu_s w_s \\ \vdots & \ddots & \vdots \\ \mu_1 w_1 & \cdots & \mu_s w_s \end{pmatrix} \begin{pmatrix} I_-^{m=0}(\tau_N, \mu_1) \\ \vdots \\ I_-^{m=0}(\tau_N, \mu_s) \end{pmatrix}
$$
$$
+ \frac{A_g}{\pi}\mu_0 S_0 \exp\left(-\frac{\tau_N}{\mu_0}\right)\delta_{0m} \begin{pmatrix} 1 \\ \vdots \\ 1 \end{pmatrix} + (1 - A_g)B_g\delta_{0m} \begin{pmatrix} 1 \\ \vdots \\ 1 \end{pmatrix}
$$
$$(4.149)$$

Substitution of the source vectors $\mathbf{J}_{+,1}^{m=0}(N + 1, N)$, $\mathbf{J}_{+,2}^{m=0}(N + 1, N)$ from (4.17) leads to the final expression for the reflected radiance

$$
\mathbf{I}_+^{m=0}(\tau_N) = 2A_g \begin{pmatrix} \mu_1 w_1 & \cdots & \mu_s w_s \\ \vdots & \ddots & \vdots \\ \mu_1 w_1 & \cdots & \mu_s w_s \end{pmatrix} \mathbf{I}_-^{m=0}(\tau_N)
$$
$$
+ \mathbf{J}_{+,1}^{m=0}(N + 1, N) + \mathbf{J}_{+,2}^{m=0}(N + 1, N) \qquad (4.150)
$$

At $\tau = \tau_N$ the upwelling radiation may be written as

$$\mathbf{I}_+^{m=0}(\tau_N) = \mathbb{r}_g^{m=0}\mathbf{I}_-^{m=0}(\tau_N) + \mathbf{J}_{+,1}^{m=0}(N+1,N) + \mathbf{J}_{+,2}^{m=0}(N+1,N) \quad (4.151)$$

Comparison of this expression with (4.150) gives the explicit form of the reflection matrix

$$\mathbb{r}_g^{m=0} = 2A_g \begin{pmatrix} \mu_1 w_1 & \cdots & \mu_s w_s \\ \vdots & \ddots & \vdots \\ \mu_1 w_1 & \cdots & \mu_s w_s \end{pmatrix} \quad (4.152)$$

as stated in (4.18).

It should be mentioned that for anisotropic ground reflection a similar form of the reflection matrix has to be derived from the reflection function of the ground for each azimuthal expansion term $m > 0$. In this context it is interesting to note that for an isotropically reflecting ground the boundary conditions for the azimuthal expansion terms of the radiance field $I_+^{m \geq 1}(\tau_N)$ are identical with the so-called *vacuum boundary conditions*. Vacuum boundary conditions mean that there exists no diffuse radiation that would enter the medium at its upper or lower boundary.

The treatment of complex reflection laws at the ground, that is the specification of the so-called *bidirectional reflection function* for diffuse radiation $\rho_d(\mu, \varphi, -\mu', \varphi')$, is rather difficult. Some information for treating anisotropic ground reflection within the radiative transfer problem can be found in Stamnes *et al.* (1988). These authors assume that ρ_d is a function only of the angle between the directions of incidence and reflection, i.e. there is no preferred direction for the scattering of radiation at the lower boundary. In this case, and in analogy to the phase function, the bidirectional reflection function can be expanded into its Fourier components

$$\rho_d(\mu, \varphi, -\mu', \varphi') = \sum_{m=0}^{2n-1}(2 - \delta_{0m})\rho_d^m(\mu, -\mu')\cos m(\varphi - \varphi') \quad (4.153)$$

The contribution of the ground to the upwelling radiance field due to diffuse anisotropic reflection of the downwelling diffuse and direct radiation can then be expressed by

$$I^m(\mu, \tau_N) = (1 + \delta_{0m})\int_0^1 \rho_d^m(\mu, -\mu')\mu' I^m(\tau_N, -\mu')\,d\mu'$$

$$+ \mu_0 S_0 \exp\left(-\frac{\tau_N}{\mu_0}\right)\rho_d^m(\mu, -\mu_0) \quad (4.154)$$

4.8 Problems

4.1: To become familiar with the terminology of the addition theorems and the interaction principle check the validity of equations (4.4) and (4.5).

4.2: Check in detail equations (4.7) and (4.8) by carrying out the comparison. Note well that this problem is not designed to keep you busy (as it may appear) but to familiarize yourself with the general procedure of the matrix operator method.

4.3: Carry out in detail the comparison to verify (4.13).

4.4: Consider an inhomogeneous atmosphere consisting of three homogeneous sublayers. Each sublayer has different optical properties. Follow the outlined procedure that summarizes the main steps of the MOM to formally obtain the vectors $\mathbf{I}_+^m(\tau = 0)$ and $\mathbf{I}_{\pm}^m(\tau_3)$ and the upward and downward radiance vectors at the sublayer boundaries. Make sure you have the required information before you continue with the next step.

4.5: Since Gaussian quadrature (Gauss–Legendre quadrature) is essential in our work, you should get an impression of the accuracy of this method. For this purpose evaluate the following integral $I = \int_{0.2}^{1.5} \exp(-x^2)dx$ by using a three-term formula. The weights w_i and the values μ_i where the function is evaluated are

w_i	μ_i
0.555 555 55	−0.774 596 67
0.888 888 88	0.000 000 00
0.555 555 55	0.774 596 67

(Four digits behind the zero are enough for your work). To bring the integral to the limits $(-1, 1)$ use the transformation $x = 1/2\,[(b - a)\mu + b + a]$.

4.6: Show that the δ-scaled phase function as defined by (4.35) is normalized.

4.7: Determine the system of algebraic equations that permit us to find $Z(\mu_i)$, see (4.54).

4.8: Check in detail the validity of (4.59) by substituting (4.56) into the relations (4.57) and (4.58).

4.9: Consider the radiance expansion in the form (4.63). For the determination of flux densities and for divergences only the term $m = 0$ is required. Substitute (4.63) with $m = 0$ into (2.76) to obtain a set of differential equations in terms of $I^{m=0}(\tau)$. Make full use of the orthogonality properties of the spherical functions. You may wish to use the recursion formula

$$\mu P_l(\mu) = \frac{l}{2l + 1}P_{l-1}(\mu) + \frac{l + 1}{2l + 1}P_{l+1}(\mu)$$

4.10:   Well equipped with the experience of the previous problem, verify equation
(4.65). You may wish to employ the recursion formula

$$\mu P_l^m(\mu) = \frac{l+m}{2l+1} P_{l-1}^m(\mu) + \frac{l-m+1}{2l+1} P_{l+1}^m(\mu)$$

4.11:   Derive equation (4.92) from (4.89) by using (4.91) and other required definitions. Only by carrying out the required derivation you will become familiar
with the notation of the finite difference method.

4.12:   Analogously to the derivation of (4.97) find (4.98).

4.13:   Use simple Cartesian vector operations to verify (4.130) and (4.131).

4.14:   Verify (4.137) by carrying out the required comparison.

5

Radiative perturbation theory

5.1 Adjoint formulation of the radiative transfer equation

Problems in radiative transfer theory can be solved by various solution methods. In the previous chapter we have discussed a number of these which we will classify as the *forward* or the *regular methods*. In addition to applying the forward solutions, it is also possible to use the so-called *adjoint solution techniques* which offer the decisive advantage that for certain types of transfer problems the numerical effort can be drastically reduced.

In this section we will formulate the adjoint technique. It will be necessary to introduce a new terminology involving expressions such as the radiative effect and the atmospheric response due to the presence of energy sources. By means of an important but simple example we will demonstrate the numerical advantage that the adjoint technique offers in comparison to the forward formulation. In Section 5.2 we will introduce the *perturbation technique* and show how to apply it to the forward as well as to the adjoint formulation.

The adjoint method originated as a purely mathematical tool for the solution of linear operator equations. Discussions on this subject can be found in textbooks on principles of applied mathematics such as Courant and Hilbert (1953), Friedman (1956) and Keener (1988). Before presenting the basic radiative transfer theory in the adjoint form, we would like to point out forcefully that this formulation is not merely another solution method to solve the RTE. It is a method of reformulating the transfer problem for maximum computational efficiency.

In the sequel, it will be of advantage to use the height z as the vertical coordinate instead of the optical thickness. We start our analysis with the RTE for the total radiation in the form, cf. (2.27)[1]

$$\frac{d}{ds}I(s, \Omega) = -k_{\text{ext}}(s)I(s, \Omega) + \frac{k_{\text{sca}}(s)}{4\pi} \int_{4\pi} \mathcal{P}(s, \Omega' \to \Omega)I(s, \Omega')d\Omega' + J^{\text{e}}(s, \Omega)$$

$$(5.1)$$

---

[1] For brevity, the index ν will again be omitted.

133

where we have used $\Omega \cdot \nabla I = dI/ds$. Assuming horizontal homogeneity of all variables the geometric increment ds may be replaced by dz/μ yielding

$$\mu \frac{\partial}{\partial z} I(z, \Omega) = -k_{ext}(z)I(z, \Omega) + \frac{k_{sca}(z)}{4\pi} \int_{4\pi} \mathcal{P}(z, \Omega' \rightarrow \Omega)I(z, \Omega')d\Omega' + J^e(z, \Omega)$$

(5.2)

We could have used the total derivative dI/dz since the variables μ and φ are treated as constants. For simplicity we ignore thermal radiation and polarization effects and consider the Sun as the only radiation source. If $z = z_t$ stands for the top of the atmosphere then the *source function for true emission* $J^e(z, \Omega)$, introduced in (2.26), may be written as

$$J^e(z, \Omega) = |\mu_0|S_0\delta(z - z_t)\delta(\mu - \mu_0)\delta(\varphi - \varphi_0) = Q(z, \Omega)$$

(5.3)

where δ denotes the Dirac δ-function. Equation (5.3) is by no means the only possible formulation of the solar source. Had we used the RTE after the direct–diffuse splitting of the radiation field a different form for the source function $Q(z, \Omega)$ would have resulted.

We are now going to introduce the *linear differential operator L*

$$L = \mu \frac{\partial}{\partial z} + k_{ext}(z) - \frac{k_{sca}(z)}{4\pi} \int_{4\pi} \mathcal{P}(z, \Omega' \rightarrow \Omega) \circ d\Omega'$$

(5.4)

which is a part of (5.2). The symbol $\circ$ occurring in the integral of this equation must be replaced by the particular function on which the operator L operates. Utilizing (5.3) and (5.4) the RTE can be written in the brief and elegant form

$$LI(z, \Omega) = Q(z, \Omega)$$

(5.5)

Before we discuss the so-called adjoint operator L^+ which is associated with L, we need to introduce the definition of the *inner product*

$$\langle f_1, f_2 \rangle = \int_0^{z_t} \int_0^{2\pi} \int_{-1}^1 f_1(z, \mu, \varphi) f_2(z, \mu, \varphi) d\mu \, d\varphi \, dz$$
$$= \int_0^{z_t} \int_{4\pi} f_1(z, \Omega) f_2(z, \Omega) d\Omega \, dz$$

(5.6)

where the bracket expression on the left-hand side is a shorthand notation for the integral expressions of the right-hand side. Here f_1 and f_2 are two arbitrary functions. We wish to emphasize that in the definition of the inner product the integrals extend over the complete range of each variable, i.e. the integration extends over the so-called *phase space* of a particular problem.

Now we will derive the adjoint formulation of the transfer problem as well as the proper boundary conditions. The theory of linear operators defines the *adjoint linear differential operator* L^+ corresponding to the linear differential operator L. The operator L^+ is uniquely defined if for any two arbitrary functions I and I^+ the following relation is valid

$$\langle I^+, LI \rangle = \langle L^+ I^+, I \rangle \quad \text{or} \quad \langle LI, I^+ \rangle = \langle I, L^+ I^+ \rangle \qquad (5.7)$$

The validity of the second equation, that is the possibility to interchange the expressions within a bracket, follows immediately from the definition of the inner product (5.6). In radiative transfer problems the two functions I and I^+ represent the radiance and the *adjoint radiance*, respectively.

Now the problem arises in which way L^+ should be determined. This could be done by imposing suitable boundary conditions on I^+ and then attempt to derive L^+. Simple examples of this type are given by Friedman (1956) and Keener (1988). At this point we recommend that the student consults Appendix 5.5.1 to this chapter. It might be more practical, however, to follow Bell and Glasstone (1970) by postulating L^+ and then determine the appropriate boundary conditions on I^+. Box *et al.* (1988) successfully used the second type of approach. Their approach will be described in this chapter. It appears that Gerstl (1982) introduced the adjoint method into the meteorological literature on radiative transfer. The adjoint formulation to solve transport problems was also successfully applied in reactor physics.

We will now introduce the adjoint linear differential operator L^+ by means of

$$L^+ = -\mu \frac{\partial}{\partial z} + k_{\text{ext}}(z) - \frac{k_{\text{sca}}(z)}{4\pi} \int_{4\pi} \mathcal{P}(z, \Omega \to \Omega') \circ d\Omega' \qquad (5.8)$$

which differs from (5.5) only in the sign of the first term, also known as the *streaming term*, and the interchange of the initial and final directions Ω and Ω' in the *phase function*. According to the discussion in Section 1.6.2 for homogeneous spherical particles the scattering process depends only on the cosine of the scattering angle $\cos \Theta = \Omega' \cdot \Omega$, see (1.43). Since the commutative law holds for the scalar product of two vectors we may write

$$\mathcal{P}(z, \Omega' \to \Omega) = \mathcal{P}(z, \Omega' \cdot \Omega) = \mathcal{P}(z, \Omega \cdot \Omega') = \mathcal{P}(z, \Omega \to \Omega') \qquad (5.9)$$

so that the only real difference between L and L^+ is the opposite sign of the streaming term.

In analogy to the forward form (5.5) of the RTE, the *adjoint form of the radiative transfer equation* can be written as

$$L^+ I^+(z, \Omega) = Q^+(z, \Omega) \qquad (5.10)$$

At this point the adjoint source Q^+ will be left completely unspecified. For mathematical convenience and also for purposes of interpretation, we introduce a function $\Psi(z, \Omega)$ according to the definition

$$\Psi(z, \Omega) = I^+(z, -\Omega) \tag{5.11}$$

Inserting this definition into (5.10), using L^+ as defined by (5.8), results in

$$Q^+(z, \Omega) = \left(-\mu\frac{\partial}{\partial z} + k_{\text{ext}}(z)\right)\Psi(z, -\Omega) - \frac{k_{\text{sca}}(z)}{4\pi}$$
$$\times \int_{4\pi} \mathcal{P}(z, \Omega \to \Omega')\Psi(z, -\Omega')d\Omega' \tag{5.12}$$

In order to transform this operator equation to the form (5.5), we use the substitutions $\Omega \to -\Omega$ and $\Omega' \to -\Omega'$. Since $\Omega = \Omega(\mu, \varphi)$ and $\varphi \geq 0$, it is obvious that the change of sign of the direction vector Ω also requires a change of sign of μ so that instead of (5.12) we obtain

$$Q^+(z, -\Omega) = \left(\mu\frac{\partial}{\partial z} + k_{\text{ext}}(z)\right)\Psi(z, \Omega) - \frac{k_{\text{sca}}(z)}{4\pi}$$
$$\times \int_{4\pi} \mathcal{P}(z, -\Omega \to -\Omega')\Psi(z, \Omega')d\Omega' \tag{5.13}$$

According to (5.9) we may write

$$\mathcal{P}(z, -\Omega \to -\Omega') = \mathcal{P}(z, \Omega \to \Omega') = \mathcal{P}(z, \Omega' \to \Omega) = \mathcal{P}(z, \Omega' \cdot \Omega) \tag{5.14}$$

Utilizing this information together with (5.13) in (5.10) yields for the adjoint form of the RTE

$$\boxed{L\Psi(z, \Omega) = Q^+(z, -\Omega)} \tag{5.15}$$

This form involves the linear differential operator L, the function $\Psi(z, \Omega)$ and the adjoint source Q^+ for the direction $-\Omega$. The term Ψ is also called the *pseudo-radiance*.

We will now briefly summarize the previous discussion by momentarily assuming that the adjoint source Q^+ is known.

(i) In order to determine the radiance $I^+(z, \Omega)$ due to the adjoint source $Q^+(z, \Omega)$, we first solve the forward RTE (5.15) for the *pseudo-source* $Q^+(z, -\Omega)$. This yields the auxiliary function $\Psi(z, \Omega)$.

(ii) Changing the direction according to (5.11) results in the adjoint radiance $I^+(z, \Omega)$.

(iii) For the determination of the auxiliary function Ψ, any standard solution method can be used to solve the RTE (5.15). Certainly, the accuracy of the adjoint radiance calculations equals the accuracy of the chosen forward procedure.

5.2 Boundary conditions

With the help of physical arguments it is a rather simple matter to specify the boundary conditions for the radiance I. It is much more difficult, in the general case, to obtain the boundary conditions for the adjoint radiance I^+ which is not readily visualized. It might be preferable to think of the function I^+ as a mathematical entity which was introduced for operational purposes and less on grounds of physical arguments. In order to obtain the boundary conditions for the adjoint radiances I^+ we apply the definition (5.7) and obtain

$$
\int_0^{z_t} \int_{4\pi} I^+(z, \Omega) \left(\mu \frac{\partial}{\partial z} I(z, \Omega) + k_{ext}(z) I(z, \Omega) \right.
$$
$$
\left. - \frac{k_{sca}(z)}{4\pi} \int P(z, \Omega' \to \Omega) I(z, \Omega') d\Omega' \right) d\Omega \, dz
$$
$$
= \int_0^{z_t} \int_{4\pi} I(z, \Omega) \left(-\mu \frac{\partial}{\partial z} I^+(z, \Omega) + k_{ext}(z) I^+(z, \Omega) \right.
$$
$$
\left. - \frac{k_{sca}(z)}{4\pi} \int_{4\pi} P(z, \Omega \to \Omega') I^+(z, \Omega') d\Omega' \right) d\Omega \, dz \tag{5.16}
$$

As will be observed, we have two sets of functions. The first set contains the radiances $I(z, \Omega)$ to which we apply the operator L and certain boundary conditions to be specified shortly. The second set consists of the adjoint radiances I^+ to which we apply the operator L^+ and some boundary conditions which may differ from the boundary conditions that apply to the radiances I.

Obviously, the second terms on each side of (5.16) drop out. After renaming the integration variables, we find that the third terms also cancel so that we obtain

$$
\int_0^{z_t} \int_{4\pi} \mu I^+(z, \Omega) \frac{\partial}{\partial z} I(z, \Omega) d\Omega \, dz = -\int_0^{z_t} \int_{4\pi} \mu \frac{\partial}{\partial z} I^+(z, \Omega) I(z, \Omega) d\Omega \, dz \tag{5.17}
$$

Performing a partial integration of the right-hand side of (5.17) over z from $z = 0$ to $z = z_t$ we obtain

$$
\int_0^{z_t} \int_{4\pi} \mu I^+(z, \Omega) \frac{\partial}{\partial z} I(z, \Omega) d\Omega \, dz = -\int_{4\pi} \mu I^+(z, \Omega) I(z, \Omega) d\Omega \Big|_{z=0}^{z=z_t}
$$
$$
+ \int_0^{z_t} \int_{4\pi} \mu I^+(z, \Omega) \frac{\partial}{\partial z} I(z, \Omega) d\Omega \, dz \tag{5.18}
$$

from which follows immediately

$$
\int_{4\pi} \mu I^+(z_t, \Omega) I(z_t, \Omega) d\Omega = \int_{4\pi} \mu I^+(0, \Omega) I(0, \Omega) d\Omega \tag{5.19}
$$

The evaluation of this equation requires the specification of the boundary conditions for I and I^+.

5.2.1 Vacuum boundary conditions

The *vacuum boundary conditions* for the radiance I state that an atmospheric layer is illuminated only by the parallel radiation of the Sun while diffuse illumination at the boundaries of the layer is not admitted. Since at the top of the atmosphere the incoming parallel solar radiation is already included in the source term Q, it does not have to be considered here. Thus, the vacuum boundary conditions can be stated in the form

$$
\begin{array}{ll}
\text{top of the atmosphere:} & I(z_t, \mu, \varphi) = 0, \quad 0 \le \varphi \le 2\pi, \quad -1 \le \mu < 0 \\
\text{Earth's surface:} & I(0, \mu, \varphi) = 0, \quad 0 \le \varphi \le 2\pi, \quad 0 < \mu \le 1
\end{array} \tag{5.20}
$$

In order to apply these equations, we split (5.19) as shown into

$$
\int_0^{2\pi} \int_{-1}^0 \mu I^+(z_t, \mu, \varphi) I(z_t, \mu, \varphi) d\mu d\varphi + \int_0^{2\pi} \int_0^1 \mu I^+(z_t, \mu, \varphi) I(z_t, \mu, \varphi) d\mu d\varphi
$$

$$
= \int_0^{2\pi} \int_{-1}^0 \mu I^+(0, \mu, \varphi) I(0, \mu, \varphi) d\mu d\varphi \tag{5.21}
$$

$$
+ \int_0^{2\pi} \int_0^1 \mu I^+(0, \mu, \varphi) I(0, \mu, \varphi) d\mu d\varphi
$$

Thus, with the help of (5.20), we recognize immediately that (5.21) reduces to

$$
\int_0^{2\pi} \int_0^1 \mu I^+(z_t, \mu, \varphi) I(z_t, \mu, \varphi) d\mu d\varphi = \int_0^{2\pi} \int_{-1}^0 \mu I^+(0, \mu, \varphi) I(0, \mu, \varphi) d\mu d\varphi \tag{5.22}
$$

We note that the radiances I and I^+ appearing in this equation are completely arbitrary and independent of each other. In the range of integration the radiances $I(z_t, \mu, \varphi)$ and $I(0, \mu, \varphi)$ differ from zero. Thus, the only way to satisfy (5.22) is to require that the boundary conditions of the adjoint radiances are given by

$$
\begin{array}{ll}
\text{top of the atmosphere:} & I^+(z_t, \mu, \varphi) = 0, \quad 0 \le \varphi \le 2\pi, \quad 0 < \mu \le 1 \\
\text{Earth's surface:} & I^+(0, \mu, \varphi) = 0, \quad 0 \le \varphi \le 2\pi, \quad -1 \le \mu < 0
\end{array}
$$

$$\tag{5.23}$$

Summarizing, the vacuum boundary conditions of the regular radiances I require that the incoming diffuse radiation is zero at the base and the top of the atmosphere while in the adjoint formulation the outgoing radiances I^+ at the boundaries must be zero. We refer to the solution of the RTE utilizing the vacuum boundary conditions as the *standard problem*. For further reference see also Chapter 3.

5.2.2 Boundary conditions for a reflecting surface

The most important extension of the standard problem is the inclusion of ground reflection. For simplicity, we will assume a *Lambertian surface* with albedo A_g. As already mentioned previously, such a surface reflects the incoming radiances from the upper hemisphere isotropically according to

$$\pi I(0, \mu, \varphi) = A_g \int_0^{2\pi} \int_{-1}^0 |\mu'| I(0, \mu', \varphi') d\mu' d\varphi', \quad 0 < \varphi \le 2\pi, \quad 0 < \mu \le 1$$

$$(5.24)$$

Since the integration is over all directions, the right-hand side of this equation is independent of any direction so that the upward radiance $I(0, \mu, \varphi)$ must also be independent of direction, that is it has the same value for every of μ and φ.

The inclusion of the surface reflection also requires a change of the lower adjoint boundary conditions in (5.23) while the upper adjoint boundary condition in (5.23) remains unaffected. Let us return to equation (5.21). Substituting the upper boundary conditions of (5.20) and (5.23) into (5.21) we find that the left-hand side vanishes. Substitution of (5.24) into the right-hand side of (5.21) yields

$$\int_0^{2\pi} \int_{-1}^0 \mu I^+(0, \mu, \varphi) I(0, \mu, \varphi) d\mu d\varphi$$

$$= -\frac{A_g}{\pi} \int_0^{2\pi} \int_{-1}^0 |\mu'| I(0, \mu', \varphi') d\mu' d\varphi' \int_0^{2\pi} \int_0^1 \mu I^+(0, \mu, \varphi) d\mu d\varphi \quad (5.25)$$

$$= \frac{A_g}{\pi} \int_0^{2\pi} \int_{-1}^0 \mu I(0, \mu, \varphi) d\mu d\varphi \int_0^{2\pi} \int_0^1 \mu' I^+(0, \mu', \varphi') d\mu' d\varphi'$$

In the integral on the left-hand side only the function $I^+(0, \mu, \varphi)$, $(-1 \le \mu < 0)$, is unknown. Since I and I^+ are independent of each other, we conclude that in this integral $I^+(0, \mu, \varphi)$ must be given by the following expression

$$\pi I^+(0, \mu, \varphi) = A_g \int_0^{2\pi} \int_0^1 \mu' I^+(0, \mu', \varphi') d\mu' d\varphi', \quad 0 < \varphi \le 2\pi, \quad -1 \le \mu < 0$$

$$(5.26)$$

By comparing this equation with (5.24) the symmetry between the forward and the adjoint boundary conditions becomes apparent. More complicated boundary formulations are possible but will not be considered here. Figure 5.1 depicts some information about the forward and the adjoint formulation showing the various directions of the regular and the adjoint radiances at the upper and lower boundary of the atmosphere.

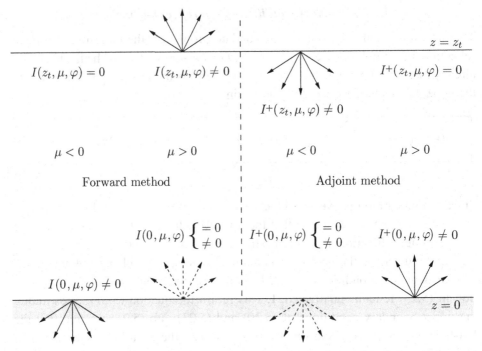

Fig. 5.1 Comparison of directions of radiances in the forward and the adjoint formulation. Dashed arrows indicate no reflection ($\ldots = 0$) or reflection at the Earth's surface ($\ldots \neq 0$).

5.2.3 Inclusion of surface reflection in the formulation of the radiances

Equation (5.26) gives the ground reflection in the adjoint formulation. Now we must find a general expression for the adjoint radiance in the presence of ground reflection which applies to an arbitrary height. This task will be accomplished by first finding a suitable expression in the forward mode which can then be transformed to the adjoint formulation.

Let $I_v(z, \mu, \varphi)$ represent the solution of the standard problem, that is the solution of the RTE with vacuum boundary conditions (index v) and source term Q

$$LI_v(z, \mu, \varphi) = Q(z, \mu, \varphi) \tag{5.27}$$

Then the downward flux density in the forward mode at the Earth's surface ($z = 0$) is given by

$$E_{v,-}(0) = \int_0^{2\pi} \int_{-1}^0 |\mu| I_v(0, \mu, \varphi) d\mu d\varphi \tag{5.28}$$

For purely mathematical purposes, in analogy to the form (5.3), we introduce an artificial surface source Q_g by means of

$$Q_g = \begin{cases} \mu \delta(z) & 0 < \mu \le 1 \\ 0 & -1 \le \mu < 0 \end{cases} \tag{5.29}$$

Now let $I_g(z, \mu, \varphi)$ represent the solution of the RTE with vacuum boundary conditions and source term Q_g

$$L I_g(z, \mu, \varphi) = Q_g(z, \mu, \varphi) \tag{5.30}$$

Owing to the special formulation of the surface source Q_g the radiance I_g is a dimensionless quantity.[2] The corresponding dimensionless downward flux density $E_{g,-}(0)$ at the surface is given by

$$E_{g,-}(0) = \int_0^{2\pi} \int_{-1}^0 |\mu| I_g(0, \mu, \varphi) d\mu d\varphi \tag{5.31}$$

In Appendix 5.5.2 we will show that $I(z, \mu, \varphi)$ may be written as function of the vacuum solution and the solution with surface reflection in the form

$$\boxed{I(z, \mu, \varphi) = I_v(z, \mu, \varphi) + E_{v,-}(0)\frac{\frac{A_g}{\pi}}{1 - \frac{A_g}{\pi}E_{g,-}(0)} I_g(z, \mu, \varphi)} \tag{5.32}$$

This expression shows in which way at an arbitrary height z the radiance $I(z, \mu, \varphi)$ is influenced by the ground albedo A_g.

The task ahead is to transform equation (5.32) to the adjoint representation. To this end we first consider the vacuum problem without any surface reflection. By repeating (5.10), adding the suffix v to I^+ (vacuum boundary conditions) for distinction, we obtain

$$L^+ I_v^+(z, \mu, \varphi) = Q^+(z, \mu, \varphi) \tag{5.33}$$

Using (5.11), we find the auxiliary function for the vacuum problem

$$\Psi_v(z, \mu, \varphi) = I_v^+(z, -\mu, \varphi) \tag{5.34}$$

According to (5.15), the auxiliary function Ψ_v satisfies the forward RTE with source $Q^+(z, -\mu, \varphi)$

$$L\Psi_v(z, \mu, \varphi) = Q^+(z, -\mu, \varphi) \tag{5.35}$$

---

[2] Note that the unit of $\delta(z)$ is (m^{-1}).

The vacuum boundary conditions for Ψ_v may be obtained by substituting (5.34) into (5.23).

For the reflecting surface we obtain from (5.26)

$$\pi \Psi(0, -\mu, \varphi) = A_g \int_0^{2\pi} \int_0^1 \mu' \Psi(0, -\mu', \varphi') d\mu' d\varphi', \quad 0 < \varphi \le 2\pi, \quad -1 \le \mu < 0$$
(5.36)

which may also be written as

$$\pi \Psi(0, \mu, \varphi) = A_g \int_0^{2\pi} \int_{-1}^0 |\mu'| \Psi(0, \mu', \varphi') d\mu' d\varphi', \quad 0 < \varphi \le 2\pi, \quad 0 < \mu \le 1$$
(5.37)

To include the surface reflection in the adjoint formulation we may proceed analogously to the forward formulation. Comparison of (5.35) with (5.5) shows that the auxiliary function $\Psi_v(z, \mu, \varphi)$ enters the RTE in the same way as the forward radiance $I(z, \mu, \varphi)$ if the source $Q(z, \mu, \varphi)$ is replaced by $Q^+(z, -\mu, \varphi)$. Furthermore, comparison of equation (5.37) with (5.24) shows the direct analogy of the lower boundary conditions of $\Psi(0, \mu, \varphi)$ and $I(0, \mu, \varphi)$. These two comparisons imply that the auxiliary function Ψ describing the radiation field including ground reflection should be given by an expression which is analogous to the form (5.32), that is

$$\Psi(z, \mu, \varphi) = \Psi_v(z, \mu, \varphi) + \tilde{E}_{v,-}(0) \frac{\frac{A_g}{\pi}}{1 - \frac{A_g}{\pi} \tilde{E}_{g,-}(0)} \Psi_g(z, \mu, \varphi)$$
(5.38)

where the terms $\tilde{E}_{v,-}(0)$ and $\tilde{E}_{g,-}(0)$ are still unspecified. The quantity $\Psi_g(z, \mu, \varphi)$ is the solution to

$$L\Psi_g(z, \mu, \varphi) = Q_g(z, \mu, \varphi)$$
(5.39)

with vacuum boundary conditions for Ψ_g, and Q_g as defined by (5.29).

We are now able to find the meanings of $\tilde{E}_{v,-}(0)$ and $\tilde{E}_{g,-}(0)$. Comparison of (5.39) with (5.30), observing (5.11), yields the following identities:

$$I_g(z, \mu, \varphi) = \Psi_g(z, \mu, \varphi) = I_g^+(z, -\mu, \varphi)$$
(5.40)

According to (5.31) and due to the similarities of the forms stated by (5.32) and (5.38), for the quantity $\tilde{E}_{g,-}(0)$ we conjecture

$$\tilde{E}_{g,-}(0) = \int_0^{2\pi} \int_{-1}^0 |\mu| \Psi_g(0, \mu, \varphi) d\mu d\varphi = E_{g,-}(0)$$
(5.41)

The form of $\tilde{E}_{v,-}(0)$ follows from (5.28) so that

$$\tilde{E}_{v,-}(0) = \int_0^{2\pi} \int_{-1}^0 |\mu| \Psi_v(0, \mu, \varphi) d\mu d\varphi = \int_0^{2\pi} \int_0^1 \mu I_v^+(0, \mu, \varphi) d\mu d\varphi = E_{v,-}^+(0)$$

(5.42)

Moreover, from (5.34) we find

$$\Psi_v(z, -\mu, \varphi) = I_v^+(z, \mu, \varphi)$$

(5.43)

Utilizing these pieces of information, we are ready to write down the desired adjoint relationship to handle surface reflection

$$I^+(z, \mu, \varphi) = I_v^+(z, \mu, \varphi) + E_{v,-}^+(0) \frac{\frac{A_g}{\pi}}{1 - \frac{A_g}{\pi} E_{g,-}(0)} I_g(z, -\mu, \varphi)$$

(5.44)

Admittedly, some of the above arguments are somewhat indirect.

In order to solve the forward as well as the adjoint problem, including ground reflection, we have to treat three individual problems:

(i) solution of $L I_v = Q$
(ii) solution of $L I_g = Q_g$
(iii) solution of $L^+ I_v^+ = Q^+$.

It should be noted that the solution for I_g plays a dual role since it is needed to solve the forward problem (5.32) as well as the adjoint problem (5.44).

5.3 Radiative effects

Many practical applications do not require the complete information contained in the radiation field which is described by the distribution of the radiances at a particular point. Often it is sufficient to extract certain integral quantities from the field such as upward and downward flux densities, net flux densities at certain heights and radiative heating rates in selected atmospheric layers. Each integral quantity is known as the *radiative effect* $\mathcal{E}$.

The reader will have noticed that the adjoint method was introduced without any particular motivation. We will now give a convincing reason why this method was introduced and point out the enormous advantage this method offers when solving certain radiative transfer problems.

In case of the forward mode, the radiance I follows from the solution of the RTE

$$LI = Q$$

(5.45)

The radiative effect $\mathcal{E}$ of the radiation field is expressed by the general definition

$$\mathcal{E} = < R, I > \qquad (5.46)$$

where $R(z, \mu, \varphi)$ is known as the corresponding *response function*. To state it more clearly: to each radiative effect belongs a response function R which expresses the response of the medium to the illumination of the atmosphere by the source Q. Soon we will give examples of how to formulate R.

Now we briefly consider the RTE of the corresponding adjoint problem

$$L^+ I^+ = Q^+ \qquad (5.47)$$

We recall that the adjoint source Q^+ was not specified previously. Since Q^+ is completely arbitrary we are free to set $Q^+ = R$. This particular choice of Q^+ results in the advantages of the adjoint method that we have mentioned above. Combining (5.46) and (5.47) with $R = Q^+$ we obtain

$$\boxed{\mathcal{E} = \langle R, I \rangle = \langle Q^+, I \rangle = \langle L^+ I^+, I \rangle = \langle I^+, LI \rangle = \langle I^+, Q \rangle} \qquad (5.48)$$

where the last two expressions have been obtained by means of (5.7) and (5.45).

To familiarize ourselves with the formulation of the response function, we consider a few very useful examples.

Example I

Suppose, at level z_0 we wish to calculate the downward flux density $E_-(z_0)$ which is a typical effect of the radiation field. The basic formula is

$$\mathcal{E} = E_-(z_0) = \int_0^{2\pi} \int_{-1}^0 |\mu| I(z_0, \mu, \varphi) d\mu d\varphi \qquad (5.49)$$

In order to state R, we make use of the *Heaviside step function U*, defined by

$$U(x - a) = \begin{cases} 1 & x > a \\ 0 & x < a \end{cases} \qquad (5.50)$$

Utilizing (5.50) equation (5.49) can also be expressed as

$$\mathcal{E} = \int_0^{z_t} \int_0^{2\pi} \int_{-1}^1 |\mu| \delta(z - z_0) U(-\mu) I(z, \mu, \varphi) d\mu \, d\varphi \, dz \qquad (5.51)$$

$$= \langle |\mu| \delta(z - z_0) U(-\mu), I \rangle$$

Comparison of this equation with (5.46) shows that the response function R is given by

$$R(z, \mu) = |\mu| \delta(z - z_0) U(-\mu) \qquad (5.52)$$

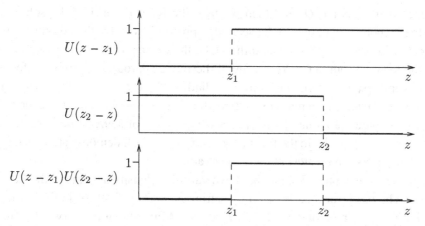

Fig. 5.2 Arrangement of the Heaviside step functions $U(z - z_1)$, $U(z_2 - z)$ and the product $U(z - z_1)U(z_2 - z)$.

Example II

Suppose we wish to find the average spectral solar heating rate within an atmospheric layer $\Delta z = z_2 - z_1$. Omitting in (2.46) the thermal emission J^e we integrate this equation over Δz yielding

$$\left.\frac{\partial T}{\partial t}\right|_{\mathrm{rad}, \Delta z} = \frac{1}{\Delta z} \int_{z_1}^{z_2} \left.\frac{\partial T}{\partial t}\right|_{\mathrm{rad}} dz = \frac{1}{\Delta z} \int_{z_1}^{z_2} \int_0^{2\pi} \int_{-1}^1 \frac{k_{\mathrm{abs}}(z)}{\rho(z)c_p} I(z, \mu, \varphi) d\mu\, d\varphi$$

(5.53)

In order to find the response function R, we use the product of two Heaviside step functions $U(z - z_1)U(z_2 - z)$, see Figure 5.2. Thus for this particular example the radiative effect $\mathcal{E}$ representing the *radiative temperature change* of layer Δz is given by

$$\mathcal{E} = \frac{1}{\Delta z} \int_0^{z_t} \int_0^{2\pi} \int_{-1}^1 \frac{k_{\mathrm{abs}}(z)}{\rho(z)c_p} U(z - z_1)U(z_2 - z)I(z, \mu, \varphi) d\mu\, d\varphi$$

$$= \left\langle \frac{k_{\mathrm{abs}}(z)}{\rho(z)c_p\Delta z} U(z - z_1)U(z_2 - z), I \right\rangle$$

(5.54)

yielding the response function as

$$R(z) = \frac{k_{\mathrm{abs}}(z)}{\rho(z)c_p\Delta z} U(z - z_1)U(z_2 - z)$$

(5.55)

Example III

From (5.48) we see that there exist two ways of computing the radiative effect $\mathcal{E}$. The first is the standard or the forward approach. We solve the RTE assuming the presence of the source Q, and then we take the inner product of the solution I with the response function R. The second approach is to solve the adjoint form of the RTE with source $Q^+ = R$, and then take the inner product of the solution I^+ with the forward source Q. Which one of these two approaches is to be preferred? If one is interested to compute a single radiative effect, corresponding to a single source Q, then it does not matter which method is given preference. However, if we wish to compute the radiative effect of a whole series of sources then one should choose very judiciously. In the following example, we will demonstrate which type of method offers the greatest numerical advantage.

Suppose we wish to calculate the downward flux density arriving at the Earth's surface. The basic formula is given by setting $z_0 = 0$ in (5.49) or in the response function (5.52). The radiance $I(0, \mu, \varphi)$ is found by solving the forward form of the RTE (5.45) utilizing the source distribution function (5.3). However, the same radiative effect can be found from (5.48), or explicitly

$$\mathcal{E} = \langle I^+, Q \rangle = \int_0^{z_t} \int_0^{2\pi} \int_{-1}^1 I^+(z, \mu, \varphi) Q(z, \mu, \varphi) d\mu \, d\varphi \, dz \qquad (5.56)$$

Inserting here for $Q(z, \mu, \varphi)$ the expression (5.3) the integration can be performed analytically and we obtain

$$\mathcal{E} = E_-(0) = |\mu_0| S_0 I^+(z_t, \mu_0, \varphi_0) \qquad (5.57)$$

This interesting result shows that the downward flux density at the ground is completely determined by the adjoint radiance I^+ at the top of the atmosphere at z_t for the special direction (μ_0, φ_0). Moreover, the source flux density is $|\mu_0| S_0$ where S_0 is the solar constant for the particular wavelength under consideration. For a broader solar band an integration over the wave number or the wavelength needs to be carried out.

Let us reconsider the classical solution (5.49) to find $E_-(0)$. For each given direction (μ_0, φ_0) of the Sun, i.e. for a fixed Q, the forward RTE (5.45) must be solved to find the corresponding distribution $I(0, \mu, \varphi)$. If the daily course of the downward directed radiative flux density at the ground is required on the basis of N solar positions, the RTE (5.45) must be solved N times, that is for each position (μ_0, φ_0) of the Sun.

If we employ the adjoint formulation, we must solve (5.47), choosing as the adjoint source $Q^+ = R$. In practice, the adjoint radiance distribution at the top of the atmosphere, $I^+(z_t, \mu, \varphi)$, is calculated for all directions (μ, φ). This includes all solar angles (μ_0, φ_0). Thus, plotting the values $|\mu_0| S_0 I^+(z_t, \mu_0, \varphi_0)$ versus μ_0 gives the diurnal course of $E_-(0)$ for all μ_0. In one-dimensional slab geometry, $\mathcal{E}$ as well as I^+ are invariant to φ_0. Hence only one single adjoint solution of the RTE is required in comparison to N necessary forward computations. This is the decisive advantage that the adjoint method offers for many interesting situations.

Additional information concerning the advantage of the adjoint formulation of the RTE can be found in Gerstl and Zardecki (1985).

5.4 Perturbation theory for radiative effects

As is well-known, the calculation of radiative effects for realistic scattering and absorbing atmospheres requires a large computer effort if exact solutions of the RTE are demanded. In contrast, the use of approximate methods reduces the computer effort substantially. The amount of accuracy which is lost due to approximation depends on the type of the chosen approach. As a first approximation method, we now introduce the so-called *perturbation theory* which greatly differs from all the other approximation procedures to be discussed in the next chapter. A number of more or less successful attempts in this direction have been made to treat aerosol scattering as a perturbation applied to a Rayleigh scattering atmosphere of fixed optical path length, see e.g. Sekera (1956), Deirmendjian (1957, 1959) and Box and Deepak (1979). A different approach is due to Fymat and Abhyankar (1969a,b) and Abhyankar and Fymat (1970a,b) who considered variations of the single scattering albedo as a perturbation to simulate aerosol scattering.

In this section we are going to introduce still another procedure due to Box *et al.* (1989a) which is a natural continuation of the theoretical developments described in the previous sections. The main idea is to first solve the RTE for a certain realistic base atmosphere. Solutions to 'neighboring' atmospheres can be obtained by perturbing the functions characterizing the base atmosphere. Since the radiative transport operator L is not self-adjoint, i.e. $L \neq L^+$, not only the forward radiance, but also the adjoint radiance will have to be accounted for. This method is not only fast but in most cases also quite accurate. It has also been shown that this type of the perturbation technique can be further developed to improve the accuracy of the procedure without losing a substantial amount of the numerical advantages characterizing the present stage of development.

Before we begin with the discussion of the perturbation technique, we would like to convince the reader that the following treatment is not simply an academic exercise. Often it is wrongly argued that with the development of ever faster computers, all approximation methods will soon be entirely obsolete. It should not be overlooked that with the advent of faster computers more realistic and thus more complex weather forecasting and climate prediction models are being developed which require the implementation of increasingly more accurate and still faster radiation transfer codes. Since the exact and frequent evaluation of the RTE for the entire spectrum is out of the question for some time to come, the development of improved approximation methods is a matter of necessity. The perturbation technique has the potential of being such a method. Thus we are not simply presenting an academic exercise but a procedure of great usefulness.

5.4.1 Basic perturbation theory

The state of the atmosphere is characterized by functions such as $k_{ext}(z)$, $k_{sca}(z)$ and $\mathcal{P}(z, \mathbf{\Omega})$. We must distinguish between the operators, the radiances and functions of the perturbed atmosphere and the corresponding quantities describing the base atmosphere. This will be done by attaching a zero as subscript to all symbols characterizing the base atmosphere. Thus, instead of (5.4), (5.8), (5.45) and (5.47) henceforth denoting the equations for the perturbed atmosphere, for the base atmosphere we write

$$\text{(a)}\quad L_0 I_0(z, \mathbf{\Omega}) = Q, \qquad \text{(b)}\quad L_0^+ I_0^+(z, \mathbf{\Omega}) = Q^+ = R \qquad (5.58)$$

with

$$\text{(a)}\quad L_0 = \mu\frac{\partial}{\partial z} + k_{ext,0}(z) - \frac{k_{sca,0}(z)}{4\pi}\int_{4\pi}\mathcal{P}_0(z, \mathbf{\Omega'} \to \mathbf{\Omega})\circ d\Omega'$$

$$\text{(b)}\quad L_0^+ = -\mu\frac{\partial}{\partial z} + k_{ext,0}(z) - \frac{k_{sca,0}(z)}{4\pi}\int_{4\pi}\mathcal{P}_0(z, \mathbf{\Omega} \to \mathbf{\Omega'})\circ d\Omega' \qquad (5.59)$$

Of course, the source Q, as defined in (5.3), is the same for any atmosphere. A little reflection shows that the radiative effect for the base atmosphere is given by

$$\mathcal{E}_0 = \langle R, I_0\rangle = \langle I_0^+, Q\rangle \qquad (5.60)$$

We will now define the perturbation quantities L, L^+, I and I^+ by means of

$$L = L_0 + \Delta L, \quad L^+ = L_0^+ + \Delta L^+, \quad I = I_0 + \Delta I, \quad I^+ = I_0^+ + \Delta I^+ \qquad (5.61)$$

so that the RTE can be written as

$$Q = LI = (L_0 + \Delta L)(I_0 + \Delta I) = L_0 I_0 + \Delta L I_0 + L\Delta I \qquad (5.62)$$

Due to (5.58a) this equation reduces to the important relation

$$\boxed{\Delta L I_0 + L\Delta I = 0} \qquad (5.63)$$

Forming the inner product of (5.63) with I_0^+ yields

$$\langle I_0^+, \Delta L I_0\rangle + \langle I_0^+, L\Delta I\rangle =$$
$$\langle I_0^+, \Delta L I_0\rangle + \langle I^+ - \Delta I^+, L\Delta I\rangle =$$
$$\langle I_0^+, \Delta L I_0\rangle + \langle I^+, L(I - I_0)\rangle - \langle \Delta I^+, L\Delta I\rangle = \qquad (5.64)$$
$$\langle I_0^+, \Delta L I_0\rangle + \langle L^+ I^+, I\rangle - \langle L^+ I^+, I_0\rangle - \langle \Delta I^+, L\Delta I\rangle =$$
$$\langle I_0^+, \Delta L I_0\rangle + \langle R, I\rangle - \langle R, I_0\rangle - \langle \Delta I^+, L\Delta I\rangle = 0$$

By using the definitions (5.46) and (5.60) this equation may be rewritten as

$$\mathcal{E} = \mathcal{E}_0 - \langle I_0^+, \Delta L I_0\rangle + \langle \Delta I^+, L\Delta I\rangle \qquad (5.65)$$

Since we consider only linear perturbation theory in this book, we omit the last term in this equation which is a second-order perturbation or a small correction to the first order and obtain finally

$$\boxed{\mathcal{E} \approx \mathcal{E}_0 - \langle I_0^+, \Delta L I_0 \rangle} \tag{5.66}$$

This equation is the basic estimate of the radiative effect formula for our work. It tells us that the radiative effect of an arbitrary atmosphere is given by the corresponding radiative effect of the base atmosphere and a correction term $\langle I_0^+, \Delta L I_0 \rangle$ which we call the *perturbation integral*. As will be seen soon, for the evaluation of the perturbation integral it is not necessary to solve the RTE again. This leads to the idea to solve the RTE for a set of different base atmospheres by means of complex exact solution methods. The results of these calculations will be stored as a data base which is then used to calculate the radiative effects of arbitrary atmospheres according to (5.66). In practice one chooses that base atmosphere of the data base which yields the smallest perturbations from the actual atmosphere so that the error of the linear approximation remains as small as possible.

Of course, for additional accuracy it is possible to include second-order perturbations by utilizing (5.65). This, however, complicates the analysis. Some progress in this direction has been made but will not be reported here.

5.4.2 An alternative formulation of the radiative effect

In order to avoid the expensive repetitive evaluation of the exact forward form or the adjoint form of the RTE, we have introduced the perturbation theory. We have also derived the important relation (5.66) to estimate the radiative effect (5.48). This is not the only way to obtain an approximate radiation effect formula. By applying the variational method, several stationary functionals have been investigated in the literature to estimate the effect of interest. In this section we present such an estimate of the effect formula in the form derived by Gerstl and Stacey (1973). This method is based on *Schwinger's variational principle*. We begin by defining the functional

$$F[I, I^+] = \langle R, I \rangle + \langle I^+, Q - LI \rangle \tag{5.67}$$

Thus, $F[I, I^+]$ consists of the radiative effect $\mathcal{E} = \langle R, I \rangle$ and a perturbation term $\langle I^+, Q - LI \rangle$. However, if I is the solution of the RTE with source Q, then the perturbation term vanishes and the functional is equal to the radiative effect.

Let us assume that equations (5.58) have been solved already, that is I_0 and I_0^+ are known, and in particular $L_0 I_0 = Q$. In contrast to this, the solutions I and I^+ to (5.45) and (5.47) are still unknown. We are now going to evaluate the functional

F in terms of the solutions I_0 and I_0^+

$$F[I_0, I_0^+] = \langle R, I_0 \rangle + \langle I_0^+, Q - LI_0 \rangle = \langle R, I_0 \rangle + \langle I_0^+, L_0 I_0 - LI_0 \rangle$$
$$= \langle R, I_0 \rangle - \langle I_0^+, \Delta L I_0 \rangle = \mathcal{E}_0 - \langle I_0^+, \Delta L I_0 \rangle \qquad (5.68)$$

where use was made of (5.60) and (5.61). By comparing (5.68) with (5.66) it is seen that $F[I_0, I_0^+]$ agrees with the linearized form of the radiative effect $\mathcal{E}$.

Now we introduce the arbitrary normalizations

$$C = \frac{I}{I_1}, \quad C^+ = \frac{I^+}{I_1^+} \qquad (5.69)$$

Substituting (5.69) into (5.67) gives

$$F[CI_1, C^+ I_1^+] = C\langle R, I_1 \rangle + C^+ \langle I_1^+, Q \rangle - CC^+ \langle I_1^+, LI_1 \rangle \qquad (5.70)$$

The normalization factors C, C^+ will be determined from the requirement that the functional $F[CI_1, C^+ I_1^+]$ is stationary with respect to arbitrary variations in the normalization factors C and C^+. Thus from the conditions

$$\boxed{\frac{\partial F}{\partial C} = 0, \quad \frac{\partial F}{\partial C^+} = 0} \qquad (5.71)$$

we find the stationary values for C and C^+. This is Schwinger's variational principle. A brief calculation gives

$$C = \frac{\langle I_1^+, Q \rangle}{\langle I_1^+, LI_1 \rangle}, \quad C^+ = \frac{\langle R, I_1 \rangle}{\langle I_1^+, LI_1 \rangle} \qquad (5.72)$$

Substituting these expressions into (5.70) leads to the *Schwinger functional*

$$\boxed{F_S[I_1, I_1^+] = F[CI_1, C^+ I_1^+] = \frac{\langle R, I_1 \rangle \langle I_1^+, Q \rangle}{\langle I_1^+, LI_1 \rangle}} \qquad (5.73)$$

We will now apply the approximate functions I_0 and I_0^+ as trial functions for I_1 and I_1^+. This results in

$$F_S[I_0, I_0^+] = \frac{\langle R, I_0 \rangle \langle I_0^+, Q \rangle}{\langle I_0^+, LI_0 \rangle} = \frac{\langle R, I_0 \rangle \langle I_0^+, Q \rangle}{\langle I_0^+, L_0 I_0 \rangle + \langle I_0^+, \Delta L I_0 \rangle}$$
$$= \frac{\langle R, I_0 \rangle}{1 + \frac{\langle I_0^+, \Delta L I_0 \rangle}{\langle I_0^+, Q \rangle}} \qquad (5.74)$$

Using the definition (5.60) for the base radiative effect $\mathcal{E}_0$, we find the radiative effect according to the Schwinger functional

$$\mathcal{E}_S = \frac{\mathcal{E}_0}{1 + \langle I_0^+, \Delta L I_0 \rangle / \mathcal{E}_0} \qquad (5.75)$$

It is a matter of interest to compare the two formulations of the radiative effect according to equations (5.66) and (5.75). Expanding (5.75) into a Taylor series we find

$$\mathcal{E}_S = \mathcal{E}_0 \left(1 - \frac{\langle I_0^+, \Delta L I_0 \rangle}{\mathcal{E}_0} + \frac{\langle I_0^+, \Delta L I_0 \rangle^2}{\mathcal{E}_0^2} \mp \cdots \right) \qquad (5.76)$$

If the ratio in the denominator of (5.75) is small in comparison to 1, we may discontinue the expansion after the linear term yielding

$$\mathcal{E}_S \approx \mathcal{E}_0 - \langle I_0^+, \Delta L I_0 \rangle \qquad (5.77)$$

in agreement with (5.66) and (5.68). For larger disturbances the higher-order terms in (5.76) contribute significantly to the Schwinger appproximation. Only detailed calculations can give information as to which one of these two approximations is to be preferred.

5.4.3 Evaluation of the perturbation integral

In the final step of the analysis we evaluate the perturbation integral

$$\Delta \mathcal{E} = \langle I_0^+, \Delta L I_0 \rangle \qquad (5.78)$$

Substituting the phase function $\mathcal{P}(z, \Omega' \to \Omega) = \mathcal{P}(z, \cos \Theta)$ in the form (2.68) into the definitions (5.4) and (5.59a) of the operators L, L^+ and applying them to I_0 gives

$$L I_0(z, \Omega) = \left(\mu \frac{\partial}{\partial z} + k_{\text{ext}}(z) \right) I_0(z, \Omega) - \frac{k_{\text{sca}}(z)}{4\pi} \int_{4\pi} \left(\sum_{m=0}^{\infty} (2 - \delta_{0m}) \right.$$

$$\left. \times \sum_{l=m}^{\infty} p_l^m(z) P_l^m(\mu) P_l^m(\mu') \cos m(\varphi - \varphi') \right) I_0(z, \Omega') d\Omega'$$

$$L_0 I_0(z, \Omega) = \left(\mu \frac{\partial}{\partial z} + k_{\text{ext},0}(z) \right) I_0(z, \Omega) - \frac{k_{\text{sca},0}(z)}{4\pi} \int_{4\pi} \left(\sum_{m=0}^{\infty} (2 - \delta_{0m}) \right.$$

$$\left. \times \sum_{l=m}^{\infty} p_{l,0}^m(z) P_l^m(\mu) P_l^m(\mu') \cos m(\varphi - \varphi') \right) I_0(z, \Omega') d\Omega' \qquad (5.79)$$

Subtracting both equations yields

$$\Delta L I_0(z, \Omega) = \Delta k_{\text{ext}}(z) I_0(z, \Omega) - \frac{1}{4\pi} \int_{4\pi} \left(\sum_{m=0}^{\infty} (2 - \delta_{0m}) \right.$$

$$\left. \times \sum_{l=m}^{\infty} \Delta \eta_l^m(z) P_l^m(\mu) P_l^m(\mu') \cos m(\varphi - \varphi') \right) I_0(z, \Omega') d\Omega'$$

(5.80)

where the abbreviations

$$\Delta k_{\text{ext}}(z) = k_{\text{ext}}(z) - k_{\text{ext},0}(z)$$
$$\Delta \eta_l^m(z) = k_{\text{sca}}(z) p_l^m(z) - k_{\text{sca},0}(z) p_{l,0}^m(z)$$

(5.81)

have been utilized. The most striking feature of (5.80) is that the partial derivative $\mu \partial / \partial z$ does not appear in the perturbation operator ΔL. We conclude that for the evaluation of the perturbation integral (5.78) it is not necessary to solve the RTE. As already mentioned this is the paramount advantage of the perturbation method. Once the radiative effect is known for a base atmosphere, the corresponding radiative effect of an arbitrary atmosphere may be obtained without solving the RTE once more. Finally, it is obvious that

$$\Delta L^+ = \Delta L \qquad (5.82)$$

In the previous formulas, for simplicity, we have omitted the wave number or wavelength subscript. Thus all formulas refer to monochromatic radiation. In order to obtain physically relevant expressions, we must integrate over the wavelength to get the radiative effect for an absorption band or even for the entire solar spectrum.

A number of physically relevant quantities such as flux densities, net flux densities and heating rates are independent of the azimuthal angle. Inspection of (5.79) and (5.80) shows that in this case in the sum over m only the term $m = 0$ must be considered which results in an important simplification.

We wish to summarize: with the help of the perturbation parameters $\Delta k_{\text{ext}}(z)$ and $\Delta \eta_l^m(z)$ we can construct all kinds of perturbations from a base atmosphere. The following procedure can be used for an efficient calculation of a radiative effect $\mathcal{E}$ for an arbitrary atmosphere in the framework of linear perturbation theory.

(1) Solve the RTE for a base atmosphere to obtain the base values I_0. Since the base values I_0 are calculated once only, they may be obtained by means of very accurate and thus elaborate solution methods of the RTE.
(2) Calculate the corresponding I_0^+ for the base solution.
(3) Calculate the base radiative effect $\mathcal{E}_0$. Use both formulations of (5.60) to check the results.
(4) Find $\mathcal{E}$ according to (5.66) or (5.75).

5.5 Appendix

5.5.1 Linear operator and its adjoint

The differential operator L is linear:

$$L(\alpha f + \beta g) = \alpha L f + \beta L g \qquad (5.83)$$

where α, β are real numbers. L is bounded:

$$||Lf|| \leq k \, ||f||, \quad k \geq 0 \qquad (5.84)$$

Every bounded linear operator L has an adjoint L^+ defined by the relation

$$\langle Lu, v \rangle = \langle u, L^+ v \rangle \qquad (5.85)$$

L is self-adjoint if

$$L = L^+ \qquad (5.86)$$

Example: $Lu = du/dx$ with boundary condition $u(0) = 2u(1)$

$$\langle u, v \rangle = \int_0^1 u(x)v(x)dx$$

$$\langle Lu, v \rangle = \int_0^1 \frac{du(x)}{dx} v(x)dx = u(1)\left[v(1) - 2v(0)\right] - \int_0^1 u(x)\frac{dv(x)}{dx}dx$$

$$L^+ v = -\frac{dv}{dx} \quad \text{with} \quad v(1) = 2v(0) \qquad (5.87)$$

5.5.2 Superposition formula for the inclusion of Lambertian surface reflection

The fundamental formula (5.32) was stated by Box *et al.* (1988) without giving a derivation. We will now derive this equation by considering a planetary atmosphere which is illuminated by parallel solar radiation assuming that no diffuse radiation is incident at the top of the atmosphere. At the lower boundary we place a Lambertian surface to simulate ground reflection. The solution $I(\tau, \mu, \varphi)$ to this type of radiative transfer problem can be obtained by solving two independent more simple problems. The first problem (i) involves the illumination of the atmosphere by parallel solar radiation and by employing vacuum boundary conditions for the diffuse radiation field. The second problem (ii) assumes that the atmosphere is illuminated from below by a purely diffuse radiation field resulting from a reflecting Lambertian ground. At the top of the atmosphere again a vacuum boundary condition is employed. To express the complete solution of the radiation problem, the solutions to problems (i) and (ii) must be superimposed in a suitable but linear fashion.

Let us first consider problem (i) as described by

$$\mu \frac{d}{d\tau} I_v(\tau, \mu, \varphi) = I_v(\tau, \mu, \varphi) - \frac{\omega_0}{4\pi} \int_0^{2\pi} \int_{-1}^1 \mathcal{P}(\tau, \mu', \varphi', \mu, \varphi) I_v(\tau, \mu', \varphi') d\mu' d\varphi'$$

$$- \frac{\omega_0}{4\pi} S_0 \mathcal{P}(\tau, \mu_0, \varphi_0, \mu, \varphi) \exp\left(-\frac{\tau}{|\mu_0|}\right) \qquad (5.88)$$

with vacuum boundary conditions (suffix v)

$$\text{top of the atmosphere:} \quad I_v(0, \mu, \varphi) = 0, \quad 0 \le \varphi \le 2\pi, \quad -1 \le \mu < 0$$
$$\text{Earth's surface:} \quad I_v(\tau_0, \mu, \varphi) = 0, \quad 0 \le \varphi \le 2\pi, \quad 0 < \mu \le 1 \qquad (5.89)$$

The solution $I(\tau, \mu, \varphi)$ to the solar radiation problem involving a Lambertian surface at $\tau = \tau_0$ with albedo A_g follows from the same type of transfer equation. While the upper boundary condition remains unchanged, the lower boundary condition will be replaced by

$$I(\tau_0, \mu, \varphi) = \frac{A_g}{\pi} \int_0^{2\pi} \int_{-1}^0 |\mu'| I(\tau_0, \mu', \varphi') d\mu' d\varphi'$$

$$+ \frac{A_g}{\pi} |\mu_0| S_0 \exp\left(-\frac{\tau_0}{|\mu_0|}\right), \quad 0 < \mu \le 1 \qquad (5.90)$$

Next we consider problem (ii) as described by the RTE in the form

$$\mu \frac{d I_d^r(\tau, \mu)}{d\tau} = I_d^r(\tau, \mu) - \frac{\omega_0}{2} \int_{-1}^1 \mathcal{P}(\tau, \mu', \mu) I_d^r(\tau, \mu') d\mu' \qquad (5.91)$$

The purely diffuse radiance $I_d^r(\tau, \mu)$ (suffix d) is generated by an isotropically reflecting surface (suffix r). Observe that the phase function is azimuthally averaged and that the primary scattering term appearing in (5.88) is absent since no parallel solar radiation is involved. At the top of the atmosphere we apply the vacuum boundary condition. To formulate the lower boundary condition we must include both the reflected flux densities resulting from the various I_v and I_d^r in addition to the flux density due to the direct solar radiation at the ground. Thus for problem (ii) the boundary conditions are given by

$$\text{top of the atmosphere:} \quad I_d^r(0, \mu) = 0, \quad\quad -1 \le \mu < 0$$
$$\text{Earth's surface:} \quad I_d^r(\tau_0, \mu) = \frac{A_g}{\pi} E_-(\tau_0), \quad 0 < \mu \le 1 \qquad (5.92)$$

where $E_-(\tau_0)$ is the total downward flux density at the Earth's surface

$$E_-(\tau_0) = E_{v,-}(\tau_0) + 2\pi \int_{-1}^0 |\mu'| I_d^r(\tau_0, \mu') d\mu' \qquad (5.93)$$

which includes the total downward flux density $E_{v,-}(\tau_0)$ resulting from the vacuum problem (i), that is

$$E_{v,-}(\tau_0) = \int_0^{2\pi} \int_{-1}^0 |\mu'| I_v(\tau_0, \mu', \varphi') d\mu' d\varphi' + |\mu_0| S_0 \exp\left(-\frac{\tau_0}{|\mu_0|}\right) \quad (5.94)$$

Note that in (5.92) the factor $1/\pi$ converts the flux density to the isotropic radiance.

Let us now consider another special radiative transfer problem in terms of a dimensionless radiance I_g assuming an isotropic radiation source of unit strength at the lower boundary and vacuum boundary condition at the top of the atmosphere. This transfer problem is defined by

$$\mu \frac{d}{d\tau} I_g(\tau, \mu) = I_g(\tau, \mu) - \frac{\omega_0}{2} \int_{-1}^1 \mathcal{P}(\tau, \mu', \mu) I_g(\tau, \mu') d\mu' \quad (5.95)$$

with the boundary conditions

$$\begin{array}{ll} \text{top of the atmosphere:} & I_g(0, \mu) = 0, \quad -1 \le \mu < 0 \\ \text{Earth's surface:} & I_g(\tau_0, \mu) = 1, \quad 0 < \mu \le 1 \end{array} \quad (5.96)$$

The solution to this problem will be employed to find the solution to the RTE (5.91). This is possible since $I_g(\tau, \mu)$ represents the response of the atmosphere due to an isotropic diffuse illumination at $\tau = \tau_0$ with flux density

$$E_{g,+}(\tau_0) = 2\pi \int_0^1 \mu I_g(\tau_0, \mu) d\mu = \pi \quad (5.97)$$

The total flux density in upward direction of the combined problem (i) and (ii) is given by

$$E_+(\tau_0) = A_g E_-(\tau_0) \quad (5.98)$$

so that

$$\frac{E_+(\tau_0)}{E_{g,+}(\tau_0)} = \frac{A_g}{\pi} E_-(\tau_0) \quad (5.99)$$

Analogously to this expression for $I_d^r(\tau, \mu)$ we may write

$$\frac{I_d^r(\tau, \mu)}{I_g(\tau, \mu)} = \frac{A_g}{\pi} E_-(\tau_0) \quad \text{or} \quad I_d^r(\tau, \mu) = \frac{A_g}{\pi} E_-(\tau_0) I_g(\tau, \mu) \quad (5.100)$$

In order to comprehend this conclusion, first of all, we recognize the similarity of the transfer equations (5.91) and (5.95) and the formal agreement between the upper boundary conditions in (5.92) and (5.96). The lower boundary conditions in (5.92) and (5.96) vary in form, but both refer to isotropic upward radiation. The step leading to (5.100) is motivated by the fact that the transfer problem for I_g is linear with respect to the isotropic boundary conditions as applied to I_g at $\tau = \tau_0$.

This implies that a multiple of the upwelling radiation illuminating the atmosphere from below results in a corresponding multiple of $I_g(\tau, \mu)$ which represents the response of the atmosphere to the lower boundary conditions in (5.92).

To complete the problem, we need to express the factor multiplying $I_g(\tau, \mu)$ in (5.100) in terms of known quantities. With the help of (5.94) and (5.100) we obtain

$$
\begin{aligned}
\frac{A_g}{\pi} E_-(\tau_0) &= \frac{A_g}{\pi} \left[E_{v,-}(\tau_0) + 2\pi \int_{-1}^{0} |\mu'| I_d^r(\tau_0, \mu') d\mu' \right] \\
&= \frac{A_g}{\pi} \left[E_{v,-}(\tau_0) + 2A_g E_-(\tau_0) \int_{-1}^{0} |\mu'| I_g(\tau_0, \mu') d\mu' \right]
\end{aligned}
\tag{5.101}
$$

By introducing the abbreviation

$$
E_{g,-}(\tau_0) = 2\pi \int_{-1}^{0} |\mu'| I_g(\tau_0, \mu') d\mu'
\tag{5.102}
$$

into (5.101) we find the required expression

$$
\begin{aligned}
E_-(\tau_0) &= \frac{1}{1 - (A_g/\pi)E_{g,-}(\tau_0)} E_{v,-}(\tau_0) \\
&= E_{v,-}(\tau_0) \left[1 + \frac{A_g}{\pi} E_{g,-}(\tau_0) + \left[\frac{A_g}{\pi} E_{g,-}(\tau_0) \right]^2 + \cdots \right]
\end{aligned}
\tag{5.103}
$$

The physical interpretation of this expression is as follows: the first term represents the downward flux density of problem (i), while terms two, three, etc., describe the additional contributions due to one, two, etc., reflection interactions between the Lambertian surface and the plane–parallel atmosphere.

Substituting (5.103) into (5.100) gives

$$
I_d^r(\tau, \mu) = E_{v,-}(\tau_0) \frac{(A_g/\pi)}{1 - (A_g/\pi)E_{g,-}(\tau_0)} I_g(\tau, \mu)
\tag{5.104}
$$

The radiance $I(\tau, \mu, \varphi)$ is the combination of the radiances $I_v(\tau, \mu, \varphi)$ and $I_d^r(\tau, \mu)$, i.e.

$$
\boxed{
\begin{aligned}
I(\tau, \mu, \varphi) &= I_v(\tau, \mu, \varphi) + I_d^r(\tau, \mu) \\
&= I_v(\tau, \mu, \varphi) + E_{v,-}(\tau_0) \frac{(A_g/\pi)}{1 - (A_g/\pi)E_{g,-}(\tau_0)} I_g(\tau, \mu)
\end{aligned}
}
\tag{5.105}
$$

While in (5.32) we have written the general form for $I_g(z, \mu, \varphi)$, here we have omitted the azimuthal dependency of this quantity which is the actual form used by Box *et al.* (1989b) in their computations.

Finally, it is noteworthy that in Chapter 4 we have included an equation similar to (5.32), see (3.48), which is due to Chandrasekhar (1960). We wish to make reference to Muldashev *et al.* (1999) who have generalized Chandrasekhar's approach by deriving the internal radiance field at an arbitrary level τ due to the presence of a Lambertian surface as described in detail above.

5.6 Problems

5.1: Show that the third terms in (5.16) cancel out.
5.2: Write down in detail all steps involved in (5.64).
5.3: Verify both parts in (5.72).
5.4: Show that (5.73) is correct.

6

Two-stream methods for the solution of the radiative transfer equation

6.1 δ-scaling of the phase function

A particular difficulty when solving the RTE for solar radiation in an atmosphere containing aerosol particles and cloud droplets is associated with the fact that the scattering phase function for such particle populations is highly peaked in the forward direction. For cloud particles the energy scattered within an angle interval of about 5° around the forward direction is four to five magnitudes larger than that part of the energy related to sideward or backward scattering. Therefore, it can be concluded that an accurate treatment of the radiation field for highly peaked phase functions requires a very large number of expansion terms when, according to (2.55), the phase function is written as a series of Legendre polynomials.

In order to incorporate the forward diffraction contribution in multiple scattering computations we may proceed as follows. Let f represent that part of the radiation which is scattered in the forward direction. In specifying this forward part we may define the so-called *δ-scaled phase function* by

$$\mathcal{P}^*(\cos \Theta) = 2f\delta(1 - \cos \Theta) + (1 - f) \sum_{k=0}^{n-1} p_k^* P_k(\cos \Theta) \tag{6.1}$$

where the first term on the right-hand side represents the forward scattering contribution. The modified expansion coefficients p_k^* will now be determined in such a way that the first n moments $p_0^*, p_1^*, \ldots, p_{n-1}^*$ of the modified phase function $\mathcal{P}^*$ are equal to the corresponding moments of the original phase function $\mathcal{P}$. With the help of (2.61) we may, therefore, write

$$\int_{-1}^{1} P_l(\cos \Theta) \mathcal{P}^*(\cos \Theta) d \cos \Theta = \int_{-1}^{1} P_l(\cos \Theta) \mathcal{P}(\cos \Theta) d \cos \Theta, \quad l = 0, \ldots, n - 1 \tag{6.2}$$

According to Morse and Feshbach (1953) the δ-function can be expressed as an infinite series of Legendre polynomials, that is

$$\delta(x - x') = \sum_{k=0}^{\infty} \frac{2k+1}{2} P_k(x) P_k(x') \qquad (6.3)$$

Utilizing this expression together with the orthogonality relations for the Legendre polynomials (2.59b), we obtain

$$\int_{-1}^{1} P_l(\cos\Theta) 2f\delta(1 - \cos\Theta) d\cos\Theta$$

$$= \int_{-1}^{1} P_l(\cos\Theta) 2f \sum_{k=0}^{\infty} \frac{2k+1}{2} P_k(1) P_k(\cos\Theta) d\cos\Theta = 2f \quad (6.4)$$

since $P_k(1) = 1$, see Abramowitz and Stegun (1972). Hence, (6.2) may be rewritten as

$$2f + (1 - f) \int_{-1}^{1} P_l(\cos\Theta) \sum_{k=0}^{n-1} p_k^* P_k(\cos\Theta) d\cos\Theta$$

$$= \int_{-1}^{1} P_l(\cos\Theta) \mathcal{P}(\cos\Theta) d\cos\Theta \quad l = 0, 1, \ldots, n-1 \quad (6.5)$$

By applying once more the orthogonality relations (2.59b) we finally obtain

$$\boxed{p_l^* = \frac{p_l - (2l+1)f}{1 - f}, \qquad l = 0, 1, \ldots, n-1} \qquad (6.6)$$

So far we have given no rule for determining f. Since the modified phase function is truncated after the $(n-1)$th term, that is $p_l^* = 0$ for $l \geq n$, we obtain from (6.6) a relation between f and the p_l

$$f = \frac{p_l}{2l+1}, \qquad l \geq n \qquad (6.7)$$

However, this expression does not determine f uniquely. In order to come up with a suitable choice we consider the difference between the original phase function $\mathcal{P}$ and the scaled phase function $\mathcal{P}^*$. In noting that the δ-function can be expanded in an infinite series of Legendre polynomials, see (6.3), the difference between $\mathcal{P}$ and $\mathcal{P}^*$ can be determined from

$$\mathcal{P}(\cos\Theta) - \mathcal{P}^*(\cos\Theta) = \sum_{l=n}^{\infty} [p_l - (2l+1)f] P_l(\cos\Theta) \qquad (6.8)$$

By choosing

$$\boxed{f = \frac{p_n}{2n+1}} \qquad (6.9)$$

we observe that the leading term in the series (6.8) vanishes, while the remaining terms of the series do not vanish. On the basis of numerical experience, Wiscombe (1977) argues that it is more important to force agreement in the phase function for the lowest order term $l = n$ than for higher order terms $l > n$. Therefore, we will use (6.9) to determine the fraction of the radiation which is scattered in the forward direction.

In the following we will demonstrate how the δ-scaling of the phase function affects the RTE in the absence of *thermal emission*. With reference to (2.29) and (2.47) we write the RTE for the total radiance I_{tot} including the parallel solar radiation as

$$\frac{d I_{tot}}{ds} = -k_{ext} I_{tot} + k_{ext} J \tag{6.10}$$

where the multiple scattering term is given by

$$J = \frac{\omega_0}{4\pi} \int_{4\pi} \mathcal{P}(\cos \Theta) I_{tot}(s, \Omega') d\Omega' \tag{6.11}$$

Introducing the δ-approximation of the phase function (6.1) into the multiple scattering term and noting that

$$\delta(1 - \cos \Theta) = 2\pi \, \delta(\mu - \mu') \delta(\varphi - \varphi') \tag{6.12}$$

(see, e.g. Jackson, 1975), we obtain

$$J = \frac{\omega_0}{4\pi} \int_0^{2\pi} \int_{-1}^{1} \left[4\pi f \delta(\mu - \mu') \delta(\varphi - \varphi') + (1 - f) \sum_{l=0}^{n-1} p_l^* P_l(\cos \Theta) \right]$$
$$\times I_{tot}(s, \mu', \varphi') d\mu' d\varphi' \tag{6.13}$$

The term containing the two δ-functions can be evaluated immediately so that

$$J = \omega_0 f I_{tot}(s, \mu, \varphi) + \frac{\omega_0(1 - f)}{4\pi} \int_0^{2\pi} \int_{-1}^{1} \sum_{l=0}^{n-1} p_l^* P_l(\cos \Theta) I_{tot}(s, \mu', \varphi') d\mu' d\varphi' \tag{6.14}$$

For the RTE we then obtain

$$\frac{d I_{tot}}{ds} = -k_{ext}^* I_{tot} + k_{ext}^* J^* \tag{6.15}$$

where the modified source term J^* is given by

$$J^* = \frac{\omega_0^*}{4\pi} \int_0^{2\pi} \int_{-1}^{1} \sum_{l=0}^{n-1} p_l^* P_l(\cos \Theta) I_{tot}(s, \mu', \varphi') d\mu' d\varphi' \tag{6.16}$$

The modified extinction coefficient k_{ext}^* and the modified single scattering albedo ω_0^* are

$$k_{ext}^* = (1 - \omega_0 f) k_{ext}, \qquad \omega_0^* = \frac{(1 - f)\omega_0}{1 - \omega_0 f} \tag{6.17}$$

By comparing (6.15) with (6.10) we observe that both types of the radiative transfer equation have the same form. This means that the δ-adjustment of the phase function leaves the RTE formally invariant. In summary, in going from the original unscaled to the scaled problem we have to make the following replacements

$$
\begin{aligned}
p_l \rightarrow p_l^* &= \frac{p_l - (2l+1)f}{1-f} \\
k_{\text{ext}} \rightarrow k_{\text{ext}}^* &= (1 - \omega_0 f)k_{\text{ext}} \\
\omega_0 \rightarrow \omega_0^* &= \frac{(1-f)\omega_0}{1 - \omega_0 f}
\end{aligned}
\tag{6.18}
$$

where the forward diffraction part f of the phase function is given by (6.9).

For a homogeneous layer with thickness Δz the optical depth $\Delta \tau = k_{\text{ext}}\Delta z$ has to be replaced by the scaled value

$$
\Delta \tau^* = k_{\text{ext}}^* \Delta z = (1 - \omega_0 f)\Delta \tau
\tag{6.19}
$$

It is important to note that in the δ-approximation for consistency the extinction of the direct solar beam is given by

$$
S(\tau^*) = S_0 \exp\left(-\frac{\tau^*}{\mu_0}\right)
\tag{6.20}
$$

Since $\tau^* \leq \tau$, the δ-scaled direct flux density is always larger than its unscaled counterpart. This means that the sum of the scaled direct beam plus the scaled diffuse radiation should be compared with the measured values of the global radiation, that is the total downward radiation.

The δ-scaling approximation for solar radiative transfer has been widely employed. In particular, this method has been applied to the so-called *two-stream methods*, see Joseph *et al.* (1976) and Wiscombe (1977). These authors have shown that the δ-scaling resulted in improved flux density and heating rate calculations.

6.2 The two-stream radiative transfer equation

It is well known that in a plane–parallel atmosphere the evaluation of the *thermodynamic heat equation* requires the knowledge of the net radiative flux densities $E_{\text{net},z} = E_{+,z} - E_{-,z}$ but does not need the complete directional dependence of the radiation field.[1] An accurate determination of these flux densities requires an integration of the azimuthally independent radiation field $I^{m=0}(\tau, \mu)$ over all directions as shown in (2.126).

In many circumstances the computational effort to determine the up- and downward flux densities in this manner is far too high so that approximate methods

---

[1] For simplicity, the subscript z occurring at the radiative flux densities will henceforth be omitted.

must be used. These are the so-called *two-stream methods* (TSM) which originated with Schuster (1905), Schwarzschild (1906), Emden (1913), and Eddington (1916). Owing to their high computational efficiency, two-stream methods are now widely used in conjunction with global or mesoscale climate models for determining the heating and cooling rates in the solar and long-wave radiative spectrum. In the following we will show how various versions of the TSM can be derived from the RTE.

In principle we could start from DOM by considering two radiation streams only. However, it is more expedient to begin with the azimuthally averaged radiation field in the absence of thermal radiation since the solar and infrared spectra may easily be separated. For convenience we repeat (2.81), omitting there the infrared radiation term

$$\mu \frac{d}{d\tau} I(\tau, \mu) = I(\tau, \mu) - \frac{\omega_0}{2} \int_{-1}^{1} \mathcal{P}(\mu, \mu') I(\tau, \mu') d\mu'$$
$$- \frac{\omega_0}{4\pi} S_0 \exp\left(-\frac{\tau}{\mu_0}\right) \mathcal{P}(\mu, -\mu_0) \tag{6.21}$$

The up- and downward radiative flux densities follow from

$$E_+ = 2\pi \int_0^1 \mu I(\tau, \mu) d\mu, \qquad E_- = 2\pi \int_0^1 \mu I(\tau, -\mu) d\mu \tag{6.22}$$

so that the total net flux density is obtained from

$$E_{\text{net}} = E_+ - E_- - \mu_0 S_0 \exp\left(-\frac{\tau}{\mu_0}\right) \tag{6.23}$$

see also Section 2.7. The last term on the right-hand side of (6.23) represents the contribution of the unscattered solar beam. Integration of (6.21) over the 2π upper hemisphere leads to

$$\frac{dE_+}{d\tau} = 2\pi \int_0^1 I(\tau, \mu) d\mu - \omega_0 \pi \int_0^1 \int_{-1}^{1} \mathcal{P}(\mu, \mu') I(\tau, \mu') d\mu' d\mu$$
$$- S_0 \exp\left(-\frac{\tau}{\mu_0}\right) \frac{\omega_0}{2} \int_0^1 \mathcal{P}(\mu, -\mu_0) d\mu \tag{6.24}$$

Let us now discuss that part of the diffuse downward light which is scattered into the backward hemisphere. From the phase function we define the so-called *backscattered fraction* or *backscattering coefficient* $b(-\mu')$ (Wiscombe and Grams, 1976) as

$$b(-\mu') = \frac{1}{2} \int_0^1 \mathcal{P}(\mu, -\mu') d\mu, \qquad \mu' > 0 \tag{6.25}$$

where, as usual, $-\mu'$ refers to the direction of the downward diffuse light as illustrated in Figure 6.1. Due to the normalization condition (4.34) and the symmetry

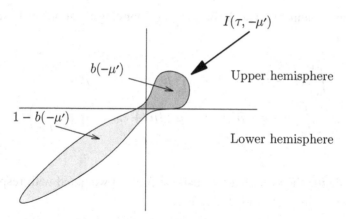

Fig. 6.1 Illustration of the backscattered fraction $b(-\mu')$ in the phase function.

properties (4.95) of $\mathcal{P}(\mu, \mu')$ it follows that

$$\frac{1}{2}\int_0^1 \mathcal{P}(-\mu, -\mu')d\mu = \frac{1}{2}\int_0^1 \mathcal{P}(\mu, \mu')d\mu = 1 - b(-\mu'), \qquad \mu' > 0 \quad (6.26)$$

With this definition we can split the multiple scattering term in (6.24) into two parts

$$\frac{1}{2}\int_0^1 \int_{-1}^1 \mathcal{P}(\mu, \mu')I(\tau, \mu')d\mu'd\mu$$

$$= \frac{1}{2}\int_0^1 \int_0^1 \mathcal{P}(\mu, \mu')I(\tau, \mu')d\mu'd\mu + \frac{1}{2}\int_0^1 \int_{-1}^0 \mathcal{P}(\mu, \mu')I(\tau, \mu')d\mu'd\mu$$

$$= \frac{1}{2}\int_0^1 \int_0^1 \mathcal{P}(\mu, \mu')I(\tau, \mu')d\mu'd\mu + \frac{1}{2}\int_0^1 \int_0^1 \mathcal{P}(\mu, -\mu')I(\tau, -\mu')d\mu'd\mu$$

$$= \int_0^1 [1 - b(-\mu')]I(\tau, \mu')d\mu' + \int_0^1 b(-\mu')I(\tau, -\mu')d\mu' \quad (6.27)$$

which motivates the definition (6.25). For the primary scattered sunlight in the backward direction we obtain analogously to (6.25)

$$b(-\mu_0) = \frac{1}{2}\int_0^1 \mathcal{P}(\mu, -\mu_0)d\mu, \qquad \mu_0 > 0 \quad (6.28)$$

Substituting (6.25)–(6.28) into (6.24) yields

$$\frac{dE_+}{d\tau} = 2\pi \int_0^1 I(\tau, \mu)d\mu - 2\pi\omega_0 \int_0^1 [1 - b(-\mu')]I(\tau, \mu')d\mu'$$

$$-2\pi\omega_0 \int_0^1 b(-\mu')I(\tau, -\mu')d\mu' - \omega_0 S_0 \exp\left(-\frac{\tau}{\mu_0}\right)b(-\mu_0) \quad (6.29)$$

In an analogous manner we can derive the corresponding equation for the downward flux density E_-

$$
\frac{dE_-}{d\tau} = -2\pi \int_0^1 I(\tau, -\mu)d\mu + 2\pi\omega_0 \int_0^1 b(-\mu')I(\tau, \mu')d\mu'
$$
$$
+ 2\pi\omega_0 \int_0^1 [1 - b(-\mu')]I(\tau, -\mu')d\mu' + \omega_0 S_0 \exp\left(-\frac{\tau}{\mu_0}\right)[1 - b(-\mu_0)]
$$

$$(6.30)$$

Next we define the average of a quantity $\psi(\tau, \mu)$ weighted with respect to the up- and downwelling intensities $I(\tau, \pm\mu)$

$$
\psi_+(\tau) = \frac{\int_0^1 \psi(\tau, \mu)I(\tau, \mu)d\mu}{\int_0^1 I(\tau, \mu)d\mu}, \qquad \psi_-(\tau) = \frac{\int_0^1 \psi(\tau, \mu)I(\tau, -\mu)d\mu}{\int_0^1 I(\tau, -\mu)d\mu} \qquad (6.31)
$$

For the particular choice $\psi = \mu$ we obtain from (6.31)

$$
\mu_+(\tau) = \frac{2\pi \int_0^1 \mu I(\tau, \mu)d\mu}{2\pi \int_0^1 I(\tau, \mu)d\mu} = \frac{E_+}{2\pi \int_0^1 I(\tau, \mu)d\mu}
$$

$$(6.32)$$

$$
\mu_-(\tau) = \frac{2\pi \int_0^1 \mu I(\tau, -\mu)d\mu}{2\pi \int_0^1 I(\tau, -\mu)d\mu} = \frac{E_-}{2\pi \int_0^1 I(\tau, -\mu)d\mu}
$$

The parameters μ_+ and μ_- can be interpreted as the mean directional cosines of the up- and downward diffuse radiation. Utilizing in (6.31) $\psi = b(-\mu)$ we obtain in the same way

$$
b_+(\tau) = \frac{2\pi \int_0^1 b(-\mu)I(\tau, \mu)d\mu}{2\pi \int_0^1 I(\tau, \mu)d\mu}, \qquad b_-(\tau) = \frac{2\pi \int_0^1 b(-\mu)I(\tau, -\mu)d\mu}{2\pi \int_0^1 I(\tau, -\mu)d\mu}
$$

$$(6.33)$$

It should be emphasized that, in general, the parameters $\mu_\pm, b_\pm$ are functions of the optical depth τ since the radiances depend on τ. Combining (6.32) and (6.33) yields

$$2\pi \int_0^1 b(-\mu)I(\tau, \mu)d\mu = \frac{b_+}{\mu_+}E_+, \qquad 2\pi \int_0^1 b(-\mu)I(\tau, -\mu)d\mu = \frac{b_-}{\mu_-}E_-$$

$$(6.34)$$

Equations (6.32) and (6.34) can now be employed to eliminate all expressions in (6.29) and (6.30) which contain the radiance $I(\tau, \mu)$. The resulting differential equations for the up- and downward flux densities can be written as a 2×2 matrix differential equation which reads

$$\boxed{\frac{d\mathbf{E}}{d\tau} = \mathbb{A} \cdot \mathbf{E} + S_0 \exp\left(-\frac{\tau}{\mu_0}\right)\mathbf{S}} \qquad (6.35)$$

where

$$\mathbf{E} = \begin{pmatrix} E_+ \\ E_- \end{pmatrix}, \qquad \mathbb{A} = \begin{pmatrix} \alpha_{11} & \alpha_{12} \\ \alpha_{21} & \alpha_{22} \end{pmatrix}, \qquad \mathbf{S} = \begin{pmatrix} -\omega_0 b(-\mu_0) \\ \omega_0[1 - b(-\mu_0)] \end{pmatrix} \qquad (6.36)$$

The coefficients α_{jk}, $(j, k = 1, 2)$, of the matrix $\mathbb{A}$ are given by

$$\alpha_{11} = \frac{1 - \omega_0(1 - b_+)}{\mu_+}, \qquad \alpha_{12} = -\frac{\omega_0 b_-}{\mu_-}$$

$$\alpha_{21} = \frac{\omega_0 b_+}{\mu_+}, \qquad \alpha_{22} = -\frac{1 - \omega_0(1 - b_-)}{\mu_-} \qquad (6.37)$$

It must be stressed that the parameters $\mu_\pm, b_\pm$ occurring in $\mathbb{A}$ are unknown within the two-stream approximation. Therefore, an ambiguity exists in specifying these values. To the best of our knowledge practically all applications of the TSM ignore the τ-dependency of both $\mu_\pm$ as well as $b_\pm$. While some authors provide different constants for the parameters for the upper and lower hemisphere, others make no distinction and, therefore, set $\mu_+ = \mu_-$, $b_+ = b_-$. In the way these parameters are chosen, slight distinctions between the different TSM schemes occur in the literature.

For a homogeneous layer $\Delta\tau_i = \tau_i - \tau_{i-1}$ the system (6.35) is a first-order differential equation with constant coefficients α_{jk} which can be solved analytically. The integration constants are determined from the boundary conditions, i.e. the downward flux density $E_-(\tau_{i-1})$ at the upper boundary and the upward flux density $E_+(\tau_i)$ at the lower boundary of the homogeneous layer.

In order to solve the two-stream equations for an inhomogeneous atmosphere we may proceed as in the DOM method, see Section 4.3.

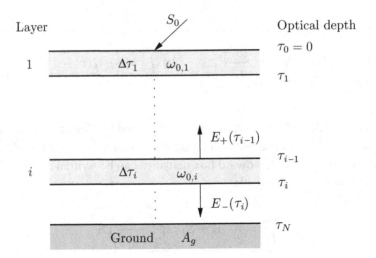

Fig. 6.2 Subdivision of an inhomogeneous atmosphere into N homogeneous sublayers. At the boundaries τ_{i-1} and τ_i the up- and downward flux densities are given by $E_+(\tau_{i-1})$ and $E_-(\tau_i)$, respectively.

(1) First, solve (6.35) for each individual homogeneous sublayer i defined by the optical depth interval $\Delta\tau_i$.
(2) The flux densities $E_\pm(\tau_i)$ are forced to be continuous at each interior interface τ_i.
(3) If τ_N is the total optical depth of the atmosphere then for a reflecting ground the flux density $E_+(\tau_N)$ is determined by the diffusely reflected flux density $E_-(\tau_N)$ plus diffuse or specular reflection of the direct solar radiation reaching the ground.
(4) Similar to DOM, a linear system of equations has to be solved to obtain the up- and downward flux densities at each level i of the model atmosphere.

Figure 6.2 illustrates the sublayering of the atmosphere as well as the up- and downward diffuse flux densities at the boundaries of an arbitrary layer i.

6.3 Different versions of two-stream methods

In this section we will present some two-stream methods which have been widely discussed in the literature.

6.3.1 Two-stream method with hemispheric isotropy

In this version of TSM it is assumed that the up- and downward diffuse radiation field is isotropic. Evaluating for this particular situation (6.32) and (6.33) yields

$$\mu_\pm = \bar{\mu} = \frac{1}{2}, \qquad b_\pm = \bar{b} = \int_0^1 b(-\mu)d\mu \qquad (6.38)$$

Truncating in (2.80) the phase function after the linear term, i.e. $P(\mu, -\mu') = 1 - p_1\mu\mu'$, and substituting this expression into (6.25) results in

$$b(-\mu') = \frac{1}{2} - \frac{3}{4}g\mu' \implies \bar{b} = \frac{1}{2} - \frac{3}{8}g \tag{6.39}$$

Here the so-called *asymmetry parameter* of the phase function g has been introduced. This quantity is defined by the first moment of the phase function

$$g = \frac{1}{2}\int_{-1}^{1}\cos\Theta P(\cos\Theta)d\cos\Theta = \frac{p_1}{3} \tag{6.40}$$

whereby the integral was evaluated by means of (2.61). Hence it is seen that for isotropic scattering $g = 0$.

At this point it is necessary to make a brief remark on the phase function as given by (2.55). If the phase function is found with the help of the rigorous electromagnetic theory, known as the *Mie–Debye theory* to be discussed later, we speak of the *Mie phase function* $P_{\text{Mie}}(\cos\Theta)$. Often it is convenient to approximate P_{Mie} with the help of the asymmetry parameter. The resulting phase function is known as the *Henyey–Greenstein phase function* defined by

$$P_{\text{HG}}(\cos\Theta) = \frac{1 - g^2}{(1 + g^2 - 2g\cos\Theta)^{3/2}} = \sum_{l=0}^{\infty} p_{l,\text{HG}} P_l(\cos\Theta) \tag{6.41}$$

$$\text{with} \quad p_{l,\text{HG}} = (2l + 1)g^l$$

6.3.2 The Eddington approximation

In the *classical Eddington approximation* the radiance is arranged by means of

$$I(\tau, \mu, \varphi) = I_0(\tau) + \mu I_1(\tau) \tag{6.42}$$

so that I is independent of the azimuthal angle φ. Substituting this equation into (6.21) gives

$$\mu\frac{d}{d\tau}[I_0(\tau) + \mu I_1(\tau)] = I_0(\tau) + \mu I_1(\tau) - \frac{\omega_0}{2}\int_{-1}^{1} P(\mu, \mu')[I_0(\tau) + \mu' I_1(\tau)]d\mu'$$

$$- \frac{\omega_0}{4\pi}S_0\exp\left(-\frac{\tau}{\mu_0}\right)P(\mu, -\mu_0) \tag{6.43}$$

The multiple scattering integral may be evaluated by means of (2.80) and the orthogonality relations for the Legendre polynomials (2.59b) yielding[2]

$$\int_{-1}^{1} \mathcal{P}(\mu, \mu')(I_0 + \mu' I_1)d\mu' = 2I_0 + 2g\mu I_1 \qquad (6.44)$$

with $g = p_1/3$. Hence, a consequence of the Eddington approximation (6.42) is that the phase function is truncated after the linear term. Integrating (6.43) over μ in the limits $[-1, 1]$ we obtain a differential equation for I_1. Multiplication of (6.43) by μ and subsequent integration over μ gives the corresponding differential equation for I_0. Thus we obtain

$$\frac{dI_1}{d\tau} = 3(1 - \omega_0)I_0 - \frac{3}{4\pi}\omega_0 S_0 \exp\left(-\frac{\tau}{\mu_0}\right)$$

$$\frac{dI_0}{d\tau} = (1 - \omega_0 g)I_1 + \frac{3}{4\pi}\omega_0 g\mu_0 S_0 \exp\left(-\frac{\tau}{\mu_0}\right) \qquad (6.45)$$

According to (6.22) the up- and downward radiative flux densities are given by

$$E_+ = \pi\left(I_0 + \frac{2}{3}I_1\right), \qquad E_- = \pi\left(I_0 - \frac{2}{3}I_1\right) \qquad (6.46)$$

Hence, I_0 and I_1 may be written as

$$I_0 = \frac{1}{2\pi}(E_+ + E_-), \qquad I_1 = \frac{3}{4\pi}(E_+ - E_-) \qquad (6.47)$$

Substituting these relations into (6.45) we obtain

$$\frac{d}{d\tau}(E_+ - E_-) = 2(1 - \omega_0)(E_+ + E_-) - \omega_0 S_0 \exp\left(-\frac{\tau}{\mu_0}\right)$$

$$\frac{d}{d\tau}(E_+ + E_-) = \frac{3}{2}(1 - \omega_0 g)(E_+ - E_-) + \frac{3}{2}\omega_0 g\mu_0 S_0 \exp\left(-\frac{\tau}{\mu_0}\right) \qquad (6.48)$$

Addition and subtraction of (6.48a,b) finally results in

$$\frac{d}{d\tau}E_+ = \left[(1 - \omega_0) + \frac{3}{4}(1 - \omega_0 g)\right]E_+$$

$$+ \left[(1 - \omega_0) - \frac{3}{4}(1 - \omega_0 g)\right]E_- - \frac{\omega_0}{2}\left(1 - \frac{3}{2}g\mu_0\right)S_0 \exp\left(-\frac{\tau}{\mu_0}\right)$$

$$\frac{d}{d\tau}E_- = -\left[(1 - \omega_0) - \frac{3}{4}(1 - \omega_0 g)\right]E_+$$

$$- \left[(1 - \omega_0) + \frac{3}{4}(1 - \omega_0 g)\right]E_- + \frac{\omega_0}{2}\left(1 + \frac{3}{2}g\mu_0\right)S_0 \exp\left(-\frac{\tau}{\mu_0}\right) \qquad (6.49)$$

---

[2] Recall that $P_0(\mu) = 1$ and $P_1(\mu) = \mu$.

As stated before, the right-hand side of (6.44) implies a linear approximation of the phase function so that $b(-\mu')$ and $\bar{b}$ are given by (6.39). According to (6.33) the quantities $b_\pm$ are independent of direction. Since $\bar{b}$ is also independent of direction, we approximate $b_\pm$ by $\bar{b}$. Assuming additionally that $\mu_\pm = 1/2$ we find from (6.37) and (6.49)

$$\alpha_{11,\mathrm{Ed}} = \alpha_{11} - \frac{1}{4}, \qquad \alpha_{21,\mathrm{Ed}} = \alpha_{21} - \frac{1}{4}$$
$$\alpha_{12,\mathrm{Ed}} = \alpha_{11} + \frac{1}{4}, \qquad \alpha_{22,\mathrm{Ed}} = \alpha_{11} + \frac{1}{4} \tag{6.50}$$

Furthermore, one may easily see that the factors multiplying the solar radiation terms in (6.49) are given by the components of $\mathbf{S}$ as defined in (6.36). The scattering problem based on the approximation (6.42) was treated in some detail by Shettle and Weinman (1970) and, among others, by Zdunkowski and Junk (1974).

Two-stream approximations often yield unsatisfactory results because in these methods the strong forward scattering peak of the phase function is not accounted for. A distinct improvement of a particular TSM is achieved by utilizing the δ-scaled phase function defined in (6.1). In the δ-two-stream approach, $\mathcal{P}^*$ reduces to the form given in (4.35).

Introducing $\mathcal{P}^*$ in the classical Eddington method yields the *δ-Eddington approximation* where, according to (6.18), the original unscaled parameters (τ, p_1, ω_0) are replaced by $(\tau^*, p_1^*, \omega_0^*)$ with

$$\tau^* = (1 - \omega_0 f)\tau, \qquad p_1^* = \frac{p_1 - 3f}{1 - f}, \qquad \omega_0^* = \frac{(1 - f)\omega_0}{1 - \omega_0 f} \tag{6.51}$$

The fraction f of radiation in the diffraction peak is determined with the help of (6.9) yielding for $n = 2$

$$\text{Mie phase function:} \qquad f = \frac{p_2}{5}$$
$$\text{Henyey–Greenstein phase function:} \qquad f = \frac{p_{2,\mathrm{HG}}}{5} = g^2 = \left(\frac{p_1}{3}\right)^2 \tag{6.52}$$

where in case of the Henyey–Greenstein phase function (6.41) has been used. Owing to the δ-scaling, the backscattered fraction of the direct solar radiation can be expressed as

$$b(-\mu_0) = \frac{1}{2}\left(1 - \frac{3}{2}g^*\mu_0\right) \qquad \text{with} \quad g^* = \frac{g - f}{1 - f} = \frac{p_1^*}{3} \tag{6.53}$$

6.3.3 Discrete ordinates formalism

Setting in the discrete ordinates method $s=1$, according to (2.90) the Gaussian weights and nodes are given by

$$w_1 = w_{-1} = 1, \qquad \mu_1 = \frac{1}{\sqrt{3}}, \qquad \mu_{-1} = -\frac{1}{\sqrt{3}} \tag{6.54}$$

Recall that the nodes μ_i are the positive zeros of the corresponding Legendre polynomials. Equation (6.54) represents the simplest possible choice for the number of streams in DOM. The upward and downward radiation streams can be interpreted as traveling along the directions μ_1 and μ_{-1}, respectively. For the backscattered fraction and for a first-order representation of the phase function we obtain from (6.25)

$$b(-\mu') = \frac{1}{2} - \frac{3}{2}g\mu_1\mu' \tag{6.55}$$

For the determination of $\bar{b}$ equation (6.38) will be evaluated by means of the Gaussian quadrature. However, it is noteworthy that the double-Gaussian quadrature leading to $\mu_1 = 1/2$ is not recommended since for $g=1$ the quadrature of $\int_0^1 b(-\mu')d\mu'$ yields an unphysical total backscattered fraction $\bar{b} = 1/8$. However, choosing $g=1$ means that the total radiation is scattered in the forward direction, i.e. $\bar{b}$ must be zero. With the ordinary Gaussian quadrature one finds indeed the physically correct value $\bar{b}=0$.

The parameters necessary for evaluating the matrix $\mathbb{A}$ in (6.36) are given by

$$\mu_{\pm} = \mu_1 = 1/\sqrt{3}, \qquad b_+ = 1 - \frac{1}{2}\mathcal{P}(\mu_1, \mu_1), \qquad b_- = \frac{1}{2}\mathcal{P}(\mu_1, -\mu_1) = b_+ \tag{6.56}$$

Here the first-order approximation of the phase function

$$\mathcal{P}(\mu_1, \pm\mu_1) = 1 \pm 3g\mu_1\mu_1 = 1 \pm g \tag{6.57}$$

has been used. In the primary scattering term the phase function and the corresponding backscattered fraction for the primary scattered sunlight are approximated as

$$\mathcal{P}(\mu_0, \pm\mu_1) = 1 \pm \sqrt{3}g\mu_0 \implies b(-\mu_0) = \frac{1}{2}\mathcal{P}(\mu_0, -\mu_1) = \frac{1}{2}\left(1 - \sqrt{3}g\mu_0\right) \tag{6.58}$$

6.3.4 Practical improved flux method

In the so-called *practical improved flux method* (PIFM) by Zdunkowski *et al.* (1982) the following parameters are employed

$$\mu_{\pm} = \bar{\mu} = \frac{1}{2}, \qquad b_{\pm} = \bar{b} = \frac{3}{8}(1 - g), \qquad b(-\mu_0) = \frac{1}{2} - \frac{3}{4}g\mu_0 \tag{6.59}$$

It is to be noted that in PIFM the δ-scaling of the phase function is applied to the primary scattering term only, that is

$$
\begin{aligned}
\frac{d}{d\tau} E_+ \bigg|_{\text{prim}} &= \omega_0(1 - f)b^*(-\mu_0)S_0 \exp\left(-\frac{(1 - \omega_0 f)\tau}{\mu_0}\right) \\
\frac{d}{d\tau} E_- \bigg|_{\text{prim}} &= \omega_0(1 - f)[1 - b^*(-\mu_0)]S_0 \exp\left(-\frac{(1 - \omega_0 f)\tau}{\mu_0}\right)
\end{aligned}
\tag{6.60}
$$

The δ-scaled backscattered fraction for primary scattered light is defined by

$$
b^*(-\mu_0) = \frac{1}{2} - \frac{3}{4}g^*\mu_0 = \frac{1}{2} - \frac{3}{4}\frac{g - f}{1 - f}\mu_0
\tag{6.61}
$$

The particular choices (6.60) replace the primary scattering term on the right-hand side of (6.35). In PIFM the transfer of the diffuse radiation still employs the unscaled optical parameters ω_0 and $\bar{b}$. For PIFM the coefficients of the matrix $\mathbb{A}$ are then given by

$$
\alpha_{11} = -\alpha_{22} = \frac{1 - \omega_0(1 - \bar{b})}{\bar{\mu}}, \qquad \alpha_{12} = -\alpha_{21} = -\frac{\omega_0 \bar{b}}{\bar{\mu}}
\tag{6.62}
$$

This selective type of scaling prevents negative flux densities which occasionally occur in case of the traditional δ-scaling.

Finally, we will state different advantages and disadvantages of the two-stream methods.

(1) Two-stream methods are computationally very fast and can be employed as a standard technique for radiative transfer calculations in climate models.
(2) It is easy to apply modifications to a particular TSM which account for partial cloudiness. Details will be given in Section 5.6 of this chapter.
(3) Two-stream methods yield sufficiently accurate results for radiative heating and cooling rates. Maximum errors are typically in the order of 10%.
(4) Application of the δ-scaled phase function yields distinct improvements of the results of the corresponding δ-two-stream methods.

The derivation of the TSM as outlined above follows mainly the work of Zdunkowski *et al.* (1980), Zdunkowski and Korb (1985), and Ceballos (1988). However, many more versions of TSMs (see, e.g. Meador and Weaver, 1980; Bott and Zdunkowski, 1983) can be found in the literature which all have their particular advantages and disadvantages.

6.4 Analytical solution of the two-stream methods
for a homogeneous layer

In all versions of TSM presented in the previous section it turned out that $\mu_+ = \mu_-$
and $b_+ = b_-$. In this case the coefficients of the matrix $\mathbb{A}$ are related by $\alpha_{11} = -\alpha_{22}$
and $\alpha_{21} = -\alpha_{12}$, see (6.37), and the RTE (6.35) reduces to

$$\begin{pmatrix} \dfrac{dE_+}{d\tau} \\[2ex] \dfrac{dE_-}{d\tau} \end{pmatrix} = \begin{pmatrix} \alpha_1 & -\alpha_2 \\ \alpha_2 & -\alpha_1 \end{pmatrix} \begin{pmatrix} E_+ \\ E_- \end{pmatrix} + \begin{pmatrix} \alpha_3 \\ \alpha_4 \end{pmatrix} S \tag{6.63}$$

Here the abbreviations

$$\alpha_1 = \alpha_{11}, \qquad \alpha_2 = \alpha_{21}, \qquad \alpha_3 = -\omega_0 b(-\mu_0), \qquad \alpha_4 = \omega_0 \left[1 - b(-\mu_0) \right] \tag{6.64}$$

have been introduced while the solar radiation $S(\tau)$ is given by the solution of the
differential equation

$$\frac{dS}{d\tau} = -\frac{1}{\mu_0} S_0 \exp\left(-\frac{\tau}{\mu_0}\right) \implies S(\tau) = S_0 \exp\left(-\frac{\tau}{\mu_0}\right) \tag{6.65}$$

Equation (6.63) describes a set of coupled ordinary differential equations for E_+
and E_-. For a homogeneous layer $\Delta\tau_i = \tau_i - \tau_{i-1}$ the coefficients α_j, $j = 1, \ldots, 4$
of the system are constant so that it is possible to obtain analytical solutions. First
we solve the homogeneous system

$$\begin{pmatrix} \dfrac{dE_+}{d\tau} \\[2ex] \dfrac{dE_-}{d\tau} \end{pmatrix} = \begin{pmatrix} \alpha_1 & -\alpha_2 \\ \alpha_2 & -\alpha_1 \end{pmatrix} \begin{pmatrix} E_+ \\ E_- \end{pmatrix} \tag{6.66}$$

Inserting the trial solutions $E_+ = A_1 \exp(\tilde{\lambda}\tau)$, $E_- = A_2 \exp(\tilde{\lambda}\tau)$ into (6.66) yields
the linear equation

$$\begin{pmatrix} \alpha_1 - \tilde{\lambda} & -\alpha_2 \\ \alpha_2 & -\alpha_1 - \tilde{\lambda} \end{pmatrix} \begin{pmatrix} E_+ \\ E_- \end{pmatrix} = 0 \tag{6.67}$$

This equation can be fulfilled only if the determinant vanishes, i.e.

$$\begin{vmatrix} \alpha_1 - \tilde{\lambda} & -\alpha_2 \\ \alpha_2 & -\alpha_1 - \tilde{\lambda} \end{vmatrix} = -(\alpha_1 - \tilde{\lambda})(\alpha_1 + \tilde{\lambda}) + \alpha_2^2 = 0 \tag{6.68}$$

From (6.68) we obtain the two eigenvalues of the system

$$\tilde{\lambda}_{1,2} = \pm\lambda \quad \text{with} \quad \lambda = \sqrt{\alpha_1^2 - \alpha_2^2} \tag{6.69}$$

A solution of the homogeneous system can be obtained by inserting the expressions

$$\begin{aligned}\tilde{\lambda}_1 : \quad E_+ &= C_1 \exp(\lambda\tau), \qquad E_- = D_1 \exp(\lambda\tau) \\ \tilde{\lambda}_2 : \quad E_+ &= C_2 \exp(-\lambda\tau), \qquad E_- = D_2 \exp(-\lambda\tau)\end{aligned} \tag{6.70}$$

into (6.67) yielding

$$\begin{aligned}\tilde{\lambda}_1 : \quad (\alpha_1 - \lambda)C_1 - \alpha_2 D_1 &= 0, \qquad \alpha_2 C_1 - (\alpha_1 + \lambda)D_1 = 0 \\ \tilde{\lambda}_2 : \quad (\alpha_1 + \lambda)C_2 - \alpha_2 D_2 &= 0, \qquad \alpha_2 C_2 - (\alpha_1 - \lambda)D_2 = 0\end{aligned} \tag{6.71}$$

From these equations the ratio of the constants can be determined as

$$\begin{aligned}D_1 &= \frac{(\alpha_1 - \lambda)}{\alpha_2}C_1 = \frac{\alpha_2}{(\alpha_1 + \lambda)}C_1 \\ D_2 &= \frac{(\alpha_1 + \lambda)}{\alpha_2}C_2 = \frac{\alpha_2}{(\alpha_1 - \lambda)}C_2\end{aligned} \tag{6.72}$$

with

$$\frac{\alpha_2}{(\alpha_1 + \lambda)} = \frac{(\alpha_1 - \lambda)}{\alpha_2} \tag{6.73}$$

which is equivalent to $\lambda = \sqrt{\alpha_1^2 - \alpha_2^2}$. The general solution of the homogeneous system, including two constants of integration, is given by a superposition of the individual solutions and may be written as

$$\begin{aligned}E_{+,\,\mathrm{h}}(\tau) &= C_1 \exp(\lambda\tau) + C_2 \exp(-\lambda\tau) \\ E_{-,\,\mathrm{h}}(\tau) &= C_1 \frac{\alpha_2}{(\alpha_1 + \lambda)} \exp(\lambda\tau) + C_2 \frac{\alpha_2}{(\alpha_1 - \lambda)} \exp(-\lambda\tau)\end{aligned} \tag{6.74}$$

In the next step we determine a particular solution of the inhomogeneous system. This can be most easily obtained by using a trial solution having the functional form of the inhomogeneous term of the differential equation (6.63), that is

$$E_{+,\,\mathrm{p}}(\tau) = \alpha_5 S_0 \exp\left(-\frac{\tau}{\mu_0}\right), \qquad E_{-,\,\mathrm{p}}(\tau) = \alpha_6 S_0 \exp\left(-\frac{\tau}{\mu_0}\right) \tag{6.75}$$

Inserting these trial solutions into the inhomogeneous system (6.63) leads to another inhomogeneous system of linear equations for α_5 and α_6

$$\begin{pmatrix} \dfrac{1}{\mu_0} + \alpha_1 & -\alpha_2 \\[2mm] \alpha_2 & \dfrac{1}{\mu_0} - \alpha_1 \end{pmatrix} \begin{pmatrix} \alpha_5 \\ \alpha_6 \end{pmatrix} + \begin{pmatrix} \alpha_3 \\ \alpha_4 \end{pmatrix} = 0 \tag{6.76}$$

with the solutions

$$\alpha_5 = \frac{\left(\alpha_1 - \frac{1}{\mu_0}\right)\alpha_3 - \alpha_2\alpha_4}{\left(\frac{1}{\mu_0}\right)^2 - \lambda^2}, \qquad \alpha_6 = \frac{\alpha_2\alpha_3 - \left(\alpha_1 + \frac{1}{\mu_0}\right)\alpha_4}{\left(\frac{1}{\mu_0}\right)^2 - \lambda^2} \tag{6.77}$$

Inspection of these equations reveals that for $\mu_0 = 1/\lambda$ the constants α_5 and α_6 become infinitely large. However, this so-called *resonance case* may be easily avoided by adding to the actual solar position μ_0 a small increment $\pm\Delta\mu_0$.

In the third step, the complete solution of the inhomogeneous system of linear differential equations (6.63) is obtained by adding the homogeneous and the particular solutions (6.74) and (6.75). Hence we obtain

$$\boxed{\begin{aligned} E_+(\tau) &= C_1\exp(\lambda\tau) + C_2\exp(-\lambda\tau) + \alpha_5 S_0\exp\left(-\frac{\tau}{\mu_0}\right) \\ E_-(\tau) &= C_1\frac{\alpha_2}{(\alpha_1 + \lambda)}\exp(\lambda\tau) + C_2\frac{\alpha_2}{(\alpha_1 - \lambda)}\exp(-\lambda\tau) + \alpha_6 S_0\exp\left(-\frac{\tau}{\mu_0}\right) \end{aligned}}$$

$$\tag{6.78}$$

Our final goal is to obtain expressions for the flux densities $E_+(\tau_{i-1})$ and $E_-(\tau_i)$ leaving the homogeneous layer $\Delta\tau_i$ in terms of the flux densities $E_-(\tau_{i-1})$ and $E_+(\tau_i)$ entering this layer. With the help of (6.78), the incident flux densities at the boundaries of the layer can be written down immediately

$$E_+(\tau_i) = C_1\exp(\lambda\tau_i) + C_2\exp(-\lambda\tau_i) + \alpha_5 S(\tau_{i-1})\exp\left(-\frac{\Delta\tau_i}{\mu_0}\right)$$

$$E_-(\tau_{i-1}) = C_1\frac{\alpha_2}{(\alpha_1 + \lambda)}\exp(\lambda\tau_{i-1}) + C_2\frac{\alpha_2}{(\alpha_1 - \lambda)}\exp(-\lambda\tau_{i-1}) + \alpha_6 S(\tau_{i-1})$$

$$\tag{6.79}$$

where $S(\tau_{i-1})$ is the solar radiative flux density incident at the top of the layer, that is

$$S(\tau_{i-1}) = S_0\exp\left(-\frac{\tau_{i-1}}{\mu_0}\right) \tag{6.80}$$

Equations (6.79) may be solved to determine the integration constants C_1 and C_2. The result is

$$\begin{aligned} C_1 &= \beta_{11}E_+(\tau_i) + \beta_{12}E_-(\tau_{i-1}) + \beta_{13}S(\tau_{i-1}) \\ C_2 &= \beta_{21}E_+(\tau_i) + \beta_{22}E_-(\tau_{i-1}) + \beta_{23}S(\tau_{i-1}) \end{aligned} \tag{6.81}$$

where the following constants have been introduced

$$\beta_{11} = A \frac{\alpha_2}{\alpha_1 - \lambda} \exp(-\lambda \tau_{i-1}), \qquad\qquad \beta_{12} = -A \exp(-\lambda \tau_i)$$

$$\beta_{21} = -A \frac{\alpha_2}{\alpha_1 + \lambda} \exp(\lambda \tau_{i-1}), \qquad\qquad \beta_{22} = A \exp(\lambda \tau_i)$$

$$\beta_{13} = -\beta_{11} \alpha_5 \exp\left(-\frac{\Delta \tau_i}{\mu_0}\right) - \beta_{12} \alpha_6, \qquad \beta_{23} = -\beta_{21} \alpha_5 \exp\left(-\frac{\Delta \tau_i}{\mu_0}\right) - \beta_{22} \alpha_6$$

$$A = \left[\frac{\alpha_2}{\alpha_1 - \lambda} \exp(\lambda \Delta \tau_i) - \frac{\alpha_2}{\alpha_1 + \lambda} \exp(-\lambda \Delta \tau_i)\right]^{-1}$$

$$(6.82)$$

According to (6.78) the radiative flux densities leaving the layer $\Delta \tau_i$ are given by

$$E_+(\tau_{i-1}) = C_1 \gamma_{11} + C_2 \gamma_{21} + \alpha_5 S(\tau_{i-1})$$

$$E_-(\tau_i) = C_1 \gamma_{12} + C_2 \gamma_{22} + \alpha_6 S(\tau_{i-1}) \exp\left(-\frac{\Delta \tau_i}{\mu_0}\right) \qquad (6.83)$$

with

$$\gamma_{11} = \exp(\lambda \tau_{i-1}), \qquad\qquad \gamma_{21} = \exp(-\lambda \tau_{i-1})$$

$$\gamma_{12} = \frac{\alpha_2}{(\alpha_1 + \lambda)} \exp(\lambda \tau_i), \qquad \gamma_{22} = \frac{\alpha_2}{(\alpha_1 - \lambda)} \exp(-\lambda \tau_i) \qquad (6.84)$$

From (6.65) and (6.80) we obtain for the solar radiation flux density at τ_i

$$S(\tau_i) = S(\tau_{i-1}) \exp\left(-\frac{\Delta \tau_i}{\mu_0}\right) \qquad (6.85)$$

Substituting (6.81) into (6.83) and combining the result with (6.85) we obtain the compact matrix notation

$$\begin{pmatrix} E_+(\tau_{i-1}) \\ E_-(\tau_i) \\ S(\tau_i) \end{pmatrix} = \begin{pmatrix} a_{11} & a_{12} & a_{13} \\ a_{21} & a_{22} & a_{23} \\ a_{31} & a_{32} & a_{33} \end{pmatrix} \begin{pmatrix} E_+(\tau_i) \\ E_-(\tau_{i-1}) \\ S(\tau_{i-1}) \end{pmatrix} \qquad (6.86)$$

with

$$a_{11} = \beta_{11} \gamma_{11} + \beta_{21} \gamma_{21}, \quad a_{12} = \beta_{12} \gamma_{11} + \beta_{22} \gamma_{21}, \quad a_{13} = \beta_{13} \gamma_{11} + \beta_{23} \gamma_{21} + \alpha_5$$

$$a_{21} = \beta_{11} \gamma_{12} + \beta_{21} \gamma_{22}, \quad a_{22} = \beta_{12} \gamma_{12} + \beta_{22} \gamma_{22}, \quad a_{23} = \beta_{13} \gamma_{12} + \beta_{23} \gamma_{22} + \alpha_6 a_{33}$$

$$a_{31} = 0, \qquad\qquad a_{32} = 0, \qquad\qquad a_{33} = \exp\left(-\frac{\Delta \tau_i}{\mu_0}\right)$$

$$(6.87)$$

From (6.82), (6.84), and (6.87) it may be easily seen that

$$a_{11} = a_{22} = A\frac{2\lambda}{\alpha_2^2}, \qquad a_{12} = a_{21} = A[\exp(\lambda\Delta\tau_i) - \exp(-\lambda\Delta\tau_i)] \qquad (6.88)$$

Thus we finally obtain the analytical solution to the TSM as

$$\begin{pmatrix} E_+(\tau_{i-1}) \\ E_-(\tau_i) \\ S(\tau_i) \end{pmatrix} = \begin{pmatrix} a_{11} & a_{12} & a_{13} \\ a_{12} & a_{11} & a_{23} \\ 0 & 0 & a_{33} \end{pmatrix} \begin{pmatrix} E_+(\tau_i) \\ E_-(\tau_{i-1}) \\ S(\tau_{i-1}) \end{pmatrix} \qquad (6.89)$$

The a_{jk} occurring in this equation are functions of the optical properties of the layer $\Delta\tau_i$. They have the following physical meaning:

a_{11}: *transmission coefficient for diffuse radiation,*
a_{12}: *reflection coefficient for diffuse radiation,*
a_{13}: *reflection coefficient for the primary scattered parallel solar radiation,*
a_{23}: *transmission coefficient for the primary scattered parallel solar radiation,*
a_{33}: *transmission coefficient for the direct parallel solar radiation.*

6.5 Approximate treatment of scattering in the infrared spectral region

It is well-known that scattering of solar radiation by cloud droplets and aerosol particles is a dominant factor affecting the Earth's planetary albedo and thus the global climate. In the previous chapters we have discussed various calculation methods to determine solar radiances. To obtain the flux densities from the radiances is a relatively simple calculational procedure. A review of the literature shows that the scattering of thermal radiation by clouds and aerosol layers is often neglected in weather and climate models for two main reasons: (i) with the exception of the *atmospheric window region* ranging from about 8–12.5 μm, long-wave radiative transfer is dominated by absorption and emission processes due to atmospheric water vapor and some other trace gases and by water droplets and ice particles; and (ii) multiple scattering calculations require a large amount of computer time in comparison to situations where the scattering part of the *source function* in the RTE can be neglected.

In the following we will show how to include multiple scattering in the infrared window region in addition to absorption and emission by modifying the two stream equation (6.63). The same method, if desired, can also be applied to any other part of the long-wave spectrum. Since the solar and the infrared spectrum can be separated, say at 4 μm, all we need to do is to replace the solar scattering term in equation (6.63) by the *Planckian emission* $B(\tau)$. Analogously to (6.63) we obtain for the infrared spectral region with multiple scattering

$$
\begin{pmatrix} \dfrac{dE_+}{d\tau} \\[2mm] \dfrac{dE_-}{d\tau} \end{pmatrix} = \begin{pmatrix} \alpha_1 & -\alpha_2 \\ \alpha_2 & -\alpha_1 \end{pmatrix} \begin{pmatrix} E_+ \\ E_- \end{pmatrix} + \alpha_3 \begin{pmatrix} -B(\tau) \\ B(\tau) \end{pmatrix} \tag{6.90}
$$

with

$$
\alpha_1 = \alpha_{11}, \qquad \alpha_2 = \alpha_{21}, \qquad \alpha_3 = \frac{\pi(1-\omega_0)}{\bar{\mu}} \tag{6.91}
$$

The terms α_{11}, α_{21} are given by (6.37) where again $b_+ = b_-$ and $\mu_+ = \mu_- = \bar{\mu}$. As before, we suppress any reference to wave number or wavelength. In order to handle the emission term in an effective way we make an Eddington type approximation for the thermal radiation, cf. (6.42)

$$
B(\tau) = B_0 + B_1\tau \tag{6.92}
$$

where B_0 and B_1 are constants.

For a homogeneous layer the complete mathematical solution to the system (6.90) is found in the same way as in the short-wave radiation case. The solution of the homogeneous part of this differential equation is given by (6.74). To determine a particular solution of the inhomogeneous system, instead of (6.75), we now substitute a trial solution of the type of the inhomogeneous term (6.92) into (6.90), i.e.

$$
E_{+,\mathrm{p}}(\tau) = \alpha_4 + \alpha_5\tau, \qquad E_{-,\mathrm{p}}(\tau) = \alpha_6 + \alpha_7\tau \tag{6.93}
$$

By comparing in the resulting equations coefficients of zero-th and first order in τ we obtain four linear equations for the unknown constants α_i, $i = 4, \ldots, 7$ which may be easily solved yielding

$$
\alpha_4 = \gamma_1 B_0 + \gamma_2 B_1, \qquad \alpha_6 = \gamma_1 B_0 - \gamma_2 B_1, \qquad \alpha_5 = \alpha_7 = \gamma_1 B_1
$$

$$
\text{with} \quad \gamma_1 = \frac{\alpha_3}{\alpha_1 - \alpha_2}, \qquad \gamma_2 = \frac{\alpha_3}{\alpha_1^2 - \alpha_2^2} = \frac{\alpha_3}{\lambda^2} \tag{6.94}
$$

Hence, the general solution of (6.90) may be written as

$$
\boxed{\begin{aligned}
E_+(\tau) &= C_1 \exp(\lambda\tau) + C_2 \exp(-\lambda\tau) + \gamma_1(B_0 + B_1\tau) + \gamma_2 B_1 \\
E_-(\tau) &= C_1 \frac{\alpha_2}{(\alpha_1 + \lambda)} \exp(\lambda\tau) + C_2 \frac{\alpha_2}{(\alpha_1 - \lambda)} \exp(-\lambda\tau) + \gamma_1(B_0 + B_1\tau) - \gamma_2 B_1
\end{aligned}}
$$

$$\tag{6.95}$$

The determination of the integration constants C_1 and C_2 is performed in the same way as in the short-wave situation. First we use (6.95) to calculate the radiative flux densities entering the homogeneous layer $\Delta\tau_i = \tau_i - \tau_{i-1}$. Introducing the abbreviations

$$\tilde{E}_+(\tau) = E_+(\tau) - \gamma_1(B_0 + B_1\tau) - \gamma_2 B_1$$
$$\tilde{E}_-(\tau) = E_-(\tau) - \gamma_1(B_0 + B_1\tau) + \gamma_2 B_1$$

(6.96)

we obtain

$$\tilde{E}_+(\tau_i) = C_1 \exp(\lambda\tau_i) + C_2 \exp(-\lambda\tau_i)$$
$$\tilde{E}_-(\tau_{i-1}) = C_1 \frac{\alpha_2}{(\alpha_1 + \lambda)} \exp(\lambda\tau_{i-1}) + C_2 \frac{\alpha_2}{(\alpha_1 - \lambda)} \exp(-\lambda\tau_{i-1})$$

(6.97)

By comparing (6.97) with (6.79) we see that formally both systems differ only in the solar radiation term which is missing in (6.97). Hence, it is an easy task to find the integration constants by dropping the solar radiation term in (6.81)

$$C_1 = \beta_{11}\tilde{E}_+(\tau_i) + \beta_{12}\tilde{E}_-(\tau_{i-1}), \qquad C_2 = \beta_{21}\tilde{E}_+(\tau_i) + \beta_{22}\tilde{E}_-(\tau_{i-1})$$

(6.98)

The constants β_{ij} follow from (6.82).

Utilizing (6.84) in (6.97) the radiative fluxes leaving the layer $\Delta\tau_i$ may be formulated as

$$\tilde{E}_+(\tau_{i-1}) = C_1\gamma_{11} + C_2\gamma_{21}, \qquad \tilde{E}_-(\tau_i) = C_1\gamma_{12} + C_2\gamma_{22}$$

(6.99)

Substitution of (6.98) into these equations results in

$$\begin{pmatrix} \tilde{E}_+(\tau_{i-1}) \\ \tilde{E}_-(\tau_i) \end{pmatrix} = \begin{pmatrix} a_{11} & a_{12} \\ a_{21} & a_{22} \end{pmatrix} \begin{pmatrix} \tilde{E}_+(\tau_i) \\ \tilde{E}_-(\tau_{i-1}) \end{pmatrix}$$

(6.100)

where the a_{ij} are given by (6.87).

Finally, we need to set up proper boundary conditions at the top of the atmosphere ($\tau = 0$) and at the Earth's surface ($\tau = \tau_g$). Denoting the albedo of the ground by A_g we obtain from (6.96)

$$\tilde{E}_+(\tau_g) = A_g E_-(\tau_g) + \pi B(T_g)(1 - A_g) - \gamma_1(B_0 + B_1\tau_g) - \gamma_2 B_1$$
$$\tilde{E}_-(0) = E_-(0) - \gamma_1 B_0 + \gamma_2 B_1$$

(6.101)

Usually we set $E_-(0) = 0$.

As in previous cases, the atmosphere will be subdivided into numerous homogeneous layers. The vertical variation of temperature is accounted for by permitting the *Planck function* to vary according to the linearization (6.92). By using a sufficiently large number N of homogeneous sublayers, we can represent with high accuracy the non-isothermal structure of the atmosphere. The upward and downward directed flux densities of each layer are obtained by constructing from (6.100) a $(2N \times 2N)$-tridiagonal equation system. This system may be solved by means of standard numerical procedures. Finally, the results for $\tilde{E}_\pm(\tau_i)$ are used in (6.96) yielding $E_\pm(\tau_i)$.

6.6 Approximations for partial cloud cover

The radiative transfer models described in the previous sections are only suitable for horizontally homogeneous layers. In numerical weather prediction models dealing with partial cloudiness within a numerical grid box, particular problems arise since in the layers which are partially filled with clouds the assumption of horizontal homogeneity is void. In addition to the horizontal inhomogeneity within a grid box, the radiation scheme has to account for the vertical distribution of fractional cloud cover within a numerical grid column.

In order to treat the situation with partial cloudiness more realistically, assumptions about the vertical distribution of the partial cloud cover have to be introduced. In combination with two-stream radiative transfer two particular approximations are widely employed to treat the overlap of contiguous cloud layers. These are the concepts of *random overlap* and *maximum overlap*. For random overlap the combined partial cloud cover of both layers is obtained by multiplying the cloud covers of each individual layer. Apparently, this concept was first used in the radiative transfer model by Manabe and Strickler (1964). The maximum overlap assumption means that the combined partial cloud cover of two vertically adjacent cloud layers is arranged in such a way that the cloudy portions of both layers overlap maximally. This scheme has first been employed in the two-stream flux transfer by Geleyn and Hollingsworth (1979). In the following we will illustrate both cases in detail.

6.6.1 Partial cloud cover with random overlap

In the random overlap concept it is assumed that the clouds of contiguous layers overlap in a random way. If in a real situation cloudy layers are separated by cloud-free regions then it seems physical to postulate that these layers are statistically independent. For simplicity let us first consider only the transmission of the downward directed diffuse radiation.

Figure 6.3 illustrates the random overlap assumption. At the bottom of the partially cloudy layer $i - 1$ the two radiative flux densities $E^c_-(\tau_{i-1})$ and $E^f_-(\tau_{i-1})$ emanate, whereby the superscripts c and f denote the cloudy and cloud-free regions, respectively. These two fluxes are added yielding the single flux $E_-(\tau_{i-1}) = E^f_-(\tau_{i-1}) + E^c_-(\tau_{i-1})$. If C_i is the cloud cover of layer i then it is assumed that the fraction $C_i E_-(\tau_{i-1})$ enters the cloudy portion of this layer, whereas the remaining fraction $(1 - C_i)E_-(\tau_{i-1})$ propagates through the clear sky portion. For the cloudy and cloud-free parts of the downward radiation at τ_i we obtain

$$E_-(\tau_i)^f = a^f_{11}(1 - C_i)E_-(\tau_{i-1}), \qquad E_-(\tau_i)^c = a^c_{11}C_i E_-(\tau_{i-1}) \qquad (6.102)$$

Here, a^f_{11} and a^c_{11} are, respectively, the transmission coefficients for diffuse radiation of the cloud-free and the cloudy part of layer i. The two relations (6.102) may be

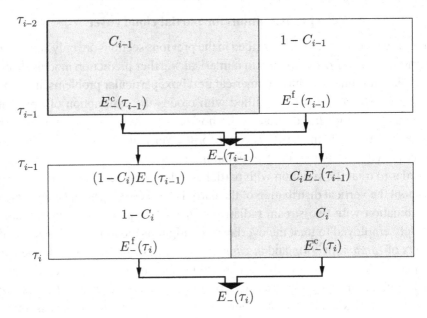

Fig. 6.3 Transfer of the downwelling diffuse radiation between two individual sublayers assuming random overlap of the partial cloud fractions C_{i-1} and C_i.

combined to a single equation for the total downward radiative flux density

$$E_-(\tau_i) = E_-^f(\tau_i) + E_-^c(\tau_i) = \left[a_{11}^f(1 - C_i) + a_{11}^c C_i\right] E_-(\tau_{i-1}) \qquad (6.103)$$

This means that in the random overlap approximation we have to replace a_{11} in (6.89) by

$$a_{11} = (1 - C_i)a_{11}^f + C_i a_{11}^c \qquad (6.104)$$

The transfer of the diffuse upward flux densities and the solar radiation can be treated in exactly the same manner so that similar relations apply to the other a_{jk} coefficients of (6.89). In summary, we obtain

$$a_{jk} = (1 - C_i)a_{jk}^f + C_i a_{jk}^c, \qquad j, k = 1, 2, 3 \qquad (6.105)$$

Consider, for example, the direct solar radiation through two adjacent cloud layers of cloudiness C_{i-1} and C_i. The emerging directly transmitted solar radiation at the base of layer i is proportional to the product $C_{i-1}C_i$ according to the rule that statistically independent probabilities multiply. This explains the name 'random'.

6.6.2 Partial cloud cover with maximum overlap

The assumption of maximum overlap of vertically adjacent clouds is reasonably well justified for a situation for which the clouds in the various layers are formed

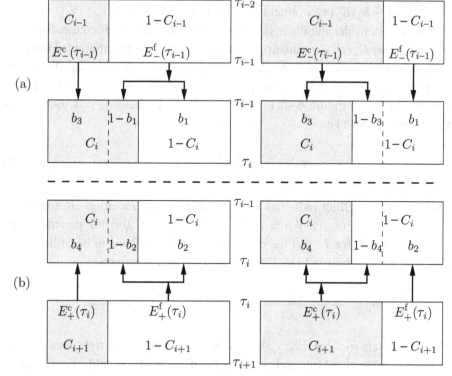

Fig. 6.4 Downward (a) and upward (b) transmission of diffuse radiation through two adjacent layers with different partial cloud cover for the maximum overlap concept.

by the same physical process. Again we first consider only the transmission of downward directed diffuse radiation in two contiguous sublayers $i-1$ and i with cloud covers C_{i-1} and C_i. The situation is illustrated in Figure 6.4(a). Two different cases must be treated as shown in the left and right panel of this figure. In one case (left panel) the lower layer has the larger cloud cover, that is $C_i > C_{i-1}$, whereas in the other case (right panel) the situation is reversed.

From Figure 6.4(a) it is seen that the transfer of the downward radiation can be formulated as follows. The cloudy part of layer $i-1$ transmits $E_-^c(\tau_{i-1})$ whereas the cloudless part transmits $E_-^f(\tau_{i-1})$. Both parts of the transmitted radiation must now be distributed between the cloudy and cloudless parts of the lower layer i. Let the parameter b_3 refer to that fraction of $E_-^c(\tau_{i-1})$ leaving the base of layer $i-1$ and entering the cloudy part of layer i. This fraction is equal to 1 if $C_i \geq C_{i-1}$ (left panel of Figure 6.4(a)) and equal to C_i/C_{i-1} if $C_i < C_{i-1}$ (right panel of the figure). Both conditions can be combined yielding

$$b_3 = \frac{\min(C_i, C_{i-1})}{C_{i-1}} \tag{6.106}$$

Inspection of the right panel of Figure 6.4(a) shows that in this particular situation the fraction $(1 - b_3)E_-^c(\tau_{i-1})$ enters the cloud free part of layer i.

Let us now consider the clear sky part in the left panel of Figure 6.4(a). The diffuse radiation $E_-^f(\tau_{i-1})$ emanating from layer $i - 1$ will be subdivided between the cloudy part $C_i - C_{i-1}$ and the clear sky region $1 - C_i$. This is done by means of the parameter b_1 which in the left panel is given by $b_1 = (1 - C_i)/(1 - C_{i-1})$. In the right panel of Figure 6.4(a) where $C_{i-1} > C_i$ we have $b_1 = 1$. Both cases can again be combined as

$$b_1 = \frac{1 - \max(C_i, C_{i-1})}{1 - C_{i-1}} \tag{6.107}$$

The diffuse upwelling radiation $E_+(\tau_i)$ and the solar radiation are treated in the same way as $E_-(\tau_{i-1})$. While for the solar radiation also the parameters b_1 and b_3 are applied, for $E_+(\tau_i)$ the coefficients b_2 and b_4 have to be utilized, see Figure 6.4(b). Analogously to (6.106) and (6.107) they are given as

$$b_2 = \frac{1 - \max(C_i, C_{i+1})}{1 - C_{i+1}}, \qquad b_4 = \frac{\min(C_i, C_{i+1})}{C_{i+1}} \tag{6.108}$$

The parameters b_j, $j = 1, \ldots, 4$ are now introduced on the right-hand side of (6.89) to obtain the cloud free and cloudy parts of the radiation field.[3] For the cloud free part we have

$$\begin{pmatrix} E_+^f(\tau_{i-1}) \\ E_-^f(\tau_i) \\ S^f(\tau_i) \end{pmatrix} = \begin{pmatrix} a_{11}^f & a_{12}^f & a_{13}^f \\ a_{12}^f & a_{11}^f & a_{23}^f \\ 0 & 0 & a_{33}^f \end{pmatrix} \begin{pmatrix} b_2 E_+^f(\tau_i) + (1 - b_4)E_+^c(\tau_i) \\ b_1 E_-^f(\tau_{i-1}) + (1 - b_3)E_-^c(\tau_{i-1}) \\ b_1 S^f(\tau_{i-1}) + (1 - b_3)S^c(\tau_{i-1}) \end{pmatrix}$$

$$\tag{6.109}$$

while the cloudy parts of the radiative fluxes are written as

$$\begin{pmatrix} E_+^c(\tau_{i-1}) \\ E_-^c(\tau_i) \\ S^c(\tau_i) \end{pmatrix} = \begin{pmatrix} a_{11}^c & a_{12}^c & a_{13}^c \\ a_{12}^c & a_{11}^c & a_{23}^c \\ 0 & 0 & a_{33}^c \end{pmatrix} \begin{pmatrix} (1 - b_2)E_+^f(\tau_i) + b_4 E_+^c(\tau_i) \\ (1 - b_1)E_-^f(\tau_{i-1}) + b_3 E_-^c(\tau_{i-1}) \\ (1 - b_1)S^f(\tau_{i-1}) + b_3 S^c(\tau_{i-1}) \end{pmatrix}$$

$$\tag{6.110}$$

---

[3] It is noteworthy that a particular coefficient b_j is set equal to 1 if an undetermined expression 0/0 occurs. This follows from physical reasoning or from applying l'Hôpital's rule.

These two systems of linear equations determine the flux densities $E_+^f(\tau_{i-1})$, $E_+^c(\tau_{i-1})$, $E_-^f(\tau_i)$, $E_-^c(\tau_i)$, $S^f(\tau_i)$, and $S^c(\tau_i)$ that emerge from the homogeneous layer $\Delta\tau_i$ as a function of the corresponding incoming flux densities $E_+^f(\tau_i)$, $E_+^c(\tau_i)$, $E_-^f(\tau_{i-1})$, $E_-^c(\tau_{i-1})$, $S^f(\tau_{i-1})$, and $S^c(\tau_{i-1})$. At an arbitrary level i the physically relevant flux densities are the sum of the individual cloudy and cloud free contributions

$$\boxed{\begin{aligned} E_+(\tau_i) &= E_+^f(\tau_i) + E_+^c(\tau_i) \\ E_-(\tau_i) &= E_-^f(\tau_i) + E_-^c(\tau_i) \\ S(\tau_i) &= S^f(\tau_i) + S^c(\tau_i) \end{aligned}} \qquad (6.111)$$

A comment is due regarding the ground reflection which, for simplicity, is assumed to be isotropic, i.e. the ground is a Lambertian reflector. In this case the diffuse upwelling flux densities are given by

$$E_+^f(\tau_N) = A_g\left[E_-^f(\tau_N) + S^f(\tau_N)\right], \qquad E_+^c(\tau_N) = A_g[E_-^c(\tau_N) + S^c(\tau_N)]$$

$$(6.112)$$

Such a situation may arise if ground fog occurs. Note that the downward flux densities leaving the lowest layer from which they emerged, after reflection enter the same section of cloudy and cloud-free air.

Due to the fact that the systems (6.109) and (6.110) are coupled, for an inhomogeneous atmosphere with N homogeneous sublayers we have to solve a $6N$-dimensional linear matrix system. This means that in case of partially cloudy layers the maximum overlap treatment costs about twice the computational time as the original TSM which assumes 100% cloud cover in each cloudy layer. In contrast to this, the numerical effort of the random overlap approach is almost the same as in the original TSM since in both cases the same number of $3N$ coupled equations has to be solved.

6.7 The classical emissivity approximation

In Section 2.6.2 we have shown that in case of a purely absorbing atmosphere it is possible to obtain an analytical solution to the RTE, see (2.123). We have already stated that in the infrared spectral region, apart from the atmospheric window, scattering processes may savely be neglected in the calculation of atmospheric radiative transfer in many cases. However, the analytic solution is still relatively elaborate since it consists of integrals of the *transmission function* over the absorbing mass. Furthermore, it is important to note that (2.123) is a spectral equation.[4] This

---

[4] Recall that in all equations of Chapter 2 for simplicity the index ν has been omitted.

means that we need to perform another integration over the entire infrared spectral region in order to obtain the total infrared radiative heating rate.

In this section we will present the *classical emissivity method* which is one of the fastest ways to find simple approximate values of the infrared radiative flux densities in a nonscattering atmosphere. The major simplification of the emissivity method consists in the decoupling of the frequency integration from the integration over the absorbing mass. To derive the classical flux-emissivity equations, we assume that the temperature and pressure dependency of the *mass absorption coefficient* $\kappa_{\mathrm{abs},\nu}(p, T)$ can be expressed as

$$\kappa_{\mathrm{abs},\nu}(p, T) = \kappa_{\mathrm{abs},\nu}(p_0, T_0)\frac{p}{p_0}\sqrt{\frac{T_0}{T}} \qquad (6.113)$$

The suffix 0 refers to standard conditions. In the following chapter we will present a detailed discussion of different forms of the absorption coefficient. There it will be shown that the so-called *Lorentz absorption coefficient* closely resembles the form (6.113).

Utilizing (6.113) the argument of the *flux-transmission function* $\mathcal{T}_{\mathrm{f},\nu}$, as given by (2.142), may be written as

$$\int_u^{u'} \kappa_{\mathrm{abs},\nu}(t)dt = \kappa_{\mathrm{abs},\nu}(p_0, T_0)\int_u^{u'} \frac{p}{p_0}\sqrt{\frac{T_0}{T}}dt = \kappa_{\mathrm{abs},\nu}(p_0, T_0)(w' - w) \quad (6.114)$$

The integral $\int_u^{u'} p/p_0\sqrt{T_0/T}\,dt$ which is symbolically written as $w' - w$ is known as the *reduced absorber mass*. This is equivalent to the definition

$$dw = \frac{p}{p_0}\sqrt{\frac{T_0}{T}}du = \frac{p}{p_0}\sqrt{\frac{T_0}{T}}\rho_{\mathrm{abs}}ds = -\frac{p}{p_0}\sqrt{\frac{T_0}{T}}\rho_{\mathrm{abs}}dz \qquad (6.115)$$

see also (2.118). With the help of (6.114) the flux-transmission function can be approximated as

$$\mathcal{T}_{\mathrm{f},\nu}(u, u') = 2E_3\left(\int_u^{u'} \kappa_{\mathrm{abs},\nu}(t)dt\right) \approx 2E_3[\kappa_{\mathrm{abs},\nu}(p_0, T_0)(w' - w)] = \mathcal{T}_{\mathrm{f},\nu}(w, w')$$

$$(6.116)$$

where $w' \geq w$. Since $E_3(0) = 1/2$ we obtain the physically correct transmission function $\mathcal{T}_{\mathrm{f}}(w, w') = 1$ if no absorption takes place.

Integration over ν yields for the upward and downward directed flux densities as given by (2.143)

(a) $E_+(w) = \pi \displaystyle\int_0^\infty [B_{v,\text{g}} - B_v(w_\text{g})]\mathcal{T}_{\text{f},v}(w, w_\text{g})dv + \pi \int_0^\infty B_v(w)dv$

$\quad + \pi \int_0^\infty \int_w^{w_\text{g}} \frac{dB_v}{dw'}\mathcal{T}_{\text{f},v}(w, w')dw'dv$

(6.117)

(b) $E_-(w) = -\pi \displaystyle\int_0^\infty B_v(w' = 0)\mathcal{T}_{\text{f},v}(0, w)dv + \pi \int_0^\infty B_v(w)dv$

$\quad - \pi \int_0^\infty \int_0^w \frac{dB_v}{dw'}\mathcal{T}_{\text{f},v}(w', w)dw'dv$

where in (6.117b) it has been assumed that $I_{-,v}(0) = 0$.

Let us consider the simple situation of an isothermal atmosphere $T = T_0$. In this case B_v is a constant so that $dB_v/dw = 0$. Furthermore, we assume that $B_{v,\text{g}} = B_v(w_\text{g})$. Then the equations for the upwelling and downwelling flux densities reduce to

$$E_+(w) = \pi \int_0^\infty B_v(T_0)dv = \sigma T_0^4$$

$$E_-(w) = \pi \int_0^\infty B_v(T_0)[1 - \mathcal{T}_{\text{f},v}(0, w)]dv$$

(6.118)

This means that in an isothermal medium of temperature $T = T_0$ the upwelling flux density is independent of altitude and is identical with the emission of a *black body* having this temperature. The situation is different for the downwelling flux density. At $w = 0$ there is no downwelling radiation. However, if $w \to \infty$ the flux transmission vanishes so that the downward flux density is identical with σT_0^4.

In order to decouple the frequency integration from the integration over the absorbing mass, we introduce the following mean transmissivities

$$\bar{\mathcal{T}}_\text{f}(T, w, w') = \frac{\pi \displaystyle\int_0^\infty B_v(T)\mathcal{T}_{\text{f},v}(w, w')dv}{\pi \displaystyle\int_0^\infty B_v(T)dv}$$

$$\hat{\mathcal{T}}_\text{f}(T, w, w') = \frac{\pi \displaystyle\int_0^\infty \frac{dB_v}{dT}\mathcal{T}_{\text{f},v}(w, w')dv}{\pi \displaystyle\int_0^\infty \frac{dB_v}{dT}dv}$$

(6.119)

with

$$\pi \int_0^\infty B_v(T)dv = \pi B(T) = \sigma T^4, \quad \pi \int_0^\infty \frac{dB_v}{dT}dv = \pi \frac{d}{dT}\left(\int_0^\infty B_v dv\right) = \pi \frac{dB}{dT}$$

(6.120)

Furthermore, we introduce the so-called *flux-emissivity function* $\varepsilon_f(T, w, w')$ by means of

$$\boxed{\varepsilon_f(T, w, w') = 1 - \bar{T}_f(T, w, w')}$$

(6.121)

where $w' \geq w$. Utilizing (6.119) and (6.120) the last integral occurring in (6.117a) may be reformulated as

$$
\begin{aligned}
\int_0^\infty \int_w^{w_g} \frac{dB_\nu}{dw'} T_{f,\nu}(w, w') dw' d\nu &= \int_w^{w_g} \frac{dT}{dw'} \int_0^\infty \frac{dB_\nu}{dT} T_{f,\nu}(w, w') d\nu dw' \\
&= \int_w^{w_g} \frac{dT}{dw'} \frac{dB}{dT} \hat{T}_f(T, w, w') dw' \\
&= \int_w^{w_g} \frac{dB}{dw'} \hat{T}_f(T, w, w') dw'
\end{aligned}
$$

(6.122)

An analogous result will be obtained for the last integral in (6.117b).

From (6.119) it is seen that, in contrast to $T_{f,\nu}(w, w')$, the mean transmissivities $\bar{T}_f(T, w, w')$ and $\hat{T}_f(T, w, w')$ explicitly depend on temperature as caused by the temperature-dependent weighting functions $B_\nu(T)$ and dB_ν/dT. This leads to the next assumption of the classical emissivity method which neglects this temperature dependence. Thus we use the following approximations

$$T_f(w, w') \approx \bar{T}_f(T_m, w, w') \approx \hat{T}_f(T_m, w, w'), \qquad \varepsilon_f(w, w') \approx \varepsilon_f(T_m, w, w')$$

(6.123)

where a suitable mean temperature T_m is used for the evaluation of the Planck function and $w' \geq w$.

Introducing the above approximations into (6.117) yields

$$
\begin{aligned}
E_+(w) &= \pi [B_g - B(w_g)] T_f(w, w_g) + \pi B(w) + \pi \int_w^{w_g} \frac{dB}{dw'} T_f(w, w') dw' \\
E_-(w) &= -\pi B(w' = 0) T_f(0, w) + \pi B(w) - \pi \int_0^w \frac{dB}{dw'} T_f(w', w) dw'
\end{aligned}
$$

(6.124)

Partial integration finally gives the so-called *flux-emissivity equations*

$$\boxed{\begin{aligned}
E_+(w) &= \pi B_g \left[1 - \varepsilon_f(w, w_g)\right] + \pi \int_w^{w_g} B(w') \frac{\partial \varepsilon_f(w, w')}{\partial w'} dw' \\
E_-(w) &= -\pi \int_0^w B(w') \frac{\partial \varepsilon_f(w', w)}{\partial w'} dw'
\end{aligned}}$$

(6.125)

Equation (6.125) may be used to calculate the infrared radiative heating rates at the reference level z_R according to

$$
\left.\frac{\partial T}{\partial t}\right|_{\text{rad}} = -\frac{1}{c_p \rho_R}\left(\frac{dE_+}{dz} - \frac{dE_-}{dz}\right)\Bigg|_{z=z_R} = -\frac{1}{c_p \rho_R}\left(\frac{dE_+}{dw} - \frac{dE_-}{dw}\right)\left.\frac{dw}{dz}\right|_{z=z_R}
$$

$$
= \frac{\rho_{\text{abs,R}}\, p_R}{c_p \rho_R\, p_0}\sqrt{\frac{T_0}{T_R}}\left(\frac{dE_+}{dw} - \frac{dE_-}{dw}\right)
$$

(6.126)

where use was made of (6.115). Differentiation of (6.125) with respect to w yields[5]

(a) $\quad\displaystyle \frac{dE_+}{dw} = -\pi B_g \frac{\partial \varepsilon_f(w, w_g)}{\partial w} + \pi \int_w^{w_g} B(w') \frac{\partial^2 \varepsilon_f(w, w')}{\partial w' \partial w}\, dw'$

$\qquad\qquad - \lim_{w' \to w} \pi B(w') \frac{\partial \varepsilon_f(w, w')}{\partial w'}$ (6.127)

(b) $\quad\displaystyle \frac{dE_-}{dw} = -\pi \int_0^w B(w') \frac{\partial^2 \varepsilon_f(w', w)}{\partial w' \partial w}\, dw' - \lim_{w' \to w} \pi B(w') \frac{\partial \varepsilon_f(w', w)}{\partial w'}$

Partial integration of these equations results in

(a) $\quad\displaystyle \frac{dE_+}{dw} = \pi[B(w_g) - B_g] \frac{\partial \varepsilon_f(w, w_g)}{\partial w} - \pi \int_w^{w_g} \frac{dB}{dw'}\frac{\partial \varepsilon_f(w, w')}{\partial w}\, dw'$

$\qquad\qquad - \lim_{w' \to w} \pi B(w') \frac{\partial \varepsilon_f(w, w')}{\partial w} - \lim_{w' \to w} \pi B(w') \frac{\partial \varepsilon_f(w, w')}{\partial w'}$ (6.128)

(b) $\quad\displaystyle \frac{dE_-}{dw} = \pi B(w = 0) \frac{\partial \varepsilon_f(0, w)}{\partial w} + \pi \int_0^w \frac{dB}{dw'}\frac{\partial \varepsilon_f(w', w)}{\partial w}\, dw'$

$\qquad\qquad - \lim_{w' \to w} \pi B(w') \frac{\partial \varepsilon_f(w', w)}{\partial w} - \lim_{w' \to w} \pi B(w') \frac{\partial \varepsilon_f(w', w)}{\partial w'}$

From the definition of ε_f one may easily see that for two arbitrary values w_1, w_2 with $w_2 > w_1$ the following relations hold

(a) $\quad\displaystyle \frac{\partial \varepsilon_f(w_1, w_2)}{\partial w_2} > 0, \qquad \frac{\partial \varepsilon_f(w_1, w_2)}{\partial w_1} < 0$

(6.129)

(b) $\quad\displaystyle \lim_{w_2 \to w_1} \frac{\partial \varepsilon_f(w_1, w_2)}{\partial w_2} = -\lim_{w_2 \to w_1} \frac{\partial \varepsilon_f(w_1, w_2)}{\partial w_1}$

Equation (6.129a) simply describes the fact that with increasing layer thickness the emissivity is increasing and vice versa. Utilizing (6.129b) we find that the last two

---

[5] Note that for the differentiation of the integrals the Leibniz rule has to be applied.

terms in (6.128a,b) cancel. Subtraction of these two equations finally yields

$$
\frac{dE_+}{dw} - \frac{dE_-}{dw} = \pi[B(w_g) - B_g]\frac{\partial \varepsilon_f(w, w_g)}{\partial w} - \pi \int_w^{w_g} \frac{dB}{dw'}\frac{\partial \varepsilon_f(w, w')}{\partial w}dw'
$$
$$
- \pi B(w = 0)\frac{\partial \varepsilon_f(0, w)}{\partial w} - \pi \int_0^w \frac{dB}{dw'}\frac{\partial \varepsilon_f(w', w)}{\partial w}dw'
$$

(6.130)

For the discussion of (6.126) we consider an atmosphere where $dT/dz < 0$. From (6.115) follows that in this case $dB/dw' > 0$. Utilizing (6.130) we see that for $B_g > B(w_g)$, that is the Earth's surface is warmer than the overlying air, the first term of (6.130) yields a positive contribution to the radiative heating rate. The same is true for the first integral of (6.130) describing the radiative energy arriving at the reference level from the warmer atmosphere below. The last two terms of (6.130) are negative resulting in radiative cooling of the reference level. The expression $\pi B(0)\partial \varepsilon(0, w)/\partial w$ sometimes is called the *cooling to space term*.

A brief discussion of the classical emissivity method is mandatory. In the middle of the twentieth century, i.e. when computers to perform radiative transfer simulations were not available, the emissivity approximation was often used as a very fast way to calculate infrared radiative flux densities and heating rates in a nonscattering atmosphere. However, there are several disadvantages to the classical emissivity method.

(1) The entire infrared spectral range is treated in one part, i.e. the infrared atmospheric window is not separated from the remainder of the spectrum. While water vapor radiation may dominate the radiation field outside the atmospheric window and outside the carbon dioxide absorption band, within the window the influence of aerosol particles and other substances is of great importance. This treatment causes substantial errors in the radiation budget because aerosol particles and hydrometeors scatter and absorb thermal radiation in the window region.
(2) Inclusion of scattering processes is not possible. To include scattering, we have to proceed as shown in a previous section.
(3) A reasonably accurate treatment of the overlap of gas absorption due to several gases is not possible because absorption by atmospheric trace gases is wavelength sensitive.
(4) Finally, it should not be overlooked that ignoring the temperature dependence of the emissivities, see (6.123), is a crude approximation.

Rodgers (1967) presented a novel approach on the use of emissivity in atmospheric radiation calculations. He introduced two different emissivity functions to approximate the upward and downward flux densities. Moreover, he also introduced the CO_2 contribution.

An efficient improvement of the classical emissivity method consists in the subdivision of the entire infrared spectral region into a certain number of frequency intervals Δv_j, $j = 1, \ldots, J$. In each of these intervals we determine mean values of the emissivity and the Planck function by setting

$$\varepsilon_{f,j}(w, w') \approx \frac{1}{B_j(T_m)} \int_{\Delta v_j} B_v(T_m) \varepsilon_{f,v}(w, w') dv$$

$$B_j(T_m) \approx \int_{\Delta v_j} B_v(T_m) dv \tag{6.131}$$

where T_m is again a suitable mean temperature. These equations imply the assumption that within Δv_j the values of B_v remain nearly constant.

The upward and downward directed radiative flux densities are now given as

$$E_+(w) = \sum_{j=1}^{J} E_{+,j}(w), \qquad E_-(w) = \sum_{j=1}^{J} E_{-,j}(w)$$

$$\text{with} \quad E_{+,j}(w) = \pi B_{g,j}[1 - \varepsilon_{f,j}(w, w_g)] + \pi \int_w^{w_g} B_j(w') \frac{\partial \varepsilon_{f,j}(w, w')}{\partial w'} dw'$$

$$E_{-,j}(w) = -\pi \int_0^w B_j(w') \frac{\partial \varepsilon_{f,j}(w', w)}{\partial w'} dw'$$

$$\tag{6.132}$$

Usually, a total number of about 10 to 20 frequency bands will be sufficient to obtain a distinct improvement of the radiative flux densities and the corresponding heating rates.

Once the subdivision of the total infrared spectral region into several subregions has been introduced in the emissivity method, it is only a simple task to include the so-called *overlap effects* of two different absorbers. By overlap effects we mean that two gases absorb in the same spectral region. Within a given frequency interval Δv_j the mean transmission resulting from two absorbers 1 and 2 is given by

$$\mathcal{T}_{f,j} = \frac{1}{\Delta v_j} \int_{\Delta v_j} \mathcal{T}_{f,j,1}(v) \mathcal{T}_{f,j,2}(v) dv \tag{6.133}$$

Let the transmission of each absorber be given by a mean transmission $\bar{\mathcal{T}}_{f,j}$ and a deviation $\mathcal{T}'_{f,j}(v)$. Then we obtain

$$\mathcal{T}_{f,j} = \frac{1}{\Delta v_j} \int_{\Delta v_j} \left[\bar{\mathcal{T}}_{f,j,1} + \mathcal{T}'_{f,j,1}(v) \right] \left[\bar{\mathcal{T}}_{f,j,2} + \mathcal{T}'_{f,j,2}(v) \right] dv = \tilde{\mathcal{T}}_{f,j} + \Delta \mathcal{T}_{f,j}$$

$$\text{with} \quad \tilde{\mathcal{T}}_{f,j} = \bar{\mathcal{T}}_{f,j,1} \bar{\mathcal{T}}_{f,j,2}, \qquad \Delta \mathcal{T}_{f,j} = \frac{1}{\Delta v_j} \int_{\Delta v_j} \mathcal{T}'_{f,j,1}(v) \mathcal{T}'_{f,j,2}(v) dv \tag{6.134}$$

If the absorption coefficients of the two gases are uncorrelated then $\Delta \mathcal{T}_{f,j}$ vanishes. If a correlation exists between the two absorption coefficients, then $\Delta \bar{\mathcal{T}}_f$ may be determined by means of detailed line-by-line calculations as will be explained in the following chapter.

A particularly simple situation arises if one of the two absorbers is a so-called *gray absorber*, that is the absorption coefficient and thus the transmission function of the absorber is independent of ν. This is, for instance, approximately the case when aerosol particles or cloud droplets are considered in the infrared radiative transfer calculations. From (6.133) it is immediately seen that now the mean transmission is given by

$$\mathcal{T}_{f,j} = \frac{\mathcal{T}_{f,j,1}}{\Delta \nu_j} \int_{\Delta \nu_j} \mathcal{T}_{f,j,2}(\nu) d\nu \qquad (6.135)$$

In the literature many other attempts have been made to obtain further improvements of the classical emissivity method, e.g. by including approximate ways to describe multiple scattering processes by aerosol particles and cloud droplets, (see e.g. Chou *et al.*, 1999, 2001). However, by considering the tremendous increase of computer power in the past decade or so, it might be a better idea to apply one of the two-stream approximations described in Section 5.3 not only to the short-wave but also to the infrared spectral region.

6.8 Radiation charts

Before the availability of large electronic computers, the exact integration of the flux density equations over the broad infrared water vapor spectrum was virtually impossible. By simplifying the spectrum in a reasonable manner, Mügge and Möller (1932) integrated the radiative transfer equation by graphical means thus inventing the first *radiation chart*. Their integration method makes it possible to easily obtain flux densities for any atmospheric sounding if pressure, temperature and humidity are known as a function of height. Möller (1943) improved the original radiation chart by using the Schnaidt model[6] of the absorption function. Moreover, Elsasser (1942) and later Yamamoto (1952) also devised radiation charts. These three radiation charts differ in their outward appearance due to transformation of coordinates, but they are equivalent in principle. For a given atmospheric sounding the three charts also yield somewhat different flux densities since different models of the absorption function are used. The literature also presents numerous simplified radiation charts which will be omitted in this discussion.

---

[6] The Schnaidt model will be presented in Section 7.2.5.

In the meteorological practice radiation charts are no longer used. Nevertheless, it is desirable to give a brief description how to use them since the calculation of flux densities can be readily visualized. The radiation chart is the pre-integrated radiative transfer equation over the wavelength domain of the infrared spectrum for all combinations of the coordinates (w, T) normally observed in the atmosphere. With the help of (6.115) we easily find $w(z)$ for an arbitrary atmospheric sounding measuring $T(z)$, $p(z)$ and $\rho_{abs}(z)$. By plotting and connecting the coordinates (w, T) by a smooth line we obtain the image of the atmospheric sounding on the radiation chart. The area under the curve can be measured by means of a planimeter which represents either the upward or downward flux density for a specified reference height.

We will now briefly discuss the underlying mathematical principle of the Möller radiation chart. According to (6.118) in an isothermal atmosphere of temperature T_0 the downward directed flux density emitted by a layer of absorber mass w and received at the reference level $w(z)$ is given by

$$E_-(T_0, w) = \pi \int_0^\infty B_\nu(T_0)[1 - T_{f,\nu}(0, w)]d\nu \qquad (6.136)$$

Obviously, the same relation holds for the upward directed flux density resulting from the emission of an atmospheric layer below the reference level having the same temperature and reduced absorber mass. Therefore, the subscripts $\pm$ are henceforth omitted by writing

$$E_+(T_0, w) = E_-(T_0, w) = E(T_0, w) \qquad (6.137)$$

Thus, from now on $E(T, w)$ describes the radiative flux density received at the reference level which has been emitted by a layer of temperature T and optical mass w.

The contribution of an elementary layer dw of fixed temperature T to the flux density at z is obtained by differentiating (6.137) with respect to w

$$\frac{dE(T, w)}{dw} = -\pi \int_0^\infty B_\nu(T)\frac{dT_{f,\nu}(0, w)}{dw}d\nu \qquad (6.138)$$

By introducing the abbreviations

$$x(w) = E(T_{\max}, w), \qquad y(T, w) = \frac{\dfrac{dE(T, w)}{dw}}{\dfrac{dE(T_{\max}, w)}{dw}} \qquad (6.139)$$

for an arbitrary atmosphere of thickness w_1 we find for the flux density

$$E(w_1) = \int_{x(0)}^{x(w_1)} y[T(w), w]dx \qquad (6.140)$$

where T_{max} is a maximum temperature, e.g. $T_{max} = 40°C$. The ratio $y(T, w)$ was calculated for all temperatures normally occurring in the atmosphere.

We will now briefly discuss the construction and the use of the radiation chart. In the troposphere and the lower stratosphere water vapor and carbon dioxide by far outweigh the influence of other radiatively active gases such as ozone. In the wavelength region ranging from 13.5–16.5 µm, the effect of carbon dioxide is much more important than that of water vapor. Thus Möller felt justified to subdivide the infrared spectrum into two parts. In the range from 13.5–16.5 µm he assumed that CO_2 acts completely independently of water vapor while in the remaining infrared spectral ranges from 4–13.5 µm and 16.5–100 µm only water vapor was assumed to be radiatively active.

To simplify the spectral integration, Möller divided the water vapor spectrum into 23 subintervals. In each of these, he replaced the numerous existing spectral lines by a single composite spectral line and used the Schnaidt model to approximately account for the overlap of neighboring spectral lines. Proceeding in this way, Möller obtained $x(w)$ and $y(T, w)$ by numerical integration as intervals on the abscissa and the ordinate of his chart.

Figure 6.5 displays schematically the Möller radiation chart whose shape is rectangular. The smaller rectangle on the left side with horizontal isotherms is the CO_2 chart while the larger part with curved isotherms depicts the water vapor radiation chart. The upper isotherm T_{max} is a straight line since $y = 1$ independent of w, see (6.139). The combined CO_2 and H_2O chart areas represent the flux density emitted by a black body of temperature T_{max}. Analogously, the combined areas of the two parts under any isotherm $T = const$ represent the *black body radiation* of that temperature. The right hand ordinate of the x-axis of each part of the chart refers $w = \infty$ where the emission of the corresponding absorber is given by the black body radiation. Owing to the strong CO_2 absorption, in the spectral section extending from 13.5–16.5 µm, black body radiation is almost emitted by $w(CO_2) = 10$ cm NTP (normal temperature and pressure). In the section of the water vapor spectrum the flux density emitted by $w(H_2O) = 100$ g cm^{-2} already approximates black body radiation. As an example, the shaded area in Figure 6.5(a) depicts the *emission of an isothermal layer* of temperature T_0 and absorber masses $w(CO_2)=1$ and $w(H_2O)=0.1$. In Figure 6.5(b) we have shown the contributions of three isothermal layers of temperatures T_i, $i = 1, 2, 3$ to the total flux density. The CO_2 and H_2O absorber masses of these layers are assumed to be $\Delta w_1 = 0.1$, $\Delta w_2 = 0.9$ and

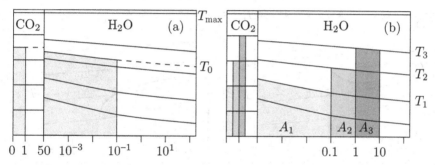

Fig. 6.5 Schematic radiation chart after Möller. Horizontal lines in the CO_2 chart and curved lines in the H_2O chart denote isotherms. Shaded areas describe the radiative flux densities. (a) Emission of an isothermal layer; (b) emission of three different isothermal layers. See also text.

$\Delta w_3 = 9$. In this example the total flux density is given by the sum of all three layers in the CO_2 and the H_2O chart.

We will now briefly show how to find the upward and downward flux densities with the help of the radiation chart which is equivalent to the integration of (6.140). Utilizing (6.115) for a given atmospheric sounding we determine the (T, w) relationship. This information will be used in (6.140) to find the curve $[x(w), y(T(w), w)]$. Inserting this curve into the radiation chart one obtains the radiation emitted by a particular layer. As an example, Figure 6.6(a) and (b) show the resulting curves for upward and downward radiation at reference level z with temperature $T(z)$. The area in Figure 6.6(a) depicts the upward radiation reaching the level from below. Starting from the point $[x(w = 0), y(T(z))]$ the curve approaches the isotherm T_g at $x(w_+)$ where w_+ is the reduced absorber mass of CO_2 or H_2O between the Earth's surface and the level z. The *black body emission of the Earth's surface* with temperature T_g is taken into account by following the curve from the point $[x(w_+), y(T_g, w_+)]$ along the T_g isotherm towards the right end of the x-axis. The area in Figure 6.6(b) describes the downward radiation at z. Starting again at the point $[x(w = 0), y(T(z))]$ the curve approaches the x-axis at w_- whereby w_- denotes the absorber masses of carbon dioxide and water vapor between the top of the atmosphere and the level z. By subtracting the areas under the curves in Figure 6.6(b) from those in Figure 6.6(a) one obtains the net radiative flux density $E_{net}(z)$. Choosing a neighboring reference layer $z + \Delta z$ one obtains the net flux density $E_{net}(z + \Delta z)$. The quantity $\Delta E_{net} = E_{net}(z + \Delta z) - E_{net}(z)$ is proportional to the infrared radiative heating rate at level z.

Finally, it is a simple task to include the effects of a cloudy atmosphere by first assuming a total cloud cover and black body emission of the cloud. Analogously to the treatment of the Earth's surface, now the corresponding (T, w) curves for upward

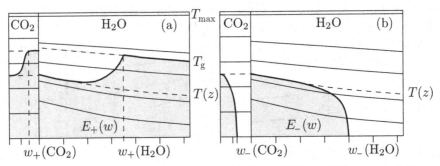

Fig. 6.6 Schematic radiation chart after Möller. (a) Upward-directed radiation, (b) downward directed radiation.

(cloud below z) or downward (cloud above z) radiation approach the isotherm of the cloud's temperature and follows from there the isotherm to the right end of the x-axis.

By calculating upward and downward flux densities for the clear and the cloud covered sky and properly weighting the individual flux densities with the fractional cloudiness, we obtain net flux densities and heating rates for cloudy atmospheres. Möller (1951) carried out numerous calculations to obtain vertical profiles of *radiative temperature changes* for various air masses and found cooling practically everywhere. To check the validity of his and other radiation charts many comparisons were made with ground based and free air measurements. In general, there was reasonable agreement. In particular, Gergen (1956, 1957, 1958) carried out numerous free air measurements of the infrared radiation with a spherical receiver. Zdunkowski (1963) constructed a radiation chart for a spherical receiver using precisely Möller's (1943, 1944) absorption data. He found good but not perfect agreement with measurements.

We conclude this section with a few additional comments on the other radiation charts. Elsasser (1942) assumed that the complete 15 μm CO_2 absorption band acts as a black absorber. Indeed, at the center of the band the absorption is almost black for very small path lengths. However, for small pressures in the higher atmosphere and at the wings of the band, the absorption is far from complete. Therefore, Yamamoto (1952) developed a special method to treat the overlap effect of carbon dioxide and water vapor. At the base of his diagram he provided an additional scale to handle the overlap effect graphically. The contribution of ozone to the flux densities was still ignored; the effect is small at low atmospheric levels. The three charts yield compatible results.

Elsasser and Culbertson (1961) presented a set of radiation tables for the absorption bands of atmospheric gases including ozone. Special tables were provided to handle overlap effects. The tables are very versatile, but at the same time very tedious to use. Bruinenberg (1946), also using Möllers absorption data, devised a numerical integration method for the calculation of radiative temperature changes. His method revealed some details of the radiative temperature change profiles that could not be obtained with the more crude graphical integration procedures. Brooks (1950) simplified this numerical method. Hales (1951) was the first to develop a graphical method to calculate radiative temperature changes. A similar procedure was given later by Yamamoto and Onishi (1953).

Nowadays practically all calculations are carried out with high-speed computers making it possible to attack many problems that were impossible to handle earlier. Nevertheless, the earlier important research was carried far enough to analyze and comprehend many of the interesting problems associated with radiative transfer.

6.9 Radiative equilibrium

In the final section of this chapter we wish to explore in some depth the concept of radiative equilibrium. Consider a horizontally homogeneous atmosphere where radiative cooling and heating is the only process to form the vertical temperature profile, that is other types of heat transfer such as heat conduction, convection and latent heat release are ignored. If this atmosphere approaches thermal equilibrium, i.e. $\partial T(z)/\partial t = 0$ at all levels z, the atmosphere has reached *radiative equilibrium*. Since we assume the existence of a horizontally homogeneous atmosphere, the condition describing radiative equilibrium implies a vanishing vertical divergence of the radiative net flux density.

With the exception of spatially very limited regions around kinks in the vertical temperature profile, long-wave radiation causes atmospheric cooling in the free atmosphere which is stronger practically everywhere than direct solar heating. Some numerical results of radiative temperature changes are given, for example, by Liou (2002). The resulting cooling by radiative transfer must be compensated in some manner since the atmospheric temperature does not decrease permanently. In the real atmosphere, such compensating heating effects are turbulent heat transport and the liberation of latent heat due to water vapor condensation. Apparently no tropospheric layer is in radiative equilibrium.

It appears that Emden (1913) was the first to investigate the atmospheric temperature profile resulting from the condition of radiative equilibrium. He found a strong superadiabatic temperature gradient in the lower troposphere and a uniform temperature of $-60°C$ in higher atmospheric layers. However, the height of the computed tropopause somewhere between 6 and 8 km was too low. Certainly, the superadiabatic lapse rate resulted from the disregard of all processes other than radiative heating. Since the calculated stratospheric temperature at the tropopause agreed remarkably well with observations it was erroneously concluded that the stratospheric temperature can be satisfactorily explained by the radiative equilibrium of water vapor alone.

The agreement between Emden's calculation and observations of the stratospheric temperature was due to a lucky choice of the magnitude of the *gray absorption coefficient* of water vapor, which was assumed to be the only atmospheric absorber. By subdividing the water vapor spectrum into two parts and approximating each region by a different gray absorption coefficient, still a very poor approximation, Möller (1941) showed that even in higher layers the temperature continues to decrease to values as low as $-100°$C. The absorption of solar radiation influences this result to a very small extent only. This should refute the idea that the stratospheric temperature can be explained by the radiative equilibrium of water vapor alone.

While it is still worthwhile to discuss Emden's model, it is more instructive for us to follow Goody's (1964a) more modern treatment, which is based on the solution of the radiative transfer equation in the form discussed earlier. We start out by repeating the monochromatic RTE for a nonscattering medium in *local thermodynamic equilibrium*

$$\mu \frac{d}{d\tau} I_\nu(\tau, \mu) = I_\nu(\tau, \mu) - B_\nu(\tau) \qquad (6.141)$$

Integration of this equation over the unit sphere yields

$$\frac{d}{d\tau} E_{\text{net},\nu}(\tau) = 2\pi [I_{+,\nu}(\tau) + I_{-,\nu}(\tau)] - 4\pi B_\nu(\tau) \qquad (6.142)$$

Here, we have introduced mean values of the upward and downward directed radiances as defined by

$$\boxed{I_{+,\nu}(\tau) = \int_0^1 I_{+,\nu}(\tau, \mu) d\mu, \qquad I_{-,\nu}(\tau) = \int_0^1 I_{-,\nu}(\tau, \mu) d\mu} \qquad (6.143)$$

which is equivalent to the assumption of isotropic radiation in the upward and downward direction. $E_{\text{net},\nu}(\tau)$ is the *monochromatic net radiative flux density*. According to (1.37c) this quantity may be expressed as

$$E_{\text{net},\nu}(\tau) = 2\pi \int_0^1 I_{+,\nu}(\tau, \mu)\mu d\mu - 2\pi \int_0^1 I_{-,\nu}(\tau, \mu)\mu d\mu = \pi [I_{+,\nu}(\tau) - I_{-,\nu}(\tau)]$$

$$(6.144)$$

whereby in the integrals the radiances have been approximated by their isotropic values.

Differentiation of (6.142) with respect to τ results in

$$\frac{d^2}{d\tau^2}E_{\text{net},\nu}(\tau) = 2\pi\left[\frac{d}{d\tau}I_{+,\nu}(\tau) + \frac{d}{d\tau}I_{-,\nu}(\tau)\right] - 4\pi\frac{d}{d\tau}B_\nu(\tau) \qquad (6.145)$$

Multiplying (6.141) by μ and integrating the result over the unit sphere gives

$$2\pi\frac{d}{d\tau}\left(\int_0^1 \mu^2 I_{+,\nu}(\tau,\mu)d\mu + \int_0^1 \mu^2 I_{-,\nu}(\tau,\mu)d\mu\right) = E_{\text{net},\nu}(\tau) \qquad (6.146)$$

In this equation we again replace the radiances $I_{\pm,\nu}(\tau,\mu)$ in the integrals by $I_{\pm,\nu}(\tau)$ and obtain

$$2\pi\frac{d}{d\tau}I_{+,\nu}(\tau) + 2\pi\frac{d}{d\tau}I_{-,\nu}(\tau) = 3E_{\text{net},\nu}(\tau) \qquad (6.147)$$

Substituting this equation into (6.145) finally results in a second-order ordinary differential equation for the net flux density

$$\boxed{\frac{d^2}{d\tau^2}E_{\text{net},\nu}(\tau) - 3E_{\text{net},\nu}(\tau) = -4\pi\frac{d}{d\tau}B_\nu(\tau)} \qquad (6.148)$$

For the solution of (6.148) we need to formulate two boundary conditions. Combination of (6.142) and (6.144) gives

$$\begin{aligned}
I_{-,\nu}(\tau) &= \frac{1}{4\pi}\frac{d}{d\tau}E_{\text{net},\nu}(\tau) - \frac{1}{2\pi}E_{\text{net},\nu}(\tau) + B_\nu(\tau) \\
I_{+,\nu}(\tau) &= \frac{1}{4\pi}\frac{d}{d\tau}E_{\text{net},\nu}(\tau) + \frac{1}{2\pi}E_{\text{net},\nu}(\tau) + B_\nu(\tau)
\end{aligned} \qquad (6.149)$$

Evaluating these equations at the upper $(\tau = 0)$ and lower $(\tau = \tau_g)$ boundary of the atmosphere yields

$$\begin{aligned}
I_{-,\nu}(0) &= \frac{1}{4\pi}\frac{d}{d\tau}E_{\text{net},\nu}(\tau)\Big|_{\tau=0} - \frac{1}{2\pi}E_{\text{net},\nu}(0) + B_\nu(0) \\
I_{+,\nu}(\tau_g) &= \frac{1}{4\pi}\frac{d}{d\tau}E_{\text{net},\nu}(\tau)\Big|_{\tau=\tau_g} + \frac{1}{2\pi}E_{\text{net},\nu}(\tau_g) + B_\nu(\tau_g)
\end{aligned} \qquad (6.150)$$

In the case of *monochromatic radiative equilibrium* the divergence of the net radiative flux density vanishes, $E_{\text{net},\nu}(\tau)$ must be constant throughout the entire atmosphere, that is

$$\frac{d}{d\tau}E_{\text{net},\nu}(\tau) = 0 \implies E_{\text{net},\nu}(\tau) = E_{\text{net},\nu} = const \qquad (6.151)$$

In this case the boundary conditions (6.150) reduce to

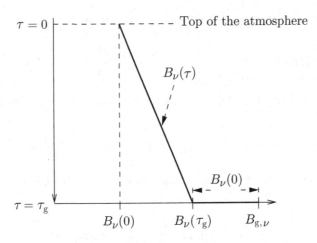

Fig. 6.7 Vertical distribution of $B_\nu(\tau)$ for monochromatic radiative equilibrium. At the top of the atmosphere the temperature adjusts itself to the net flux density.

(a) $E_{\text{net},\nu} = 2\pi B_\nu(0)$

$\qquad\qquad\qquad\qquad\qquad\qquad\qquad\qquad\qquad\qquad\qquad\qquad$ (6.152)

(b) $I_{+,\nu}(\tau_g) = B_{g,\nu} = B_\nu(\tau_g) + \dfrac{1}{2\pi} E_{\text{net},\nu} = B_\nu(\tau_g) + B_\nu(0)$

Equation (6.152a) implies the assumption that no radiation is incident at the top of the atmosphere, i.e. $I_{-,\nu}(0) = 0$.

Application of the equilibrium condition (6.151) to (6.148) results in

$$\frac{d}{d\tau} B_\nu(\tau) = \frac{3}{4\pi} E_{\text{net},\nu} \qquad\qquad\qquad (6.153)$$

This differential equation may be easily integated yielding $B_\nu(\tau)$ as a linear function of τ

$$\boxed{\begin{aligned} B_\nu(\tau) &= \left(1 + \frac{3\tau}{2}\right) \frac{E_{\text{net},\nu}}{2\pi} = \left(1 + \frac{3\tau}{2}\right) B_\nu(0) \\[2mm] B_{g,\nu} &= B_\nu(\tau_g) + B_\nu(0) = \left(2 + \frac{3\tau_g}{2}\right) B_\nu(0) \end{aligned}}$$

$\qquad\qquad\qquad\qquad\qquad\qquad\qquad\qquad\qquad\qquad\qquad$ (6.154)

where use was made of the boundary conditions (6.152). Figure 6.7 depicts the vertical profile of $B_\nu(\tau)$ for monochromatic radiative equilibrium together with the corresponding values at the boundaries of the atmosphere. At the Earth's surface we observe a discontinuity of the curve expressing a temperature jump ΔT_g between the surface temperature T_g and the lowest atmospheric layer $T(\tau_g)$, i.e.

$$\Delta T_g = T_g - T(\tau_g) > 0 \qquad\qquad\qquad (6.155)$$

Emden's investigation pertains to gray absorption since he approximated the strongly varying water vapor spectrum by a single absorption coefficient. In order to formally pass from the monochromatic to the gray absorption, we have to omit the subscript ν in the previous equations. In this case the vertical temperature profile of the atmosphere in radiative equilibrium as well as the temperature of the ground are obtained from

$$T(\tau) = \left[\frac{\pi}{\sigma}B(\tau)\right]^{1/4}, \qquad T_g = \left\{\frac{\pi}{\sigma}[B(\tau_g) + B(w = 0)]\right\}^{1/4}$$

$$\text{with} \quad B(\tau) = \int_0^\infty B_\nu(\tau)d\nu$$

(6.156)

For the temperature of the Earth's surface we may also write

$$T_g = \left[T^4(\tau_g) + T^4(0)\right]^{1/4} \qquad (6.157)$$

Figure 6.8 depicts qualitatively different vertical temperature profiles resulting from different choices of the total optical thickness τ_g. By considering only gray absorption of water vapor in the troposphere, the height of the tropopause is given by the optical top of the atmosphere, i.e. at $\tau = 0$. Furthermore, in the stratosphere the temperature distribution remains constant with height. The dotted curve shows the situation with very weak absorption in the troposphere, that is $\tau_{g,1} \to 0$. In this case $B(\tau)$ and, thus, the tropospheric temperature remain nearly constant with height. At the same time the temperature jump $\Delta T_{g,1}$ is largest. Evaluating for this particular situation (6.157) yields a temperature jump of more than 47 K and 37 K for $T(\tau_{g,1}) = T(0) = 250$ K and 200 K, respectively. This means that ΔT_g decreases with decreasing temperature at the tropopause.

From (6.154) we see that for a given $B_\nu(0)$ with increasing total optical thickness τ_g the quantity $B_{g,\nu}$ and, therefore, also $T(\tau_g)$ are increasing. According to (6.157) the values of T_g are also affected thereof. To give a numerical example, for $T(0) = 200$ K and $T(\tau_g) = 250$ K we obtain $T_g = 272.4$ K, that is $\Delta T_g = 22.4$ K. Choosing $T(0) = 200$ K and $T(\tau_g) = 270$ K yields $T_g = 288.4$ K or $\Delta T_g = 18.4$ K. Hence, with increasing total optical thickness of the atmosphere the temperature is increasing in the lower atmospheric layers while at the same time the temperature jump is decreasing. In the limit of a very opaque atmosphere with an unrealistically low temperature $T(0) = 150$ K and $T(\tau_g) = 300$ K, ΔT_g would be less than 5 K.

Goody (1964b) applied the radiative equilibrium model of the gray absorber by assuming that the gray absorption coefficient and the absorber mass follow the same height distribution. He also used a realistic temperature at the tropopause which he estimated from a simple heat balance consideration assuming a mean global albedo

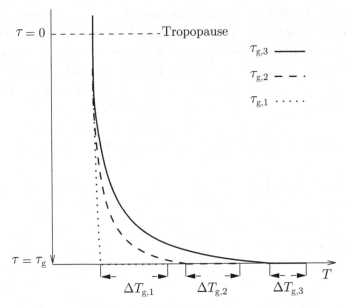

Fig. 6.8 Schematic vertical temperature distributions in case of a gray absorbing troposphere for different values of the total absorber mass with $\tau_{g,1} \ll \tau_{g,2} < \tau_{g,3}$ and $\Delta T_{g,1} > \Delta T_{g,2} > \Delta T_{g,3}$.

of 0.4. For $\tau_g = 1$ he found $T(\tau_g) = 259.7$ K, $\Delta T_g = 22.8$ K, while $\tau_g = 4$ resulted in $T(\tau_g) = 335.9$ K and $\Delta T_g = 11.3$ K. At the same time the temperature gradients at the ground were $-\partial T/\partial z = 19.5$ and 36.0 K km^{-1}, respectively.

From the above findings we may conclude that the radiative equilibrium model with gray absorption in the troposphere yields satisfactory vertical temperature distributions in terms of lapse rates which are decreasing with height. Moreover, by a fortuitous choice of all model parameters it is possible to obtain relatively good values of the stratospheric temperature and the tropopause height. However, the model also shows many deficiencies. The most important are listed here.

(1) The temperature jump occurring at the Earth's surface is unrealistically large.
(2) The lapse rates in the lowest atmospheric layers are distinctly too high in comparison to the observed value of about 6.5 K km^{-1}.
(3) The temperature increase observed in the real stratosphere cannot be simulated with the model.
(4) The stratospheric temperature $T(0)$ is positively correlated with T_g, see (6.157).

Certainly, there are many reasons why Emden's model fails to produce better results. Here, we mention the following shortcomings.

(1) The assumption of a gray absorbing atmosphere.
(2) The missing absorption of ozone in the stratosphere.
(3) The disregard of dynamic and thermodynamic processes in the lower troposphere.

Paltridge and Platt (1976) used simple arguments to show that even the subdivision of the infrared spectral region into two subintervals, i.e. the almost transparent atmospheric window and the remaining gray absorbing region, is sufficient to achieve a decoupling of the stratospheric temperature from the temperature of the Earth's surface. Hence, with this simple treatment it is possible to obtain warm temperatures at the ground, but at the same time cold stratospheric temperatures and vice versa. This is in accordance with observations in the tropics and higher geographic latitudes.

Möller and Manabe (1961) successfully used the emissivity method to handle the transfer problem. In addition to the water vapor absorption they included the absorption effects of CO_2 and O_3 yielding more realistic temperature profiles with temperatures increasing with height in the stratosphere. Manabe and Möller (1961) carried out detailed calculations with a more refined model. They used a time-marching procedure as the solution method starting out with an isothermal atmosphere with the observed temperature at the Earth's surface. However, this method is very time consuming. Manabe and Strickler (1964) pointed out that it is necessary to include various aspects of large-scale dynamics to obtain better results. Apart from the inclusion of carbon dioxide and ozone they used a simple convective adjustment scheme to account for the convective mixing and the latent heat release in the lower troposphere. The resulting *radiative–convective equilibrium* showed much more realistic temperature profiles than those produced by a pure radiative equilibrium model.

6.10 Problems

6.1: Verify equation (6.8).

6.2: Verify equation (6.26).

6.3: In detail follow the steps from (6.92) to (6.95).

6.4: Consider the approximate treatment of scattering (Section 6.5) in the infrared spectral region. For the conservative case $\omega_0 = 1$ find the solutions $E_+(\tau)$ and $E_-(\tau)$ for a cloud layer of optical thickness τ_c. The boundary conditions are given by $E_-(\tau=0)=E_-(0)$ and $E_+(\tau=\tau_c)=E_+(\tau_c)$.

6.5: (a) Find the integration constants C_1 and C_2 in (6.95) by assuming the boundary conditions $E_-(\tau = 0) = E_-(0)$ and $E_+(\tau = \tau_c) = E_+(\tau_c)$, where τ_c is the cloud layer optical thickness.

(b) Find the divergence of the net flux density, i.e. $d(E_+ - E_-)/d\tau$.

6.6: Again we consider the approximative scattering in the infrared spectrum as discussed in Section 6.5. For a cloud model consisting of three layers, find expressions for the upward and downward directed flux density. Each layer of optical thickness T_i, $i = 1, 2, 3$ is homogeneous but of different optical properties. Set up the required linear system which permits you to determine the six integration constants $C_{1,i}$, $C_{2,i}$, $i = 1, 2, 3$. The required boundary conditions and the continuity statements at the layer boundaries are

$$E_-(\tau_1 = 0) = E_-(0)$$
$$E_-(\tau_1 = T_1) = E_-(\tau_2 = 0)$$
$$E_+(\tau_1 = T_1) = E_+(\tau_2 = 0)$$
$$E_-(\tau_2 = T_2) = E_-(\tau_3 = 0)$$
$$E_+(\tau_2 = T_2) = E_+(\tau_3 = 0)$$
$$E_+(\tau_3 = T_3) = E_+(T_1 + T_2 + T_3)$$

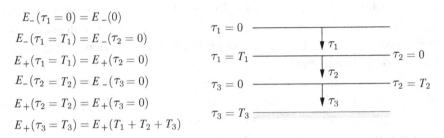

6.7: In case of an isothermal and nonscattering atmosphere equation (6.95) reduces to the so-called *Schwarzschild equation*. Carry out the reduction assuming the boundary conditions $E_-(\tau = 0) = E_-(0)$ and $E_+(\tau = \tau_c) = E_+(\tau_c)$.

Hint: Use *Robert's approximation*, that is $\exp(-\tau/\bar{\mu}) \approx 2E_3(\tau)$, see (2.142), where $\bar{\mu}$ is an average value.

6.8: Carry out the required integration to show that (6.125) follows from (6.124).

6.9: The Sun may be treated as a black body emitting the largest amount of energy at the wavelength of 0.5 μm. The Earth may also behave as a black body in the spectral region of infrared emission. Which temperature will the Earth assume if the incoming solar radiation and the outgoing infrared radiation are balanced? Compare your result with the measured mean temperature of 14°C at the Earth's surface (average value over all latitudes and seasons) and explain the difference.

Required information: Sun's radius, 695 300 km; distance Sun–Earth, 149 600 000 km; Earth's radius, 6371 km; mean albedo of the Earth, 30%.

6.10: Set up the radiation budget at the top of the atmosphere and at the Earth's surface. The following quantities are given: $\bar{A}$, mean absorption of solar radiation by the atmosphere; A_{tot}, global albedo (albedo of the entire system); T_s, radiation temperature of the black body surface of the Earth; T_a, radiation temperature of the atmosphere assumed to be gray; ε, thermal emissivity of the gray atmosphere; and S_0; solar constant.

(a) Find T_s^4 and T_a^4 from the budget equations.

(b) Assume $\bar{A}=0.26$, $A_{tot}=0.31$ and $\varepsilon=0.8$ to find numerical values of T_s and T_a.

(c) Now suppose that the atmosphere is completely transparent to solar radiation but completely opaque ($\varepsilon=1$) to thermal radiation. Find the new values of T_s and T_a.

(d) Suppose that T_s is fixed at 283 K. For a given $\bar{A}=0.2$ and $A_{tot}=0.30$ find the thermal emissivity and T_a.

7

Transmission in individual spectral lines and in bands of lines

High resolution spectroscopy reveals that the absorption and emission of radiation by atmospheric gases is not continuously distributed over the entire spectral range. In fact, the absorption and emission spectra are composed of numerous *spectral lines* of different strength. Molecules have three different forms of internal energy E_{int}: rotational, vibrational and electronic. These energy forms are quantized and are expressed by one or more quantum numbers. If a molecule absorbs or emits radiation, a transition from one energy level to another takes place. During an absorption process the molecule captures a photon thus reaching a higher level of internal energy. Hence the molecule is said to be in an *excited state*. *Emission* of radiation occurs if the molecule releases a photon resulting in a transition to a lower energy level, that is, the molecule leaves the excited state. Both processes yield spectral absorption and emission lines which are characteristic for a particular molecule. According to (1.15) the change of internal energy $\Delta E_{int}(\nu)$ of a molecule resulting from the uptake or release of a photon is given by *Planck's relation*

$$\Delta E_{int}(\nu) = \pm h\nu \tag{7.1}$$

where h is *Planck's constant* and ν is the frequency of the absorbed or emitted energy.

From these considerations one expects that each individual line has an infinitely small width expressing the *monochromatic absorption* and *emission* of radiation with frequency ν. In nature, however, it is observed that individual lines do not have a zero line width, but they are broadened over a narrow frequency range. This *line broadening* is caused by external influences affecting the molecule during the absorption and emission process. There are mainly three effects which are responsible for the broadening of spectral lines. These are listed below.

(1) *Natural broadening*: owing to the finite natural lifetime of a molecule in an excited state, according to Heisenberg's uncertainty principle the emitted energy is distributed over a narrow frequency interval $\Delta \nu$.

(2) *Collision broadening*: during the emission of radiation the molecule will collide with other molecules. This interaction disturbs the emission process resulting in a broadening of the emission line. This process is also called *pressure broadening*.

(3) *Doppler broadening*: the *Doppler effect* caused by the thermal motion of the molecule yields a broadening of the line. Often it is also called *thermal broadening*.

The natural lifetime of the excited state of a molecule is of the order 10^{-2} to 10^{-1} s, (Houghton and Smith, 1966). This is much larger than the time between collisions in a gas at normal atmospheric pressures. Therefore, the first effect is much smaller than the second and the third so that we are justified to disregard it in our discussion. In the lowest 30 km of the atmosphere the line broadening due to molecular collisions is much more important than the Doppler broadening. At altitudes higher than about 50 km, however, the Doppler broadening becomes more and more important as compared to the collision broadening. Certainly, this is caused by the vertical decrease of the air density yielding a reduction of the number of molecular collisions with height while at the same time the mean free path length of the molecules is increasing. In a region of about 30–50 km the collision broadening as well as the Doppler effect should be taken into account.

7.1 The shape of single spectral lines

In this section we will determine the shape of the mass absorption coefficient $\kappa_{abs,\nu}$ of an absorbing gas resulting from the line broadening by collisions of molecules and from the Doppler effect. We start with the description of the isolated effects. At the end of this section the collision and Doppler broadening effects will be combined yielding the so-called *Voigt profile*. For ease of notation, the mass absorption coefficient will henceforth also be denoted by k_ν, that is $k_\nu = \kappa_{abs,\nu}$ as given in (1.41). Thus we omit the reference to the density of the absorbing gas. Furthermore, k_ν will simply be called the absorption coefficient.

7.1.1 The Lorentz line

The simplest approach describing the collision broadening effect is due to *Lorentz* who assumed that at each collision the interaction of radiation with a molecule is momentarily halted and a random phase change is introduced. This is called a *strong encounter*. First we will proceed to give a mathematical description of the collision or *Lorentz broadening*.

Let us consider an electromagnetic wave with circular frequency ω_0 which is incident on an absorbing atmospheric gas molecule. The time interval during which this molecule absorbs the wave is given by $-t_0/2 \leq t < t_0/2$. The time signal of this wave can be expressed by

$$f(t) = \begin{cases} \exp(-i\omega_0 t) & -t_0/2 \leq t < t_0/2 \\ 0 & \text{otherwise} \end{cases} \tag{7.2}$$

Using the Fourier transformation we may switch from the time domain to the frequency domain of the wave. If $g(\omega)$ denotes the Fourier transform of $f(t)$ then $g(\omega)$ and $f(t)$ are related by

$$g(\omega) = \frac{1}{2\pi} \int_{-\infty}^{\infty} f(t) \exp(i\omega t) \, dt, \qquad f(t) = \int_{-\infty}^{\infty} g(\omega) \exp(-i\omega t) \, d\omega \tag{7.3}$$

Inserting (7.2) into the first equation of (7.3) we obtain

$$g(\omega) = \frac{1}{2\pi} \int_{-t_0/2}^{t_0/2} \exp[i(\omega - \omega_0)t] \, dt = \frac{\sin\left[\frac{(\omega - \omega_0)t_0}{2}\right]}{\pi(\omega - \omega_0)} \tag{7.4}$$

Instead of ω we may also use the frequency since $\omega = 2\pi\nu$ so that

$$g(\nu) = \frac{\sin[\pi(\nu - \nu_0)t_0]}{2\pi^2(\nu - \nu_0)} \tag{7.5}$$

The modulus or the absolute value of $g(\nu)$ is called the *amplitude spectrum*. Since the square of the amplitude of a wave is proportional to the energy of the oscillation, the quantity $G(\nu) = |g(\nu)|^2$ is the so-called *power spectrum*. Figure 7.1 illustrates the time signal $f(t)$ of the real part of the finite wave, its Fourier transform and its power spectrum.

Let us now briefly discuss how the shape of a spectral line is influenced by the collisions with other molecules. Let p_c stand for the number of collisions per unit time so that $p_c \Delta t$ is the number of collisions within the time increment Δt. Thus, $q = 1 - p_c \Delta t$ is the probability that the molecule experiences no collisions within Δt. Let the wave have a duration of $t_0 = n\Delta t$. Assuming that the collisions during each time increment Δt are independent of each other, we obtain the probability that the molecule experiences no collisions within the time span t_0 as

$$q^n = (1 - p_c \Delta t)^n = (1 - p_c \Delta t)^{t_0/\Delta t} = [(1 - p_c \Delta t)^{-1/(p_c \Delta t)}]^{-p_c t_0} \tag{7.6}$$

Since $\lim_{x \to 0}(1 - x)^{-1/x} = e$, for $\Delta t \to 0$ we obtain from (7.6)

$$\lim_{\Delta t \to 0} q^n = \exp(-p_c t_0) = \exp\left(-\frac{t_0}{\bar{\tau}}\right) \tag{7.7}$$

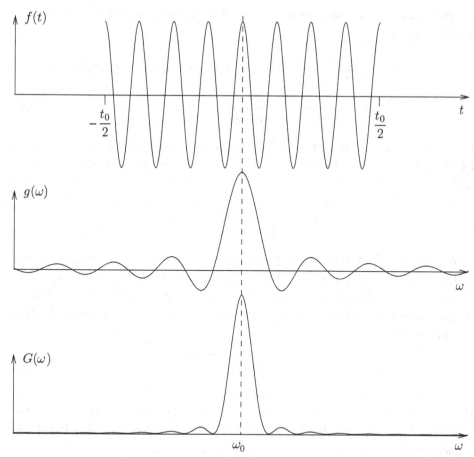

Fig. 7.1 Illustration of a finite wave $f(t)$ in the time domain, its Fourier transform $g(\omega)$ and its power spectrum $G(\omega)$.

Here, $\bar{\tau} = 1/p_c$ is the average time between two collisions and $t_0/\bar{\tau}$ is the total number of collisions within t_0.

The absorption coefficient of the Lorentz line is proportional to the probability $\exp(-p_c t_0)$ for the time interval t_0 between collisions and to the power spectrum $|g(\nu)|^2$ describing the spectral energy distribution. Furthermore, in order to obtain the total absorption, we need to integrate over all possible time spans t_0 during which the molecule absorbs radiation. Hence $k_{\nu,\mathrm{L}}$ may be written in the form

$$k_{\nu,\mathrm{L}} = A \int_0^\infty |g(\nu)|^2 \exp(-p_c t_0)\, dt_0 \qquad (7.8)$$

where A is a constant. Substituting (7.5) into this equation the definite integral may be evaluated yielding

$$k_{v,L} = \frac{A}{4\pi^4(v-v_0)^2} \int_0^\infty \sin^2\left[\pi(v-v_0)t_0\right]\exp\left(-p_c t_0\right)dt_0$$

$$= \frac{A}{2\pi^2 p_c\left[p_c^2 + 4\pi^2(v-v_0)^2\right]} \tag{7.9}$$

From this equation it is seen that the absorption coefficient is largest at $v = v_0$. Denoting $k_{v_0,L} = k_0$ we find from (7.9) that $A = k_0 2\pi^2 p_c^3$. Substituting this expression into (7.9) yields

$$k_{v,L} = \frac{p_c^2 k_0}{p_c^2 + 4\pi^2(v-v_0)^2} \tag{7.10}$$

We will now introduce the *half-width of the Lorentz line* α_L. In general, the half-width of a spectral line is defined by the distance from the line center $v = v_0$ to the points $v_{1,2}$ where the absorption coefficient has decreased to one-half of its maximum value. Hence we may write $\alpha_L^2 = (v_{1,2} - v_0)^2$. Evaluating (7.10) at $v = v_{1,2}$ yields

$$\alpha_L = \frac{p_c}{2\pi} = \frac{1}{2\pi\,\bar{t}} \tag{7.11}$$

showing that the Lorentz half-width is inversely proportional to the average time between collisions. Replacing p_c^2 in (7.10) by means of (7.11) we obtain immediately

$$k_{v,L} = \frac{\alpha_L^2 k_0}{\alpha_L^2 + (v-v_0)^2} \tag{7.12}$$

Finally, we introduce the *line intensity* or the *line strength S* from the definition

$$S = \int_{-\infty}^\infty k_v\,dv \tag{7.13}$$

Substitution of (7.12) into this equation yields the *intensity of the Lorentz line*

$$S = \pi\alpha_L k_0 \tag{7.14}$$

so that the *absorption coefficient of the Lorentz line* can be written as

$$k_{v,L} = \frac{1}{\pi}\frac{\alpha_L S}{\alpha_L^2 + (v-v_0)^2} \tag{7.15}$$

This is the form of the absorption coefficient which is normally used in radiative transfer calculations.

For any line shape the absorption coefficient can be formulated as

$$k_\nu = Sf(\nu - \nu_0)$$
(7.16)

where $f(\nu - \nu_0)$ is the so-called *line-shape factor*. According to (7.13) the line-shape factor is normalized, that is

$$\int_{-\infty}^{\infty} f(\nu - \nu_0) d(\nu - \nu_0) = 1$$
(7.17)

From (7.15) we see that the *line-shape factor of the Lorentz line* is given by

$$f_L(\nu - \nu_0) = \frac{1}{\pi} \frac{\alpha_L}{\alpha_L^2 + (\nu - \nu_0)^2}$$
(7.18)

Integrating this equation over all frequencies according to (7.17) shows that the Lorentz line-shape factor is normalized.

It is customary in spectroscopy to introduce the wave number $\tilde{\nu}$ instead of the frequency ν. If λ represents the wavelength then $\tilde{\nu}$ is defined as $\tilde{\nu} = 1/\lambda$ (cm^{-1}). Using the basic relation $\nu = c/\lambda = c\tilde{\nu}$, where c is the speed of light in a vacuum, we may replace $k_{\nu,L}$ by $k_{\tilde{\nu},L}$. However, since c cancels out in (7.12), the form of the absorption coefficient remains the same. As is common usage, we will not replace $k_{\nu,L}$ by $k_{\tilde{\nu},L}$ but simply continue to write $k_{\nu,L}$. If ν represents the wave number expressed in units of cm^{-1} then α_L must also be expressed in units of cm^{-1} and S in cm^{-2} so that the absorption coefficient k_ν has units of cm^{-1}. Otherwise, if ν represents the frequency then α_L must be expressed in s^{-1} and S in units of cm^{-1} s^{-1}. Whenever a question arises about the set of units used, it is usually not difficult to decide if we work with the frequency or the wave number system. Unfortunately, some writers even call the wave number simply the frequency. The problem associated with the introduction of the wave number is discussed in more detail by Goody (1964a).

The shape of the Lorentz line is shown in Figure 7.2. Note that per definition the area under the line profile is S.

Since the collision of molecules depends on their number density and on their velocity, it is expected that the Lorentz half-width is a function of pressure and temperature. In the following we will discuss a simplified collision model by assuming the existence of preferably elastic spheres. Let us consider identical molecules with radius r which are frozen in position with the exception of one individual molecule that is moving along an irregular zigzag path with an average velocity $\bar{v}$. At the instant of collision the center-to-center distance of the colliding molecules

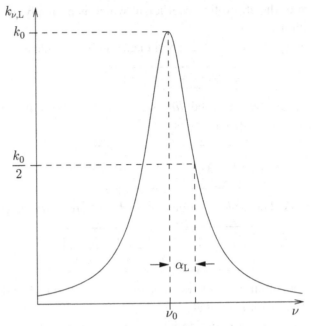

Fig. 7.2 Shape of the Lorentz form of the spectral line.

is $2r$. Thus the *collision cross section* σ_c of this molecule is given by

$$\sigma_c = \pi(r+r)^2 = 4\pi r^2 \qquad (7.19)$$

During the time t the molecule has moved the distance $\bar{v}t$. If n is the number of molecules at rest per unit volume, during t the individual molecule experiences a total number of N_c collisions with

$$N_c = \sigma_c n \bar{v} t \qquad (7.20)$$

The collision frequency p_c, that is the number of collisions per unit time, is given by

$$p_c = \frac{1}{\bar{\tau}} = \sigma_c n \bar{v} \qquad (7.21)$$

where, as before, $\bar{\tau}$ is the average time between two successive collisions. From the kinetic gas theory it is known that $\bar{v}$ depends on the temperature T and on the mass m of the molecules as expressed by

$$\bar{v} = \sqrt{\frac{8kT}{\pi m}} \qquad (7.22)$$

where k is the *Boltzmann constant*. For a brief derivation of $\bar{v}$ see Appendix 7.6.1. The pressure p of an ideal gas is given by

$$p = nkT \tag{7.23}$$

Substituting (7.22) and (7.23) into (7.21) we obtain for the reciprocal of the average collision time

$$\frac{1}{\bar{\tau}} = \sigma_c \frac{p}{kT} \sqrt{\frac{8kT}{\pi m}} = C \frac{p}{\sqrt{T}} \tag{7.24}$$

The constant C depends on the gas.

We are now ready to give an expression for the pressure and temperature dependence of the half-width of the Lorentz line. Since $\alpha_L = 1/2\pi\bar{\tau}$ we find

$$\alpha_L(p, T) = \frac{C}{2\pi} \frac{p}{\sqrt{T}} \tag{7.25}$$

For reference values of the pressure and temperature (p_0, T_0) we denote the half-width by

$$\alpha_{L,0} = \alpha_L(p_0, T_0) = \frac{C}{2\pi} \frac{p_0}{\sqrt{T_0}} \tag{7.26}$$

Utilizing this equation in (7.25) gives

$$\boxed{\alpha_L(p, T) = \alpha_{L,0} \frac{p}{p_0} \sqrt{\frac{T_0}{T}}} \tag{7.27}$$

Thus it is seen that the Lorentz half-width depends linearly on pressure and thus decreases with height. Furthermore, the temperature dependence of α_L is relatively weak as compared to the pressure dependence so that it is sometimes completely ignored. The standard half-width $\alpha_{L,0}$ can be determined by experiment or from quantum theory. Values of $\alpha_{L,0}$ for atmospheric gases can be found in tabulated form in the literature.

As we have seen, the average time $\bar{\tau}$ can be determined from classical gas kinetic theory enabling an estimate of α_L. However, such an estimate is several times too small. Although the absolute magnitude of α_L cannot be found accurately, the pressure dependence given in (7.27) is accurately followed. A more detailed discussion on this subject is given, for example, in Goody (1964a).

7.1.2 The thermal Doppler line

At high altitudes in the atmosphere, that is at low pressure, the collision broadening becomes less important while at the same time the *Doppler shift* in frequency

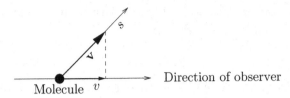

Fig. 7.3 Molecule moving with velocity component v towards the observer.

becomes the main line broadening effect. Let us assume that a molecule which is moving with velocity **v** along a path s has the velocity component v towards the observer, see Figure 7.3. If the frequency of the radiation emitted by the stationary molecule is ν_0, then the observed frequency ν_0' is given by

$$\nu_0' = \nu_0 \left(1 + \frac{v}{c}\right), \qquad |v| \ll c \tag{7.28}$$

where c is the speed of light. Hence, the Doppler shift in frequency is

$$\Delta\nu_0' = \nu_0' - \nu_0 = \nu_0 \frac{v}{c} \tag{7.29}$$

The number concentration of molecules, dn, belonging to the velocity interval $(v, v + dv)$ can be obtained from Maxwell's one-dimensional velocity distribution, see also Appendix 7.5.1

$$dn = n\sqrt{\frac{m}{2\pi kT}} \exp\left(-\frac{mv^2}{2kT}\right) dv$$

$$= \frac{n}{\sqrt{\pi}\, v_0} \exp\left[-\left(\frac{v}{v_0}\right)^2\right] dv \quad \text{with} \quad v_0 = \sqrt{\frac{2kT}{m}} \tag{7.30}$$

Due to the square of the velocity in the exponent the component v can either be positive (towards the observer) or negative (away from the observer). In order to account for the Doppler effect of arbitrary v we have to integrate over the Maxwell distribution as illustrated in Figure 7.4. Then the average of the velocity squared is given by

$$\overline{v^2} = \frac{1}{n} \int_{-\infty}^{\infty} v^2 dn = \frac{1}{\sqrt{\pi}\, v_0} \int_{-\infty}^{\infty} v^2 \exp\left[-\left(\frac{v}{v_0}\right)^2\right] dv = \frac{v_0^2}{2} \tag{7.31}$$

First we consider the case of pure Doppler broadening, that is we neglect the effect of both natural as well as *pressure broadening*. In the following section we will discuss the combined effect of pressure and Doppler broadening. If the

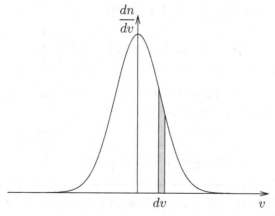

Fig. 7.4 Maxwell's one-dimensional velocity distribution for the component v.

molecule is at rest, then the absorption coefficient is proportional to the Dirac δ-function

$$k_v = S\delta(v - v_0) \tag{7.32}$$

Thus, according to (7.16) the *line-shape factor for monochromatic emission* is given by $\delta(v - v_0)$. If we take the Doppler effect due to a particular velocity v into account we obtain

$$k_v = S\delta\left(v - v_0'\right) \tag{7.33}$$

Note that this equation corresponds to the emission or absorption by a non-broadened *monochromatic line* with frequency v_0'. From (7.28) we see that instead of integrating over all possible velocities v it is equivalent to integrate over all corresponding frequencies v_0'. The right-hand side of (7.33) corresponds to a situation where only a single frequency shift occurs. Therefore, the absorption coefficient for all possible frequency shifts can be obtained from

$$k_v = S \int_{-\infty}^{\infty} P\left(v_0'\right)\delta\left(v - v_0'\right)dv_0' \tag{7.34}$$

where the probability distribution function $P(v_0')$ follows directly from the Maxwell distribution

$$P\left(v_0'\right)dv_0' = \frac{dn}{n} = \frac{1}{\sqrt{\pi}\,v_0}\exp\left[-\left(\frac{v}{v_0}\right)^2\right]dv \tag{7.35}$$

From (7.28) we have $dv'_0 = (v_0/c)dv$ so that

$$P(v'_0) = \frac{c}{\sqrt{\pi}\, v_0 v_0} \exp\left[-\left(\frac{v}{v_0}\right)^2\right] \tag{7.36}$$

Substituting this expression into (7.34) and evaluating the integral results in the *absorption coefficient of the Doppler line*

$$k_{v,\mathrm{D}} = \frac{Sc}{\sqrt{\pi}\, v_0 v_0} \exp\left[-\left(\frac{(v - v_0)c}{v_0 v_0}\right)^2\right] \tag{7.37}$$

The maximum of the absorption coefficient occurs at the line center where $v = v_0$ and is given by

$$k_0 = \frac{Sc}{\sqrt{\pi}\, v_0 v_0} \tag{7.38}$$

Finally, we determine the particular frequency $v = v_0 \pm \alpha_\mathrm{D}$ where $k_{v,\mathrm{D}} = k_0/2$. This leads to the *half-width of the Doppler line* which has the form

$$\alpha_\mathrm{D} = \sqrt{\ln 2}\, \frac{v_0 v_0}{c} = \sqrt{\ln 2}\, \frac{v_0}{c}\sqrt{\frac{2kT}{m}} \tag{7.39}$$

It is noteworthy that the Doppler half-width depends on temperature only but not on pressure.

By comparing (7.16) with (7.37) one may easily see that the *line-shape factor of the Doppler line* is given by

$$f_\mathrm{D}(v - v_0) = \frac{\sqrt{\ln 2}}{\sqrt{\pi}\,\alpha_\mathrm{D}} \exp\left[-\left(\frac{(v - v_0)\sqrt{\ln 2}}{\alpha_\mathrm{D}}\right)^2\right] \tag{7.40}$$

where use was made of (7.39). Finally, it is not difficult to verify that, in accordance with (7.17), the Doppler line-shape factor is also normalized.

7.1.3 The Voigt profile

A comparison of broadening effects due to molecular collisions and the Doppler effect reveals that pressure broadening dominates in the troposphere and lower stratosphere while Doppler broadening is most important in atmospheric layers above 50 km. At sea level the Doppler half-width α_D is about two orders of magnitude smaller than α_L. In an altitude range from about 30 to 50 km, however, both mechanisms need to be considered since they are in the same order of magnitude. As already mentioned, in the meteorologically relevant part of the Earth's atmosphere natural line broadening remains negligibly small in comparison to collisional and *thermal broadening*.

In order to combine the pressure and Doppler broadening effect we apply Doppler broadening to a line which has already been broadened by the pressure effect. This is achieved by replacing in (7.32)–(7.34) the Dirac δ-function, expressing the monochromatic emission, by the Lorentz line-shape factor, expressing the pressure broadening. Instead of (7.34) we then obtain

$$k_\nu = S \int_{-\infty}^{\infty} P(\nu_0') f_L(\nu - \nu_0') d\nu_0' \tag{7.41}$$

By comparing (7.36) with (7.40) one may easily verify that

$$P(\nu_0') = f_D(\nu_0' - \nu_0) \tag{7.42}$$

Substituting this expression into (7.41) yields the *absorption coefficient of the Voigt line*

$$\boxed{k_{\nu,V} = S \int_{-\infty}^{\infty} f_D(\nu_0' - \nu_0) f_L(\nu - \nu_0') d\nu_0' = S f_V(\nu - \nu_0)} \tag{7.43}$$

Here, $f_V(\nu - \nu_0)$ is the *line-shape factor of the Voigt line* which is given by the convolution of the Lorentz and the Doppler line-shape factors, i.e.

$$f_V(\nu - \nu_0) = f_V(y) = \int_{-\infty}^{\infty} f_D(x) f_L(y - x) dx \tag{7.44}$$

with $x = \nu_0' - \nu_0$ and $y = \nu - \nu_0$. This equation reflects the fact that the signal of the product of two spectra is equivalent to the convolution of the individual signals. Substituting (7.18) and (7.40) into (7.44) yields for the Voigt line-shape factor

$$\boxed{f_V(\nu - \nu_0) = \frac{\sqrt{\ln 2}}{\sqrt{\pi^3} \alpha_D} \int_{-\infty}^{\infty} \exp\left[-\left(\frac{x\sqrt{\ln 2}}{\alpha_D}\right)^2\right] \frac{\alpha_L}{\alpha_L^2 + (y - x)^2} dx} \tag{7.45}$$

Integration of this equation over all frequencies shows that the Voigt line-shape factor is also normalized. Furthermore, it may be easily verified that in the limits $\alpha_D \to 0$ and $\alpha_L \to 0$ the Voigt line approaches the Lorentz and Doppler line, respectively.

Unfortunately, it is not possible to give an analytical solution to the integral in (7.45). However, it causes no problems to find a numerical solution. Nevertheless, in the literature there exist several analytical approximations for the Voigt line (e.g. Penner, 1959; Fels, 1979). Fomichev and Shved (1985) suggest the following formula

$$
\begin{aligned}
f_V(\nu - \nu_0) = & \sqrt{\frac{\ln 2}{\pi}} \frac{1 - \xi}{\alpha_V} \exp(-\ln 2\eta^2) + \frac{\xi}{\pi \alpha_V (1 + \eta^2)} \\
& - \frac{\xi(1 - \xi)}{\pi \alpha_V} \left(\frac{3}{2 \ln 2} + 1 + \xi \right) \\
& \times \left(0.066 \exp(-0.4\eta^2) - \frac{1}{40 - 5.5\eta^2 + \eta^4} \right)
\end{aligned}
\tag{7.46}
$$

with $\xi = \alpha_L/\alpha_V$ and $\eta = (\nu - \nu_0)/\alpha_V$. The term α_V is the *half-width of the Voigt line* and is given by

$$
\alpha_V = 0.5 \left(\alpha_L + \sqrt{\alpha_L^2 + 4\alpha_D^2} \right) + 0.05\alpha_L \left(1 - \frac{2\alpha_L}{\alpha_L + \sqrt{\alpha_L^2 + 4\alpha_D^2}} \right)
\tag{7.47}
$$

Generally, this approximation yields results with an accuracy of less than 3%. Numerical evaluation of (7.46) reveals that the approximate form of the Voigt line-shape factor is also normalized and it approaches the Doppler and Lorentz line for $\alpha_L \to 0$ and $\alpha_D \to 0$.

As an example, Figure 7.5 shows the three line types for the half-widths $\alpha_L = \alpha_D$. The curves are plotted as function of $x = (\nu - \nu_0)/\alpha_L$ and they are normalized with respect to $f_D(0)$. From this figure we conclude that collisional line broadening dominates in the line wings, whereas near the line center $f_D > f_L$. The effect of the Voigt line is to increase the absorption in the wings and to decrease it in the center of the line as compared to the Lorentz and the Doppler line. The same behavior of the Voigt line is also observed for other choices of the half-widths with $\alpha_L \lessgtr \alpha_D$.

However, if there is practically total absorption near the center of the spectral line, the simpler Lorentz shape can still be used instead of the more complicated Voigt shape. This is due to the fact that, for very strong absorption, the particular shape of the line near its center is of secondary importance. A detailed discussion of the Voigt profile can be found, for example, in Unsöld (1968).

7.2 Band models

Inspection of the absorption spectrum in the short-wave and long-wave spectral region reveals that there exist many absorption lines for different molecular species. With increasing spectral resolution the structure of the absorption spectrum

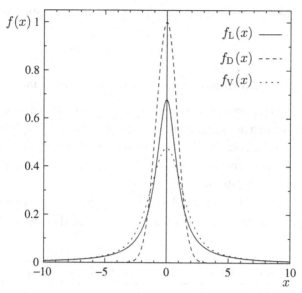

Fig. 7.5 Comparison of the Lorentz, the Doppler and the Voigt line-shape factors with $\alpha_L = \alpha_D$ and $x = (\nu - \nu_0)/\alpha_L$.

increases in complexity. For a particular gas the positions and the strengths of the spectral lines are calculable with the help of the quantum theory of radiation. It is found that there exist approximately 50 000 relevant lines for water vapor which have to be considered in detailed spectral calculations for atmospheric problems. Truly exact calculations of the absorption spectrum are carried out by the so-called *line-by-line models*, meaning that the absorption coefficient must be determined along the shape of each individual spectral line. In order to determine the radiation field these calculations have to be performed for many altitudes within the atmosphere because gaseous concentrations, half-widths and line intensities depend on pressure and temperature.

One can easily imagine that radiative transfer by the exact line-by-line method requires an extraordinary computational effort. Such benchmark calculations are normally carried out for a few representative model atmospheres only. For routine purposes, however, one needs to design simpler approaches which do not treat the spectroscopic structure in detail, but still take the main features of the absorption spectrum into account. These are the so-called *band models*.

A view of the absorption spectrum gives the impression that the positions of the lines are randomly placed and that the line intensities are distributed according to some statistical law. In reality, the spectral arrangement of the lines is not completely random since line positions and intensities can be determined via quantum mechanical laws. Nevertheless, random line position and statistical distribution laws for line intensities may be used to determine the absorption of simple band models.

7.2.1 Mean absorption in a single Lorentz line

In the following we will first discuss the simple but important case of absorption by a single Lorentz line in a homogeneous atmosphere. The resulting mean absorption of such a spectral line will be used to find the limiting forms of very weak and very strong absorption. These results in turn will be needed as closure assumptions of various band models to be discussed later.

The *monochromatic absorption* $\mathcal{A}_\nu(u)$ in a single line depends on the spectral absorption coefficient and the optical mass u of the gas in the following way

$$\boxed{\mathcal{A}_\nu(u) = 1 - \exp(-k_\nu u) = 1 - \mathcal{T}_\nu(u) \quad \text{with} \quad \mathcal{T}_\nu(u) = \exp(-k_\nu u)} \quad (7.48)$$

where u was defined in (2.118). $\mathcal{T}_\nu(u)$ is the monochromatic transmission of the single line. This fundamental quantity represents that part of the incident radiation which remains after passage of the optical mass.

The single line model can be extended to more general situations involving the absorption in several well-separated lines so that no appreciable *line overlap* ocurs. If there exist N different absorbers with absorption coefficients $k_{\nu,i}$ and optical masses u_i the total monochromatic transmission is given by

$$\boxed{\mathcal{T}_\nu = \exp\left(-\sum_{i=1}^{N} k_{\nu,i} u_i\right) = \prod_{i=1}^{N} \mathcal{T}_{\nu,i}(u_i)} \quad (7.49)$$

indicating that the individual monochromatic transmissions can be multiplied to yield the total monochromatic effect. It should be noted, however, that this *multiplication property of the transmissions* is not strictly true for the transmission of band intervals.

A convenient way to define the spectrally integrated absorption is by means of the *equivalent width* $W(u)$

$$W(u) = \int_{-\infty}^{\infty} \mathcal{A}_\nu(u) d\nu = \int_{-\infty}^{\infty} [1 - \exp(-k_\nu u)] d\nu \quad (7.50)$$

The relationship between W and u is also called the *curve of growth*. The notion of the equivalent width stems from the fact that a rectangular line with width W, whose line center is completely absorbed ($\mathcal{A}_\nu = 1$), gives rise to the same area under $\mathcal{A}_\nu$.

Instead of $W(u)$, which has the dimension of a frequency, one defines the average absorption $A(u)$ of a single line over a frequency interval Δv taken sufficiently wide as[1]

$$A(u) = \frac{W(u)}{\Delta v} = \frac{1}{\Delta v} \int_{-\infty}^{\infty} A_v(u) dv = \frac{1}{\Delta v} \int_{-\infty}^{\infty} [1 - \exp(-k_v u)] dv \qquad (7.51)$$

Introducing in this equation the absorption coefficient of the Lorentz line (7.15) and using the substitutions

$$\bar{u} = \frac{Su}{2\pi \alpha_L}, \qquad x = \frac{v - v_0}{\alpha_L} \qquad (7.52)$$

we obtain the expression

$$A(\bar{u}) = \frac{\alpha_L}{\Delta v} \int_{-\infty}^{\infty} \left[1 - \exp\left(-\frac{2\bar{u}}{(1 + x^2)} \right) \right] dx \qquad (7.53)$$

Integration by parts results in

$$A(\bar{u}) = \frac{4\bar{u}\alpha_L}{\Delta v} \int_{-\infty}^{\infty} \frac{x^2 \exp(-2\bar{u}/(1 + x^2))}{(1 + x^2)^2} dx$$

$$= \frac{4\bar{u}\alpha_L}{\Delta v} \int_{-\infty}^{\infty} \frac{\exp(-2\bar{u}/(1 + x^2))}{(1 + x^2)} dx - \frac{4\bar{u}\alpha_L}{\Delta v} \int_{-\infty}^{\infty} \frac{\exp(-2\bar{u}/(1 + x^2))}{(1 + x^2)^2} dx$$

$$(7.54)$$

In the next step we may treat the variable $\bar{u}$ as a parameter in the integral of (7.53). Thus, the first and second derivative of A with respect to $\bar{u}$ are

$$\frac{dA}{d\bar{u}} = \frac{2\alpha_L}{\Delta v} \int_{-\infty}^{\infty} \frac{\exp(-2\bar{u}/(1 + x^2))}{(1 + x^2)} dx,$$

$$(7.55)$$

$$\frac{d^2 A}{d\bar{u}^2} = -\frac{4\alpha_L}{\Delta v} \int_{-\infty}^{\infty} \frac{\exp(-2\bar{u}/(1 + x^2))}{(1 + x^2)^2} dx$$

Substituting these expressions into (7.54) gives an ordinary linear differential equation of second order for $A(\bar{u})$

$$\bar{u}\frac{d^2 A}{d\bar{u}^2} + 2\bar{u}\frac{dA}{d\bar{u}} - A(\bar{u}) = 0 \qquad (7.56)$$

This differential equation can be easily solved by means of Laplace transforms. Let us define the Laplace transform of $A(\bar{u})$ as

$$\mathcal{L}[A(\bar{u})] = \int_0^{\infty} \exp(-s\bar{u}) A(\bar{u}) d\bar{u} = y(s) \qquad (7.57)$$

---

[1] Many authors denote the average absorption by $\bar{A}(u)$. This notation is unnecessarily complicated since averaging implies an integration over v which is just as well indicated by $A(u)$.

Then the properties of the Laplace transform yield

$$\mathcal{L}\left[\frac{d\mathcal{A}}{d\bar{u}}\right] = sy(s) - \mathcal{A}(0), \quad \mathcal{L}\left[\frac{d^2\mathcal{A}}{d\bar{u}^2}\right] = s^2 y(s) - s\mathcal{A}(0) - \frac{d\mathcal{A}}{d\bar{u}}(0) \quad (7.58)$$

The two boundary conditions $\mathcal{A}(0)$ and $(d\mathcal{A}/d\bar{u})(0)$ to evaluate the Laplace transforms can be obtained very easily. From physical reasoning we must have $\mathcal{A}(0) = 0$. To obtain the first derivative of $\mathcal{A}$ at $\bar{u} = 0$ we expand (7.51) in a power series. For small values of u we find

$$u \ll 1: \quad \mathcal{A}(u)\Delta v = Su, \quad \mathcal{A}(\bar{u}) = \frac{2\pi\alpha_L\bar{u}}{\Delta v}, \quad \frac{d\mathcal{A}}{d\bar{u}}(0) = \frac{2\pi\alpha_L}{\Delta v} \quad (7.59)$$

The additional transform rule

$$\mathcal{L}[\bar{u}f(\bar{u})] = -\frac{df}{ds} \quad \text{with} \quad \mathcal{L}[f(\bar{u})] = f(s) \quad (7.60)$$

can then be used to formulate (7.56) in transformed space as

$$\frac{dy}{ds} + \frac{2y}{s+2} + \frac{3y}{s(s+2)} = 0 \quad (7.61)$$

It may be easily verified that the solution to this differential equation is given by

$$y(s) = \frac{C}{\sqrt{(s+2)s^3}} \quad (7.62)$$

where C is a constant of integration.

The inverse Laplace transform of $y(s)$ may be found in transform tables and is given by

$$\mathcal{L}^{-1}[y(s)] = \mathcal{A}(\bar{u}) = C\bar{u}\exp(-\bar{u})[I_0(\bar{u}) + I_1(\bar{u})] \quad (7.63)$$

where I_0, I_1 are the *modified Bessel functions of the first kind*. The constant C is evaluated with the help of (7.59) and by observing that

$$I_0(0) = 1, \quad I_1(0) = I_2(0) = 0$$
$$\frac{dI_0(\bar{u})}{d\bar{u}} = I_1(\bar{u}), \quad 2\frac{dI_1(\bar{u})}{d\bar{u}} = I_0(\bar{u}) + I_2(\bar{u}) \quad (7.64)$$

This results in

$$C = \frac{2\pi\alpha_L}{\Delta v} \quad (7.65)$$

which is required to evaluate (7.63). As final result we obtain the average absorption of an isolated Lorentz line

$$\boxed{\mathcal{A}(\bar{u}) = \frac{2\pi\alpha_L}{\Delta v}\bar{u}\exp(-\bar{u})[I_0(\bar{u}) + I_1(\bar{u})]} \quad (7.66)$$

This important result is due to Ladenburg and Reiche (1913) who originally used the *Bessel functions of the first kind* J_0 and J_1 with pure imaginary argument to express $A(\bar{u})$. In the literature the *Ladenburg and Reiche function* $L(\bar{u})$ is defined as

$$\boxed{L(\bar{u}) = \bar{u}\exp{(-\bar{u})}\left[I_0(\bar{u}) + I_1(\bar{u})\right]} \qquad (7.67)$$

so that

$$\boxed{A(\bar{u}) = \frac{2\pi\alpha_L}{\Delta\nu}L(\bar{u})} \qquad (7.68)$$

The Laplace transform method to find the Ladenburg and Reiche function is due to Zdunkowski (1974). The original method to obtain $L(\bar{u})$ is outlined in Appendix 7.6.2.

Let us discuss the average absorption of an isolated Lorentz line for the two important cases $\bar{u} \ll 1$ and $\bar{u} \gg 1$. For small $\bar{u}$ the I_n may be expressed as infinite series of the form (Abramowitz and Stegun, 1972)

$$I_n(x) = \sum_{k=0}^{\infty} \frac{1}{k!\,\Gamma(n+k+1)}\left(\frac{x}{2}\right)^{2k+n} \qquad (7.69)$$

where $\Gamma(x)$ is the Gamma function which is defined by the definite integral

$$\Gamma(x) = \int_0^{\infty} \exp{(-t)}\,t^{x-1}dt \qquad (7.70)$$

For a positive integer n one finds $\Gamma(n+1) = n!$ The leading terms of the power series for I_0 and I_1 are then given by

$$I_0(\bar{u}) = 1 + \left(\frac{\bar{u}}{2}\right)^2 + \frac{1}{4}\left(\frac{\bar{u}}{2}\right)^4 + \cdots$$
$$I_1(\bar{u}) = \frac{\bar{u}}{2} + \frac{1}{2}\left(\frac{\bar{u}}{2}\right)^3 + \frac{1}{12}\left(\frac{\bar{u}}{2}\right)^5 + \cdots \qquad (7.71)$$

For $\bar{u} \ll 1$ it is sufficiently accurate if we retain only the first two terms of these series. Furthermore, since $\exp{(-\bar{u})} \approx 1 - \bar{u}$ the third-order expansion of the average absorption is found to be

$$A(\bar{u}) \approx \frac{2\pi\alpha_L}{\Delta\nu}\left(\bar{u} - \frac{\bar{u}^2}{2} - \frac{\bar{u}^3}{4}\right) \qquad (7.72)$$

The so-called *weak line approximation* is defined as the linear part of the absorption law (7.72)

$$\boxed{A_w(\bar{u}) \approx \frac{2\pi\alpha_L}{\Delta\nu}\bar{u} \quad \text{or} \quad A_w(u) \approx \frac{Su}{\Delta\nu}} \qquad (7.73)$$

This very important result shows that very weak absorption varies linearly with u. Moreover, it is found that the absorption is independent of the half-width of the spectral line. Since in (7.51) no special form of k_v is involved, it may be easily seen that for $u \ll 1$ the general form of the weak line approximation (7.73) is valid for arbitrary line shapes.

The second important case is known as strong absorption, i.e. $\bar{u} \gg 1$. Now we introduce the asymptotic expressions for the modified Bessel functions (Abramowitz and Stegun, 1972)

$$I_n(x) = \frac{\exp(x)}{\sqrt{2\pi x}} \left[1 + \mathcal{O}\left(\frac{1}{x}\right) \right] \tag{7.74}$$

We immediately find the so-called *strong line approximation* or *square-root law*

$$\boxed{ \mathcal{A}_s(\bar{u}) \approx \frac{2\alpha_L}{\Delta v} \sqrt{2\pi \bar{u}} \quad \text{or} \quad \mathcal{A}_s(u) = \frac{2}{\Delta v} \sqrt{S\alpha_L u} } \tag{7.75}$$

It should be noted that, contrary to the weak line law, the strong line approximation for certain combinations of S, α_L and u might give an unphysical value $\mathcal{A} > 1$. Thus, one should correctly limit the average absorption in (7.75) by 1. In contrast, the weak line limit is always bounded.

In the following we will seek a physical interpretation for these limiting values of the average absorption. In a pressure-broadened spectral line the monochromatic transmission reads

$$\mathcal{T}_v(\bar{u}) = \exp\left(-\frac{2\alpha_L^2 \bar{u}}{v^2 + \alpha_L^2} \right) \tag{7.76}$$

Figure 7.6 depicts the absorption and transmission by a single Lorentz line for different values of the parameter $\bar{u}$. It can be seen that near the line center $\mathcal{T}_v$ gradually approaches zero for increasing $\bar{u}$. For $\bar{u} = 5$ the absorption is already complete ($\mathcal{A}_v = 1$) as long as v stays within a distance of one half-width from the line center.

For $v \gg \alpha_L$ we observe that in the denominator of $\mathcal{T}_v$ in (7.76) the term α_L^2 can safely be neglected in comparison to the term v^2. In fact, already for $v > 10\alpha_L$ this gives rather accurate approximations. Furthermore, if we assume that $\bar{u}$ is very large, then the absorption near the line center is complete. Neglecting α_L^2 in the denominator of (7.76) does not appreciably change the monochromatic absorption for $\bar{u} \gg 1$ and for any $\bar{u}$ far from the line center. In summary, for these two cases we find the following approximate form for the monochromatic transmission

$$\mathcal{T}_v(\bar{u}) \approx \exp\left(-\frac{2\alpha_L^2 \bar{u}}{v^2} \right) \tag{7.77}$$

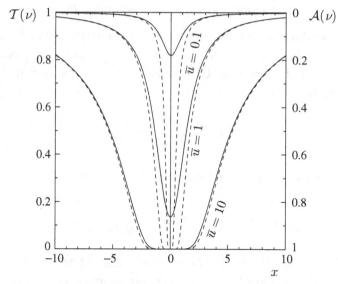

Fig. 7.6 Comparison of the transmission function (7.76) (full curves) with the approximate form (7.77) (dashed curves) for different values of the variable $\bar{u}$ with $x = \nu/\alpha_L$.

The accuracy of this approximation can be inferred from the dashed curves in Figure 7.6. As expected for large $\bar{u}$ and for ν far away from the center line the approximation is very good. For small $\bar{u}$, however, extremely large errors are observed near the center line.

Using (7.52) and (7.53) it is easy to find an expression for the average absorption of an isolated Lorentz line for large $\bar{u}$. Setting $\nu_0 = 0$ and introducing the substitution $\xi = 2\alpha_L^2 \bar{u}/\nu^2$ we again obtain the strong line approximation

$$A(\bar{u}) \approx \frac{\alpha_L}{\Delta\nu}\sqrt{2\bar{u}} \int_0^\infty (1 - \exp(-\xi))\xi^{-3/2}d\xi = \frac{2\alpha_L}{\Delta\nu}\sqrt{2\pi\bar{u}} \tag{7.78}$$

In case of very weak absorption the linear expansion of the exponential (7.76) leads to

$$\mathcal{T}_\nu(\bar{u}) \approx 1 - \frac{2\alpha_L^2\bar{u}}{\nu^2 + \alpha_L^2} \tag{7.79}$$

For $\nu/\alpha_L \gg 1$ and small values for $\bar{u}$ we obtain $\mathcal{T}_\nu(\bar{u}) \approx 1$ in accordance with Figure 7.6.

7.2.2 Band model for nonoverlapping lines

The simplest band model to be treated is a spectral interval $\Delta \nu = N\delta$ containing N nonoverlapping lines of mean line spacing δ. Thus the average equivalent width $W(u)$ of all lines in the band can be obtained by summation over the individual equivalent widths W_i, i.e.

$$W(u) = \frac{1}{N} \sum_{i=1}^{N} W_i(u) \tag{7.80}$$

The average band absorption then follows from

$$\mathcal{A}(u) = \frac{1}{N\delta} \sum_{i=1}^{N} W_i(u) = \frac{1}{N} \sum_{i=1}^{N} \mathcal{A}_i(u) \tag{7.81}$$

It is useful to consider the two limiting cases of weak and strong absorption. In the weak line limit the individual lines have the average absorption $\mathcal{A}_{w,i}(u) = S_i u / \delta$. Thus, for N nonoverlapping lines we obtain the weak line limit

$$\boxed{\mathcal{A}_w(u) = \frac{1}{N\delta} \sum_{i=1}^{N} S_i u} \tag{7.82}$$

In this limit $\mathcal{A}$ is independent of the line shape. For the strong line limit of N nonoverlapping Lorentz lines we similarly obtain

$$\boxed{\mathcal{A}_s(u) = \frac{2}{N\delta} \sum_{i=1}^{N} \sqrt{S_i \alpha_{L,i} u}} \tag{7.83}$$

In general, the band model for nonoverlapping spectral lines does not apply to atmospheric conditions. Nevertheless, the model results will be needed in the construction of realistic atmospheric *transmission functions*.

7.2.3 Random band models

For triatomic molecules such as H_2O, the band structure is very complex and gives the impression that the spectral lines are randomly spaced. We will use several steps to construct a general random model.

First we consider an infinite array of identical lines of unspecified line shape. Suppose that N lines are distributed randomly in the interval between $-N\delta/2$ and $N\delta/2$ where δ is the mean line spacing. If a line is centered at $\nu = \nu_i$ within this interval, the contribution of this line to the absorption coefficient at the center of the interval at $\nu = 0$ is $k_{\nu,i}$. The contribution from all N lines in the spectral interval is $\sum_{i=1}^{N} k_{\nu,i}$. Therefore, the resultant transmission is given by

$$\mathcal{T}_{\nu=0}(u) = \exp\left(-\sum_{i=1}^{N} k_{\nu,i} u\right) = \prod_{i=1}^{N} \exp(-k_{\nu,i} u) \tag{7.84}$$

We have assumed that the lines in this superficial spectrum are randomly spaced. Due to the random spacing, the probability for a line to be located in the interval ν_i to $\nu_i + d\nu_i$ is $d\nu_i/\delta$. Consequently, the joint probability for lines to be located in the intervals extending from ν_1 to $\nu_1 + d\nu_1$, ν_2 to $\nu_2 + d\nu_2$, ..., ν_i to $\nu_i + d\nu_i$, ..., ν_N to $\nu_N + d\nu_N$ is given by $\prod_{i=1}^{N} d\nu_i/\delta$.

In order to allow for all possible arrangements of lines we must permit each line to be located anywhere in the interval $\Delta\nu$ extending from $-N\delta/2$ to $N\delta/2$. Therefore, the average value of (7.84) over the interval is given by

$$\mathcal{T}(u) = \frac{\displaystyle\prod_{i=1}^{N} \frac{1}{\delta} \int_{-N\delta/2}^{N\delta/2} \exp(-k_{\nu,i} u) d\nu_i}{\displaystyle\prod_{i=1}^{N} \frac{1}{\delta} \int_{-N\delta/2}^{N\delta/2} d\nu_i} \tag{7.85}$$

Assuming that each arrangement of lines is equally probable then (7.85) can be simplified substantially. Instead of the product $\prod_{i=1}^{N}$ we obtain the N-th power so that (7.85) can be written as

$$\mathcal{T}(u) = \left[\frac{1}{N\delta} \int_{-N\delta/2}^{N\delta/2} \exp\left(-k_\nu u\right) d\nu\right]^{N} = \left[1 - \frac{1}{N\delta} \int_{-N\delta/2}^{N\delta/2} [1 - \exp(-k_\nu u)] d\nu\right]^{N} \tag{7.86}$$

If we permit the number of lines to be very large ($N \to \infty$) then this equation becomes the exponential function

$$\mathcal{T}(u) = \exp\left[-\frac{1}{\delta} \int_{-\infty}^{\infty} [1 - \exp(-k_\nu u)] d\nu\right] = \exp\left(-\frac{W(u)}{\delta}\right) \tag{7.87}$$

where we have used the definition (7.50) of the equivalent width. It should be noted that no particular line profile has been specified in the derivation of (7.87). In general, we will use the Lorentz profile to specify the absorption coefficient.

The artificial situation of lines with equal intensity will now be generalized to a more realistic spectrum consisting of randomly distributed lines with different intensities. Let us consider a certain frequency range containing N lines. Each of these is assumed to belong to an infinite array of randomly spaced lines of equal intensity and mean spacing $N\delta$. According to (7.87) the mean transmission of array i is given by

$$T_i(u) = \exp\left(-\frac{W_i(u)}{N\delta}\right) \tag{7.88}$$

Due to the assumption of randomness in each array we may expect that the multiplication property (7.49) also applies so that

$$T(u) = \prod_{i=1}^{N} T_i(u) = \exp\left(-\frac{1}{N\delta}\sum_{i=1}^{N} W_i(u)\right) \tag{7.89}$$

or

$$\boxed{T_G(u) = \exp\left(-\frac{W(u)}{\delta}\right) = \exp[-\mathcal{A}(u)]} \tag{7.90}$$

The transmission function $T_G(u)$ is due to Goody (1952). Sometimes (7.90) is also attributed to Mayer (1947) and one speaks of the *Mayer–Goody model*.

In the following we will show how to find $\mathcal{A}(u)$. Since (7.90) refers to a realistic spectrum, the line intensities of the N lines vary in strength. In order to account for this situation various probability distributions $p(S)$ for the distributions of S have been proposed in the literature. We are going to discuss three prominent models.

Goody's exponential model

Goody (1952) proposed the following distribution for the line intensities

$$p(S) = \frac{1}{\sigma}\exp(-S/\sigma) \tag{7.91}$$

where σ is the average value of the line intensity in the given spectral interval. The quantity $p(S)dS$ then gives the percentage of lines in the spectral interval which belong to the line intensity interval $(S, S+dS)$. It is easily seen that the probability distribution (7.91) fulfils the normalization condition

$$\int_0^\infty p(S)dS = 1 \tag{7.92}$$

The expectation value for the absorption $\mathcal{A}(u)$ due to the exponential model is obtained from

$$\mathcal{A}(u) = \int_0^\infty p(S)\mathcal{A}(S,u)dS$$

$$\text{with} \quad \mathcal{A}(S,u) = \frac{1}{\delta}\int_{-\infty}^\infty \left[1 - \exp\left(-f_v Su\right)\right]dv \tag{7.93}$$

Introducing (7.91) into (7.93) yields

$$A(u) = \int_0^\infty \frac{1}{\sigma} \exp\left(-\frac{S}{\sigma}\right) \frac{1}{\delta} \int_{-\infty}^\infty [1 - \exp(-f_\nu S u)] d\nu dS = \frac{1}{\delta} \int_{-\infty}^\infty \frac{f_\nu u \sigma}{1 + f_\nu u \sigma} d\nu \tag{7.94}$$

In the following we will use the line-shape factor of the Lorentz line (7.18) for $\nu_0 = 0$. Employing the substitutions

$$\bar{u} = \frac{\sigma u}{2\pi \alpha_L}, \qquad y = \frac{\alpha_L}{\delta} \tag{7.95}$$

we obtain for the average absorption of the exponential model

$$A(u) = \int_{-\infty}^\infty \frac{2\bar{u}\alpha_L y}{\nu^2 + \alpha_L^2(1 + 2\bar{u})} d\nu = \frac{2\pi \bar{u} y}{\sqrt{1 + 2\bar{u}}} \tag{7.96}$$

The parameters σ and α_L, or equivalently $\bar{u}$ and y, have not been specified so far. They can be determined by the so-called *matching principle*, i.e. one forces the random band model to obey the limiting forms of the weak and the strong line approximation (7.82) and (7.83) of the band model with non-overlapping lines. Thus the matching principle can also be viewed as the closure assumption of the random band models. For the weak line and strong line approximation we find using (7.96)

$$\bar{u} \ll 1: \qquad 2\pi y \bar{u} = \frac{1}{N\delta} \sum_{i=1}^N S_i u$$

$$\bar{u} \gg 1: \qquad \pi y \sqrt{2\bar{u}} = \frac{2}{N\delta} \sum_{i=1}^N \sqrt{S_i \alpha_{L,i} u} \tag{7.97}$$

These two equations can be used to determine the unknown parameters $\bar{u}$ and y. The result is

$$\bar{u} = \frac{u}{8} \frac{C^2}{D^2}, \qquad y = \frac{4}{\pi N\delta} \frac{D^2}{C}$$

$$\text{with} \quad C = \sum_{i=1}^N S_i, \qquad D = \sum_{i=1}^N \sqrt{S_i \alpha_{L,i}} \tag{7.98}$$

The exponential distribution for S as assumed in the Goody model involves a range of line intensities. It is clear that the strongest lines will be most important for very short path lengths, while the very weak line intensities dominate the transmission of radiation over very long path lengths. Therefore, in order to correctly describe the mean transmission for arbitrary path lengths, the distribution function for S must be carefully chosen. While the exponential distribution leads to a simple expression, in several cases the model provides rather inaccurate representations of the line intensities in atmospheric bands. For example, in the case of water vapor, the exponential distribution fails to account for the large number of weak lines in the bands between 50 and 100 μm wavelength. For more details see Goody and Yung (1989).

Godson's inverse power model

Godson (1955) applied an inverse power law to describe the distribution of line intensities as

$$p(S) = \begin{cases} \dfrac{\kappa}{S} & \text{for } S_0 < S \leq S_1 \\ 0 & \text{for } S > S_1 \end{cases} \tag{7.99}$$

The constant κ is a model parameter which is used to normalize the probability distribution according to (7.92). The expectation value for absorption, employing the inverse power law, is given by

$$A(u) = \lim_{S_0 \to 0} \frac{1}{\delta} \int_{S_0}^{S_1} \frac{\kappa}{S} \int_{-\infty}^{\infty} [1 - \exp(-f_v Su)] dv dS \tag{7.100}$$

Since $p(S)$ approaches infinity if $S_0 \to 0$, the limiting value of the integral must be considered. Using the substitutions

$$\bar{u} = \frac{Su}{2\pi\alpha}, \quad x = \frac{\kappa v}{\delta}, \quad y = \frac{\kappa\alpha}{\delta} \tag{7.101}$$

we obtain for a Lorentz line the intermediate result

$$A(\bar{u}_1) = \lim_{\bar{u}_0 \to 0} \int_{\bar{u}_0}^{\bar{u}_1} \frac{1}{\bar{u}} \int_{-\infty}^{\infty} \left[1 - \exp\left(\frac{-2\bar{u} y^2}{x^2 + y^2} \right) \right] dx d\bar{u} \tag{7.102}$$

where the upper bound $\bar{u}_1$ implicitly contains the dependence on u and S_1. It is convenient to carry out the integration over x first. From Appendix 7.5.2 we know

$$\int_{-\infty}^{\infty} \left[1 - \exp\left(-\frac{2\bar{u} y^2}{(x^2 + y^2)} \right) \right] dx = 2\pi y\bar{u} \exp(-\bar{u}) [I_0(\bar{u}) + I_1(\bar{u})] \tag{7.103}$$

Inserting this expression in (7.102) yields

$$A(\bar{u}_1) = 2\pi y \lim_{\bar{u}_0 \to 0} \int_{\bar{u}_0}^{\bar{u}_1} \exp(-\bar{u}) [I_0(\bar{u}) + I_1(\bar{u})] d\bar{u} \tag{7.104}$$

The modified Bessel functions fulfill the following functional relations

(a) $\dfrac{d}{dx}[x^{-\nu}I_\nu(x)] = x^{-\nu}I_{\nu+1}(x)$

(b) $\dfrac{d}{dx}[x\exp(-x)(I_0(x) + I_1(x))] = \exp(-x)I_0(x)$

$$(7.105)$$

where ν is a fixed real number. Integration by parts of the following integral gives

$$\int_{\bar{u}_0}^{\bar{u}_1}\exp(-\bar{u})I_1(\bar{u})d\bar{u} = \int_{\bar{u}_0}^{\bar{u}_1}\exp(-\bar{u})I_0(\bar{u})d\bar{u} + \exp(-\bar{u})I_0(\bar{u})\Big|_{\bar{u}_0}^{\bar{u}_1} \quad (7.106)$$

where (7.105a) with $\nu = 0$ has been used. Substituting this equation into (7.104) we obtain

$$\mathcal{A}(\bar{u}_1) = 2\pi y \lim_{\bar{u}_0 \to 0}\left(\int_{\bar{u}_0}^{\bar{u}_1} 2\exp(-\bar{u})I_0(\bar{u})d\bar{u} + 2\pi y\exp(-\bar{u})I_0(\bar{u})\Big|_{\bar{u}_0}^{\bar{u}_1}\right) \quad (7.107)$$

The integral on the right-hand side of this equation may be evaluated by means of (7.105b) yielding

$$\mathcal{A}(\bar{u}_1) = 2\pi y[2\bar{u}_1\exp(-\bar{u}_1)[I_0(\bar{u}_1) + I_1(\bar{u}_1)] + \exp(-\bar{u}_1)I_0(\bar{u}_1)]$$
$$-2\pi y\lim_{\bar{u}_0 \to 0}[2\bar{u}_0\exp(-\bar{u}_0)[I_0(\bar{u}_0) + I_1(\bar{u}_0)] + \exp(-\bar{u}_0)I_0(\bar{u}_0)] \quad (7.108)$$

and hence

$$\boxed{\mathcal{A}(\bar{u}_1) = 2\pi y[2\bar{u}_1\exp(-\bar{u}_1)[I_0(\bar{u}_1) + I_1(\bar{u}_1)] + \exp(-\bar{u}_1)I_0(\bar{u}_1)] - 2\pi y}$$

$$(7.109)$$

Here, use was made of $I_0(0) = 1$ and $I_1(0) = 0$, see (7.71). This is the final form of the mean absorption of Godson's model.

The unknown parameters y and $\bar{u}_1$ will again be determined by requiring in the weak and strong line limit agreement between the Godson model and the band model for nonoverlapping lines

$$\bar{u}_1 \ll 1: \quad 2\pi y\bar{u}_1 = \frac{1}{N\delta}\sum_{i=1}^{N} S_i u$$

$$\bar{u}_1 \gg 1: \quad 4y\sqrt{2\pi\bar{u}_1} = \frac{2}{N\delta}\sum_{i=1}^{N}\sqrt{S_i\alpha_i u} \quad (7.110)$$

Solving these two equations gives

$$\bar{u}_1 = \frac{2u}{\pi} \frac{C^2}{D^2}, \qquad y = \frac{1}{4N\delta} \frac{D^2}{C} \tag{7.111}$$

where C and D are given by (7.98). The Godson model allows for a good representation of weak lines in certain absorption bands and is usually more accurate than the exponential law.

The Malkmus model

So far the most successful statistical model is due to Malkmus (1967). This model is a combination of the Goody and the Godson model. Malkmus also observed that the exponential distribution (7.91) substantially underestimates the number of weak lines. If we consider only the Boltzmann factor in the line intensity formula to be discussed in a later chapter, then $S \sim \exp[-E/kT]$, where E represents the lower energy level in a molecular transition. From this relation he concludes that $dE/dS \sim S^{-1}$. In many cases the number density of lines, n, is approximately equally spaced with respect to variations in E, that is $dn/dE \sim const$. The probability $p(S)dS$ to find lines with intensity S must be proportional to the change of n versus S. Therefore, we obtain the relationship

$$p(S) \sim \frac{dn}{dS} \sim \frac{dn}{dE} \frac{dE}{dS} \sim \frac{dE}{dS} \sim \frac{1}{S} \tag{7.112}$$

which shows that $p(S)$ should vary as S^{-1}. Indeed, it is this dominating influence which determines the accuracy of the average band absorption. For this reason Malkmus proposes a multiplicative combination of Goody's and Godson's statistical models, that is

$$p(S) = \frac{1}{S} \exp\left(-\frac{S}{\sigma}\right) \tag{7.113}$$

where σ is the average value of the line intensities. For $S = 0$ the function $p(S)$ of the Godson and the Malkmus model is not defined. Nevertheless, both models can be applied by using a limiting procedure.

The average absorption $A(u)$ due to the Malkmus model is now given by

$$A(u) = \lim_{\varepsilon \to 0} \int_\varepsilon^\infty \frac{1}{S} \exp\left(-\frac{S}{\sigma}\right) \frac{1}{\delta} \int_{-\infty}^\infty [1 - \exp(-f_\nu Su)] d\nu dS \tag{7.114}$$

The integral over S will be evaluated by means of

$$\lim_{\varepsilon \to 0} \int_\varepsilon^\infty \frac{1}{S} \exp\left(-\frac{S}{\sigma}\right)[1 - \exp(-f_v S u)]dS$$

$$= \lim_{\varepsilon \to 0} \int_\varepsilon^\infty \frac{1}{S} \exp\left(-\frac{S}{\sigma}\right)dS - \lim_{\varepsilon' \to 0} \int_{\varepsilon'}^\infty \frac{1}{S'} \exp\left(-\frac{S'}{\sigma}\right)dS'$$

$$= -\lim_{\varepsilon \to 0}\left[\ln \varepsilon - \frac{\varepsilon}{\sigma} + \frac{1}{4}\left(\frac{\varepsilon}{\sigma}\right)^2 - \frac{1}{18}\left(\frac{\varepsilon}{\sigma}\right)^3 \pm \cdots\right] \tag{7.115}$$

$$+ \lim_{\varepsilon' \to 0}\left[\ln \varepsilon' - \frac{\varepsilon'}{\sigma} + \frac{1}{4}\left(\frac{\varepsilon'}{\sigma}\right)^2 - \frac{1}{18}\left(\frac{\varepsilon'}{\sigma}\right)^3 \pm \cdots\right]$$

$$= \ln(1 + \sigma f_v u)$$

where the substitutions $S' = S(1 + \sigma f_v u)$ and $\varepsilon' = \varepsilon(1 + \sigma f_v u)$ have been introduced. Furthermore, it should be noted that the contributions of the upper limit of the integrals over S and S' cancel so that only the difference of the lower integral limits remains.

Substituting this result into (7.114) and applying the Lorentz line-shape factor gives the average absorption of the Malkmus model

$$\boxed{A(u) = \frac{1}{\delta} \int_{-\infty}^\infty \ln(1 + \sigma f_v u)dv = 2\pi y(\sqrt{1 + 2\bar{u}} - 1)} \tag{7.116}$$

with $y = \alpha_L/\delta$ and $\bar{u} = \sigma u/(2\pi \alpha_L)$. Finally, with the help of the matching procedure we obtain

$$\bar{u} \ll 1: \qquad 2\pi y \bar{u} = \frac{1}{\Delta v} \sum_{i=1}^N S_i u$$

$$\bar{u} \gg 1: \qquad 2\pi y \sqrt{2\bar{u}} = \frac{2}{\Delta v} \sum_{i=1}^N \sqrt{S_i \alpha_i u} \tag{7.117}$$

so that the unknown parameters y and $\bar{u}$ are given as

$$\boxed{\bar{u} = \frac{u}{2}\frac{C^2}{D^2}, \qquad y = \frac{1}{\pi \Delta v}\frac{D^2}{C}} \tag{7.118}$$

The terms C and D are again given by (7.98). Applications of the Malkmus model can be found, for example, in Crisp *et al.* (1986) and Lacis and Oinas (1991).

It should be observed that the absorption equations (7.96), (7.109) and (7.116) have been derived on the assumption that only the intensities vary from line to line. The model parameters $\bar{u}$ and y employ the complete spectral data so that the variation of the half-width from line to line is also taken into account.

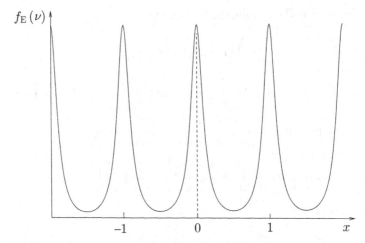

Fig. 7.7 Line-shape factor of Elsasser's regular band model with $x = \nu/\delta$.

The model parameters $\bar{u}$ and y have been determined for the three statistical models using different formulations of $p(S)$. We are now in the position to employ the Mayer–Goody transmission formula (7.90) with some confidence. The spectral data required to evaluate $\bar{u}$ and y are given in Goody (1964a) or in Goody and Yung (1989). This part of the sections on band models largely follow Goody's (1964a) excellent text.

7.2.4 Elsasser's regular model

Spectra of linear molecules such as CO_2 often appear as a superposition of arrays of more or less regularly spaced lines with nearly identical line shape. Due to this observation Elsasser (1942) constructed a band model consisting of an infinite number of evenly spaced identical Lorentz lines. The absorption coefficient is given by the superposition of all lines, that is

$$k_{\nu,E} = \sum_{n=-\infty}^{\infty} \frac{S}{\pi} \frac{\alpha_L}{(\nu - n\delta)^2 + \alpha_L^2} \qquad (7.119)$$

Figure 7.7 illustrates the line-shape factor $f_E(\nu) = k_{\nu,E}/S$ of the Elsasser band model. The spectral lines are separated by the distance δ.

The function $k_{\nu,E}$ possesses an infinite number of simple poles at $\nu = j\delta \pm i\alpha_L$. With the help of the Mittag–Leffler theorem, see Appendix 7.6.3, the infinite sum can be stated as a closed expression involving periodic and hyperbolic functions. This leads to the absorption coefficient

$$k(s) = \frac{S}{\delta} \frac{\sinh \beta}{\cosh \beta - \cos s} \qquad (7.120)$$

where $s = 2\pi \nu/\delta$ and $\beta = 2\pi \alpha_L/\delta$. Figure 7.7 suggests that k_ν varies periodically within the bounds $k_{min} \leq k_\nu \leq k_{max}$. From (7.120) we immediately find

$$k_{min} = \frac{S}{\delta} \frac{\sinh \beta}{\cosh \beta + 1}, \qquad k_{max} = \frac{S}{\delta} \frac{\sinh \beta}{\cosh \beta - 1} \qquad (7.121)$$

Next we will compute the average transmission of the regular band. Due to the periodicity it is sufficient to average over the interval $[-\delta/2, \delta/2]$, i.e.

$$T(u) = \frac{1}{\delta} \int_{-\delta/2}^{\delta/2} \exp(-k_\nu u) \, d\nu = \frac{1}{2\pi} \int_{-\pi}^{\pi} \exp[-k(s)u] \, ds \qquad (7.122)$$

It is convenient to determine the first derivative of T with respect to u

$$\frac{dT}{du} = -\frac{1}{2\pi} \int_{-\pi}^{\pi} k(s) \exp[-k(s)u] \, ds \qquad (7.123)$$

Substituting

$$\cos \phi = \frac{1 - \cosh \beta \cos s}{\cosh \beta - \cos s} \qquad (7.124)$$

gives

$$d\phi = -\frac{\cosh \beta - \cos \phi}{\sinh \beta} ds = -k(\phi) \frac{\delta}{S} ds \qquad (7.125)$$

The latter expression is valid since by means of simple trigonometric manipulations it can be shown that

$$\frac{\cosh \beta - \cos \phi}{\sinh \beta} = \frac{\sinh \beta}{\cosh \beta - \cos s} \qquad (7.126)$$

Using the above relations (7.123) may be rewritten as

$$\frac{dT}{du} = \frac{S}{2\pi \delta} \int_0^{2\pi} \exp\left(-\frac{Su}{\delta} \frac{\cosh \beta - \cos \phi}{\sinh \beta}\right) d\phi \qquad (7.127)$$

If we substitute in the argument of the exponential function the expression

$$y = \frac{Su}{\delta \sinh \beta} \qquad (7.128)$$

we obtain

$$\frac{dT}{dy} = \frac{\sinh \beta}{2\pi} \exp(-y \cosh \beta) \int_0^{2\pi} \exp(y \cos \phi) \, d\phi \qquad (7.129)$$

The Bessel functions J_n of the first kind can be expressed by the following definite integral (Watson, 1980)

$$J_n(x) = \frac{(-i)^n}{2\pi} \int_{-\pi}^{\pi} \cos(nw) \exp(ix \cos w) \, dw, \qquad n \in \mathcal{N}_0 \qquad (7.130)$$

Substitution of $\phi = \pi - w$ yields

$$J_n(x) = \frac{(-i)^n}{2\pi} \int_{0}^{2\pi} \cos(n\pi - n\phi) \exp[ix \cos(\pi - \phi)] \, d\phi \qquad (7.131)$$

which for $n = 0$ and $y = -ix$ turns into

$$J_0(iy) = \frac{1}{2\pi} \int_{0}^{2\pi} \exp(y \cos \phi) \, d\phi \qquad (7.132)$$

Hence (7.129) can be expressed as

$$\frac{dT}{dy} = \sinh \beta \exp(-y \cosh \beta) J_0(iy) \qquad (7.133)$$

A final substitution of $Y = y \sinh \beta = Su/\delta$ then leads to a formula for the average transmission of the Elsasser model

$$T(Y) = \int_{Y}^{\infty} \exp(-Y' \coth \beta) J_0 \left(\frac{iY'}{\sinh \beta} \right) dY' \qquad (7.134)$$

There are no analytical solutions to this integral. However, results can be found by means of numerical integration.

From Figure 7.7 we conclude that there exists an average value $\bar{k}$ for the absorption coefficient k_ν. For the rapidly varying absorption coefficients of real spectra it is usually impossible to give a reliable value of $\bar{k}$. Formally we have

$$\bar{k} = \frac{1}{\delta} \int_{-\delta/2}^{\delta/2} \frac{S}{\delta} \frac{\sinh \beta}{\cosh \beta - \cos(2\pi \nu/\delta)} \, d\nu \qquad (7.135)$$

Using the indefinite integral

$$\int \frac{dx}{b + c \cos ax} = \frac{2}{a\sqrt{b^2 - c^2}} \tan^{-1} \left[\frac{(b-c) \tan(ax/2)}{\sqrt{b^2 - c^2}} \right] \qquad \text{for} \quad b^2 > c^2 \qquad (7.136)$$

listed in integration tables, we find the expected result

$$\bar{k} = \frac{S}{\delta} \qquad (7.137)$$

This relation could have been guessed from Figure 7.7, since within the periodic pattern of Lorentz lines the area under each line is S. Despite the fact that within the interval $[-\delta/2, \delta/2]$ the central line does contribute to this area less than S, say $S - \Delta S$, the missing part is contributed by the infinite number of neighboring lines whose wings give exactly rise to the contribution ΔS.

We will conclude this section by discussing two special cases of the Elsasser band model and the resulting consequences.

(i) **Large values of β** Large β-values imply small distances between spectral lines since $\beta = 2\pi\alpha_L/\delta$. In this case $\coth\beta$ approaches 1 while $\sinh\beta$ becomes a fairly large number. For a fixed β-value the contribution of the product of functions under the integral sign of (7.134) to the value of the integral decreases rapidly with increasing Y'. Saying it slightly differently, the contribution of the product of functions to the integral is dominated by smaller values of Y' due to the rapidly decreasing exponential function with increasing Y'. Thus for large β and not too large values of Y', the fraction $Y'/\sinh\beta$ is a small number also so that the argument of $J_0(iY'/\sinh\beta) = I_0(Y'/\sinh\beta)$ is a small number and may be approximated by $I_0 \approx 1$. For this special case the transmission function $T(Y)$, originally given by (7.134), may be approximated by

$$T(Y) = \int_Y^\infty \exp(-Y')dY' = \exp\left(-\frac{Su}{\delta}\right) \tag{7.138}$$

This is already a reasonably good approximation for $\beta = 2$ which is equivalent to $\delta = \pi\alpha_L$. The fraction S/δ may be considered as an absorption coefficient of a continuous spectrum. The spectral lines strongly overlap and no line structure is observed. The special case of the weak line approximation, see (7.73), is included in (7.138) by setting $\delta = \Delta\nu$ and $Su/\delta \ll 1$.

(ii) **Small values of β** For small β-values ($\delta \gg \alpha_L$) the distance between line centers considerably exceeds the half-width. This implies that the overlap effect is negligible for weak lines. However, for strong lines considerable overlap may still take place in the line wings. In this case $\cosh\beta \approx 1$ and $\sinh\beta \approx \beta$. Thus equation (7.120) simplifies to

$$k(s) = \frac{S\beta}{\delta}\frac{1}{1 - \cos s} = \frac{S\beta}{2\delta}\frac{1}{\sin^2(s/2)} \tag{7.139}$$

Setting $m = Su\beta/2\delta$ and $\sin^2(s/2) = 1/y$, the transmission function (7.122) can be written as

$$T = \frac{1}{\pi}\int_0^\pi \exp[-k(s)u]\,ds = \frac{1}{\pi}\int_1^\infty \frac{\exp(-my)}{y\sqrt{y-1}}dy \tag{7.140}$$

In order to write this expression in terms of a tabulated function, we differentiate (7.140) with respect to m and obtain

$$\frac{dT}{dm} = -\frac{1}{\pi}\int_1^\infty \frac{\exp(-my)}{\sqrt{y-1}}dy = \frac{\exp(-m)}{\pi}\int_0^\infty \frac{\exp(-m\xi)}{\sqrt{\xi}}d\xi = \frac{\exp(-m)}{\sqrt{\pi m}} \tag{7.141}$$

from which follows

$$T = \frac{1}{\sqrt{\pi}} \int_m^{m_1} \frac{\exp(-m')}{\sqrt{m'}} dm' = \frac{2}{\sqrt{\pi}} \int_{\sqrt{m}}^{\sqrt{m_1}} \exp(-x^2) dx \qquad (7.142)$$

The upper limit $\sqrt{m_1}$ of the second integral can be found from the condition $T = 1$ when $m = 0$ (m is proportional to u) by observing the identity

$$\int_0^{\infty} \exp(-x^2) dx = \frac{\sqrt{\pi}}{2} \qquad (7.143)$$

so that $\sqrt{m_1} = \infty$.

For the evaluation of $T(u)$ we will now introduce the tabulated error function $\phi(x)$ which is defined by

$$\phi(x) = \frac{2}{\sqrt{\pi}} \int_0^x \exp(-s^2) ds \qquad (7.144)$$

Replacing β in m we find

$$T(u) = 1 - \phi(\sqrt{m}) = 1 - \phi\left(\frac{1}{\delta}\sqrt{\pi S \alpha_L u}\right) = 1 - \phi\left(\frac{\sqrt{\pi}}{2}\frac{W_s}{\delta}\right) \qquad (7.145)$$

Here we have also used the equivalent width of the strong line approximation $W_s = A_s(u)\Delta v = 2\sqrt{S\alpha_L u}$, see (7.75). As a matter of notation Elsasser also introduced the *generalized absorption coefficient* $l = 2\pi S\alpha_L/\delta^2$ so that the transmission function can be written in the form

$$T(u) = 1 - \phi(\sqrt{lu/2}) \qquad (7.146)$$

Finally, let us consider the series expansion of the error function

$$\phi(x) = \frac{2}{\sqrt{\pi}}\left(x - \frac{x^3}{3} + \cdots\right) \qquad (7.147)$$

Hence, for small values of the argument we obtain from (7.145)

$$T(u) = 1 - \frac{W_s}{\delta} \qquad (7.148)$$

which is the correct form of the *transmission function for nonoverlapping lines*. Additional information can be found in Goody and Yung (1989), Kondrat'yev (1965) and elsewhere.

7.2.5 The Schnaidt model

There is another formulation of the absorption function $A(u)$ which is due to Schnaidt (1939). He assumed that the effect of line overlap was to terminate each line at a distance $\delta/2$ from the line center. Thus, instead of (7.51) the average absorption is now written as

$$\mathcal{A}(u) = \frac{1}{\Delta \nu} \int_{-\delta/2}^{\delta/2} [1 - \exp(-k_\nu u)] d\nu \qquad (7.149)$$

Since the integral is not taken over the interval $[-\infty, \infty]$ as in (7.51), the weak line limit (7.75) does not follow for $u \to 0$. It will be observed that the termination of the integral at $[-\delta/2, \delta/2]$ excludes the influence of any spectral lines which are located outside this range.

7.3 The fitting of transmission functions

At the beginning of the previous section we have already stated that, owing to the strongly selective absorption of atmospheric gases, the spectrum must be highly resolved for a sufficiently accurate frequency integration of the radiative transfer equation. In the most accurate *line-by-line integration* the RTE is evaluated along the profile of each spectral line taking due account of the contribution of neighboring and distant strong spectral lines. However, the line-by-line integration is far too expensive for practical applications such as the computation of radiative heating rates in climate or weather prediction models. The band models discussed in the previous section provide one way to obtain less costly yet sufficiently reliable solutions to the RTE.

Another possibility to reduce the computational effort of the line-by-line method is to use certain parameterization techniques describing the mean transmission of a particular gas in a given $\Delta \nu$ interval. Two highly successful methods are the *exponential sum-fitting of transmissions* and the *correlated k-distribution method* which will be discussed in the following sections.

7.3.1 Exponential sum-fitting of transmissions

Let us approximate the *transmission function* $\mathcal{T}_{\Delta \nu}(u)$ for a given interval $\Delta \nu$ by an exponential expression of the form

$$\mathcal{T}_{\Delta \nu}(u) \approx E_{\Delta \nu}(u) = \sum_{i=1}^{m} a_i \exp(-b_i u) \qquad (7.150)$$

where, as usual, u is the gas absorber amount. The task ahead is to find by some method reliable values of the coefficient pairs (a_i, b_i), $i = 1, \ldots, m$. The coefficient a_i is a dimensionless positive weight while b_i must be interpreted as the corresponding gray absorption coefficient. Thus the mean transmission over the spectral interval $\Delta \nu$ in approximated form can be considered as the sum of partial transmissions. Each subband i is described in terms of the dimensionless 'width' a_i and the gray absorption coefficient b_i. Clearly, if $u = 0$ then the transmission equals 1. In summary, we search for an exponential expression of the transmission function that meets the following constraints

$$\sum_{i=1}^{m} a_i = 1, \quad a_i > 0, \quad b_i \geq 0 \qquad \text{for all } i \tag{7.151}$$

Now we briefly describe the highly successful least-square technique formulated by Wiscombe and Evans (1977) to find reliable values of (a_i, b_i). For a uniform grid of absorber mass increments Δu we have

$$u_n = n\Delta u, \qquad n = 0, \ldots, N \tag{7.152}$$

where a total number of $N + 1$ grid points is used.[2] For the u_n the exponential fit (7.150) then reads

$$E_{\Delta \nu}(u_n) = \sum_{i=1}^{m} a_i \exp\left(-b_i n\Delta u\right) = \sum_{i=1}^{m} a_i \theta_i^n \quad \text{with} \quad \theta_i = \exp(-b_i \Delta u) \tag{7.153}$$

The 'exact' values of the transmission function $\mathcal{T}_{\Delta \nu}(u_n)$ for all pathlengths may be found by employing a high resolution line-by-line integration technique or by using a very accurate band model.

Using a set of equally spaced arguments, the 'distance' between $\mathcal{T}_{\Delta \nu}(u_n)$ and $E_{\Delta \nu}(u_n)$ is expressed by the least squares residual

$$R_0 = \sum_{n=1}^{N} w_n \left[\mathcal{T}_{\Delta \nu}(u_n) - E_{\Delta \nu}(u_n)\right]^2 \tag{7.154}$$

where the $w_n \geq 0$ are the least square weights and $E_{\Delta \nu}(u_n)$ is the sum of powers as expressed in (7.153) with $0 \leq \theta_i \leq 1$ when $b_i \geq 0$. The best fit is defined as the one that minimizes R_0 over all permissible values of a_i, b_i and m. Considering the θ_i in (7.153) as known, the standard linear least squares normal equations for $a_1, \ldots a_m$ are given by

$$P(\theta_i) = \frac{\partial R_0}{\partial a_i} = 0, \qquad i = 1, \ldots, m \tag{7.155}$$

where

$$P(\theta) = 2 \sum_{n=1}^{N} p_n \theta^n \tag{7.156}$$

---

[2] In general, an arbitrary nonuniform grid spacing can also be used. This makes it possible, for example, to obtain very accurate fits for values of u for which the mean transmission is close to 1.

is known as the residual polynomial. The coefficients

$$p_n = w_n \left[E_{\Delta \nu}(u_n) - T_{\Delta \nu}(u_n) \right], \qquad n = 1, \ldots, N \tag{7.157}$$

are weighted point-by-point differences between the exponential fit and the 'exact' data. The set a_i satisfying (7.155) minimizes R_0 for fixed θ_i. We call $P(\theta)$ the residual polynomial whose central role in the approximation technique will be recognized from the following theorem:

A best fit $(a_i > 0, b_i \geq 0)$ to the original data provided by $T_{\Delta \nu}(u)$ has been achieved if and only if the residual polynomial satisfies the conditions

$$\begin{align}
\text{(a)} \quad & P(\theta_i) = 0, \qquad i = 1, \ldots, m \\
\text{(b)} \quad & P(\theta) \geq 0, \qquad 0 \leq \theta \leq 1
\end{align} \tag{7.158}$$

It should be noted that condition (7.158a) is exactly equation (7.155). The method iterates back and forth between solving condition (7.158a) for the coefficients a_i and improving toward condition (7.158b) by adding a new exponential factor θ_i. For details of the method the reader should consult the original paper.

Let us demonstrate the exponential sum fitting method by means of a simple example. Suppose that the functional form we wish to approximate is given by

$$T(u) = 0.6 \exp(-0.1u) + 0.3 \exp(-0.01u) + 0.1 \exp(-0.001u) \tag{7.159}$$

The exponential sum fit shall have the same analytical form with the unknown coefficients (a_i, b_i), $i = 1, 2, 3$, i.e.

$$E(u) = a_1 \exp(-b_1 u) + a_2 \exp(-b_2 u) + a_3 \exp(-b_3 u) \tag{7.160}$$

The algorithm of Wiscombe and Evans (1977) will be used on a grid consisting of 20 grid points with

$$\begin{align}
u_n = \{ & 0, \ 1, \ 2, \ 3, \ 4, \ 5, \ 10, \ 30, \ 60, \ 150, \ 300, \ 400, \ 500, \\
& 1000, \ 1500, \ 2000, \ 3000, \ 4000, \ 5000, \ 6000 \} \\
w_n = & \frac{1}{T(u_n)}, \qquad n = 0, \ldots, 19
\end{align} \tag{7.161}$$

Note that the minimization problem involving the linear and the nonlinear part as given above, can be solved by an iterative process. For details the reader is referred to Wiscombe and Evans (1977). For this particularly simple example the result to five significant digits is listed next

$$\begin{align}
a_1 &= 0.59955, & b_1 &= 0.098566 \\
a_2 &= 0.29980, & b_2 &= 0.0099970 \\
a_3 &= 0.099916, & b_3 &= 0.00099870
\end{align} \tag{7.162}$$

By comparing these numbers with (7.159) it is seen that the exponential sum fitting method is capable of retrieving the given nonlinear function with high accuracy for as many as 20 data points. Further examples for the exponential fits of transmission functions for atmospheric gases and intercomparisons with band model expressions can be found in their article.

7.3.2 The k-distribution method

An alternative approach to obtain rather accurate expressions for the average transmission in a particular wave number band is the so-called *k-distribution method*. This method is considerably faster than the band model approach described in the previous sections. It also allows for a self-consistent treatment of multiple scattering in an absorbing atmosphere. The approach makes use of the fact that for a homogeneous atmosphere, the transmission within a relatively wide wave number interval is independent of the ordering of the values of the absorption coefficient k_ν. This means that the fractional absorption caused by absorption coefficients belonging to the interval $(k_\nu, k_\nu + dk_\nu)$ is associated with the number of instances for which k_ν attains values for this particular k_ν interval. This leads us to the conclusion that we must determine the probability density function $f(k)$ for the k values in the interval $(k_\nu, k_\nu + dk_\nu)$. More generally speaking, the absorption associated with a particular value of $k = k_\nu$ is proportional to the expression $f(k)dk$. In essence, this probability treatment means that we transform the transmission computation from wave number space (ν-space) into the probability space of k values (k-space).

The k-distribution method is based on an idea of grouping frequency intervals according to the line strengths as described by Ambartsumian (1936). This procedure has been employed by Chou and Arking (1980) to compute infrared cooling rates. The same authors carried out heating rate computations in the solar spectral region, see Chou and Arking (1981). The interested reader is referred to the more recent treatments by Lacis and Oinas (1991) and Fu and Liou (1992).

Similar to band models, the k-distribution approach is first developed for homogeneous absorber paths. For nonhomogeneous paths one uses the so-called *correlated k-distribution* (CKD) approach first introduced by Lacis *et al.* (1979). The correlated assumption means that the vertical inhomogeneity of the atmosphere is accurately accounted for by assuming the existence of a simple correlation of the k-distributions for different temperatures and pressures. Moreover, the CKD approach allows us to fully treat the complicated Voigt line shape. The CKD method can be used for thermal and solar radiative transfer likewise. As will be explained later, the treatment of multiple scattering in a realistic atmosphere containing aerosol particles and water droplets or ice crystals can also be straightforwardly done with the CKD approach. A detailed discussion of this CKD method will be given below.

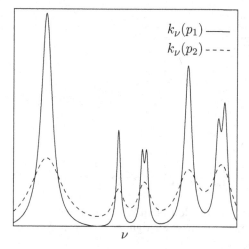
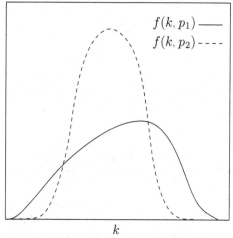

Fig. 7.8 Schematic illustration of the k-distribution method. The left panel shows the variation of the absorption coefficient in a spectral interval for two different pressures p_1 (full curve) and $p_2 = 3p_1$ (dashed curve). The corresponding probability density functions $f(k)$ are shown in the right panel.

Basic illustration of the k-distribution method

In the k-distribution method it is not important to specify at which particular position within the interval $\Delta \nu$ a certain value of k_ν occurs. As soon as we know the probability $f(k)dk$ of the occurrence of absorption coefficients belonging to $(k, k + dk)$, the average transmission can be obtained by an integration over k

$$\mathcal{T}_{\Delta\nu}(u) = \frac{1}{\Delta\nu} \int_{\Delta\nu} \exp(-k_\nu u)\, d\nu = \int_0^\infty f(k) \exp(-ku)\, dk \qquad (7.163)$$

From this definition it can be inferred that the average transmission is the Laplace transform of the probability density function $f(k)$. Vice versa, $f(k)$ can be obtained from the inverse Laplace transformation of the average transmission function, i.e.

$$\mathcal{T}_{\Delta\nu}(u) = \mathcal{L}[f(k)], \qquad f(k) = \mathcal{L}^{-1}[\mathcal{T}_{\Delta\nu}(u)] \qquad (7.164)$$

From these equations it may be concluded that only in very few cases it is possible to obtain $f(k)$.

Figure 7.8 depicts schematically the essence of the k-distribution approach. The left panel of the figure shows the absorption line spectrum consisting of seven Lorentz lines with different half-widths and line intensities. The full curve is plotted for an arbitrary pressure p_1 whereas the dashed curve is plotted for pressure $p_2 = 3p_1$. Owing to the pressure dependence of α_L, see (7.27), the *line broadening* is

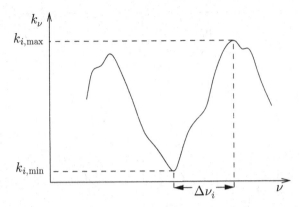

Fig. 7.9 Definition of the subinterval Δv_i whose left boundary is given at the location where k_v attains are local minimum. Likewise the right boundary of Δv_i is given by the next local maximum of k_v.

much stronger for p_2 than for p_1 so that the high peaks occurring in the $k_v(p_1)$-curve are strongly damped in the $k_v(p_2)$-curve. As a consequence of this, $f(k, p_2)$ covers a smaller range of k-values expressing the missing small and large absorption coefficients as compared to the $f(k, p_1)$-curve.

We conclude the following: with the k-distribution approach one essentially replaces the integration of k_v in wave number space by a similar integration of the product of the spectral transmission with $f(k)$ in k-space. Figure 7.8 also suggests that in k-space the integrand may have a much smoother shape which eases the effort for numerical quadrature.

Algorithm for computing the k-distribution

An algorithm for computing the probability density function $f(k)$ consists of the following steps.

(1) Subdivide the entire spectral range Δv into suitably chosen subintervals Δv_i. Figure 7.9 depicts how the left and right boundaries have to be chosen so that the inverse function $v = v(k)$ of the absorption coefficient k_v is uniquely defined. Note that within Δv_i there exist no other local minima and maxima of the absorption coefficient, that is $k_{i,\min} \leq k_v < k_{i,\max}$.

(2) Next we define the transformation which maps the differential dv into k-space

$$\frac{dv}{\Delta v} \rightarrow \frac{dk}{\Delta v} \left| \frac{dv}{dk_v} \right|_{\Delta v_i}, \qquad i = 1, 2, \ldots, N \tag{7.165}$$

Here $|dv/dk_v|_{\Delta v_i}$ means that this derivative has to be provided for the subinterval Δv_i.

(3) If $\Psi = \Psi(k_v, u)$ is an arbitrary function of the spectral absorption coefficient, analogously to (7.163) we may define

$$\bar{\Psi}(u) = \frac{1}{\Delta v} \int_{\Delta v} \Psi(k_v, u)dv = \int_0^\infty f(k)\Psi(k, u)dk \qquad (7.166)$$

where $f(k)$ is given by

$$f(k) = \sum_{i=1}^N \frac{1}{\Delta v} \left|\frac{dv}{dk_v}\right|_{\Delta v_i} [U(k - k_{i,\min}) - U(k - k_{i,\max})] \qquad (7.167)$$

and U is the Heaviside step function as defined by (5.50). Note that the difference of the Heaviside step functions in (7.167) picks out the required range of k values for which the first derivative of the inverse relationship $v = v(k)$ has to be applied.

Equation (7.167) can be applied in two different ways.

(1) Use the complete listing of the spectral information on the absorption coefficient for a particular gas, e.g. use the Air Force Geophysics Laboratory (AFGL) CD-ROM database HITRAN for the spectral line compilation. This database contains the required information on a line-by-line basis for all important atmospheric absorber gases. Now use (7.167) to count the occurrence of all spectral lines within the subinterval Δv_i which fall into a particular range of absorption coefficients $(k_j, k_j + \Delta k_j)$. Try different choices for the binning width Δk_j in order to cover a large range of possible k_j values. However, a too coarse stepping Δk_j might result in the unwanted effect that the frequency distribution exhibits holes at certain points. Some experimentation might be necessary to yield the best results. After repeating this procedure for all subintervals in the spectral band under consideration, the probability density function $f(k)$ can be derived.
(2) For certain analytic band models, explicit expressions for k_v as a function of wave number exist. Under certain circumstances, the derivative $|dv/dk_v|^{-1}$ can be calculated explicitly.

In the following we will give an illustrative example, how $f(k)$ can be calculated for the regular Elsasser band model which has already been discussed in the previous section. Owing to the symmetry of the absorption coefficient, we can confine the discussion to the spectral interval $[0, \delta/2]$. For the probability density function we then obtain

$$f(k) = \frac{2}{\delta} \left|\frac{dv}{dk_v}\right| \quad \text{for } 0 \le v \le \delta/2 \qquad (7.168)$$

According to (7.120) the absorption coefficient for the regular Elsasser band can be expressed in a closed analytical form. The minimum and the maximum values of k_v are given by (7.121). Clearly, k_v is strictly monotone increasing in the given interval so that

$$k_{\min} \leq k_\nu \leq k_{\max} \quad \text{for } 0 \leq \nu \leq \delta/2 \tag{7.169}$$

Differentiation of (7.120) with respect to ν yields

$$\frac{dk_\nu}{d\nu} = -\frac{2\pi \, k_\nu^2 \sin s}{S \, \sinh \beta} \implies \left|\frac{d\nu}{dk_\nu}\right| = \frac{S \, \sinh \beta}{2\pi \, k_\nu^2 \sin s} \tag{7.170}$$

For the product of k and $f(k)$ we thus obtain

$$k f(k) = \frac{S \, \sinh \beta}{\pi \delta \, k \sin s} \tag{7.171}$$

An expression for $\sin s$ can be derived from (7.120)

$$\sin^2 s = 1 - \cos^2 s = \sinh^2 \beta \left[\frac{2\bar{k}}{k} \coth \beta - 1 - \left(\frac{\bar{k}}{k}\right)^2 \right] \tag{7.172}$$

where $\bar{k} = S/\delta$, see (7.137). Using the above expressions, after a few simple steps we find the k-distribution for the regular Elsasser model in analytic form as

$$\boxed{k f(k) = \frac{1}{\pi} \left[2\frac{k}{\bar{k}} \coth \beta - 1 - \left(\frac{k}{\bar{k}}\right)^2 \right]^{-1/2}} \tag{7.173}$$

Finally, from (7.121) we may see that $k/\bar{k}$ is bounded by

$$\frac{\sinh \beta}{\cosh \beta + 1} \leq \frac{k}{\bar{k}} \leq \frac{\sinh \beta}{\cosh \beta - 1} \tag{7.174}$$

The cumulative k-distribution

We have seen that the k-distribution approach replaces the wave number integration by an integration over k-space. For convenience, let us set the minimum and maximum value for the absorption coefficient to $k_{\min} \to 0$ and $k_{\max} \to \infty$. Then we obtain for the mean transmission in the band $\Delta \nu$

$$\mathcal{T}_{\Delta\nu}(u) = \int_{\Delta\nu} \exp\left(-k_\nu u\right) \frac{d\nu}{\Delta\nu} = \int_0^\infty f(k) \exp\left(-ku\right) dk \tag{7.175}$$

Note that $f(k)$ is a normalized probability density function, i.e.

$$\int_0^\infty f(k)dk = 1 \tag{7.176}$$

Having found the k-distribution, we may further define the related cumulative probability density function $g(k)$ via

$$g(k) = \int_0^k f(k')dk' \tag{7.177}$$

In particular we have

$$g(0) = 0, \qquad g(k \to \infty) = 1, \qquad dg(k) = f(k)dk \tag{7.178}$$

By definition $g(k)$ is a monotone increasing function. Moreover, while for many gases the spectral absorption coefficient is a highly variable function in ν-space, its probability density $f(k)$ exhibits much less variation. The reader will not be surprised when going from $f(k)$ to $g(k)$ that the variability decreases even more thus leading to a rather smooth cumulative probability density $g(k)$. Having found $g(k)$, the mean transmission can be obtained from the basic equation (7.163) so that

$$\mathcal{T}_{\Delta\nu}(u) = \int_{\Delta\nu} \exp\left(-k_\nu u\right) \frac{d\nu}{\Delta\nu} = \int_0^1 \exp\left[-k(g)u\right] dg \tag{7.179}$$

It should be noted that $k = k(g)$ is the inverse function of $g = g(k)$.

Due to the fact that the cumulative probability density function is rather smooth, the integration over g in (7.179) can be computed very accurately. Often one employs Gaussian quadrature which means that certain g_j and w_j are used for the abscissa and weights of the quadrature rule. This yields

$$\mathcal{T}_{\Delta\nu}(u) = \int_0^1 \exp\left[-k(g)u\right] dg \approx \sum_{j=1}^J w_j \exp[-k(g_j)u] \tag{7.180}$$

where J is the total number of quadrature abscissa. Depending on the required accuracy of the transmission values one may use, for example, four to fifteen quadrature nodes.

For demonstration purposes we will now discuss a realistic situation by considering a spectral interval within the vibration–rotation water vapor band. First we calculate the absorption spectrum, then we find the frequency distribution $f(k)$ and finally the cumulative distribution $g(k)$. Figure 7.10 depicts the spectral absorption coefficient as calculated by means of line-by-line computations using the spectroscopic data for water vapor from the HITRAN database (Rothman et al. 1987, 1992). These computations employ the Voigt profile for the shape of the spectral lines. In the HITRAN database one can find, among other information, the spectral position of each spectral line, the line intensity and the half-width for standard temperature and pressure. Furthermore, one has to describe a cutoff limit beyond which the contributions of neighboring spectral lines can be neglected. Putting the information together line-by-line, one arrives at the graph of Figure 7.10.

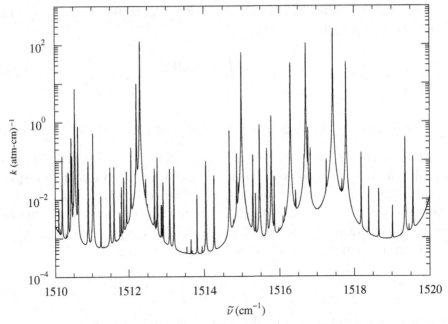

Fig. 7.10 Line-by-line calculations of the absorption coefficient for the spectral range extending from 1510–1520 cm^{-1}, $p = 10$ hPa, $T = 240$ K. This interval is located within the vibration–rotation water vapor band.

The units (atm-cm)$^{-1}$ of the ordinate are obtained by multiplying the absorption cross-section for given coordinates (p, T) by *Loschmidt's number* $N_L = 2.6867 \times 10^{19}$ cm^{-3}. This is the number of molecules per cm^3 of the absorbing gas under standard temperature and pressure conditions. The spectral lines have been computed on a wave number grid with an extremely fine resolution of 10^{-4} cm^{-1}. The influence of neighboring lines from outside the selected interval (1510–1520) cm^{-1} has been accounted for by setting the cutoff wave number to 25 cm^{-1}.

The frequency distribution $f(k)$ corresponding to Figure 7.10 was computed by binning the k-values into a discrete k-grid employing a total of 100 logarithmically equidistant k-values per decade of k, see Figure 7.11. A sufficiently fine resolution (10^{-4} cm^{-1} in our case) is mandatory in order to obtain a satisfactory $f(k)$ distribution. A coarser k-binning grid yields the physically unrealistic occurrence of gaps in the discrete frequency distribution.

Finally, the results of Figure 7.11 were used to compute the cumulative distribution $g(k)$, as stated in (7.178), see Figure 7.12. While the distribution $f(k)$ exhibits some structure, the cumulative distribution $g(k)$ is very smooth. The solid line refers to $p = 10$ hPa and $T = 240$ K. For comparison purposes an additional calculation

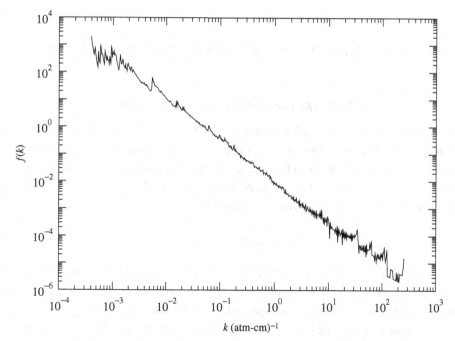

Fig. 7.11 Frequency distribution $f(k)$ of the absorption spectrum shown in Figure 7.10.

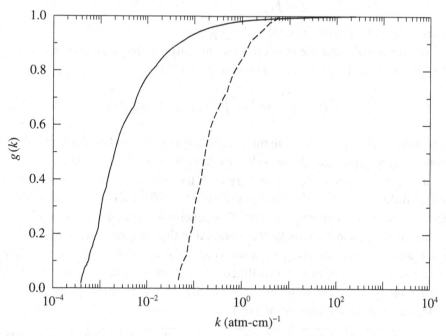

Fig. 7.12 Cumulative frequency distribution $g(k)$ for two combinations of (p, T). Solid line: $p = 10$ hPa, $T = 240$ K, dashed line: $p = 1000$ hPa, $T = 296$ K.

was carried out for $p = 1000$ hPa, $T = 296$ K which is displayed by the dashed curve.

7.3.3 The correlated k-distribution method

So far our derivations for the k-distribution method are valid only for homogeneous atmospheres. The extension to a realistic, inhomogeneous atmosphere requires some careful considerations. Here we follow the discussion of Fu and Liou (1992) and introduce the concept of correlated k-distributions. First, we have to recall the dependence of k_ν on temperature and pressure

$$k_\nu = \sum_i S_i(T) f_i(\nu, p, T) \tag{7.181}$$

where the summation indicates that several spectral lines may contribute to the value of the absorption coefficient at a particular wave number ν. The S_i and f_i are the line intensities and the line-shape factors of the lines, respectively.

If we want to apply the k-distribution method to an inhomogeneous atmosphere, we will consider an atmospheric layer defined by the heights z_1 and $z_2 > z_1$. The mean transmission for this layer can be obtained from

$$T_{\Delta\nu}(u) = \frac{1}{\Delta\nu} \int_{\Delta\nu} \exp\left[-\int_{z_1}^{z_2} k_\nu(p, T)\rho_{\text{abs}}dz\right] d\nu \tag{7.182}$$

where ρ_{abs} is the density of the absorbing gas.

Now we will discuss the mathematical and physical requirements under which (7.182) can be replaced by a form similar to (7.179), i.e.

$$T_{\Delta\nu}(u) = \int_0^1 \exp\left[-\int_{z_1}^{z_2} k(g, p, T)\rho_{\text{abs}}dz\right] dg \tag{7.183}$$

If the mean transmission is computed according to (7.183), then the k-distribution approach is referred to as the *correlated k-distribution* (CKD) *method*.

Since pressure and temperature vary over the layer (z_1, z_2), we should expect that for different levels in the atmosphere there are different $g = g(k)$ relationships for the same wave number ν. In fact, this general behavior is not what the CKD method assumes. Here we make the assumption that despite varying pressure and temperature there is only one g value for a particular ν at all levels of the inhomogeneous atmosphere. For this to be fulfilled, we first must assume that the absorption coefficients at two wave numbers ν_1 and ν_2 are the same for any p and T, if they are the same at the reference state (p_r, T_r)

$$\boxed{k(\nu_1, p_r, T_r) = k(\nu_2, p_r, T_r) \implies k(\nu_1, p, T) = k(\nu_2, p, T)} \tag{7.184}$$

We will call this the first requirement for the CKD method.

If (7.184) is assumed to be valid for an arbitrary wave number v we may conclude that the spectral absorption coefficient for arbitrary v, p and T may be cast into the functional form

$$k(v, p, T) = \chi \, [k_r(v), p, T] \quad \text{with} \quad k_r(v) = k(v, p_r, T_r) \tag{7.185}$$

where the function χ consists of a v-dependent part which can be separated from the (p, T)-dependence. Since in (7.185) there appears only a single reference function $k_r(v)$, we may compute the corresponding probability density function $f_r(k)$. Introducing (7.185) into (7.182) and transforming to the k-space, we obtain

$$\mathcal{T}_{\Delta v}(u) = \int_0^\infty \exp\left[-\int_{z_1}^{z_2} \chi(k_r, p, T)\rho_{abs}dz\right] f(k_r)dk_r \tag{7.186}$$

Assuming that for the reference condition $g_r(k_r)$ is a monotonic function of k_r, we may also compute the relationship $k_r = k_r(g_r)$. Thus, in g_r-space the function χ may be identified with another function β which depends on g_r

$$\chi \, [k_r(g_r), p, T] = \beta \, (g_r, p, T) \tag{7.187}$$

Using the last definition and (7.178), the mean transmission for the layer (z_1, z_2) can be computed from

$$\mathcal{T}_{\Delta v}(u) = \int_0^1 \exp\left[-\int_{z_1}^{z_2} \beta(g_r, p, T)\rho_{abs}dz\right] dg_r \tag{7.188}$$

As the second requirement for the CKD approach we postulate that

$$\boxed{k(v_i, p_r, T_r) > k(v_j, p_r, T_r) \implies k(v_i, p, T) > k(v_j, p, T)} \tag{7.189}$$

This assumption has an important consequence. The ordering of the spectral absorption coefficients, as required for the computation of the cumulative probability distribution, is independent of the actual values of pressure and temperature. In other words, there exists only one g for a given wave number v at different atmospheric levels, that is

$$g_r \, [k_r(v)] = g \, [k(v, p, T), p, T] \tag{7.190}$$

This last relation can be used to find the mean transmission for layer (z_1, z_2) by means of a single g function

$$\mathcal{T}_{\Delta v}(u) = \int_0^1 \exp\left[-\int_{z_1}^{z_2} \beta(g, p, T)\rho_{abs}dz\right] dg \tag{7.191}$$

By using equations (7.185), (7.187) and (7.190) we may conclude that the two requirements of the CKD method lead to the relationship

$$k = \beta [g(k)] \tag{7.192}$$

If k is a monotone function in g-space, then $\beta(g) = k(g)$ as follows from differentiating (7.192) with respect to k. This last identity makes (7.191) to be equivalent to (7.183).

The first requirement of the CKD method provides the basis for using the k-distribution method at the reference state (p_r, T_r). The second requirement relates the cumulative probability function of the reference state to any other level, i.e. for other values of p, T.

7.3.4 The k-distribution method for special situations

Two overlapping gases

For certain gases it is necessary to treat overlap effects in radiative transfer calculations. Due to the fact that computational speed is a very important issue in radiative transfer modeling, in particular for vertically inhomogeneous scattering and absorbing atmospheres, a fast method must be provided allowing the treatment of overlapping effects in an efficient manner.

The mean transmission of two different gases 1 and 2 in a wave number interval of width $\Delta \nu$ is defined as

$$\mathcal{T}_{\Delta\nu}(1, 2) = \frac{1}{\Delta \nu} \int_{\Delta\nu} \mathcal{T}_\nu(1)\mathcal{T}_\nu(2)d\nu \tag{7.193}$$

To simplify this expression we assume that the spectral transmissions of the gases are uncorrelated. This implies that the mean of the product of the individual transmissions equals the product of the corresponding mean transmissions, i.e.

$$\mathcal{T}_{\Delta\nu}(1, 2) = \mathcal{T}_{\Delta\nu}(1)\mathcal{T}_{\Delta\nu}(2) \tag{7.194}$$

Expressing the individual transmissions for the layer (z_1, z_2) in g-space as

$$\mathcal{T}_{\Delta\nu}(i) = \int_0^1 \exp\left(-\int_{z_1}^{z_2} k_i \rho_{\text{abs},i} dz\right) dg_i, \qquad i = 1, 2 \tag{7.195}$$

the mean transmission of the two overlapping gases can be formulated as

$$\mathcal{T}_{\Delta\nu}(1, 2) = \int_0^1 \int_0^1 \exp\left(-\int_{z_1}^{z_2} [k_1 \rho_{\text{abs},1} + k_2 \rho_{\text{abs},2}]dz\right) dg_1 dg_2 \tag{7.196}$$

$$\approx \sum_{m=1}^{M} \sum_{n=1}^{N} \exp(-\tau_{mn}) \Delta g_{1m} \Delta g_{2n}$$

where $\rho_{\text{abs},1}$, $\rho_{\text{abs},2}$ and k_1, k_2 are the density and the absorption coefficient of gas 1 and gas 2, respectively. The optical depth of the two gases in the layer is defined by

$$\tau_{mn} = \int_{z_1}^{z_2} [k_1 \rho_{\text{abs},1} + k_2 \rho_{\text{abs},2}] dz \qquad (7.197)$$

In summary we conclude, that for obtaining the absorption effect of two gases which overlap in a particular wave number interval, we must carry out $M \times N$ quasi-spectral radiative transfer calculations for each vertical atmospheric column. To give an example of the amount of computations involved, let us consider the overlap of CO_2 and H_2O in the spectral region 540–800 cm^{-1}. Fu and Liou (1992) recommend in their CKD method a subdivision of this region into the two subregions 540–670 cm^{-1} and 670–800 cm^{-1}. In the first subregion they employ for H_2O and CO_2 an expansion with 5 and 10 terms, respectively. In the second subinterval the corresponding numbers are 4 (H_2O) and 8 (CO_2). Hence a total of $5 \times 10 + 4 \times 8 = 82$ quasi-monochromatic radiative transfer computations has to be performed. This illustrates the increased computational effort involved when treating gas overlap effects in atmospheric radiative transfer. Additional simplifications can be used to reduce computational efforts. For more details the reader is invited to consult the original papers.

Gray absorption coefficient

We consider the wave number range $\Delta \nu$, in which the absorption coefficient takes on the constant value $k_\nu = \bar{k}$. According to (7.163) the k-distribution is then given by

$$f(k) = \delta(k - \bar{k}) \qquad (7.198)$$

where $\delta(k - \bar{k})$ is the Dirac δ-function.

Regular band of nonoverlapping rectangular lines

The next example is a regular band of an infinite number of nonoverlapping rectangular lines. Let b represent the half-width of the rectangular line and define $\alpha = b/\delta$. The absorption coefficient is then given by

$$k_\nu = \begin{cases} k_1 & \text{for } \alpha \leq \nu/\delta \leq 1/2 \\ k_2 & \text{for } 0 \leq \nu/\delta \leq \alpha \end{cases} \qquad (7.199)$$

For this spectral arrangement we read from Figure 7.13

$$f(k) = (1 - 2\alpha)\delta(k - k_1) + 2\alpha\delta(k - k_2) \qquad (7.200)$$

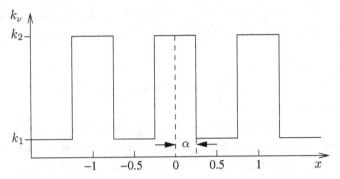

Fig. 7.13 A regular band of nonoverlapping rectangular lines with $x = v/\delta$.

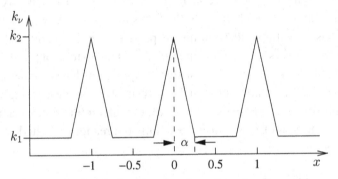

Fig. 7.14 A regular band of nonoverlapping triangular lines with $x = v/\delta$.

For the mean absorption coefficient we obtain by inspection of Figure 7.13 the relation

$$\bar{k} = k_1 + 2\alpha(k_2 - k_1) \tag{7.201}$$

Alternatively, we can calculate the mean absorption coefficient from the k-distribution

$$\bar{k} = \int_0^\infty k f(k) dk = \int_0^\infty k(1 - 2\alpha)\delta(k - k_1)dk + \int_0^\infty 2\alpha k \delta(k - k_2)dk$$

$$= (1 - 2\alpha)k_1 + 2\alpha k_2 = k_1 + 2\alpha(k_2 - k_1) \tag{7.202}$$

Regular band of triangular lines

In this case the absorption coefficient is given by, see Figure 7.14

$$k_v = \begin{cases} k_2 - \dfrac{v}{\alpha}(k_2 - k_1) & \text{for } 0 \le v/\delta \le \alpha \\ k_1 & \text{for } \alpha \le v/\delta \le 1/2 \end{cases} \tag{7.203}$$

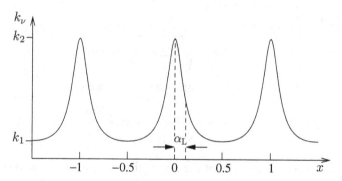

Fig. 7.15 A regular band of nonoverlapping Lorentz lines with $x = v/\delta$.

A few simple algebraic steps lead to

$$f(k) = (1 - 2\alpha)\delta(k - k_1) + \begin{cases} \dfrac{2\alpha}{k_2 - k_1} & \text{for } k_1 \leq k \leq k_2 \\ 0 & \text{for } k < k_1 \text{ or } k > k_2 \end{cases} \qquad (7.204)$$

For the average absorption coefficient we obtain

$$\bar{k} = \int_0^\infty kf(k)dk = k_1(1 - 2\alpha) + \frac{2\alpha}{k_2 - k_1}\frac{k^2}{2}\Big|_{k_1}^{k_2} = k_1 + \alpha(k_2 - k_1) \qquad (7.205)$$

This result also follows directly from inspection of Figure 7.14.

Regular band of nonoverlapping Lorentz lines

As a final example we discuss a situation which is similar to the regular Elsasser model. However, there is one exception, namely that the individual identical Lorentz lines are cut off at the distance $\delta/2$ from each center line, see Figure 7.15.
   In this case we have

$$k_v = \frac{S\alpha_L}{\pi\left(v^2 + \alpha_L^2\right)} \qquad \text{for } 0 \leq v \leq \delta/2 \qquad (7.206)$$

With this information we employ (7.168) to find $f(k)$ by differentiation

$$f(k) = \frac{1}{\delta}\left(\left|\frac{dv}{dk}\right|_{\Delta v_1} + \left|\frac{dv}{dk}\right|_{\Delta v_2}\right) = \frac{2}{\delta}\left|\frac{dv}{dk}\right|_{\Delta v_2} \qquad (7.207)$$

where $\Delta v_1 = (-\delta/2, 0)$ and $\Delta v_2 = (0, \delta/2)$. The last step follows from symmetry. The expression for the spectral absorption coefficient can be directly solved for v yielding

$$v = \sqrt{\frac{S\alpha_L}{\pi k} - \alpha_L^2} \tag{7.208}$$

Computing the first derivative of v with respect to k, we obtain

$$\frac{dv}{dk} = -\frac{1}{2\sqrt{S\alpha_L/(\pi k) - \alpha_L^2}} \frac{S\alpha_L}{\pi k^2} \tag{7.209}$$

Substituting the maximum value of the absorption coefficient $k_2 = S/\pi\alpha_L$ gives

$$\left|\frac{dv}{dk}\right| = \frac{1}{2} \frac{k_2\alpha_L}{k^{3/2}\sqrt{k_2 - k}} \tag{7.210}$$

Thus we find the k-distribution of the cut-off Lorentz band in the form

$$f(k) = \frac{k_2\alpha_L}{8k^{3/2}\sqrt{k_2 - k}} \tag{7.211}$$

From $\bar{k} = \int kf(k)dk$ we obtain after some tedious but straightforward steps

$$\bar{k} = \frac{2k_2\alpha_L}{\delta} \tan^{-1}\left(\frac{\delta}{2\alpha_L}\right) \tag{7.212}$$

Details in the derivation will be left up to the exercises.

Single scattering properties for inhomogeneous atmospheres

While gaseous atmospheric absorption is described by the spectral absorption coefficient, extinction by air molecules, aerosol particles and hydrometeors requires additional information on the extinction coefficient, the *single scattering albedo* and the *scattering phase function*. Below we will describe how these radiative properties are combined to yield the scattering and absorption properties of an individual homogeneous atmospheric layer embedded in an otherwise inhomogeneous atmosphere.

We will assume that the inhomogeneous atmosphere consists of a set of distinct homogeneous layers. For a certain wave number band Δv the total optical depth of a layer Δz consists of *Rayleigh scattering* $\Delta\tau_R$, *Mie scattering* due to atmospheric aerosol particles and hydrometeors $\Delta\tau_M$, and the contribution $\Delta\tau_G$ due to gas absorption

$$\Delta\tau = \Delta\tau_R + \Delta\tau_M + \Delta\tau_G \tag{7.213}$$

Absorption by a gas with density ρ_{abs} may be described by means of the CKD approach yielding

$$\Delta\tau_G(g) = k(g)\rho_{abs}\Delta z \tag{7.214}$$

where g is the cumulative probability density for the absorbing gas. For simplicity, we have omitted in $\tau_G(g)$ the pressure and temperature dependence. The contribution to the total optical depth in layer Δz resulting from Mie scattering is[3]

$$\Delta \tau_M = \Delta \tau_{\text{sca,M}} + \Delta \tau_{\text{abs,M}} \tag{7.215}$$

The single scattering albedo in the homogeneous layer is defined by

$$\omega_0(g) = \frac{\Delta \tau_R + \Delta \tau_{\text{sca,M}}}{\Delta \tau_R + \Delta \tau_M + \Delta \tau_G(g)} \tag{7.216}$$

Let the Legendre expansion coefficients for the Rayleigh and Mie scattering be given by $p_{l,R}$ and $p_{l,M}$. The Legendre expansion coefficients for the *phase function* for the combined action of air molecules and particle scattering is defined by a linear weighting of the expansion coefficients of the individual components. The weighting factor is the scattering optical depth of the respective material in layer Δz, i.e.

$$p_l = \frac{\Delta \tau_{\text{sca,M}} p_{l,M} + \Delta \tau_R p_{l,R}}{\Delta \tau_{\text{sca,M}} + \Delta \tau_R} \tag{7.217}$$

For many cases the *Mie phase function* can be approximated by the *Henyey–Greenstein phase function*, which for particle scattering depends on the asymmetry factor only, see (6.41). Thus we obtain for the phase function the two relations

$$\mathcal{P}(\cos \Theta) = \sum_{l=0}^{M} p_l P_l(\cos \Theta) = \begin{cases} \dfrac{3}{4}(1 + \cos^2 \Theta) & \text{Rayleigh scattering} \\[2mm] \displaystyle\sum_{l=0}^{M}(2l+1)g^l P_l(\cos \Theta) & \text{Henyey–Greenstein} \end{cases} \tag{7.218}$$

In these examples we have seen that the k-distribution method can be applied to various simple models. It is difficult to conceive any situation where this method would fail.

7.4 Transmission in inhomogeneous atmospheres

In the previous sections we have derived various transmission functions for model spectra of absorbing atmospheres. These transmission functions applying to homogeneous gaseous layers can be extended to simulate the transmission through inhomogeneous atmospheres. Ordinarily this involves coupled integrations over the atmospheric path and the wave number. In order to avoid such extremely laborious integrations to handle the transfer problem, it is customary to adopt approximate

---

[3] A detailed description of the Mie theory will be given in a later chapter.

techniques by means of scaling procedures which attempt to decouple these two types of integrations. In earlier years radiation charts and the emissivity method were used to study the radiative properties of the atmosphere. These procedures already provided reasonably accurate profiles of the vertical flux density and the radiative cooling rate for various types of air masses as was verified by comparison with measurements. These methods required that the integration over the wave number was carried out once and for all. Then followed the integration over the inhomogeneous path of the atmosphere either using a graphical or a numerical procedure. This was done with the help of a one-parameter scaling technique which will be discussed below. In order to improve the accuracy of the one-parameter procedure, the *Curtis–Godson approximation* was introduced which is a two-parameter scaling technique. Higher order scaling techniques are possible but usually they are quite difficult to apply. In the following we will describe one- and two-parameter models.

7.4.1 One-parameter scaling

The most simple scaling procedure is the so-called *one-parameter scaling*. Fortunately, this method is sufficiently accurate to handle various transfer problems. The transmission function for a non-homogeneous vertical atmospheric path may be expressed by

$$
\begin{aligned}
\mathcal{T}(u) &= \frac{1}{\Delta v} \int_{\Delta v} \exp\left(-\int k_{v,\mathrm{L}}(p,T)du\right) dv \\
&= \frac{1}{\Delta v} \int_{\Delta v} \exp\left(-\int \sum_i \frac{S_i}{\pi} \frac{\alpha_{\mathrm{L},i}}{(v-v_{0,i})^2 + \alpha_{\mathrm{L},i}^2} du\right) dv
\end{aligned}
\tag{7.219}
$$

which applies to a small spectral interval containing numerous spectral lines. To be specific we will assume that these lines have Lorentzian profiles, but for brevity henceforth we omit the index L.

For simplicity let us first consider the technique as it applies to a single spectral line. The formal extension to include neighboring lines is simple. We have shown in (7.27) that the Lorentzian half-width is proportional to the linear pressure reduction (p/p_0) while the temperature dependency varies according to $\sqrt{T_0/T}$. There is sufficient empirical and some theoretical evidence that the square root law to handle the temperature dependency is not always sufficient. Hence we will express the temperature dependency by a more general law as stated by

$$
\alpha = \alpha_0 \frac{p}{p_0}\left(\frac{T_0}{T}\right)^n
\tag{7.220}
$$

leaving the exponent n unspecified. Since (p_0, T_0) refer to the reference pressure and temperature, the ratio of the absorption coefficients $k_\nu(p, T)/k_\nu(p_0, T_0)$ is given by

$$\frac{k_\nu(p, T)}{k_\nu(p_0, T_0)} = \frac{\dfrac{S(T)}{\pi} \dfrac{\alpha}{(\nu - \nu_0)^2 + \alpha^2}}{\dfrac{S(T_0)}{\pi} \dfrac{\alpha_0}{(\nu - \nu_0)^2 + \alpha_0^2}} \approx \frac{S(T)}{S(T_0)} \frac{p}{p_0} \left(\frac{T_0}{T}\right)^n \approx \frac{p}{p_0} \left(\frac{T_0}{T}\right)^n \quad (7.221)$$

Here, we have assumed that the strong-line approximation is sufficiently accurate for the one-parameter scaling procedure. Thus, by ignoring the square of the half-width in the denominator of the spectral line, the wave number-dependent parts of the absorption coefficient can be factored out and cancel. Furthermore, we have assumed that the ratio $S(T)/S(T_0)$ is approximately equal to 1. Hence we may write

$$\int k_\nu(p, T) du = k_\nu(p_0, T_0) \int \frac{p}{p_0} \left(\frac{T_0}{T}\right)^n du = k_\nu(p_0, T_0)\tilde{u} \quad (7.222)$$

where the quantity

$$\boxed{\tilde{u} = \int \frac{p}{p_0} \left(\frac{T_0}{T}\right)^n du} \quad (7.223)$$

is known as the *scaling parameter* or the *scaling path length*. In case that there are many spectral lines in the interval, the approximation used in (7.221) must be replaced by

$$\frac{k_\nu(p, T)}{k_\nu(p_0, T_0)} \approx \frac{\displaystyle\sum_i \frac{S_i(T)\alpha_{0,i}}{(\nu - \nu_{0,i})^2} \frac{p}{p_0} \left(\frac{T_0}{T}\right)^n}{\displaystyle\sum_i \frac{S_i(T_0)\alpha_{0,i}}{(\nu - \nu_{0,i})^2}} \approx \frac{p}{p_0} \left(\frac{T_0}{T}\right)^n \quad (7.224)$$

However, the same scaling parameter $\tilde{u}$ is used whether we are scaling the absorption path of a single line or a group of lines.

As stated above, most early radiative transfer calculations were carried out with the help of radiation charts or the emissivity method. These methods were designed to employ the scaling parameter $\tilde{u}$ which can also be successfully employed in connection with the rotational water vapor band, see Chou and Arking (1980). In the older literature the parameter $\tilde{u}$ was called the *reduced absorber mass*.

The extension to an inclined path, assuming horizontally homogeneous sublayers, is quite simple as we have seen in our previous work. All we need to do is to introduce the factor $1/\mu$ in the exponent of (7.219).

7.4.2 The two-parameter scaling technique of Curtis and Godson

To simplify the transmission calculations, the approximation replaces an inhomogeneous by a more or less equivalent homogeneous layer. This will be accomplished by adjusting the parameters appearing in the absorption coefficient which is assumed to have the Lorentzian shape. The mean transmission in correct form is given by (7.219). For simplicity let us again begin with a single spectral line by omitting the summation sign. In the limit of the strong line approximation $T_s(u)$ the half-width in the denominator may be ignored. Instead of neglecting α^2 altogether, we replace it by a suitable mean value $\tilde{a}^2$ which is independent of the atmospheric path

$$T_s(u) = \frac{1}{\Delta v} \int_{\Delta v} \exp\left(-\int \frac{S(T)}{\pi} \frac{\alpha(p, T)}{(v - v_0)^2 + \tilde{a}^2} du\right) dv \qquad (7.225)$$

By expanding the exponent and discontinuing the expansion after the second term, we obtain the weak line limit $T_w(u)$ of the transmittance. Since the interval Δv is assumed to be much larger than the half-width of the line, we may extend the wave number integration to infinity yielding

$$T_w(u) = 1 - \frac{1}{\Delta v} \int \frac{S(T)\alpha(p, T)}{\pi \tilde{a}} du \int_{-\infty}^{\infty} \frac{1}{x^2 + 1} dx = 1 - \frac{1}{\Delta v} \int \frac{S(T)\alpha(p, T)}{\tilde{a}} du \qquad (7.226)$$

Here we have used the simple transformation $x = (v - v_0)/\tilde{a}$. In order to evaluate $\tilde{a}$, we also expand the exponent of (7.219) and obtain the approximation

$$T_w = 1 - \frac{1}{\Delta v} \int S(T) du \qquad (7.227)$$

which, of course, is identical with the result obtained from the Ladenburg–Reiche formula (7.73). Forcing agreement between (7.226) and (7.227) we obtain the scaling half-width $\tilde{a}$

$$\boxed{\tilde{a} = \frac{\int S(T)\alpha(p, T) du}{\int S(T) du}} \qquad (7.228)$$

which is the first scaling parameter of the Curtis–Godson approximation. By ignoring the small temperature dependence of the half-width, that is

$$\frac{\alpha}{\tilde{a}} \approx \frac{p}{\tilde{p}} \qquad (7.229)$$

we obtain the so-called *pressure scaling factor* $\tilde{p}$

$$\boxed{\tilde{p} = \frac{\int S(T) p \, du}{\int S(T) du}} \qquad (7.230)$$

To simulate the transmission through an inhomogeneous atmosphere by an equivalent homogeneous layer, using Goody's (1964b) notation, we employ the adjusted parameters $\tilde{S}, \tilde{a}, \tilde{u}$. Thus, instead of (7.219) (omitting the summation sign), we write

$$\mathcal{T} = \frac{1}{\Delta \nu} \int_{\Delta \nu} \exp\left(-\frac{\tilde{S}\tilde{a}\tilde{u}}{\pi[(\nu - \nu_0)^2 + \tilde{a}^2]}\right) d\nu \qquad (7.231)$$

We now wish to obtain explicit expressions for $\tilde{S}$, $\tilde{a}$ and $\tilde{u}$. Comparing (7.231) with (7.225) and utilizing (7.228) gives

$$\boxed{\tilde{S}\tilde{a}\tilde{u} = \int S a \, du, \qquad \tilde{S}\tilde{u} = \int S \, du, \qquad \tilde{u} = \frac{\int S \, du}{\tilde{S}}} \qquad (7.232)$$

For a moment one is tempted to conclude that we have a three-parameter approximation. However, only the parameters $\tilde{a}$ and $\tilde{u}$ are needed. The pressure variation of the inhomogeneous atmosphere is included in (7.228) or (7.230) while the temperature variation is modeled by (7.232). The mean line intensity $\tilde{S}$ can be evaluated at any specified temperature since the effect of this temperature cancels. Thus the two scaling parameters are $\tilde{a}$ (or $\tilde{p}$) and $\tilde{u}$.

For a system of many lines the principle of the Curtis–Godson approximation is the same as for a single line. We repeat equation (7.219) but we introduce the line-shape factor of the Lorentzian line according to (7.18)

$$\mathcal{T}(u) = \frac{1}{\Delta \nu} \int_{\Delta \nu} \exp\left(-\int \sum_i S_i f(\nu - \nu_{0,i}, \alpha_i) du\right) d\nu \qquad (7.233)$$

In case of the weak line approximation this equation assumes the form

$$\mathcal{T}_{\mathrm{w}}(u) = 1 - \frac{1}{\Delta \nu} \int \sum_i S_i \int_{\Delta \nu} f(\nu - \nu_{0,i}, \alpha_i) d\nu \, du = 1 - \frac{1}{\Delta \nu} \int \sum_i S_i d\nu \qquad (7.234)$$

where we have assumed that the normalization condition (7.17) of the line-shape factor is valid in the frequency interval $\Delta \nu$. In order to introduce the strong line approximation, as before, in the denominator of (7.219) we replace the half-width by the constant $\tilde{a}_i$ yielding

$$\mathcal{T}_{\mathrm{s}}(u) = \frac{1}{\Delta \nu} \int_{\Delta \nu} \exp\left(-\int \sum_i \frac{S_i}{\pi} \frac{\alpha_i}{(\nu - \nu_{0,i})^2 + \tilde{a}_i^2} du\right) d\nu \qquad (7.235)$$

To obtain the proper value for $\tilde{a}$ that fits the strong line and the weak line approximation, we expand the exponent and find

$$T_w(u) = 1 - \frac{1}{\Delta v} \int \sum_i \frac{S_i \alpha_i}{\tilde{\alpha}_i} \int_{\Delta v} f(v - v_{0,i}, \tilde{\alpha}_i) dv \, du = 1 - \frac{1}{\Delta v} \int \sum_i \frac{S_i \alpha_i}{\tilde{\alpha}_i} du$$

(7.236)

We proceed analogously to the single line case by comparing (7.234) and (7.236). This gives

$$\int \sum_i S_i du = \int \sum_i \frac{S_i p}{\tilde{p}} du$$

(7.237)

Since the pressure dependence of each line is identical, we obtain the first scaling parameter

$$\boxed{\tilde{p} = \frac{\int \sigma p du}{\int \sigma du}}$$

(7.238)

where $\sigma = 1/N \sum_i S_i$ is the mean line intensity. In case of a single line this equation, as it should, reduces to (7.230).

Now we need to find the second scaling parameter. Let us reconsider equation (7.219) whose exact analogy to a homogeneous path can be written as

$$T(u) = \frac{1}{\Delta v} \int_{\Delta v} \exp \left(-\sum_i \frac{\tilde{S}_i \tilde{\alpha}_i \tilde{u}}{\pi \left[(v - v_{0,i})^2 + \tilde{\alpha}_i^2 \right]} \right) dv$$

(7.239)

Comparison of (7.235) with (7.239) gives the second scaling parameter

$$\tilde{u} = \frac{\int \sum_i \frac{S_i \alpha_i}{(v - v_{0,i})^2 + \tilde{\alpha}_i^2} du}{\sum_i \frac{\tilde{S}_i \tilde{\alpha}_i}{(v - v_{0,i})^2 + \tilde{\alpha}_i^2}}$$

(7.240)

Recalling (7.228) and applying it to line i, after introducing the line-shape factors, cf. (7.18), results in

$$\boxed{\tilde{u} = \frac{\sum_i f(v - v_{0,i}, \tilde{\alpha}_i) \int S_i du}{\sum_i f(v - v_{0,i}, \tilde{\alpha}_i) \tilde{S}_i}}$$

(7.241)

which is the scaled absorber mass. The line intensity $\tilde{S}_i$, as explained before, can be evaluated at any specified temperature.

The quality of the Curtis–Godson approximation for spectral intervals containing many spectral lines has been tested by Walshaw and Rodgers (1963). They performed extensive cooling rate calculations using a line-by-line integration technique. For the water vapor rotational band and the 15 μm CO_2 band they found errors less than a few percent. Various authors have investigated the accuracy of the Curtis–Godson approximation. For example, Zdunkowski and Raymond (1970) investigated the transmission in small spectral intervals in the 1.9 and 6.3 μm water vapor bands. They found excellent agreement between the exact calculations and the Curtis–Godson approximation. The Curtis–Godson approximation may also be applied to other line shapes than the Lorentz line, but we omit any discussion.

While in many situations the Curtis–Godson approximation provides satisfactory results for atmospheric water vapor and carbon dioxide distributions, the method is less satisfactory for the 9.6 μm ozone band. Goody (1964b) employed van de Hulst's (1945) rather general technique to handle the transmission calculations pertaining to inhomogeneous atmospheres. This technique is based on the series expansion of the Fourier cosine transform permitting the formulation of three scaling parameters. The extension of the Curtis–Godson scaling method resulted in a significant improvement of the transmission calculations. It is also possible to use additional scaling parameters. This, however, greatly complicates the calculation procedure. For more details the reader is referred to Goody's original work.

7.5 Results

In this section we will briefly consider some typical vertical profiles of solar and infrared flux densities as well as radiative temperature changes which have been obtained by means of the radiation transfer model DISORT (Stamnes *et al.*, 1988) with a total of four discrete streams. To handle the spectrally dependent absorption by atmospheric gases we have used the correlated k-distribution parameterization of Fu and Liou (1992). Presently this appears to be the most efficient yet sufficiently accurate way to handle the spectral integration. The calculations specify the ground albedo and the solar zenith angle as $A_g = 0.1$ and $\theta_0 = 30°$ for the solar spectrum and a ground emissivity $\varepsilon_g = 1$ for the entire thermal emission spectrum. Furthermore, the calculations assume clear sky conditions (no clouds and no aerosol) in a mid-latitude summer atmosphere with typical vertical distributions of the radiatively relevant atmospheric trace gases. Figure 7.16 shows the vertical distributions of temperature, relative humidity and those of the atmospheric trace gases which have been used in the radiation calculations. The height constant CO_2 volume mixing ratio has been set to 350 ppmv where 1 ppmv $= 10^{-6}$. The integrated or total amount of ozone in a vertical column above the surface of the Earth is expressed in atmosphere centimeters (atm-cm) which is the height of the resulting volume

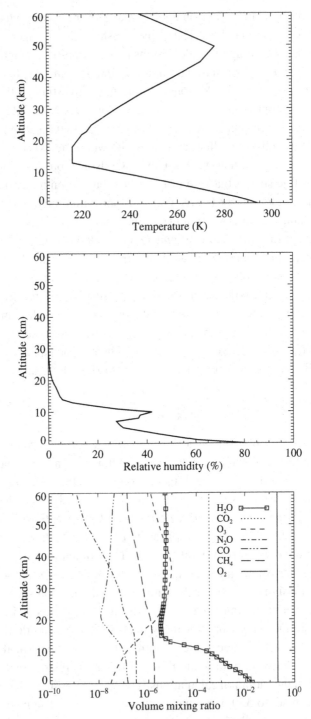

Fig. 7.16 Vertical distributions of temperature, relative humidity and the radiatively active trace gases for a mid-latitude summer atmosphere.

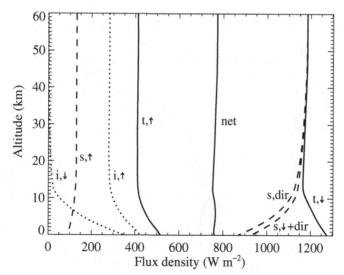

Fig. 7.17 Distribution of upward ($\uparrow$), downward ($\downarrow$) and net flux densities labeled as s (solar), i (infrared), t (total), s,dir (direct downward solar radiation) and net (net radiation).

if all the ozone in the column of unit area were brought to normal pressure and temperature (NTP). Currently, the total amount of atmospheric ozone is expressed in terms of Dobson units (DU) where 1 DU $= 10^{-3}$ atm-cm NTP. The total ozone column shown in Figure 7.16 amounts to 330 DU.

The absorption bands of the gases have been fully accounted for in the solar and in the infrared spectrum. All transfer calculations pertaining to the solar spectrum use a solar constant of 1368 W m^{-2} while the infrared calculations assume a black-body ground emission.

Figure 7.17 depicts the upward and downward directed solar and infrared flux densities and the corresponding net flux density. The value of the upward solar flux density at the ground is found by multiplying the value of total downward solar radiation reaching the ground (s,$\downarrow$ +dir) by $A_g = 0.1$. The flux densities at the ground are needed to formulate energy boundary conditions in the thermodynamic parts of weather and climate prediction models. Of particular interest is the radiative flux divergence of the combined solar and infrared spectrum, which is needed to evaluate thermodynamic equations of the type (2.44).

Radiative heating rates corresponding to the radiative flux densities of Figure 7.17 are shown in Figure 7.18. While the vertical flux density profiles are rather smooth, radiative temperature changes vary quite rapidly with height. This variation results from the vertical structure of the volume mixing ratio of the trace

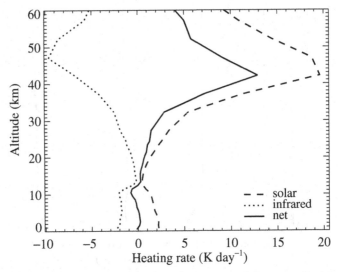

Fig. 7.18 Solar heating rates, infrared cooling rates and the net radiative temperature changes expressed in K day^{-1}.

gases and from the atmospheric vertical temperature distribution. Inspection of the curves shows that solar heating at the ground amounts to about 2 K day^{-1}. Had a normal aerosol distribution been included, we would have calculated an additional heating of about 0.1 K day^{-1}. This is a small but not entirely negligible effect. Since aerosol concentrations usually decrease with height, the aerosol contribution to the heating rate is negligible at some height above the atmospheric boundary layer. Solar heating, mainly caused by the presence of water vapor, decreases rather rapidly with height in the troposphere due to the vertically decreasing water vapor concentration. In the stratosphere strong solar heating is observed resulting from an increase of the ozone volume mixing ratio with height. At a height of 42 km where the ozone concentration begins to decrease, a maximum value of nearly 20 K day^{-1} is obtained. For a smaller value of μ_0 maximum heating would have occurred at a somewhat smaller height. For increasing values of μ_0 solar heating is most effective in the upper part of the O$_3$ layer. As stands to reason, maximum heating will then be less intense due to the smaller O$_3$ concentrations existing there.

Let us now consider the infrared cooling rates which strongly depend on the concentration of the absorbing and emitting gases and also on the vertical temperature distribution. For the midlatitude summer model, the tropospheric cooling rate of 2 K day^{-1}, mostly due to water vapor absorption bands, is nearly height constant. Additional calculations (not shown) reveal that larger tropospheric cooling will be found in the moist tropical and smaller cooling in arctic air masses. Moreover, the cooling rates will be strongly modified by the presence of clouds.

Further inspection of Figure 7.18 shows that infrared cooling decreases in the lower stratosphere and then increases to about $-10\,\mathrm{K\,day}^{-1}$ just below 50 km. Even at a height of 60 km, cooling rates still amount to about $-5\,\mathrm{K\,day}^{-1}$. The strong radiative cooling in the layer extending from 30 to 60 km is mainly caused by CO_2. In the height range from 30–50 km radiative cooling is intensified by the presence of ozone and to a smaller extent by water vapor. The strongest cooling should take place at the top of the stratospheric O_3 layer located between 20 and 40 km since shielding effects of still higher absorbing layers are absent. For a height of about 40 km Plass (1956) calculated an ozone cooling rate of about $-2.5\,\mathrm{K\,day}^{-1}$. This result was verified by more recent investigations.

Finally, we will investigate the net radiative temperature change. Within the entire troposphere the net radiative heating is almost zero. Above the troposphere radiation causes a net warming of the atmosphere. The strongest radiative heating is found in the height range from 25–50 km which roughly coincides with the stratospheric temperature increase in this height interval. The fact that the atmospheric temperature decreases above 50 km indicates that physical processes other than radiation cause atmospheric cooling in this region.

7.6 Appendix

7.6.1 Maxwell's velocity distribution and the mean molecular velocity

Let $\mathbf{v} = (v_x, v_y, v_z)$ be the vector of the thermal velocity of the air molecules. Temperature is just another measure for the average kinetic energy of the molecules. If we limit our discussion to translational energy, the relation between temperature and mean kinetic energy is given by

$$\frac{1}{2}m\overline{v^2} = \frac{3}{2}kT \tag{7.242}$$

In total there are three degrees of freedom for the translational motion. Along each of the three Cartesian axes the average kinetic energy contributes $\frac{1}{2}kT$ to the total thermal energy. Hence, we obtain

$$\frac{1}{2}m\overline{v_x^2} = \frac{1}{2}m\overline{v_y^2} = \frac{1}{2}m\overline{v_z^2} = \frac{1}{2}kT \tag{7.243}$$

Maxwell's velocity distribution $F(v_x, v_y, v_z)$ is given by

$$F(v_x, v_y, v_z) = \left(\frac{m}{2\pi kT}\right)^{3/2} \exp\left(-\frac{m\left(v_x^2 + v_y^2 + v_z^2\right)}{2kT}\right) \tag{7.244}$$

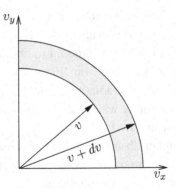

Fig. 7.19 Two-dimensional illustration of a spherical shell volume for the derivation of Maxwell's velocity distribution $f(v)$.

Maxwell's distribution has spherical symmetry and depends only on the magnitude of $\mathbf{v}$, but not on the direction of $\mathbf{v}$. To be a proper probability distribution F must be normalized

$$\int_{-\infty}^{\infty} F \, dv_x dv_y dv_z = 1 \qquad (7.245)$$

Furthermore, each translational degree of freedom must contribute $\frac{1}{2}kT$ to its average kinetic energy, see (7.243).

Due to the fact that F only depends on $v^2 = |\mathbf{v}|^2 = v_x^2 + v_y^2 + v_z^2$, integration of F over the Cartesian volume element $dv_x dv_y dv_z$ can be replaced by an integration over spherical shells having the volume $4\pi v^2 dv$, see Figure 7.19. Hence, we may write

$$f(v)dv = 4\pi n F(v)v^2 dv = 4\pi n \left(\frac{m}{2\pi kT}\right)^{3/2} \exp\left(-\frac{mv^2}{2kT}\right) v^2 dv \qquad (7.246)$$

The function $f(v)$ is Maxwell's velocity distribution for the magnitude v of the velocity. Note that the factor 4π stems from the integration over azimuth from 0 to 2π and over the cosine of the polar angle from -1 to 1. Furthermore, the factor n is necessary to account for the contributions of each molecule contained in a unit volume of space.

The most probable velocity $\hat{v}$ is defined by that velocity for which $f(v)$ attains its maximum, i.e.

$$\hat{v} = \sqrt{\frac{2kT}{m}} \qquad (7.247)$$

Likewise, the average velocity $\bar{v}$ can be computed from

$$\bar{v} = \frac{1}{n} \int_0^\infty v f(v) dv = \sqrt{\frac{8kT}{\pi m}} \tag{7.248}$$

and the root-mean-square velocity is given by

$$(\overline{v^2})^{1/2} = \left(\frac{1}{n} \int_0^\infty v^2 f(v) dv \right)^{1/2} = \sqrt{\frac{3kT}{m}} \tag{7.249}$$

Note that the factor $1/n$ in the last two expressions is required for normalization.

7.6.2 Original derivation of the Ladenburg and Reiche function

We start with the following expression for the average absorption of an isolated Lorentz line

$$A(\bar{u}) = \int_{-\infty}^\infty \left[1 - \exp\left(-\frac{2\bar{u} y^2}{x^2 + y^2} \right) \right] dx \tag{7.250}$$

$$\text{with} \quad \bar{u} = \frac{Su}{2\pi \alpha_L}, \quad x = \frac{v}{\Delta v}, \quad y = \frac{\alpha_L}{\Delta v}$$

It is convenient to introduce the substitution

$$x = y \tan(s/2) \implies dx = y \frac{d}{ds} \tan(s/2) ds \tag{7.251}$$

Using this substitution, the argument of the exponential function in (7.250) can be transformed to

$$\frac{2\bar{u} y^2}{x^2 + y^2} = \frac{2\bar{u} y^2}{y^2[1 + \tan^2(s/2)]} = 2\bar{u} \cos^2(s/2) = \bar{u}(1 + \cos s) \tag{7.252}$$

so that the average absorption may be rewritten as

$$A(\bar{u}) = y \int_{-\pi}^\pi \left[1 - \exp\left[-\bar{u}(1 + \cos s) \right] \frac{d}{ds} \tan(s/2) \right] ds \tag{7.253}$$

This expression can be further simplified by means of integration by parts

$$A(\bar{u}) = y \frac{1 - \exp\left[-\bar{u}(1 + \cos s) \right]}{\cot(s/2)} \Big|_{-\pi}^\pi$$

$$+ y\bar{u} \int_{-\pi}^\pi \exp\left[-\bar{u}(1 + \cos s) \right] \tan(s/2) \sin s\, ds \tag{7.254}$$

The first term on the right-hand side is of the form 0/0. Utilizing l'Hôpital's rule one can easily verify that this term vanishes. Using

$$\tan(s/2) = \frac{1 - \cos s}{\sin s} \tag{7.255}$$

the second term on the right-hand side of (7.254) can be rewritten yielding

$$A(\bar{u}) = y\bar{u}\exp(-\bar{u})\left(\int_{-\pi}^{\pi}\exp(-\bar{u}\cos s)\,ds - \int_{-\pi}^{\pi}\cos s\exp(-\bar{u}\cos s)\,ds\right) \tag{7.256}$$

The two integrals appearing in (7.256) are proportional to the Bessel functions of the first kind of order 0 and 1. From mathematical tables (Abramowitz and Stegun, 1972) we find

$$J_n(\rho) = \frac{(-i)^n}{2\pi}\int_{-\pi}^{\pi}\exp(i\rho\cos s)\cos(ns)\,ds \tag{7.257}$$

Hence, by setting $\rho = i\bar{u}$, we can write for (7.256)

$$A(\bar{u}) = 2\pi y\bar{u}\exp(-\bar{u})[J_0(i\bar{u}) - iJ_1(i\bar{u})] \tag{7.258}$$

The modified Bessel functions I_n are real functions and are related to the Bessel functions by

$$J_n(ix) = i^n I_n(x) \tag{7.259}$$

This completes the alternative derivation of the Ladenburg and Reiche function, that is

$$\boxed{A(\bar{u}) = 2\pi y\bar{u}\exp(-\bar{u})[I_0(\bar{u}) + I_1(\bar{u})] = 2\pi yL(\bar{u})} \tag{7.260}$$

7.6.3 The Mittag–Leffler theorem and the Elsasser model

Before stating the Mittag–Leffler theorem it might be useful for the reader to review some terminology from complex variable theory. Undoubtedly, the student was exposed to the theory of residues which is an integral part of any course on the mathematics of physics and engineering. By necessity our review will be very brief.

Consider the complex function $f(z)$ where z is the complex variable $z = x + iy$. Let $f(z)$ be single-valued and analytic inside and on a circle C except at the point $z = a$ which is taken as the center of C. Then $f(z)$ can be expanded as a *Laurent series* about $z = a$ as given by

$$f(z) = \sum_{n=-\infty}^{\infty} b_n (z - a)^n \tag{7.261}$$

The part $\sum_{n=0}^{\infty} b_n (z - a)^n$ of $f(z)$ is called the analytic part while the remainder of the Laurent series $\sum_{n=-\infty}^{-1} b_n (z - a)^n$ is known as the principal part. The expansion coefficients can be calculated with the help of

$$b_n = \frac{1}{i2\pi} \oint_C \frac{f(z)}{(z-a)^{n+1}} dz, \qquad n = 0, \pm 1, \pm 2, \ldots \tag{7.262}$$

For the special case that $n = -1$ we find

$$\oint f(z) dz = i2\pi b_{-1} \tag{7.263}$$

The latter equation can be evaluated by termwise integrating (7.261), utilizing

$$\oint \frac{1}{(z-a)^p} dz = \begin{cases} i2\pi & \text{for } p = 1 \\ 0 & \text{for } p \neq 1, \end{cases} \qquad p \text{ an integer number} \tag{7.264}$$

Since (7.263) only involves the expansion coefficient b_{-1} it is called the residue of $f(z)$ at the point $z = a$. If the principal part of $f(z)$ has only a finite number of terms

$$f(z)_{\text{principal part}} = \frac{b_{-1}}{z-a} + \frac{b_{-2}}{(z-a)^2} + \cdots + \frac{b_{-n}}{(z-a)^n} \tag{7.265}$$

where $b_{-n} \neq 0$, then $z = a$ is a pole of order n. In case that $n = 1$ we have a pole of order 1 or a simple pole. If $z = a$ is a pole of order k then the residue b_{-1} can be obtained from

$$b_{-1} = \lim_{z \to a} \frac{1}{(k-1)!} \frac{d^{k-1}}{dz^{k-1}} [(z-a)^k f(z)] \tag{7.266}$$

In case of a simple pole (7.266) simplifies to

$$b_{-1} = \lim_{z \to a} (z - a) f(z) \tag{7.267}$$

We are now ready to state the Mittag–Leffler theorem. Suppose the only singularities of $f(z)$ in the finite z-plane are the simple poles $a_1, a_2, \ldots$. Let the residues of $f(z)$ at these poles be given by $b_1, b_2, \ldots$. Then the Mittag–Leffler theorem can be expressed as

$$f(z) = f(0) + \sum_{n=1}^{\infty} b_n \left(\frac{1}{z - a_n} + \frac{1}{a_n} \right) \tag{7.268}$$

The special function $f(z) = \cot z - 1/z$ is fundamental in showing that the regular Elsasser spectral band model (7.119) can be written in the form (7.120). We will now show this by performing two major steps.

Step 1

By writing

$$f(z) = \cot z - \frac{1}{z} = \frac{z\cos z - \sin z}{z\sin z} \tag{7.269}$$

we recognize that $f(z)$ has simple poles at $z = n\pi$ with $n = \pm 1, \pm 2, \ldots$ so that the residues at the poles are

$$b_n = \lim_{z \to n\pi} (z - n\pi)\frac{z\cos z - \sin z}{z\sin z} = \lim_{z \to n\pi}\frac{z - n\pi}{\sin z}\lim_{z \to n\pi}\frac{z\cos z - \sin z}{z} = 1 \tag{7.270}$$

In this equation we have made use of the fact that the limit of a product is equal to the product of the limits. Using l'Hôpital's rule we find that at $z = 0$ the function has a removable singularity since

$$\lim_{z \to 0}\frac{z\cos z - \sin z}{z\sin z} = 0 \tag{7.271}$$

Defining $f(0) = 0$ we find from (7.268) together with (7.270)

$$f(z) = \cot z - \frac{1}{z} = \sum_n \left(\frac{1}{z - n\pi} + \frac{1}{n\pi}\right), \qquad n = \pm 1, \pm 2, \ldots$$

$$= \lim_{N \to \infty}\left[\sum_{n=-N}^{-1}\left(\frac{1}{z - n\pi} + \frac{1}{n\pi}\right) + \sum_{n=1}^{N}\left(\frac{1}{z - n\pi} + \frac{1}{n\pi}\right)\right] \tag{7.272}$$

To accept this result we would still have to show that $f(z)$ is bounded on circles C_N having their center at the origin and radius $R_N = (N + 1/2)\pi$ which is part of the requirements used to derive the Mittag–Leffler theorem. We omit this part of the proof.

Step 2

We split $\cot z$ into the real and imaginary part. According to (7.269)

$$\cot z = \frac{1}{z} + f(z) = \frac{1}{x + iy} + \sum_n \left(\frac{1}{x + iy - n\pi} + \frac{1}{n\pi}\right), \qquad n = \pm 1, \pm 2, \ldots \tag{7.273}$$

After a few simple steps we find

$$\cot z = \frac{x}{x^2 + y^2} + \sum_n \frac{x^2 + y^2 - n\pi x}{n\pi[(x - n\pi)^2 + y^2]} - i\left(\frac{y}{x^2 + y^2} + \sum_n \frac{y}{(x - n\pi)^2 + y^2}\right)$$

$$= \Re(\cot z) + i\Im(\cot z) \tag{7.274}$$

It is a simple matter to show that we may also write

$$\cot z = \frac{\cos(x + iy)}{\sin(x + iy)} = \frac{\sin(2x)}{\cosh(2y) - \cos(2x)} - i\frac{\sinh(2y)}{\cosh(2y) - \cos(2x)} \tag{7.275}$$

A comparison of the latter two equations shows that for the imaginary part we must have

$$\frac{\sinh(2y)}{\cosh(2y) - \cos(2x)} = \frac{y}{x^2 + y^2} + \sum_n \frac{y}{(x - n\pi)^2 + y^2}, \qquad n = \pm 1, \pm 2, \dots \tag{7.276}$$

Setting in (7.276) $x = \pi\nu/\delta$ and $y = \pi\alpha_L/\delta$ we obtain

$$\frac{S}{\delta} \frac{\sinh(2\pi\alpha_L/\delta)}{\cosh(2\pi\alpha_L/\delta) - \cos(2\pi\nu/\delta)} = \sum_{n=-\infty}^{\infty} \frac{S}{\pi} \frac{\alpha_L}{(\nu - n\delta)^2 + \alpha_L^2} = k_{\nu,E} \tag{7.277}$$

Finally, by introducing in (7.277) the abbreviations $\beta = 2\pi\alpha_L/\delta$ and $s = 2\pi\nu/\delta$ we find

$$k_{\nu,E} = \frac{S}{\delta} \frac{\sinh\beta}{\cosh\beta - \cos s} = k(s) \tag{7.278}$$

in accordance with (7.120).

The proof of the Mittag–Leffler theorem can be found in various textbooks such as Whittaker and Watson (1915). The theory of residues is nicely discussed, for example, in Wylie (1966) and Spiegel (1964). The latter author also proves the Mittag–Leffler theorem.

7.7 Problems

7.1: Show by direct integration that the Lorentz line-shape factor $f_L(\nu - \nu_0)$ is normalized.

7.2: Estimate the pressure levels for which the half-width α_D of the Doppler line equals the half-width α_L of the Lorentz line for a

(a) CO_2 line located at 667 cm^{-1},

(b) H_2O line at 1600 cm^{-1}. Assume that $\alpha_L(p_0 = 1013.25 \text{ hPa}) = 0.1 \text{ cm}^{-1}$. You may ignore the temperature dependence of the Lorentz half-width. Use the following model atmosphere. At the Earth's surface: $z = 0$ km, $p_0 = 1013.25$ hPa, $T = 288.15$ K,

Height level (km)	Lapse rate (K km^{-1})
0–11	6.5
11–25	0.0
25–47	−3.0
47–53	0.0
53–80	4.4
80–90	0.0

7.3: The centers of two identical Lorentz lines are separated by a distance of eight half-widths. Calculate the transmission between the lines at a wave number of three half-widths away from the line center. $S = \pi$ cm$^{-2}, \alpha_L = 0.1$ cm^{-1}, $u = 1$ cm.

7.4: Carry out the integration (7.149) of the Schnaidt model.
Hint: First assume that the line stretches to infinity on both sides of the line center and then subtract the remaining parts. Make use of the approximation that in the atmosphere $y = \alpha_L/\delta \ll 1$.

7.5: Consider an isothermal atmospheric layer so that the line intensity is constant. Also assume that the specific humidity q is constant within this layer. Show that for an Elsasser band the monochromatic transmission of this layer is given by

$$T = \left(\frac{\cosh \beta_2 - \cos s}{\cosh \beta_1 - \cos s} \right)^\eta, \qquad \eta = \frac{q S p_1 \sec \vartheta}{2\pi g \alpha_{L,1}}$$

where ϑ is the zenith angle of the radiation, $s = 2\pi \nu/\delta$, and $\beta_i = 2\pi \alpha_i/\delta$, $i = 1, 2$ are taken at the lower and upper boundary of the layer where the pressure is p_1 and p_2, respectively.

7.6: Determine the vertical transmission of radiation within the spectral range of a Lorentz line originating at the surface of the Earth where the pressure is $p_0 = 1000$ hPa. Ignore the temperature dependence of the line intensity and of the half-width α_L. The specific humidity is distributed according to $q(p) = q(p_0)(p/p_0)^{2.5}$. Assume that $S = \pi$ cm^{-2}, $\alpha_{L,0} = 0.08$ cm^{-1}, $\delta = 20\alpha_{L,0}$, and $q(p_0) = 5$ g kg^{-1}.
Hint: Use the Curtis–Godson approximation.

7.7: Consider a vertical transmission path between pressure levels p_1 and $p_2 < p_1$. The concentration c_{gas} of an absorbing gas of density ρ_{gas} is homogeneously distributed.

(a) Assuming that the line intensity is height independent, by using the Curtis–Godson approximation, show that $\tilde{p} = (p_1 + p_2)/2$.
(b) For a Lorentz line show that the equivalent widths of the strong line and weak line approximations are given by

$$\text{strong line approximation:} \quad W = \sqrt{\frac{2S\alpha_{L,0}c}{gp_0}\left(p_1^2 - p_2^2\right)}$$

$$\text{weak line approximation:} \quad W = \frac{Sc}{g}(p_1 - p_2)$$

where p_0 is the normal pressure.

7.8: For a uniformly distributed absorbing gas the height dependence of the absorption coefficient is assumed to be given by $k_\nu = k_{\nu,0}\sqrt{p/p_0}$ where p_0 is the surface pressure. The incoming solar radiation at the zenith angle ϑ_0 at the top of the atmosphere is $S_{\nu,0}$ (W m^{-2}). If the Sun's position remains fixed, find the parallel solar energy density $E_{0,\Delta t}$ within 1 hour per square meter at the Earth's surface whose albedo is $A_g = 0.31$.

7.9: Show that the absorption of a Doppler line can be expressed by

$$\mathcal{A}_D = \frac{Su}{\delta}\sum_{n=0}^{\infty}\frac{(-1)^n b^n}{(n+1)!\sqrt{n+1}} \quad \text{with} \quad b = Su\sqrt{\frac{\ln 2}{\pi\alpha_D^2}}$$

For convenience use the wave number notation.

7.10: A cylindrical absorption cell of length l is filled with an absorbing gas. The pressure within the cell can be varied. The cell is illuminated by a parallel infrared light beam parallel to the axis of the cylinder. Show that the absorption at the line center of a Lorentz line is independent of pressure.

7.11: A hypothetical spectral interval contains 100 spectral lines of equal half-width α_L. The line positions are statistically distributed. Each decade of lines is represented by a line of mean intensity S_0 with

Number of lines	S_0 (cm^{-2})
1–10	0.05
11–20	0.10
21–30	0.15
31–40	0.20
41–50	0.25
51–60	0.30
61–70	0.35
71–80	0.40
81–90	0.45
91–100	0.50

For $\alpha_L = 0.1$ cm^{-1}, $\delta = 10\alpha_L$ and $u = 1$ cm find the transmission according to the exponential model.

7.12:   Schnaidt expresses the average absorption as

$$A = \frac{1}{\delta} \int_{-\delta/2}^{\delta/2} [1 - \exp(-k_\nu u)] d\nu$$

This expression ignores the contribution of neighboring lines outside the spectral range $[-\delta/2, \delta/2]$. Calculate the distribution function for this model.

7.13:   Given is the following tropical model atmosphere.

Height (km)	Pressure (hPa)	Temperature (°C)	Relative humidity (%)	Specific humidity (g/kg)
0	1011	27.7	82	18.76
1.0	902	22.8	68	13.03
1.5	858	20.6	63	11.09
1.8	830	19.2	60	10.00
2.0	808	18.2	59	9.50
2.3	780	16.9	57	8.75
3.0	710	13.3	52	6.96
4.0	636	7.8	48	4.96
4.4	604	5.5	46	4.28
5.0	562	2.0	44	3.43
6.0	495	−3.5	41	2.36
7.0	435	−10.0	39	1.46
8.0	385	−16.6	38	0.885
9.0	340	−23.2	37	0.518
9.8	302	−28.5	38	0.351
10.0	291	−30.0	38	0.312
11.0	253	−38.6	40	0.148
12.0	217	−47.2	43	0.0679
13.0	187	−55.7	49	0.0309
14.0	156	−62.3	58	0.0180
15.0	131	−69.0	70	0.0100
16.0	112	−75.5	90	0.0054

Calculate the radiative temperature change at 0, 1 and 3 km height for the water vapor window (8–13) μm. The gray absorption coefficient k may be traced back to the dimer-water vapor molecule. k is temperature and pressure dependent according to

$$k(z) = k_0 \left[1 + \alpha_2(T(z) - T_0)\right] \frac{p^1(z)}{p_0}$$

with $k_0 = 12$ cm^2 g^{-1}, $\alpha_2 = -0.02$ K^{-1}, $T_0 = 278$ K, $p_0 = 1013.25$ hPa. p^1 is the partial pressure of water vapor. For the calculation of the Planck radiation use

$$B = \frac{\kappa \sigma T^4}{\pi} \qquad \text{with} \qquad \kappa = 0.25, \quad \sigma = 5.6697 \times 10^{-8} \text{ Wm}^{-2} \text{ K}^{-4}.$$

7.14: Consider two idealized lines in triangular form as shown in the figure.

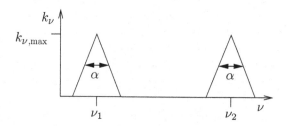

The center lines are located at ν_1 and ν_2. The distance between the lines exceeds the half-width. (Note that here the half-width is twice as large as in the usual definition.) Find an analytic representation of the spectral absorption coefficient, the equivalent width W_1, W_2 of each line and the equivalent width W of the line spectrum. Moreover, consider the limiting cases $u \to 0$ and $u \to \infty$.

7.15: In an isothermal atmospheric layer an absorbing gas is homogeneously distributed, that is the concentration c_{gas} is constant. Calculate the equivalent width of a spectral line for a vertical path in the region of the weak line approximation. Assume the Lorentz line shape.

8

Absorption by gases

8.1 Introduction

In this chapter we are going to discuss some of the more elementary ideas in connection with the absorption spectra of gases. The energy E of a molecule may be expressed as the sum of the *rotational energy* E_{rot}, *vibrational energy* E_{vib} and *electronic energy* E_{el}. Of these three types of energy, E_{rot} is generally the smallest, typically a few hundredths of an electron Volt.[1] Vibrational energies are of the order of a few tenths of one electron Volt, while the largest energies are of the electronic type which generally amount to a few electron Volts.

The absorption (emission) spectrum arising from the rotational and vibrational motion of a molecule which is not electronically excited will be located in the infrared region. In infrared absorption experiments light from a suitable source penetrates an absorption chamber containing the gas to be studied and then enters a spectrograph. If the instrument is of low resolving power, a series of wide bands is observed which correspond to the *vibrational transitions*. If an instrument of high resolving power is used, these bands are seen to consist of numerous spectral lines resulting from the energy levels of rotation.

In the next section we are going to discuss the vibrational motion of two relatively simple molecules (CO_2 and H_2O) which are particularly important in the study of radiative transfer in the atmosphere. The forces between the atoms making up the molecule may be crudely approximated by forces exerted by weightless springs which hold the atoms relative to each other in the neighborhood of certain configurations. The forces due to stretching or compression of the springs are assumed to follow *Hooke's law*.

If a molecule contains n atoms, there are $3n$ modes of motion. Of these, three correspond to translation, and three to rotation (or two for a linear molecule). The

---

[1] If an electron falls through a potential difference of one Volt it attains a kinetic energy of 1.602×10^{-19} joule which is used as the definition of one electron volt, i.e. 1 electron volt $= 1.602 \times 10^{-19}$ joule.

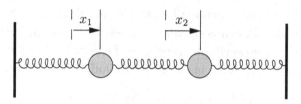

Fig. 8.1 Two coupled harmonic oscillators with equilibrium positions at $x_1 = 0$, $x_2 = 0$.

remaining $3n - 6$ (or $3n - 5$) correspond to the normal vibrational modes of the molecule.

8.2 Molecular vibrations

8.2.1 Two coupled harmonic oscillators

Before studying the vibrational motion of the carbon dioxide and water molecules, we are going to discuss a fairly simple example by considering two particles, each of mass m, connected by light springs of stiffness k, see Figure 8.1. The particles are constrained to move in a straight line. The distances x_1 and x_2 stand for the displacements of particles 1 and 2 from their equilibrium positions.

The equation of motion of each particle can be obtained quite easily by using *Lagrange's equation of motion*. The *Lagrangian function L* is defined as the difference of the kinetic energy K and the potential energy V of the system, that is

$$
\begin{aligned}
\text{(a)} \quad & L = K - V \\
\text{(b)} \quad & K = \frac{1}{2}m\dot{x}_1^2 + \frac{1}{2}m\dot{x}_2^2 \\
\text{(c)} \quad & V = \frac{1}{2}kx_1^2 + \frac{1}{2}k(x_1 - x_2)^2 + \frac{1}{2}kx_2^2
\end{aligned}
\tag{8.1}
$$

The second term on the right-hand side of (8.1c) refers to the potential energy stored in the spring connecting the two masses. If q_k and $\dot{q}_k$ stand for the *generalized coordinate* and the *generalized velocity* of the particles, then Lagrange's equation of motion of a conservative system is given by

$$
\frac{d}{dt}\frac{\partial L}{\partial \dot{q}_k} = \frac{\partial L}{\partial q_k}
\tag{8.2}
$$

Substituting for $q_k = x_1, x_2$ equation (8.1) into (8.2) yields the equations of motion for the two particles

$$
\begin{aligned}
\frac{d}{dt}\frac{\partial L}{\partial \dot{x}_1} &= \frac{\partial L}{\partial x_1} \quad \text{or} \quad m\ddot{x}_1 = -kx_1 + k(x_2 - x_1) \\
\frac{d}{dt}\frac{\partial L}{\partial \dot{x}_2} &= \frac{\partial L}{\partial x_2} \quad \text{or} \quad m\ddot{x}_2 = -kx_2 + k(x_1 - x_2)
\end{aligned}
\tag{8.3}
$$

Let us find the possible common frequencies of vibration of the two particles. These frequencies are known as *eigenfrequencies*. The associated vibrational states are the corresponding *eigenvibrations* or *normal modes of vibrations*. We are going to solve the system (8.3) by means of the trial solutions

$$x_1 = A_1 \cos \omega t, \qquad x_2 = A_2 \cos \omega t \qquad (8.4)$$

which requires that both particles vibrate with the frequency ω. Had we used a sine function or the combination of a cosine and a sine function we would still obtain the same equations expressing the conditions for the frequency. Substitution of (8.4) into (8.3) gives two linear homogeneous equations for the amplitudes A_1 and A_2

$$(-m\omega^2 + 2k)A_1 - kA_2 = 0$$
$$-kA_1 + (-m\omega^2 + 2k)A_2 = 0 \qquad (8.5)$$

Nontrivial solutions for the amplitudes exist only if the determinant of the coefficients vanishes. The expansion of the determinant results in the frequency equation

$$\begin{vmatrix} -m\omega^2 + 2k & -k \\ -k & -m\omega^2 + 2k \end{vmatrix} = (-m\omega^2 + 2k)^2 - k^2 = 0 \qquad (8.6)$$

The positive roots

$$\omega_1 = \sqrt{\frac{3k}{m}}, \qquad \omega_2 = \sqrt{\frac{k}{m}} \qquad (8.7)$$

are the eigenfrequencies of the system.

In order to get some idea about the type of the normal vibrations, we substitute (8.7) into the system (8.5) and find the conditions for the symmetric and the antisymmetric modes

$$\begin{aligned} \text{antisymmetric mode:} \quad & A_1 = -A_2 \quad \text{for} \quad \omega_1 \Longrightarrow x_1 = -x_2 \\ \text{symmetric mode:} \quad & A_1 = A_2 \quad \text{for} \quad \omega_2 \Longrightarrow x_1 = x_2 \end{aligned} \qquad (8.8)$$

The number of normal vibrations is equal to the number of coordinates which are required for a complete description of the system. The equally large amplitudes of this example result from the assumption that the two masses are equally large. The general motion of the mass points will be given by superimposing the normal vibrations with different amplitudes and phases.

8.2.2 Review of physical principles

Theoretical calculations of molecular vibrations often are carried out in a coordinate system in which the center of mass of the particles is at rest. To prevent the molecule

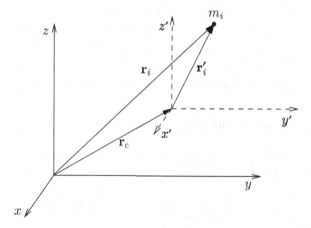

Fig. 8.2 Center of mass system.

from rotating we must require that the angular momentum of the molecule vanishes. A brief review of the physical principles involved in the solution technique now follows.

We ignore the translational motion of the system as a whole since for spectroscopic considerations only the motion of the atoms relative to the center of mass is of importance. As shown in Figure 8.2, the origin of the primed coordinate system is the center of mass whose position in the unprimed system is given by the vector $\mathbf{r}_c$.

The position of the particle m_i in the primed and unprimed (laboratory) system is $\mathbf{r}_i'$ and $\mathbf{r}_i$, respectively. In general, the center of mass is defined by

$$\mathbf{r}_c = \frac{1}{M} \sum_{i=1}^{n} m_i \mathbf{r}_i \quad \text{with} \quad M = \sum_{i=1}^{n} m_i \tag{8.9}$$

where M is the total mass. Replacing $\mathbf{r}_i$ by the vector sum $\mathbf{r}_c + \mathbf{r}_i'$ results in

$$M\mathbf{r}_c = \sum_{i=1}^{n} m_i \left(\mathbf{r}_c + \mathbf{r}_i'\right) = M\mathbf{r}_c + \sum_{i=1}^{n} m_i \mathbf{r}_i' \tag{8.10}$$

From this equation it follows that

$$\sum_{i=1}^{n} m_i \mathbf{r}_i' = 0, \quad \sum_{i=1}^{n} m_i \mathbf{v}_i' = 0 \tag{8.11}$$

whereby the second equation results from the time differentiation of the first one. Hence in the center of mass system the sum of the mass moments $m_i \mathbf{r}_i'$ as well as the sum of the linear moments $m_i \mathbf{v}_i'$ vanish.

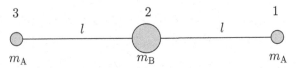

Fig. 8.3 Linear symmetric triatomic molecule in equilibrium position.

The total angular momentum of the system is defined by

$$\mathbf{J} = \sum_{i=1}^{n} m_i \left(\mathbf{r}_i \times \mathbf{v}_i\right) = \sum_{i=1}^{n} m_i \left(\mathbf{r}_c + \mathbf{r}'_i\right) \times \left(\mathbf{v}_c + \mathbf{v}'_i\right)$$

$$= M\left(\mathbf{r}_c \times \mathbf{v}_c\right) + \sum_{i=1}^{n} m_i \left(\mathbf{r}'_i \times \mathbf{v}'_i\right) = \mathbf{J}_c + \sum_{i=1}^{n} \mathbf{J}'_i \qquad (8.12)$$

showing that the angular momentum $\mathbf{J}$ can be expressed as the sum of two terms. The first term represents the angular momentum of the center of mass $\mathbf{J}_c$ having mass M, while the second term gives the sum of the angular momenta of the individual particles about the center of mass.

8.2.3 Linear triatomic molecules

In the following we calculate the normal vibrations of the linear symmetric CO_2 molecule whose equilibrium position is shown in Figure 8.3. The atoms m_A are separated from the atom m_B by the equilibrium distance l. We assume that the potential energy of the molecule depends only on the distances between the atoms and on the angle of bending. The motion of the atoms takes place in the (x, y)-plane. First we are going to discuss longitudinal vibrations.

If the vector $\mathbf{x}_i$ with components (x_i, y_i) stands for the displacement of atom i from its equilibrium position $\mathbf{r}_{0,i}$ then the momentary position of this atom is given by

$$\mathbf{r}_i = \mathbf{r}_{0,i} + \mathbf{x}_i \qquad (8.13)$$

The forces holding the atoms together, in first approximation, follow Hooke's law. For longitudinal vibrations the Lagrangian function L of the system is expressed by

$$L = K - V = \frac{m_A}{2}\left(\dot{x}_1^2 + \dot{x}_3^2\right) + \frac{m_B}{2}\dot{x}_2^2 - \frac{k_l}{2}[(x_1 - x_2)^2 + (x_3 - x_2)^2] \qquad (8.14)$$

where k_l is the spring constant for the longitudinal motion. To eliminate one coordinate, say x_2, we make use of

$$\sum_{i=1}^{n} m_i \mathbf{r}_i = \sum_{i=1}^{n} m_i \mathbf{r}_{0,i} \qquad (8.15)$$

which is a statement for the conservation of the center of mass. Application of (8.13) and (8.15) yields

$$m_A(x_1 + x_3) + m_B x_2 = 0 \tag{8.16}$$

so that L is given by

$$
\begin{aligned}
L = {} & \frac{m_A}{2}\left(\dot{x}_1^2 + \dot{x}_3^2\right) + \frac{m_A^2}{2m_B}(\dot{x}_1 + \dot{x}_3)^2 \\
& - \frac{k_l}{2}\left[\left(x_1^2 + x_3^2\right) + \frac{2m_A}{m_B}(x_1 + x_3)^2 + \frac{2m_A^2}{m_B^2}(x_1 + x_3)^2\right]
\end{aligned}
\tag{8.17}
$$

We introduce the new set of coordinates (η, ξ) by means of

$$
\eta = x_1 - x_3, \qquad \xi = x_1 + x_3 \implies
$$
$$
x_1 = \frac{\xi + \eta}{2}, \qquad x_3 = \frac{\xi - \eta}{2}
\tag{8.18}
$$

Utilizing the new coordinates the Lagrangian function may be written as

$$
L = \frac{m_A}{4}\dot{\eta}^2 + \frac{m_A m_T}{4m_B}\dot{\xi}^2 - \frac{k_l}{4}\eta^2 - \frac{k_l m_T^2}{4m_B^2}\xi^2
\tag{8.19}
$$

with $m_T = 2m_A + m_B$. In contrast to (8.17), in this form L does not contain any cross terms.

Differentiation of the Lagrange function (8.19) according to (8.2) yields

$$
\frac{d}{dt}\frac{\partial L}{\partial \dot{\xi}} = \frac{m_A m_T}{2m_B}\ddot{\xi}, \qquad \frac{\partial L}{\partial \xi} = -\frac{k_l m_T^2}{2m_B^2}\xi
\tag{8.20}
$$

so that the equations of motion for the (η, ξ) coordinates are given by

$$
\ddot{\xi} + \frac{k_l m_T}{m_A m_B}\xi = 0, \qquad \ddot{\eta} + \frac{k_l}{m_A}\eta = 0
\tag{8.21}
$$

From these equations it is seen that due to the introduction of the coordinates (η, ξ) the differential equations of motion are decoupled which was not the case in the example of two coupled harmonic oscillators, see (8.3). The coordinates (η, ξ) are known as the *normal coordinates*. It is a characteristic feature of normal coordinates that the differential equations of motion are automatically separated, there being one differential equation for each normal coordinate.

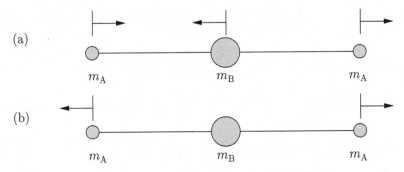

Fig. 8.4 (a) Antisymmetric mode of the longitudinal vibrations, $\eta = 0$, ν_3-vibration. (b) Symmetric mode of the longitudinal vibrations, $\xi = 0$, ν_1-vibration.

Substituting the trial solution $\exp(i\omega t)$ into (8.21) we immediately find the frequencies

$$\text{antisymmetric vibration: } \omega_a = \sqrt{\frac{k_l m_T}{m_A m_B}}$$

$$\text{symmetric vibration: } \omega_s = \sqrt{\frac{k_l}{m_A}}$$

(8.22)

Let us examine the vibrations more closely and consider case (a) of Figure 8.4. If $x_1 = x_3$ then $\eta = 0$ resulting in the antisymmetric vibrational mode. This is the reason we have added the suffix a to the circular frequency in equation (8.22). This type of motion is usually called the ν_3-*vibration*. In case (b) we set $x_1 = -x_3$ so that $\xi = 0$ resulting in the symmetric ν_1-*vibration* (suffix s in (8.22)).

So far we have restricted the motion of the atoms to one direction. A nonrigid triatomic molecule, such as CO_2, vibrates not only longitudinally but also transversally as shown in Figure 8.5. In this case the Lagrangian function is given by

$$L = \frac{m_A}{2}(\dot{y}_1^2 + \dot{y}_3^2) + \frac{m_B}{2}\dot{y}_2^2 - \frac{k_T}{2}(l\delta)^2$$

(8.23)

where the constant k_T of the potential energy part of L refers to the transversal displacement. The angle δ stands for the deviation from $180°$. The meaning of the angles α_1 and α_2 follows from the figure. Since the angle δ is assumed to be very small, we may replace it by the sine functions as shown in

$$\delta = \left(\frac{\pi}{2} - \alpha_1\right) + \left(\frac{\pi}{2} - \alpha_2\right) \approx \sin\left(\frac{\pi}{2} - \alpha_1\right) + \sin\left(\frac{\pi}{2} - \alpha_2\right)$$

$$= \cos\alpha_1 + \cos\alpha_2 = \left(\frac{y_2 - y_1}{l}\right) + \left(\frac{y_2 - y_3}{l}\right)$$

(8.24)

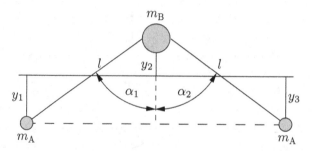

Fig. 8.5 Coordinates of the transversal vibration of a linear triatomic molecule.

As before, we eliminate one coordinate by using the conservation of the center of mass. The result is

$$m_A(y_1 + y_3) + m_B y_2 = 0 \tag{8.25}$$

In order to exclude the rotation of the molecule, the total angular momentum must vanish. The mathematical form of the angular momentum is given by (8.12). Since the deviation from the equilibrium position is small, we may approximate the vector $\mathbf{r}_i$ by $\mathbf{r}_{0,i}$. Thus the total angular momentum is approximately given by

$$\mathbf{J} = \sum_{i=1}^{3} m_i(\mathbf{r}_i \times \mathbf{v}_i) \approx \sum_{i=1}^{3} m_i(\mathbf{r}_{0,i} \times \mathbf{v}_i) = \frac{d}{dt} \sum_{i=1}^{3} m_i(\mathbf{r}_{0,i} \times \mathbf{x}_i) = 0 \tag{8.26}$$

which is satisfied by

$$\sum_{i=1}^{3} m_i(\mathbf{r}_{0,i} \times \mathbf{x}_i) = 0 \tag{8.27}$$

This equation can be used to show that $y_1 = y_3$ so that the Lagrangian function may be written as

$$L = \frac{m_A m_B}{4 m_T}(l\dot{\delta})^2 - \frac{k_T}{2}(l\delta)^2 \tag{8.28}$$

After a few steps we find the eigenfrequency of the transversal vibration

$$\omega_T = \sqrt{\frac{2 k_T m_T}{m_A m_B}} \tag{8.29}$$

Details of the derivations will be left to the exercises.

The transversal vibration shown in Figure 8.6(a) is called v_2-*vibration*. At the beginning of the chapter we have stated that this molecule should have four vibrational modes. Our calculations, however, provided only three of these. The fourth vibrational mode of the CO_2 molecule results from a twofold degeneracy, i.e. the direction of one vibrational mode is perpendicular to that of the other as shown in

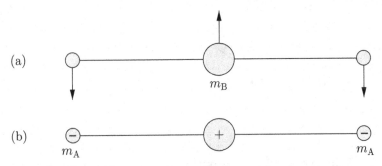

Fig. 8.6 Transversal vibration of a linear triatomic molecule.

Figure 8.6(b). This does not provide any new information. The $+$ attached to m_B and the $-$ attached to m_A simply indicate that they vibrate in opposite directions in a plane perpendicular to the plane of the paper.

As will be shown later, the interaction of an electromagnetic wave with the molecule to produce absorption or emission results from the interaction of the electric field vector $\mathbf{E}$ with the (variable) *dipole moment* $\mathbf{M}$ of the system. The dipole moment of an electrically neutral molecule is a vector whose direction is along the line joining the center of charge of the negative charges to the center of charge of the positive charges. The magnitude of the dipole moment is the length of that line multiplied by the total negative or positive charge, these being equal. An atom or a molecule is said to be *polarized* by an electric field when the displacements of the charges caused by the electric field produce or alter the dipole moment.

In a Cartesian coordinate system the components of $\mathbf{M}$ are given by

$$M_x = \sum_k e_k x_k, \quad M_y = \sum_k e_k y_k, \quad M_z = \sum_k e_k z_k \qquad (8.30)$$

where e_k is the charge of the particle k at the position (x_k, y_k, z_k). If the particles are the atoms of a molecule, the charges e_k must be considered as effective charges.

Some molecules have a permanent dipole moment such as the heteronuclear diatomic molecule CO. The dipole moment results from the asymmetric charge distribution. In contrast, homonuclear diatomic molecules such as N_2 have no electric dipole moment due to the symmetric charge distribution. Similarly, in the equilibrium configuration the CO_2 molecule has no permanent dipole moment due to the symmetric distribution of charges. Further details may be found, for example, in Wilson *et al.* (1955).

Let us re-examine Figure 8.4. The symmetric longitudinal stretching of the CO_2 molecule, usually called the ν_1-vibration, does not produce any dipole moment so that this type of vibration is inactive in the infrared spectrum. The remaining vibrations are classified as *parallel* or *perpendicular* according as the change

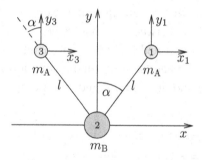

Fig. 8.7 Coordinates of the triangular molecule H_2O.

of the dipole moment takes place along or perpendicular to the axis of symmetry of the molecule which is the *internuclear axis*. The antisymmetric longitudinal or the ν_3-vibration induces a dipole moment parallel to the axis of symmetry. Thus ν_2- and ν_3-vibrations are active in the infrared spectrum. The most important CO_2 band in the infrared spectral range is centered at the wave number $\tilde{\nu} = 667.40\,\text{cm}^{-1}$ ($\lambda = 15\,\mu\text{m}$) which results from the ν_2-vibration. In contrast to the CO_2 molecule, the diatomic gases N_2 and O_2, abundantly occurring in the Earth's atmosphere, do not possess permanent electric dipole moments and, therefore, do not exhibit infrared absorption bands.

We have now examined in some detail the normal vibrations of a linear triatomic molecule. The actual internal vibrations of a semi-rigid system (small amplitude vibrations) may be very complicated. However, the motion can always be decomposed into a sum of elementary motions described by the normal vibrational modes. The frequency equations (8.22) and (8.29) are in exact agreement with the results shown in the standard reference book *Infrared and Raman Spectra* by Herzberg (1964b) where many details can be found.

8.2.4 Nonlinear triatomic molecules

It is not always as simple as in the case of the linear triatomic molecule to find the normal coordinates. This will become apparent in the discussion of the nonlinear triatomic H_2O, see Figure 8.7. The three masses are labelled as 1, 2 and 3. The equilibrium distance from the central mass m_B to the masses m_A is l.

Employing the conservation of the center of mass according to (8.15) we find

$$m_A(x_1 + x_3) + m_B x_2 = 0, \qquad m_A(y_1 + y_3) + m_B y_2 = 0 \qquad (8.31)$$

If we place ourselves in the resting position of the atom m_B so that $\mathbf{r}_{0,2} = 0$, noting that $|\mathbf{r}_{0,1}| = |\mathbf{r}_{0,3}|$, we find that the conservation of angular momentum is given by

$$(y_1 - y_3)\sin\alpha = (x_1 + x_3)\cos\alpha \qquad (8.32)$$

Details of the calculation will be left up to the exercises. Now we must find the change of l due to stretching between the masses m_A (mass point 1) and m_B (mass point 2) and between m_B (2) and m_A (3). The changes l_1 and l_3 are found by projecting the vectors $x_1 - x_2$ and $x_3 - x_2$ on the directions of the lines connecting the masses m_A (point 1) and m_B and m_B and m_A (point 3). A simple calculation gives

$$\delta l_1 = (x_1 - x_2)\sin\alpha + (y_1 - y_2)\cos\alpha$$
$$\delta l_3 = -(x_3 - x_2)\sin\alpha + (y_3 - y_2)\cos\alpha \tag{8.33}$$

The change of the angle 2α is found by projecting the vectors $x_1 - x_2$ and $x_3 - x_2$ on the directions perpendicular to the lines connecting the points 1 and 2 and 2 and 3. The result is given by

$$\delta = \frac{1}{l}[(x_1 - x_2)\cos\alpha - (y_1 - y_2)\sin\alpha - (x_3 - x_2)\cos\alpha - (y_3 - y_2)\sin\alpha] \tag{8.34}$$

Details of the calculations are left to the exercises.

The Lagrangian function is found as

$$L = \frac{m_A}{2}\left(\dot{x}_1^2 + \dot{x}_3^2\right) + \frac{m_B}{2}\dot{x}_2^2 - \frac{k_1}{2}\left[(\delta l_1)^2 + (\delta l_2)^2\right] - \frac{k_2}{2}(l\delta)^2 \tag{8.35}$$

where $\dot{x}_i = \dot{x}_i\mathbf{i} + \dot{y}_i\mathbf{j}$, etc. The third and fourth right-hand side terms describe the potential energy of the extension of the springs connecting the masses and of the bending of the molecule.

As motivated by the previous example, we introduce the new coordinates

$$q_a = x_1 + x_3, \qquad q_{s,1} = x_1 - x_3, \qquad q_{s,2} = y_1 + y_3 \implies$$
$$x_1 = \frac{1}{2}(q_a + q_{s,1}), \qquad x_2 = -\frac{m_A}{m_B}q_a, \qquad x_3 = \frac{1}{2}(q_a - q_{s,1}) \tag{8.36}$$
$$y_1 = \frac{1}{2}(q_{s,2} + q_a\cot\alpha), \qquad y_2 = -\frac{m_A}{m_B}q_{s,2}, \qquad y_3 = \frac{1}{2}(q_{s,2} - q_a\cot\alpha)$$

where we have employed the conservation of the center of mass. A simple but tedious calculation gives the Lagrangian function of the system

$$L = \frac{m_A}{4}\left(\frac{2m_A}{m_B} + \frac{1}{\sin^2\alpha}\right)\dot{q}_a^2 + \frac{m_A}{4}\dot{q}_{s,1}^2 + \frac{m_A m_T}{4m_B}\dot{q}_{s,2}^2$$
$$- q_a^2\frac{k_1}{4}\left(\frac{2m_A}{m_B} + \frac{1}{\sin^2\alpha}\right)\left(1 + \frac{2m_A}{m_B}\sin^2\alpha\right) - \frac{q_{s,1}^2}{4}(k_1\sin^2\alpha + 2k_2\cos^2\alpha)$$
$$- q_{s,2}^2\frac{m_T^2}{4m_B^2}(k_1\cos^2\alpha + 2k_2\sin^2\alpha) + q_{s,1}q_{s,2}\frac{m_T}{2m_B}(2k_2 - k_1)\sin\alpha\cos\alpha$$

$$\tag{8.37}$$

We immediately recognize that q_a is a normal coordinate since no cross-term occurs with the other coordinates. The coordinates $q_{s,1}$ and $q_{s,2}$ are not normal coordinates due to the appearance of a cross-term. As we will see later, the suffixes a and s stand for antisymmetric and symmetric vibrations. We omit the simple details in the calculation of the equation of motion involving the coordinates $\dot{q}_a$ and q_a which is decoupled from the remainder of the system. The solution of the q_a equation of motion is given by

$$\omega_a^2 = \frac{k_1}{m_A}\left(1 + \frac{2m_A}{m_B}\sin^2\alpha\right) \tag{8.38}$$

Using again the Lagrangian form of the equation of motion, after a few steps we obtain

$$\ddot{q}_{s,1} + A_1 q_{s,1} + A_2 q_{s,2} = 0$$
$$\ddot{q}_{s,2} + B_1 q_{s,2} + B_2 q_{s,1} = 0 \tag{8.39}$$

with

$$A_1 = \frac{1}{m_A}(k_1 \sin^2\alpha + 2k_2 \cos^2\alpha), \qquad A_2 = -\frac{m_T}{m_A m_B}(2k_2 - k_1)\sin\alpha\cos\alpha$$

$$B_1 = \frac{m_T}{m_A m_B}(k_1 \cos^2\alpha + 2k_2 \sin^2\alpha), \qquad B_2 = -\frac{1}{m_A}(2k_2 - k_1)\sin\alpha\cos\alpha \tag{8.40}$$

and $m_T = 2m_A + m_B$. It is seen that (8.39) is a coupled system of two second-order linear differential equations. The solution to this system can be found by any one of the standard methods. The operator method is particularly easy to apply since the two equations are decoupled almost immediately. We leave it to the exercises to verify that the characteristic equation is given by

$$\omega^4 - \omega^2(A_1 + B_1) + (A_1 B_1 - A_2 B_2) = 0 \tag{8.41}$$

permitting us to determine the eigenfrequencies $\omega_{s,1}$ and $\omega_{s,2}$ of the normal vibrations $q_{s,1}$ and $q_{s,2}$.

We will now briefly discuss the normal modes of vibration. First we consider the antisymmetric vibration described by (8.38) of the H_2O molecule. Pure q_a vibrations exist if $x_1 = x_3$ and $y_1 = -y_3$. Thus q_a describes antisymmetric vibrations with respect to the y-axis. This case is shown in Figure 8.8(a).

Inspection reveals that the vibrations corresponding to the coordinate $q_{s,1}$ and $q_{s,2}$ are symmetric with respect to the y-axis as shown in parts (b) and (c) of the figure. We set $q_a = 0$ and find $x_1 = -x_3$ and from (8.36) follows that $y_1 = y_3$.

A more exact but also more involved analysis is described in Herzberg (1964a,b) who introduces an additional interaction coefficient. This changes slightly the frequency equations (8.38) and (8.41). Setting this small interaction coefficient equal

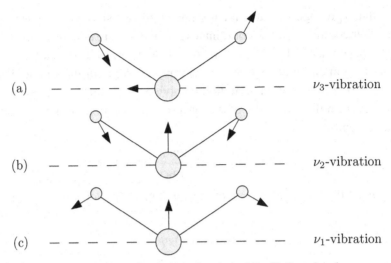

(a) ν_3-vibration

(b) ν_2-vibration

(c) ν_1-vibration

Fig. 8.8 Normal modes of vibration of the H_2O molecule.

to zero results in the equations we have derived above. The inclusion of the interaction coefficient implies that the normal coordinates shown in Figure 8.8 also
change slightly. Particularly the ν_3-*vibration* (Figure 8.8(a)), must be modified in
such a way that the arrows extending from the masses m_A are nearly parallel to the
lines connecting them with m_B. The remaining parts of the figure are qualitatively
correct. Part (b) is called the ν_2-*vibration* and part (c) the ν_1-*vibration*. None of
the normal modes of vibration we have shown is drawn to scale. The most important absorption band in the infrared spectral range is centered at the wave number
$\tilde{\nu} = 1594.78$ cm^{-1} ($\lambda = 6.27$ μm) which results from the ν_2-vibration.

The famous books *Molecular Vibrations* by Wilson *et al.* (1955) and *Infrared
and Raman Spectra* by Herzberg (1964b) describe in great detail the theory of
vibrational spectra. Discussion on normal coordinates are given in any textbook on
theoretical mechanics. Our reference goes to Greiner (1989), *Mechanik, Volume 2*,
who gives a number of examples on normal vibrations. A wealth of information
relevant to radiative transfer in the atmosphere is summarized in Goody (1964a).

8.3 Some basic principles from quantum mechanics

We shall not attempt a rigorous development of quantum mechanics, but we shall
merely state some of the basic results as they apply to the individual atom or atomic
systems. The quantum mechanical description of atomic or molecular systems
is carried out with the help of the *wave function* or the *state function* Ψ. This
function, in general, is a complex number and considered to be a function of all of
the configurational coordinates including time.

According to the basic postulates of quantum mechanics, the square of the absolute value $|\Psi|^2$ of the wave function is a measure of the probability that the considered system is located at the configuration corresponding to the particular values of the coordinates. Sometimes the product $\Psi^*\Psi$ or $|\Psi|^2$ is called the *probability distribution function* or the *probability density*. For example, if the system consists of a single electron, then the probability that the electron is located somewhere between x, y, z and $x + dx, y + dy, z + dz$ is given by $\Psi^*(x, y, z)\Psi(x, y, z)\,dx\,dy\,dz$.

From the interpretation of the wave function it follows that we cannot be certain that the electron is located at any particular place. Only the probability of being there within certain limits can be known. This interpretation is consistent with *Heisenberg's uncertainty principle*. Since the electron has to be somewhere in space the total probability has to be unity as stated in

$$\int_{-\infty}^{\infty}\int_{-\infty}^{\infty}\int_{-\infty}^{\infty} \Psi^*\Psi \, dx \, dy \, dz = 1 \tag{8.42}$$

Functions satisfying (8.42) are classified as *quadratically integrable normalized functions*.

8.3.1 Stationary and coherent states

We will now briefly discuss two particular states which are known as *stationary* and *coherent states*.

(i) Stationary states

An *eigenstate* or *characteristic state* corresponds to a perfectly defined energy. A given system may have many eigenstates each possessing, in general, a different energy. If E_n denotes the particular energy of one of its eigenstates, the complete wave function can be written as

$$\Psi_n(x, y, z, t) = \psi_n(x, y, z)\exp\left(-\frac{iE_n t}{\hbar}\right) \quad \text{with} \quad \hbar = h/(2\pi) \tag{8.43}$$

The first factor $\psi_n(x, y, z,)$ depends on the space coordinates only while the second factor gives the time dependency. The parameter h is *Planck's constant*. Multiplication of the wave function by its conjugate yields

$$\Psi_n^*\Psi_n = \psi_n^*\psi_n \tag{8.44}$$

which indicates that the probability density is constant in time or stationary in the sense that no changes at all are taking place with respect to the external surroundings. Thus an eigenstate is also a stationary state. In this situation the system does not radiate.

(ii) Coherent states

Suppose the system is in the process of changing from eigenstate Ψ_1 to Ψ_2. During the transition the state function is a linear combination of the two state functions as shown in

$$\Psi = C_1 \psi_1 \exp\left(-\frac{iE_1 t}{\hbar}\right) + C_2 \psi_2 \exp\left(-\frac{iE_2 t}{\hbar}\right) \qquad (8.45)$$

The time variation of the parameters C_1 and C_2 is slow in comparison with the time variation of the exponential factors. A state of the type (8.45) is called a *coherent state*. Inspection of this formula shows that the energy of a coherent state is not well defined since two energies are involved. In contrast to the coherent state, the energy of a stationary state is well defined. The probability density of the coherent state is given by

$$\Psi^* \Psi = C_1^* C_1 \psi_1^* \psi_1 + C_2^* C_2 \psi_2^* \psi_2 + C_1^* C_2 \psi_1^* \psi_2 \exp(i\omega t) + C_2^* C_1 \psi_2^* \psi_1 \exp(-i\omega t) \qquad (8.46)$$

with $\quad \omega = 2\pi \nu = (E_1 - E_2)/\hbar$

The quantum mechanical description of a radiating atom may be stated in the following way. During the change from one quantum state to another, the probability distribution of the electron becomes coherent and oscillates sinusoidally. This oscillation is associated with an oscillating electromagnetic field which constitutes the radiation.

8.3.2 The Schrödinger equation

So far we have not given any information in which way we might find the wave function Ψ. From classical physics we know that the time-dependent wave equation can be written in the form

$$\frac{\partial^2 \Psi}{\partial t^2} = v^2 \nabla^2 \Psi \qquad (8.47)$$

where Ψ represents any wave function. Substituting the trial solution

$$\Psi = \psi(x, y, z) \exp(i\omega t) \qquad (8.48)$$

assuming the usual sinusoidal time dependency of Ψ into (8.47) we obtain the time-independent wave equation

$$\nabla^2 \psi + \left(\frac{2\pi}{\lambda}\right)^2 \psi = 0 \qquad (8.49)$$

where λ is the wavelength.

In order to find *Schrödinger's equation*, we make use of *Einstein's mass–energy relation*

$$hv = mc^2$$ (8.50)

where c is the speed of light in empty space and m is the mass of the photon. The linear momentum of the photon is denoted by

$$p = mc = \frac{hv}{c} = \frac{h}{\lambda}$$ (8.51)

Just as light exhibits both wavelike and particle properties, De Broglie assumed that a material particle m moving with speed v, also possesses a wave-like character as stated in

$$\lambda = \frac{h}{p} = \frac{h}{mv}$$ (8.52)

This assumption was later confirmed experimentally.

Replacing in (8.49) the wavelength λ by means of (8.52) we find

$$\nabla^2 \psi + \left(\frac{2\pi p}{h}\right)^2 \psi = 0$$ (8.53)

If E stands for the total energy of the particle as given by $E = mv^2/2 + V$, V is the potential energy, the linear momentum may be expressed by

$$p^2 = 2m(E - V)$$ (8.54)

Substitution of (8.54) into (8.53) yields the famous Schrödinger equation

$$\nabla^2 \psi + \frac{8\pi^2 m}{h^2}(E - V)\psi = 0$$ (8.55)

permitting us to find the wave function ψ. In case that the physical system consists of N particles, (8.55) must be replaced by

$$\sum_{k=1}^{N} \frac{1}{m_k}\left(\frac{\partial^2 \psi}{\partial x_k^2} + \frac{\partial^2 \psi}{\partial y_k^2} + \frac{\partial^2 \psi}{\partial z_k^2}\right) + \frac{8\pi^2}{h^2}(E - V)\psi = 0$$ (8.56)

if Cartesian coordinates are used.

To obtain the wave function ψ from the Schrödinger equation can be very difficult. The solution procedure requires the specification of the potential function $V(x, y, z)$ of the physical system. Not all mathematical solutions of the Schrödinger equation are acceptable since they may not be physically meaningful. To be an acceptable solution, the function ψ must tend to zero for infinite values of the coordinates in such a way that it is quadratically integrable. This requirement leads

Table 8.1 *Examples of quantum mechanical operators*

Variable	Operator
Position	
x	x
y	y
z	z
Linear momentum	
p_x	$\dfrac{\hbar}{i}\dfrac{\partial}{\partial x}$
p_y	$\dfrac{\hbar}{i}\dfrac{\partial}{\partial y}$
p_z	$\dfrac{\hbar}{i}\dfrac{\partial}{\partial z}$
Kinetic energy	
$\dfrac{1}{2}m\left(v_x^2 + v_y^2 + v_z^2\right)$	$-\dfrac{\hbar^2}{2m}\left(\dfrac{\partial^2}{\partial x^2} + \dfrac{\partial^2}{\partial y^2} + \dfrac{\partial^2}{\partial z^2}\right)$
Potential energy	
$V(x, y, z)$	$V(x, y, z)$
Total energy	
E	$-\dfrac{\hbar}{i}\dfrac{\partial}{\partial t}$

to the result that acceptable functions can exist only if the energy E has definite values. These allowed values of E, known as eigenvalues, are *characteristic energy levels* of the system. The corresponding solutions are the eigensolutions. Simple examples will be given later.

In quantum mechanics every variable, such as position or momentum is associated with an operator. If one of these variables is denoted by g and the corresponding operator by G, then the operation on the wave function of the system by G gives, in some cases, the value for the variable g times ψ, i.e.

$$G\psi = g\psi \tag{8.57}$$

Examples of some quantum mechanical operators are given in Table 8.1.

It is often convenient to employ a form of the Schrödinger equation that makes use of a formal analogy between classical and quantum mechanics. From classical mechanics (see Appendix 8.8.1) we know that for a conservative dynamical system the sum of the kinetic energy K and the potential energy V is equal to the constant E, i.e.

$$H = K + V = E \tag{8.58}$$

The sum of K and V is called the *Hamilton function H*. If (x_k, y_k, z_k) are the coordinates of particle k and $(p_{x_k}, p_{y_k}, p_{z_k})$ the components of the linear momentum of this particle, the Hamiltonian function H can be written in the form

$$H = \sum_{k=1}^{N} \frac{1}{2m_k} \left(p_{x_k}^2 + p_{y_k}^2 + p_{z_k}^2 \right) + V(x_1, y_1, z_1, \ldots x_k, y_k, z_k, \ldots) \qquad (8.59)$$

Replacing the momenta and E according to Table 8.1 and introducing the wave function Ψ on which the operator is applied, we find the quantum mechanical analogy. This gives the *time-dependent Schrödinger equation*

$$-\frac{\hbar^2}{2} \sum_{k=1}^{N} \frac{1}{m_k} \left(\frac{\partial^2 \Psi}{\partial x_k^2} + \frac{\partial^2 \Psi}{\partial y_k^2} + \frac{\partial^2 \Psi}{\partial z_k^2} \right) + V\Psi = -\frac{\hbar}{i} \frac{\partial \Psi}{\partial t} \qquad (8.60)$$

Utilizing the information listed in Table 8.1, the Schrödinger equation can also be written as

$$\mathcal{H}\Psi = E\Psi = -\frac{\hbar}{i} \frac{\partial \Psi}{\partial t} \qquad (8.61)$$

While we denote the Hamiltonian function by the symbol H we will use the calligraphic print $\mathcal{H}$ to designate the analogous *quantum mechanical Hamilton operator*.

The general solution to (8.61) is given by

$$\Psi = \sum_n a_n \Psi_n = \sum_n a_n \psi_n \exp\left(-\frac{i E_n t}{\hbar} \right) \qquad (8.62)$$

In textbooks on quantum mechanics it is shown that any two eigenfunctions of an atomic system belonging to different eigenvalues are orthogonal. Assuming that the eigenfunctions are normalized we may write

$$\int_{-\infty}^{\infty} \psi_m^* \psi_n d\tau = \delta_{m,n} \quad \text{with} \quad d\tau = dx\, dy\, dz \qquad (8.63)$$

If the wave function Ψ is normalized to 1, then the following relation must be valid

$$\sum_n a_n^* a_n = \sum_n |a_n|^2 = 1 \qquad (8.64)$$

The latter equation implies that the product $|a_n|^2 = a_n^* a_n$ represents the probability of finding the system in a state of energy E_n at time t.

In order to discuss radiation theory we need to find a suitable expression for the Hamiltonian operator for a charged particle in an electromagnetic field. As before, we begin our discussion with the classical Hamiltonian function.

8.3.3 *Hamilton operator for a charged particle in an electromagnetic field*

The reader may wish to refer to Appendix 2 of this chapter where we have briefly summarized those relationships from electromagnetic theory which are needed later. The interaction of a charged particle of mass m with an electromagnetic field is described by the well-known *Lorentz force* equation. If $\mathbf{E}$ and $\mathbf{B}$ represent the *electric* and the *magnetic field vector*, e and $\mathbf{v}$ the charge and the velocity of the particle, then the Lorentz force equation can be written in the form

$$\boxed{\mathbf{F} = e\mathbf{E} + e\mathbf{v} \times \mathbf{B}} \tag{8.65}$$

In deriving the classical Hamiltonian function it is more convenient to use the vector potential $\mathbf{A}$ and the scalar potential ϕ rather than the field vectors themselves. The basic relationships are

$$\mathbf{E} = -\nabla\phi - \frac{\partial \mathbf{A}}{\partial t}, \qquad \mathbf{B} = \nabla \times \mathbf{A} \tag{8.66}$$

Hence the force equation assumes the form

$$\mathbf{F} = -e\frac{\partial \mathbf{A}}{\partial t} - e\nabla\phi + e\mathbf{v} \times (\nabla \times \mathbf{A}) \tag{8.67}$$

We will now briefly show that equation (8.67) can also be derived with the help of Lagrange's equation of motion if L is given by

$$
\begin{aligned}
L &= \frac{1}{2}m\mathbf{v}^2 + e(\mathbf{v} \cdot \mathbf{A}) - e\phi \\
&= \frac{1}{2}m(\dot{x}^2 + \dot{y}^2 + \dot{z}^2) + e(\dot{x}A_x + \dot{y}A_y + \dot{z}A_z) - e\phi
\end{aligned} \tag{8.68}
$$

where we have used Cartesian coordinates. The use of the Lagrangian equation requires that we treat (x, y, z) and $(\dot{x}, \dot{y}, \dot{z})$ as independent variables. For the x-component we obtain

$$
\begin{aligned}
\frac{\partial L}{\partial x} &= e\left(\dot{x}\frac{\partial A_x}{\partial x} + \dot{y}\frac{\partial A_y}{\partial x} + \dot{z}\frac{\partial A_z}{\partial x}\right) - e\frac{\partial \phi}{\partial x} \\
\frac{\partial L}{\partial \dot{x}} &= m\dot{x} + eA_x = p_x
\end{aligned} \tag{8.69}
$$

so that Lagrange's equation of motion can be generalized to

$$\frac{d}{dt}(m\mathbf{v} + e\mathbf{A}) = \nabla L \tag{8.70}$$

Using the vector identity

$$\mathbf{v} \times (\nabla \times \mathbf{A}) = (\nabla\mathbf{A}) \cdot \mathbf{v} - \mathbf{v} \cdot (\nabla\mathbf{A}) \tag{8.71}$$

the gradient of L assumes the form

$$\nabla L = e\mathbf{v} \times (\nabla \times \mathbf{A}) + e\mathbf{v} \cdot \nabla \mathbf{A} - e\nabla \phi \tag{8.72}$$

Application of the Euler expansion

$$\frac{d\mathbf{A}}{dt} = \frac{\partial \mathbf{A}}{\partial t} + \mathbf{v} \cdot \nabla \mathbf{A} \tag{8.73}$$

yields

$$\frac{d}{dt}(m\mathbf{v}) = \mathbf{F} = -e\frac{\partial \mathbf{A}}{\partial t} - e\nabla \phi + e\mathbf{v} \times (\nabla \times \mathbf{A}) \tag{8.74}$$

in accordance with (8.67)

In Appendix 8.8.1 it will be shown that the Hamiltonian function can be written as

$$H = \sum_{k=1}^{n} p_k \dot{q}_k - L = \mathbf{p} \cdot \mathbf{v} - L \tag{8.75}$$

Since the momentum of the charged particle is given by

$$\mathbf{p} = m\mathbf{v} + e\mathbf{A} \tag{8.76}$$

the classical expression for H assumes the form

$$H = \mathbf{p} \cdot \mathbf{v} - L = \frac{1}{2}m\mathbf{v}^2 + e\phi \tag{8.77}$$

Using (8.76) H can also be written as

$$H = \frac{1}{2m}(\mathbf{p} - e\mathbf{A})^2 + e\phi \tag{8.78}$$

For an electromagnetic wave such as that associated with a light wave the Maxwell conditions

$$\nabla \cdot \mathbf{A} = 0, \qquad \phi = 0 \tag{8.79}$$

apply, so that the Hamilton function will simplify. As shown in textbooks on quantum mechanics the vectors $\mathbf{p}$ and $\mathbf{A}$, in general, do not commute but follow the rule

$$\mathbf{p} \cdot \mathbf{A} - \mathbf{A} \cdot \mathbf{p} = i\hbar\nabla \cdot \mathbf{A} \tag{8.80}$$

Thus, by using (8.79), the Hamilton operator can be written as

$$\boxed{\mathcal{H} = \frac{\mathbf{p}^2}{2m} - \frac{e}{m}\mathbf{A} \cdot \mathbf{p} + \frac{e^2}{2m}\mathbf{A}^2} \tag{8.81}$$

Since the perturbation of an atomic system by a light wave will be small, in discussing radiation we may neglect the last term on the right-hand side of (8.81). This term, however, cannot be neglected when discussing the perturbations due to strong magnetic fields. The first term in (8.81) refers to the Hamiltonian of the particle in the absence of a radiation field while the second term accounts for the perturbation of the system due to an electromagnetic field. The perturbation part $\mathcal{H}'$ of the operator $\mathcal{H}$ is expressed by

$$\mathcal{H}' = -\frac{e}{m}\mathbf{A}\cdot\mathbf{p} = -\frac{e\hbar}{mi}\mathbf{A}\cdot\nabla \qquad (8.82)$$

where we have also replaced the linear momentum by the corresponding quantum mechanical operator as defined in Table 8.1. The perturbation part $\mathcal{H}'$ of the Hamiltonian is also called the *interaction Hamiltonian*.

8.3.4 The interaction Hamiltonian

Now we consider a molecular system subjected to the perturbation $\mathcal{H}'$ of an electromagnetic field of a light wave. Since the molecular dimensions are much smaller than the wavelength of the infrared light, we may consider $\mathbf{A}$ to be a constant over the molecule. For simplicity, we will assume that the field is that of a plane polarized light wave traveling in the z-direction. With $E_x = E_0 \exp[i(\omega t - kz)]$ and $E_y = E_z = 0$ we find from (8.66)

$$A_x = -\frac{E_x}{i\omega}, \quad A_y = 0, \quad A_z = 0 \qquad (8.83)$$

which enables us to obtain the interaction Hamiltonian.

By using in (8.82) the expression (8.76) of the momentum $\mathbf{p}$ we would again obtain an expression that is proportional to $\mathbf{A}^2$ which we neglect in first-order perturbation theory. Thus we simply write for the x-component of the momentum $p_x = m\dot{x}$. For a forced harmonic oscillation of frequency ω, p_x also varies sinusoidally as stated in

$$p_x = m\frac{dx}{dt} = m\frac{d}{dt}[x_0 \exp(i\omega t)] = mi\omega x \qquad (8.84)$$

so that the perturbation Hamiltonian is given by

$$\begin{aligned}
H' &= -\frac{e}{m}\mathbf{A}\cdot\mathbf{p} = exE_x = exE_{0,x}\cos\omega t \\
&= \frac{1}{2}exE_{0,x}[\exp(i\omega t) + \exp(-i\omega t)]
\end{aligned} \qquad (8.85)$$

The quantity exE_x is the product of the dipole moment and the electric field in the x-direction. Similarly we may consider the remaining directions. If the dipole moment of the system is denoted by **M** whose components are

$$M_x = \sum_k e_k x_k, \quad M_y = \sum_k e_k y_k, \quad M_z = \sum_k e_k z_k \qquad (8.86)$$

where the suffix k refers to particle k, then the interaction energy can be expressed by $\mathbf{M} \cdot \mathbf{E}$.

Any transition for which the probability can be calculated using the form (8.85) is called an *electric dipole moment transition*. The effect of the magnetic field has been ignored. The discussion we have carried out so far is a mixture of classical and quantum mechanics which is known as a *semi-classical treatment*.

8.3.5 Computation of transition probabilities

We begin our discussion by restating the wave equation in the form

$$\mathcal{H}\Psi = i\hbar \frac{\partial \Psi}{\partial t} \qquad (8.87)$$

where Ψ is the complete wave function. The Hamiltonian operator may be expressed as $\mathcal{H} = \mathcal{H}_0 + \mathcal{H}'$. The part $\mathcal{H}_0$ is independent of time while $\mathcal{H}'$ is a time-dependent perturbation. The unperturbed eigenfunctions Ψ^0 satisfy

$$\mathcal{H}_0\Psi^0 = i\hbar \frac{\partial \Psi^0}{\partial t} \qquad (8.88)$$

In order to obtain a solution to (8.87) we expand the function Ψ in terms of the unperturbed eigenfunctions Ψ^0 permitting the expansion coefficients to vary with time. Substituting

$$\Psi = \sum_n a_n(t)\Psi_n^0 \quad \text{with} \quad \Psi_n^0 = \psi_n^0 \exp\left(-\frac{iE_n t}{\hbar}\right) \qquad (8.89)$$

into (8.87) we find the following equation

$$\sum_n a_n(t)\mathcal{H}_0\Psi_n^0 + \sum_n a_n(t)\mathcal{H}'\Psi_n^0 = i\hbar \sum_n \frac{da_n}{dt}\Psi_n^0 + i\hbar \sum_n a_n(t)\frac{\partial \Psi_n^0}{\partial t} \qquad (8.90)$$

Since the unperturbed eigenfunctions satisfy (8.88), equation (8.90) immediately reduces to

$$\sum_n a_n(t)\mathcal{H}'\Psi_n^0 = i\hbar \sum_n \frac{da_n}{dt}\Psi_n^0 \qquad (8.91)$$

Now we multiply both sides of equation (8.91) by Ψ_m^{0*} and then integrate over the coordinate space yielding

$$\sum_n a_n(t) \int \Psi_m^{0*} \mathcal{H}' \Psi_n^0 \, d\tau = i\hbar \sum_n \frac{da_n}{dt} \int \Psi_m^{0*} \Psi_n^0 \, d\tau = i\hbar \frac{da_m}{dt} \qquad (8.92)$$

Hence, due to the orthogonality of the wave functions we immediately obtain the result[2]

$$\frac{da_m}{dt} = -\frac{i}{\hbar} \sum_n a_n(t) \int \Psi_m^{0*} \mathcal{H}' \Psi_n^0 \, d\tau = -\frac{i}{\hbar} \sum_n a_n(t) \left(\Psi_m^{0*} | \mathcal{H}' | \Psi_n^0 \right) \qquad (8.93)$$

Often the integral is written in the operator form as shown in the final expression of this equation. For any particular problem we have to solve a set of differential equations to get explicit expressions for the a_m.

Temporarily we consider the perturbation for a single frequency. We assume the simple situation that the system originally at $t = 0$ was in the state n so that $a_n(0) = 1$ and all the other a_k are zero at time $t = 0$. For a sufficiently short time so that all a_k are negligible except a_n, we find the following approximate expression

$$a_m(t) = -\frac{i}{\hbar} \int_0^t \int \Psi_m^{0*} \mathcal{H}' \Psi_n^0 \, d\tau \, dt' \quad \text{with} \quad a_m(0) = 0 \qquad (8.94)$$

Making use of $\Psi_n^0 = \psi_n^0 \exp(-i E_n t / \hbar)$ where the ψ_n^0 depend on space only, and replacing the perturbation Hamiltonian by (8.85), upon integration we obtain

$$a_m(t) = \frac{1}{2} E_{0,x} X_{nm} \left(\frac{1 - \exp\left[\frac{it}{\hbar}(E_m - E_n + h\nu)\right]}{E_m - E_n + h\nu} + \frac{1 - \exp\left[\frac{it}{\hbar}(E_m - E_n - h\nu)\right]}{E_m - E_n - h\nu} \right)$$

$$(8.95)$$

The expression

$$X_{nm} = \int \psi_m^{0*} M_x \psi_n^0 \, d\tau \qquad (8.96)$$

is called the x-component of the matrix element of the dipole moment for the transition n to m.

Let us consider the case $E_m > E_n$ so that the transition corresponds to absorption. The coefficient a_m will be large only if $E_m - E_n$ is approximately equal to $h\nu$ so that the denominator of the second term on the right-hand side of (8.95) is nearly

---

[2] Even shorter is Dirac's notation which is written as $\langle m | \mathcal{H}' | n \rangle$ where the terms $\langle m | = \Psi_m^{0*}$ and $| n \rangle = \Psi_n^0$ are known as (bra) and (ket), respectively. The integration over all space is implied.

zero. The first term can be ignored so that the product $a_m a_m^*$ is given with excellent approximation by

$$a_m(t) a_m^*(t) = \frac{E_{0,x}^2 |X_{nm}|^2 \sin^2\left[\frac{\pi t}{h}(E_m - E_n - h\nu)\right]}{(E_m - E_n - h\nu)^2} \tag{8.97}$$

In case of emission $E_n > E_m$ the first expression will be the dominant term and a similar expression will be found for $a_m a_m^*$.

Inspection of (8.97) shows that for small t the transition probability varies according to t^2 which is an unexpected result. This difficulty is due to the fact that so far we have considered only a single frequency. From experience we know that we never deal with strictly monochromatic radiation but always with a range of frequencies and with radiation fields having components in all three directions.

The energy density of the radiation field in the frequency interval ν to $\nu + d\nu$ will be denoted by $\hat{u}_\nu d\nu$. From electromagnetic theory it is known that the energy density and the electric field are related by

$$\hat{u}_\nu = \varepsilon_0 \overline{E^2(\nu)} \tag{8.98}$$

where the overbar represents an average value. For isotropic radiation the following equation is valid

$$\frac{1}{3}\overline{E^2(\nu)} = \overline{E_x^2(\nu)} = \overline{E_y^2(\nu)} = \overline{E_z^2(\nu)} \tag{8.99}$$

with

$$\overline{E_x^2(\nu)} = \overline{E_{0,x}^2(\nu) \cos^2 2\pi \nu t} = \frac{1}{2} E_{0,x}(\nu)^2 \tag{8.100}$$

Hence the energy density can be expressed by the following relation

$$\hat{u}_\nu = \frac{3}{2}\varepsilon_0 E_{0,x}^2 \tag{8.101}$$

Assuming that $\hat{u}_\nu$ is constant over the frequency range ν to $\nu + d\nu$ we may integrate (8.97) yielding

$$a_m(t) a_m^*(t) = \frac{2}{3\varepsilon_0} |X_{nm}|^2 \hat{u}_\nu \int_{-\infty}^{\infty} \frac{\sin^2\left[\frac{\pi t}{h}(E_m - E_n - h\nu)\right]}{(E_m - E_n - h\nu)^2} d\nu \tag{8.102}$$

At first glance it seems to be inconsistent to treat $\hat{u}_\nu$ as a constant and then integrate over the complete frequency range from $-\infty$ to $+\infty$. Nevertheless, the approximation is entirely satisfactory since $a_m a_m^*$ is very small except for such frequencies that $E_m - E_n = h\nu$. By observing that $\int_{-\infty}^{\infty} \sin^2 \alpha / \alpha^2 d\alpha = \pi$ we may carry out the integration and find the result

$$a_m(t)a_m^*(t) = \frac{2\pi^2}{3\varepsilon_0 h^2}|X_{nm}|^2 \hat{u}_\nu\, t \tag{8.103}$$

This relation shows that now $a_m a_m^*$ varies linearly with time t as should be expected. Corresponding expressions can be obtained for the other directions.

Thus we find for the total transition probability per unit time the expression

$$\frac{a_m(t)a_m^*(t)}{t} = \frac{2\pi^2}{3\varepsilon_0 h^2}|R_{nm}|^2 \hat{u}_\nu \tag{8.104}$$

with

$$|R_{nm}|^2 = |X_{nm}|^2 + |Y_{nm}|^2 + |Z_{nm}|^2 = \left| \int \psi_m^{0*} \mathbf{M} \psi_n^0 \, d\tau \right|^2 \tag{8.105}$$

The term R_{nm} is the total matrix element for the transition n to m.

So far our discussion includes only transitions between the so-called *non-degenerate energy levels*. Frequently, however, there are several different orthogonal eigenfunctions associated with one and the same eigenvalue so that we are dealing with a *degenerate state*. The frequency of this occurrence, i.e. the number of eigenfunctions corresponding to this state, is the so-called *statistical weight*.

Suppose that the levels n and m are degenerate with statistical weights g_n and g_m. The probability of transition per unit time from one of its lower states n_i to the upper level is given by

$$\frac{a_m(t)a_m^*(t)}{t} = \frac{2\pi^2}{3\varepsilon_0 h^2}\hat{u}_\nu \sum_k |R(n_i m_k)|^2 \tag{8.106}$$

where the summation must be carried out over all states belonging to the upper level. Let N_n represent the number populating the lower level n. Assuming that the lower states will be equally distributed between the g_n states, then each state will have a population N_n/g_n. Now the transition probability per unit time is given by

$$\frac{a_m(t)a_m^*(t)}{t} = \frac{2\pi^2}{3\varepsilon_0 h^2 g_n}\hat{u}_\nu \sum_i \sum_k |R(n_i m_k)|^2 \tag{8.107}$$

8.3.6 Einstein transition probabilities

So far we have discussed the process of absorption and emission in the presence of an electromagnetic field. Since a system in an excited state can emit radiation even in the absence of an electromagnetic field, the completion of the theory of radiation requires the calculation of the transition probability of *spontaneous emission*. The direct quantum-mechanical calculation of this quantity is a matter of great difficulty.

Fortunately, Einstein has shown how to tackle the problem of spontaneous emission by using thermodynamic reasoning.

We investigate the equilibrium between two states of different energy. As stated in equation (8.46) the transition between two states is always accompanied by the absorption or emission of radiation. Let us now consider the radiation density in an enclosure having opaque walls of uniform temperature T containing a large number of quantized systems which can interact with the radiation. Two states m and n of these systems have the respective energies E_m and E_n. Any transition between these states will be accompanied by absorption or emission of radiation. For a wave, considering absorption with $E_m > E_n$, we have

$$h\nu_{mn} = E_m - E_n \qquad (8.108)$$

We will denote the energy density in the spectral interval ranging from ν_{mn} to $\nu_{mn} + d\nu_{mn}$ by $\hat{u}(\nu_{mn})d\nu_{mn}$. For ease of identification we will momentarily call l the lower and u the upper energy levels. The probability p_{abs} that a system in state l absorbing a quantum of radiative energy and undergoing a transition to state u in unit time is given by

$$p_{abs}(l \rightarrow u) = B_{l \rightarrow u}\hat{u}(\nu_{lu}) \qquad (8.109)$$

where the coefficient $B_{l \rightarrow u}$ is known as the *Einstein coefficient of absorption*. The transition in the opposite direction is given by

$$p_{em}(u \rightarrow l) = A_{u \rightarrow l} + B_{u \rightarrow l}\hat{u}(\nu_{lu}) \qquad (8.110)$$

where the coefficient $B_{u \rightarrow l}$ is the *Einstein coefficient of induced emission* which is *stimulated emission* in the presence of the radiation field of volume density $\hat{u}(\nu)$. In addition spontaneous emission is taking place which is described by the *Einstein coefficient of spontaneous emission* $A_{u \rightarrow l}$.[3]

Within the enclosure the systems will have various energy states. We denote the number of systems of energy E_l by N_l, and the number of systems E_u by N_u, then in equilibrium the number of transitions from u $\rightarrow$ l must be equal to the number of transitions from l $\rightarrow$ u. Therefore, we must have the equilibrium statement

$$N_l B_{l \rightarrow u}\hat{u}(\nu_{lu}) = N_u \left[A_{u \rightarrow l} + B_{u \rightarrow l}\hat{u}(\nu_{lu})\right] \qquad (8.111)$$

Since N_l, N_u are numbers per unit volume, each term expresses the number of transitions in unit time per unit volume. Equation (8.111) can be solved to give the

---

[3] The quantities $\hat{u}(\nu_{lu})$ and $B_{l \rightarrow u}$ are respectively expressed in J s m^{-3} and m^3 J^{-1} s^{-2} while $A_{u \rightarrow l}$ is expressed in s^{-1}.

ratio

$$\frac{N_u}{N_l} = \frac{B_{l \rightarrow u} \hat{u}(\nu_{lu})}{A_{u \rightarrow l} + B_{u \rightarrow l} \hat{u}(\nu_{lu})} \tag{8.112}$$

In an enclosure at temperature T we may also express the ratio (8.112) with the help of the *Boltzmann distribution*. The number of systems having an energy E_i above the ground state is given by

$$\boxed{N_i = \frac{N}{Z} g_i \exp\left(-\frac{E_i}{kT}\right)} \tag{8.113}$$

where N is the total number of systems and Z the partition function, that is the sum over all states. The quantity g_i is the statistical weight of the level of energy E_i. Therefore, at equilibrium the following expression is valid

$$\frac{N_u}{N_l} = \frac{g_u}{g_l} \exp\left(-\frac{E_u - E_l}{kT}\right) = \frac{g_u}{g_l} \exp\left(-\frac{h\nu_{lu}}{kT}\right) \tag{8.114}$$

Substituting (8.114) into (8.112) and solving for the energy density $\hat{u}(\nu_{lu})$, we obtain

$$\hat{u}(\nu_{lu}) = \frac{g_u A_{u \rightarrow l}}{g_l B_{l \rightarrow u} \exp\left(h\nu_{lu}/kT\right) - g_u B_{u \rightarrow l}} \tag{8.115}$$

Within the enclosure the energy density must also be given by *Planck's radiation law*

$$\hat{u}(\nu_{lu}) = \frac{8\pi h \nu_{lu}^3}{c^3} \frac{1}{\exp\left(h\nu_{lu}/kT\right) - 1} \tag{8.116}$$

Comparison of (8.115) and (8.116) gives the important relations

$$\boxed{A_{u \rightarrow l} = \frac{8\pi h \nu_{lu}^3}{c^3} B_{u \rightarrow l}, \qquad g_u B_{u \rightarrow l} = g_l B_{l \rightarrow u}} \tag{8.117}$$

From time-dependent perturbation theory we obtained equation (8.107) which expresses the Einstein probability for absorption. Equation (8.109) also states this probability, but the Einstein coefficient remained undefined. We observe that in (8.107) the level m refers to the upper level. By setting (8.107) equal to (8.109), replacing n by l we find the important equation for the Einstein coefficient for absorption

$$\boxed{B_{l \rightarrow u} = \frac{2\pi^2}{3\varepsilon_0 h^2 g_l} \sum_i \sum_k |R(l_i u_k)|^2} \tag{8.118}$$

This formula will be needed when we derive the line intensity equation for the *spectral absorption coefficient*. Using (8.117) and (8.118) yields the coefficient for spontaneous emission

$$A_{u \to l} = \frac{16\pi^3 \nu_{lu}^3}{3h\varepsilon_0 c^3 g_u} \sum_i \sum_k |R(l_i u_k)|^2 \qquad (8.119)$$

To the degree of approximation we have used above, the Einstein coefficients depend mainly on the matrix element for the electric dipole moment between the two states.

If the variation of the field over the molecule is not neglected additional terms will appear. The first two of these correspond to magnetic dipole and electric quadrupole radiation. In comparison to the transition probability of electric dipole radiation the contributions of these additional terms may be ignored in most cases. Often these terms are loosely called *forbidden transitions*.

8.3.7 Line intensities

We start the discussion with *Beer's law* as given by

$$dI_\nu = -k_\nu I_\nu \, du \qquad (8.120)$$

Here, k_ν is the *monochromatic absorption coefficient* and du the *differential absorbing mass*. As we know, a spectral line is not infinitely sharp but it is broadenend. Over the small frequency interval occupied by a spectral line, the radiative energy from a nearly parallel beam varies so little that I_ν may be treated as a constant. Thus we may perform the frequency integration and obtain

$$dI_\nu = -I_\nu \, du \int_{-\infty}^{\infty} k_\nu \, d\nu = -SI_\nu \, du \qquad (8.121)$$

where we have used the definition (7.13) for the line intensity S. To find a theoretical expression for S we need to relate this quantity to the net number of transitions N_{tr} from the lower energy level E_l to the upper energy level E_u. The number N_{tr} induced by a radiation field of energy density per unit volume $\hat{u}(\nu_{lu})$ is given by

$$N_{tr} = (N_l B_{l \to u} - N_u B_{u \to l}) \, \hat{u}(\nu_{lu}) \qquad (8.122)$$

Due to (8.117) we may rewrite this expression as

$$N_{tr} = B_{l \to u} \left(N_l - N_u \frac{g_l}{g_u} \right) \hat{u}(\nu_{lu}) \qquad (8.123)$$

The energy absorbed in each transition is $h\nu_{lu}$ so that the decrease dI_ν in the beam may be expressed as

$$dI_\nu = -\frac{h\nu_{lu}}{c} B_{l \to u} \left(N_l - N_u \frac{g_l}{g_u} \right) I_{\nu_{lu}} \, du \qquad (8.124)$$

Here we have substituted I_ν/c for $\hat{u}_\nu$. Comparing the latter equation with (8.121) we find

$$S = \frac{h\nu_{lu}}{c} B_{l\to u}\left(N_l - N_u\frac{g_l}{g_u}\right) \tag{8.125}$$

With the help of (8.118) we now replace $B_{l\to u}$ by $B_{m\to n}$ and obtain

$$S = \frac{2\pi^2\nu_{mn}}{3h\varepsilon_0 c}\sum_i\sum_k|R(m_i n_k)|^2\left(\frac{N_m}{g_m} - \frac{N_n}{g_n}\right) \tag{8.126}$$

$$\text{with}\quad R(m_i n_k) = \int \psi^{0*}(m_i)\mathbf{M}\psi^0(n_k)\,d\tau$$

The indices i and k number the degenerate wave functions belonging to the energy levels m and n, respectively. The summation must be carried out over all possible combinations of wave functions of the upper state with wave functions of the lower state.

In case of thermal equilibrium, using the Boltzmann distribution (8.113), the line intensity can also be expressed as

$$\boxed{S = \frac{2\pi^2\nu_{mn}}{3h\varepsilon_0 c}\sum_i\sum_k|R(m_i n_k)|^2\frac{N}{Z}\left[1 - \exp\left(-\frac{h\nu_{nm}}{kT}\right)\right]\exp\left(-\frac{E_m}{kT}\right)}$$

$$\tag{8.127}$$

The frequency ν_{nm} has previously been called ν_0 which refers to the frequency defining the position of the center of the absorption line. The above line intensity formulas ignore the effect of *nuclear spin* which is responsible for the existence of the so-called *hyperfine structure* of the spectrum, see Rothman *et al.* (1987, 1992).

Very few problems have exact quantum mechanical solutions. To this class of problems belong the harmonic oscillator and the rigid rotator for which exact solutions can be given. More complicated problems require approximate solutions.

8.4 Vibrations and rotations of molecules

8.4.1 The harmonic oscillator

Harmonic oscillation is of considerable importance in quantum mechanics. The model of the simple harmonic oscillator is used to understand the vibrations of diatomic and polyatomic molecules. A *harmonic oscillator* is a particle of mass m moving in a straight line (say along the x-axis) subject to the potential $V = kx^2/2$ where k is Hooke's constant. According to (8.58) the classical Hamiltonian of the system is given by

$$H = \frac{p^2}{2m} + \frac{1}{2}kx^2 \tag{8.128}$$

where p is the linear momentum. Using Table 8.1, the Hamiltonian operator is expressed by

$$\mathcal{H} = -\frac{h^2}{8\pi^2 m}\frac{d^2}{dx^2} + \frac{1}{2}kx^2 \tag{8.129}$$

and the wave equation by

$$\frac{d^2\psi}{dx^2} + \frac{8\pi^2 m}{h^2}\left(E - \frac{1}{2}kx^2\right)\psi = 0 \tag{8.130}$$

Introducing the abbreviations

$$\alpha = \frac{8\pi^2 m}{h^2}E, \quad \beta = \frac{2\pi\sqrt{mk}}{h} \tag{8.131}$$

(8.130) assumes the form

$$\frac{d^2\psi}{dx^2} + (\alpha - \beta x^2)\psi = 0 \tag{8.132}$$

Changing the variable according to

$$\xi = \sqrt{\beta}x, \quad \frac{d^2}{dx^2} = \beta\frac{d^2}{d\xi^2} \tag{8.133}$$

we find

$$\frac{d^2\psi}{d\xi^2} + \left(\frac{\alpha}{\beta} - \xi^2\right)\psi = 0 \tag{8.134}$$

Now we investigate which form ψ must have in order to be an acceptable wave function for large values of ξ. For sufficiently large values of ξ the ratio α/β can be neglected in comparison to ξ^2 so that we get the approximate equation

$$\frac{d^2\psi}{d\xi^2} - \xi^2\psi = 0 \tag{8.135}$$

This equation is approximately satisfied by

$$\psi = C\exp\left(\pm\frac{\xi^2}{2}\right) \quad \text{with} \quad \frac{d^2}{d\xi^2}\exp\left(\pm\frac{\xi^2}{2}\right) = \exp\left(\pm\frac{\xi^2}{2}\right)(\xi^2 \pm 1) \tag{8.136}$$

since ± 1 may be neglected in the region of large ξ^2. For obvious reasons we cannot use the solution $\exp(+\xi^2/2)$, but $\exp(-\xi^2/2)$ behaves satisfactorily at large values of ξ. Thus we are led to the trial solution

$$\psi(\xi) = u(\xi) \exp\left(-\frac{\xi^2}{2}\right) \tag{8.137}$$

Substitution of (8.137) into (8.134) yields

$$\frac{d^2u}{d\xi^2} - 2\xi \frac{du}{d\xi} + \left(\frac{\alpha}{\beta} - 1\right)u = 0 \tag{8.138}$$

By setting $\alpha/\beta - 1 = 2n$, we obtain *Hermite's differential equation* whose solution are the well-known *Hermite polynomials* H_n, $n = 0, 1, \ldots$

$$H_n(\xi) = (-1)^n \exp(\xi^2) \frac{d^n}{d\xi^n} [\exp(-\xi^2)] \tag{8.139}$$

A few low-order expressions of these polynomials are listed next

$$H_0(\xi) = 1, \quad H_1(\xi) = 2\xi, \quad H_2(\xi) = 4\xi^2 - 2 \tag{8.140}$$

A particular solution to equation (8.138) can be written as

$$u(\xi) = H_n(\xi) \tag{8.141}$$

so that the wave function is given by

$$\psi(\xi) = C H_n(\xi) \exp\left(-\frac{\xi^2}{2}\right), \quad C = const \tag{8.142}$$

Using equation (8.131) we find that the energy E is quantized and must be written as

$$E = \frac{h}{2\pi}\left(n + \frac{1}{2}\right)\sqrt{\frac{k}{m}} = \left(n + \frac{1}{2}\right)h\nu, \quad n = 0, 1, \ldots \tag{8.143}$$

where ν is the frequency of the classical harmonic oscillator. The state with $n = 0$ is the *vibrational ground state* whose vibrational energy is not zero. The residual energy is known as the *zero point energy*. This is in agreement with *Heisenberg's uncertainty principle* which states that one can never precisely know the position and the momentum of a particle. If the oscillator had zero energy it would have zero momentum and would be located exactly at the position of the minimum potential energy.

Finally, we have to find the normalization constant C by considering the condition (8.142). For the present situation we obtain

$$\int_{-\infty}^{\infty} \psi_n^* \psi_n \, dx = \frac{C^2}{\sqrt{\beta}} \int_{-\infty}^{\infty} H_n(\xi)^2 \exp(-\xi^2) \, d\xi = 1 \quad \text{with} \quad C = \frac{\beta^{1/4}}{\sqrt{2^n n! \sqrt{\pi}}}$$

$$\tag{8.144}$$

Omitting details we find for the normalized wave function the following expression

$$\psi_n(\xi) = N_n H_n(\xi) \exp\left(-\frac{\xi^2}{2}\right) \quad \text{with} \quad N_n = \left(\sqrt{\frac{\beta}{\pi}} \frac{1}{2^n n!}\right)^{1/2} \qquad (8.145)$$

with $\xi = \sqrt{\beta} x$.

The solution of the harmonic oscillator problem outlines the common approach to solve the wave equation. We convert the wave equation into one of the standard differential equations whose solutions are known. This technique will also be used in connection with the rigid rotator problem.

8.4.2 Vibration of diatomic molecules

We consider the vibrations of the two atoms relative to each other. The simplest form of vibrations in a diatomic molecule is that each atom moves toward or away from the other in simple harmonic motion. For a molecule consisting of two like atoms such as N_2, O_2 (homonuclear) the dipole moment is zero, and therefore no transitions between the different vibrational levels will be observed. This means that no infrared absorption or emission occurs. In contrast to this, diatomic molecules such as HCl (heteronuclear) do have a permanent dipole moment.

The harmonic oscillator potential energy curve is not particularly accurate when considering actual molecules. A more satisfactory procedure is to assume some appropriate analytical expression for the potential energy curve such as

$$V(r) = D\left(1 - \exp\left[-\beta(r - r_e)\right]\right)^2 \qquad (8.146)$$

This is the so-called *Morse function*. Here D is the dissociation energy of the molecule which is obtained if $r \to \infty$. The quantity β is a constant which differs from molecule to molecule. If the distance r is equal to the equilibrium value r_e the potential energy has the minimum value $V = 0$. For $r = 0$ the potential energy approaches a large but finite value. (8.146) is satisfactory for many situations. However, for reasons of mathematical simplicity we will presently continue to use the harmonic oscillator approximation.

To determine the vibrational energy states it will be useful to introduce the concept of the reduced mass. If r_1 and r_2 are the respective distances of the atoms from the center of mass of the molecule, we obtain

$$m_1 r_1 = m_2 r_2 \qquad (8.147)$$

If $r = r_1 + r_2$ then r_1 and r_2 are given by

$$r_1 = \frac{m_2}{m_1 + m_2} r, \qquad r_2 = \frac{m_1}{m_1 + m_2} r \qquad (8.148)$$

The kinetic energy K of the diatomic system is expressed by

$$K = \frac{1}{2}m_1\dot{r}_1^2 + \frac{1}{2}m_2\dot{r}_2^2 = \frac{1}{2}\mu\dot{r}^2 \quad \text{with} \quad \mu = \frac{m_1 m_2}{m_1 + m_2} \tag{8.149}$$

where μ is the reduced mass. Thus the classical Hamilton function is given by

$$H = \frac{1}{2}\mu\dot{r}^2 + V(r) \quad \text{with} \quad V(r) = \frac{1}{2}k(r - r_e)^2 \tag{8.150}$$

so that Schrödinger's equation follows immediately

$$\frac{d^2\psi}{dr^2} + \frac{8\pi^2\mu}{h^2}[E - V(r)]\psi = 0 \tag{8.151}$$

Setting $r - r_e = x$, the latter equation has the form (8.130). Thus, according to (8.145) and (8.143), we find that the wave function and the energy of the diatomic system are given by

$$\psi_n(x) = \left(\frac{\sqrt{\beta/\pi}}{2^n n!}\right)^{1/2} H_n(\sqrt{\beta}x)e^{-\frac{1}{2}\beta x^2}, \qquad \beta = 2\pi\sqrt{\mu k}/h \tag{8.152}$$

and

$$E = \frac{h}{2\pi}\left(v + \frac{1}{2}\right)\sqrt{\frac{k}{\mu}} = \left(v + \frac{1}{2}\right)h\nu \tag{8.153}$$

As is customary in spectroscopy, we have used the symbol v for the vibrational quantum number. It will be observed that the vibrational spectrum of a diatomic molecule considered as a harmonic oscillator consists of one frequency. The selection rule $\Delta v = \pm 1$ will be derived later.

8.4.3 Vibration of polyatomic molecules

As stated before, a molecule consisting of n atoms has $3n$ degrees of freedom. Three coordinates are required to describe the translational motion of the entire system considered as concentrated at the center of mass, and three degrees are required, in general, to describe the rotational motion of the system about its center of mass. Thus $3n - 6$ degrees of freedom are left to describe the vibrational motion of the nuclei of the atoms relative to the axes with origin at the center of mass. For the linear CO_2 molecule ($3n - 5$ vibrational degrees of freedom) and for the water vapor molecule we have shown how to find the normal coordinates and normal frequencies. Any vibrational motion of the molecule may be constructed from the superposition of the normal vibrations.

In textbooks on theoretical mechanics, see also the previous examples, it is shown that in terms of the normal coordinates the kinetic and the potential vibrational energy assume the simple form

$$K = \frac{1}{2}\sum_{k=1}^{3n-6} \dot{q}_k^2, \qquad V = \frac{1}{2}\sum_{k=1}^{3n-6} \lambda_k q_k^2 \tag{8.154}$$

For the linear molecule the summation extends to $3n - 5$. In V there are no terms present involving the cross-products of the coordinates q_k. In the Hooke's law approximation the λ_k are constants. By naively treating the q_k as if they were ordinary Cartesian coordinates, ignoring any interaction with rotational motion, the wave equation can be written as

$$\sum_{k=1}^{3n-6}\frac{\partial^2 \psi}{\partial q_k^2} + \frac{8\pi^2}{h^2}\left(E - \frac{1}{2}\sum_{k=1}^{3n-6}\lambda_k q_k^2\right)\psi = 0 \tag{8.155}$$

This equation is separable into $3n - 6$ (or $3n - 5$) equations by using the substitution

$$\psi = \psi_1(q_1)\psi_2(q_2)\cdots\psi_{3n-6}(q_{3n-6}) \tag{8.156}$$

This leads to wave equations of the type

$$\frac{d^2\psi_k(q_k)}{dq_k^2} + \frac{8\pi^2}{h^2}\left(E_k - \frac{1}{2}\lambda_k q_k\right)\psi_k(q_k) = 0 \tag{8.157}$$

The total energy E is the sum of the energies E_k associated with each normal coordinate, i.e.

$$E = \sum_{k=1}^{3n-6} E_k \tag{8.158}$$

Here each equation is an ordinary differential equation in one variable. From a comparison with the wave equation for the linear harmonic oscillator (8.151) and the energy relation (8.153) we find that the eigenvalues for this problem are given by

$$E_k = h\nu_k\left(\upsilon_k + \frac{1}{2}\right), \qquad \upsilon_k = 0, 1, \ldots \tag{8.159}$$

and the ν_k follow from

$$\nu_k = \frac{1}{2\pi}\sqrt{\lambda_k} \tag{8.160}$$

This is the classical oscillation frequency of the normal vibration k and υ_k is the *vibrational quantum number*. The total vibrational energy of the system of n

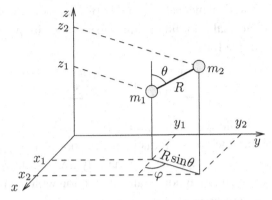

Fig. 8.9 Coordinates of the rigid rotator.

particles is found by introducing (8.159) into (8.158). For further details see, for example, Eyring *et al.* (1965).

8.4.4 Rotation of diatomic molecules

As an idealization of a diatomic molecule, we assume that the molecule consists of two atoms rigidly connected by a weightless link of constant length R. In place of the space coordinates (x_1, y_1, z_1) and (x_2, y_2, z_2) of the two point masses m_1 and m_2 we are going to introduce the center of mass coordinates (x, y, z) of the system and the spherical coordinates (r, θ, φ) of one particle referred to the other as origin. The spherical coordinates are given by

$$x_2 - x_1 = R \sin \theta \cos \varphi, \quad y_2 - y_1 = R \sin \theta \sin \varphi, \quad z_2 - z_1 = R \cos \theta \tag{8.161}$$

were θ is the polar angle and φ the azimuth angle of the system, see also Figure 8.9. The center of mass coordinates of the diatomic system are given by

$$x = \frac{m_1 x_1 + m_2 x_2}{m_1 + m_2}, \quad y = \frac{m_1 y_1 + m_2 y_2}{m_1 + m_2}, \quad z = \frac{m_1 z_1 + m_2 z_2}{m_1 + m_2} \tag{8.162}$$

For brevity of notation we introduce the symbols a and b

$$a = \frac{m_2}{m_1 + m_2} R, \quad b = \frac{m_1}{m_1 + m_2} R \tag{8.163}$$

By eliminating the coordinates (x_2, y_2, z_2) and then (x_1, y_1, z_1) from (8.161) and (8.162), we find

$$x_1 = x - a \sin \theta \cos \varphi, \quad y_1 = y - a \sin \theta \sin \varphi, \quad z_1 = z - a \cos \theta$$
$$x_2 = x + b \sin \theta \cos \varphi, \quad y_2 = y + b \sin \theta \sin \varphi, \quad z_2 = z + b \cos \theta \tag{8.164}$$

Thus the center of mass coordinates have been separated from the spherical coordinates. Since we are not interested in the translational motion of the molecule in space, we may regard the center of mass as fixed. In terms of the original coordinates (x_1, y_1, z_1) and (x_2, y_2, z_2) the kinetic energy of the system is given by

$$K = \frac{m_1}{2}\left(\dot{x}_1^2 + \dot{y}_1^2 + \dot{z}_1^2\right) + \frac{m_2}{2}\left(\dot{x}_2^2 + \dot{y}_2^2 + \dot{z}_2^2\right) \tag{8.165}$$

Using (8.164) the kinetic energy can also be written as

$$K = \frac{m_1 a^2 + m_2 b^2}{2}\left[\left(\frac{d\theta}{dt}\right)^2 + \sin^2\theta\left(\frac{d\varphi}{dt}\right)^2\right] \tag{8.166}$$

Setting the fixed center of mass coordinates (x, y, z) equal to zero, the moment of inertia I about an axis through the center of mass and perpendicular to the molecular axis is $I = m_1 a^2 + m_2 b^2$ so that the kinetic energy can be written as

$$K = \frac{I}{2}\left[\left(\frac{d\theta}{dt}\right)^2 + \sin^2\theta\left(\frac{d\varphi}{dt}\right)^2\right] \tag{8.167}$$

The quantum mechanical operator for the kinetic energy, see Table 8.1, can be written down by replacing the Laplacian in rectangular coordinates by the Laplacian in spherical coordinates. Thus the quantum mechanical Hamiltonian is given by

$$\mathcal{H} = -\frac{h^2}{8\pi^2 m}\left[\frac{1}{r^2}\frac{\partial}{\partial r}\left(r^2\frac{\partial}{\partial r}\right) + \frac{1}{r^2\sin\theta}\frac{\partial}{\partial\theta}\left(\sin\theta\frac{\partial}{\partial\theta}\right) + \frac{1}{r^2\sin^2\theta}\frac{\partial^2}{\partial\varphi^2}\right] + V \tag{8.168}$$

Since no external forces are acting on the rotator we may set the potential energy $V = 0$. Moreover, setting $r = 1$, $mr^2 = m = I$, we find that the Schrödinger equation is given by

$$\frac{1}{\sin\theta}\frac{\partial}{\partial\theta}\left(\sin\theta\frac{\partial\psi}{\partial\theta}\right) + \frac{1}{\sin^2\theta}\frac{\partial^2\psi}{\partial\varphi^2} + \frac{8\pi^2 IE}{h^2}\psi = 0 \tag{8.169}$$

which is a partial differential equation with two independent variables. We attempt to solve this equation by separating the variables, i.e. we are looking for a solution in the form

$$\psi = \Theta(\theta)\Phi(\varphi) \tag{8.170}$$

Substituting (8.170) into (8.169) gives

$$\frac{\sin\theta}{\Theta}\frac{\partial}{\partial\theta}\left(\sin\theta\frac{\partial\psi}{\partial\theta}\right) + \frac{8\pi^2 IE}{h^2}\sin^2\theta = -\frac{1}{\Phi}\frac{\partial^2\Phi}{\partial\varphi^2} \tag{8.171}$$

Since the left-hand side of this equation depends on the variable θ, the right-hand side on the variable φ, both sides must be equal to a constant, say M^2. Thus we obtain the two differential equations

$$\frac{d^2\Phi}{d\varphi^2} = -M^2\Phi \tag{8.172}$$

and

$$\frac{1}{\sin\theta}\frac{\partial}{\partial\theta}\left(\sin\theta\frac{\partial\Theta}{\partial\theta}\right) - \frac{M^2\Theta}{\sin^2\theta} + \frac{8\pi^2 IE}{h^2}\Theta = 0 \tag{8.173}$$

From (8.172) we immediately obtain the solution

$$\Phi(\varphi) = C\exp(\pm iM\varphi) \tag{8.174}$$

where C is an integration constant. This is an acceptable solution provided that M is an integer. This condition arises because the function $\Phi(\varphi)$ must be single-valued which implies

$$\Phi(\varphi) = \Phi(\varphi + 2\pi) \quad \text{or} \quad \exp(iM\varphi) = \exp[iM(\varphi + 2\pi)] \tag{8.175}$$

This requires that $\exp(i2\pi M)$ is unity which is possible only if M is an integer. It is easy to show that the normalized function Φ_M is given by

$$\Phi(\varphi) = \Phi_M(\varphi) = \frac{1}{\sqrt{2\pi}}\exp(\pm iM\varphi), \qquad M = 0, 1, \ldots \tag{8.176}$$

In order to solve equation (8.173) we set $x = \cos\theta$ and introduce the derivatives $d/d\theta$ and $d^2/d\theta^2$ according to

$$x = \cos\theta, \quad \frac{d}{d\theta} = -\sin\theta\frac{d}{dx}, \quad \frac{d^2}{d\theta^2} = \sin^2\theta\frac{d^2}{dx^2} - \cos\theta\frac{d}{dx} \tag{8.177}$$

After some simple rearrangements we obtain

$$(1-x^2)\frac{d^2\Theta}{dx^2} - 2x\frac{d\Theta}{dx} + \left(\frac{8\pi^2 IE}{h^2} - \frac{M^2}{1-x^2}\right)\Theta = 0 \tag{8.178}$$

This equation has the form

$$(1-x^2)\frac{d^2u}{dx^2} - 2x\frac{du}{dx} + \left(J(J+1) - \frac{M^2}{1-x^2}\right)u = 0 \tag{8.179}$$

which is known as the *associated Legendre equation* having the solution

$$u = P_J^M(x) \quad \text{where} \quad P_J^M(x) = (1-x^2)^{M/2}\frac{d^M}{dx^M}P_J(x) \tag{8.180}$$

The $P_J^M(x)$ are the *associated Legendre polynomials* and the $P_J(x)$ the ordinary Legendre polynomials. Whenever $M > J$ the function $P_J^M(x) = 0$. We have discussed the orthogonality properties of the Legendre polynomials in Section 2.4 when treating the scattering problem. Equations (8.178) and (8.179) are identical if

$$\frac{8\pi^2 I E}{h^2} = J(J+1) \quad \text{or} \quad E = \frac{h^2}{8\pi^2 I}J(J+1), \qquad J = 0, 1, \ldots \quad (8.181)$$

Since M enters equation (8.178) only as M^2 we must have $\Theta_{J,M} = \Theta_{J,-M}$. Solution functions that are finite, have integrable squares and are single valued, exist only for conditions that J is zero or a positive integer and $J \geq |M|$. Thus the correct normalized solution function is given by

$$\Theta(\theta) = \Theta_J^{|M|}(\theta) = \sqrt{\frac{(2J+1)}{2}\frac{(J-|M|)!}{(J+|M|)!}} P_J^{|M|}(\cos\theta) \qquad (8.182)$$

and the complete wave function by

$$\psi_{J,M} = \Phi_M(\varphi)\Theta_J^{|M|}(\theta) \qquad (8.183)$$

The allowed wave functions depend on the quantum numbers J and M which are known as the *rotational* and *magnetic quantum numbers*, respectively. For every value of J, there will be $2J+1$ values of M. For example, if $J = 2$, M can have the values $0, \pm 1, \pm 2$. This is called a $(2J+1)$ *degeneracy*. In the presence of an electric or magnetic field, this degeneracy is removed if the molecule has an electric or magnetic dipole moment, and the energy of the state will depend on M also. We will return to this topic later.

In order that radiation may interact with the molecule to produce rotation or that a rotating molecule may emit or absorb radiation, it is necessary that the molecule possesses an electric moment implying that the molecule must have a dipole moment. For this reason homonuclear molecules (having a symmetrical charge distribution about their center of mass) do not have a pure rotation spectrum.

We will show later that the selection rule governing rotational transitions is given by

$$\Delta J = J' - J'' = \pm 1 \qquad (8.184)$$

Thus the frequencies ν_R absorbed or emitted by a rotating molecule correspond to energy differences between adjacent energy levels. Denoting the rotational quantum numbers J' and J'', with $J' > J''$, we find from equation (8.181) the following relation

$$\nu_R = \frac{h}{8\pi^2 I}[J'(J'+1) - J''(J''+1)] = \frac{h}{4\pi^2 I}(J''+1), \qquad J' > J'' \quad (8.185)$$

or $\tilde{v}_R = v_R/c$, that is division of the frequency v by the speed of light c gives the wave number $\tilde{v}$.

8.4.5 Vibration–rotation of diatomic molecules

If a molecule absorbs electromagnetic energy of sufficiently high frequency, both vibration and rotation may occur simultaneously. While the rigid rotator consists of two mass points connected by a massless bar, the nonrigid (vibrating) rotator consists of two mass points which are connected by a massless spring. In order to describe the nonrigid rotator mathematically, we make use of the Hamiltonian operator (8.168) and find the following equation

$$\frac{1}{r^2}\frac{\partial}{\partial r}\left(r^2\frac{\partial\psi}{\partial r}\right) + \frac{1}{r^2\sin\theta}\frac{\partial}{\partial\theta}\left(\sin\theta\frac{\partial\psi}{\partial\theta}\right) + \frac{1}{r^2\sin^2\theta}\frac{\partial^2\psi}{\partial\varphi^2}$$
$$+ \frac{8\pi^2\mu}{h^2}[E - V(r)]\psi = 0 \tag{8.186}$$

where μ is the reduced mass. By writing the wave function as the product $\psi = \psi_r(r)\,\Theta(\theta)\,\Phi(\varphi)$, (8.186) may be separated into three ordinary differential equations. In the interest of brevity we omit mathematical details which can be found in Eyring *et al.* (1965), Pauling and Wilson (1935), and in many other modern textbooks on quantum mechanics.

As before, solving the Θ and Φ equations leads to the introduction of the rotational quantum number J. This is the Schrödinger equation for the r-component of the wave function

$$\frac{1}{r^2}\frac{\partial}{\partial r}\left(r^2\frac{\partial\psi_r}{\partial r}\right) + \left(\frac{8\pi^2\mu}{h^2}[E - V(r)] - \frac{J(J+1)}{r^2}\right)\psi_r = 0 \tag{8.187}$$

Introducing into this equation the potential energy function $V(r)$ of the harmonic oscillator, we find that the energy of the system is given by

$$E = \left(v + \frac{1}{2}\right)hv_e + J(J+1)\frac{h^2}{8\pi^2 I} - \frac{J^2(J+1)^2 h^4}{128\pi^6 v_e^2 I^2} \quad \text{with} \quad v_e = \frac{1}{2\pi}\sqrt{\frac{k}{\mu}} \tag{8.188}$$

where μ, as before, is the reduced mass. The subscript e on the frequency symbol refers to the equilibrium position of the molecule. Inspection of equation (8.188) shows that the first term describes the vibrational energy of the molecule. The second term is the energy of rotation, assuming that the molecule is rigid, while the third term introduces a correction taking into account the stretching of the actual nonrigid molecule due to the rotation.

It turns out that the harmonic oscillator approximation is a good approximation only in the neighborhood of the equilibrium value of r. A more refined approximation is the Morse function as introduced in (8.146). Substituting (8.146) into (8.187) yields a new solution for the frequency

$$\tilde{v} = \frac{E}{hc} = \tilde{v}_e \left(v + \frac{1}{2} \right) - x_e \tilde{v}_e \left(v + \frac{1}{2} \right)^2 + J(J+1)B_e$$

$$- J^2(J+1)^2 D_e - \alpha_e \left(v + \frac{1}{2} \right) J(J+1) \qquad (8.189)$$

where

$$\tilde{v}_e = \frac{\beta}{2\pi c} \left(\frac{2D}{\mu} \right)^{1/2}, \qquad D_e = \frac{h^3}{128\pi^6 \mu^3 \tilde{v}_e^2 c^3 r_e^6}, \qquad B_e = \frac{h}{8\pi^2 Ic}$$

$$x_e = \frac{h \tilde{v}_e c}{4D}, \qquad \alpha_e = \frac{3h^2 \tilde{v}_e}{16\pi^2 \mu r_e^2 D} \left(\frac{1}{\beta r_e} - \frac{1}{\beta^2 r_e^2} \right) \qquad (8.190)$$

Since we have divided the energy E by hc, (8.189) is expressed in wave numbers. For most diatomic molecules this equation gives rather accurate values of the energy levels.

The first term of (8.189) represents the harmonic oscillator approximation, the second is the correction for the anharmonicity due to the Morse function. The third and the fourth terms describe the rotational part of the energy. The fifth term takes the interaction of vibration and rotation into account. If still greater accuracy is desired, a term proportional to $(v + 1/2)^3$ or even higher powers may be introduced. The selection rule for J, that is $\Delta J = \pm 1$, is still obeyed in this more complicated model. The vibrational transitions, however, are not restricted to $\Delta v = \pm 1$ but may also differ by larger integral amounts. The transitions due to $\Delta v = \pm 2$, $\Delta v = \pm 3, \ldots$ are very weak. As a matter of terminology, transitions for which $\Delta v = \pm 1$ are known as the fundamental transitions, $\Delta v = \pm 2$ as the first overtone transition or second harmonic, $\Delta v = \pm 3$ as the second overtone or third harmonics and so on.

Now we discuss an energy level diagram of the diatomic molecule on the assumption that the molecular motion can be approximated as a harmonic oscillator and a rigid rotator. First we introduce the definition

$$\tilde{v}_0 = \tilde{v}_e (v' - v'') - x_e \tilde{v}_e \left[\left(v' + \frac{1}{2} \right)^2 - \left(v'' + \frac{1}{2} \right)^2 \right], \qquad v' > v'' \qquad (8.191)$$

describing the vibrational energy difference due to the transition v'' to v' with $v' > v''$. We consider an absorption or emission process between levels having quantum numbers v'' and J'' and v' and J' where v' and v'' have fixed values. For $J' - J'' = 1$ we find

$$\tilde{v}_R = \tilde{v}_0 + 2(J+1)B_e \qquad (8.192)$$

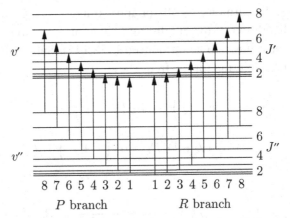

Fig. 8.10 Schematic representation of the first few P and R transitions of the vibration–rotation spectrum (rigid rotator and harmonic oscillator approximation) of a diatomic molecule. After Houghton and Smith (1966).

where J'' has been replaced by J for simplicity. The final two terms in (8.189) have been omitted for simplicity. This transition results in the so-called *R branch* as shown in Figure 8.10. If the rotational transition is $J' - J'' = -1$ then we find

$$\tilde{\nu}_P = \tilde{\nu}_0 - 2J B_e \qquad (8.193)$$

describing the *P branch*.

The fine structure consists of nearly equally spaced lines on each side of the band center where a line is missing since $\Delta J \neq 0$. The line spacing has been exaggerated compared with the spacing of the vibrational levels.

In a very few molecules transitions corresponding to $\Delta J = 0$ are also allowed giving a group of lines which is called the *Q branch*. When discussing rotation of the molecule we have implicitly assumed that there is no angular momentum about the internuclear axis. It is possible, however, that the electrons surrounding the nucleus possess angular momentum about this axis resulting in the selection rule $\Delta J = 0$ so that the *Q* branch occurs. Since all lines with $\Delta J = 0$ are located at nearly the same frequency, a strong line is produced at the center of the band. For additional details see, for example, Houghton and Smith (1966) and Herzberg (1964a,b).

8.5 Matrix elements, selection rules and line intensities

Without derivations we have previously given the selection rules for the one-dimensional harmonic oscillator and the rigid rotator. We will now show in which way these may be obtained.

8.5.1 The harmonic oscillator

In the harmonic oscillator approximation the dipole moment varies linearly with the internuclear distance

$$M = M_0 + M_1(r - r_e) = M_0 + M_1 x = M_0 + \frac{M_1}{\sqrt{\beta}}\xi \qquad (8.194)$$

Here, M_0 is the dipole moment in equilibrium position while $x = r - r_e$ is the change of the internuclear distance. The quantity M_1 is the rate of change of the dipole moment with internuclear distance and $\xi = \sqrt{\beta}x$. The defining relation (8.105) for the matrix element is rewritten as

$$R(v'', v') = M_0 \int_{-\infty}^{\infty} \psi_{v''}^{0*}(x)\psi_{v'}^{0}(x)dx + M_1 \int_{-\infty}^{\infty} x\psi_{v''}^{0*}(x)\psi_{v'}^{0}(x)\,dx \qquad (8.195)$$

now using the vibrational quantum numbers (v'', v'). From (8.145) we repeat the complete time-independent one-dimensional harmonic oscillator wave function

$$\psi_v(\xi) = N_v H_v(\xi)\exp\left(-\frac{\xi^2}{2}\right) \quad \text{with} \quad N_v = \left(\sqrt{\frac{\beta}{\pi}}\frac{1}{2^v v!}\right)^{1/2} \qquad (8.196)$$

This wave function will be introduced into (8.195) in place of the unperturbed wave functions. These form a complete orthonormal set. Thus we find

$$R(v'', v') = M_0 N_{v''} N_{v'} \int_{-\infty}^{\infty} \exp(-\xi^2)H_{v''}(\xi)H_{v'}(\xi)\frac{d\xi}{\sqrt{\beta}}$$
$$+ M_1 \frac{N_{v''} N_{v'}}{\sqrt{\beta}} \int_{-\infty}^{\infty} \xi \exp(-\xi^2)H_{v''}(\xi)H_{v'}(\xi)\frac{d\xi}{\sqrt{\beta}} \qquad (8.197)$$

Due to the orthogonality relations

$$\int_{-\infty}^{\infty} \psi_{v'}^{0*}\psi_{v''}^{0}\,dx = \delta_{v',v''} \qquad (8.198)$$

the first integral vanishes if $v' \neq v''$. In order to evaluate the second integral we introduce the *recursion formula for the Hermite polynomials* as given by

$$\xi H_v(\xi) = v H_{v-1}(\xi) + \frac{1}{2}H_{v+1}(\xi) \qquad (8.199)$$

Due to the orthogonality of the wave functions we immediately find the selection rules

$$\boxed{v' = v'' + 1, \quad v' = v'' - 1, \quad \text{or} \quad v' - v'' = \Delta v = \pm 1} \qquad (8.200)$$

Details of the calculations are left to the exercises.

We are now ready to calculate the matrix elements for the two transitions $\Delta v = \pm 1$. We recall from (8.126) that the line intensity is proportional to the square of the matrix elements. For the one-dimensional harmonic oscillator they are given by the following two equations

$$|R(v'', v')|^2 = M_1^2 \begin{cases} \dfrac{v'' + 1}{2\beta} & \text{for} \quad \Delta v = +1 \\ \dfrac{v''}{2\beta} & \text{for} \quad \Delta v = -1 \end{cases} \tag{8.201}$$

More complicated problems do not have an exact solution and approximate methods must be used. By considering, for example, the harmonic oscillator wave functions as the unperturbed functions it is possible to tackle the anharmonic oscillator as a perturbation problem. We are not going to discuss the procedure which is explained in textbooks on quantum mechanics.

8.5.2 The rigid rotator

For rotational transitions the matrix element may be expressed in the form

$$\mathbf{R}(J'', M'', J', M') = \int \psi_{J'',M''}^{0*} \mathbf{M} \psi_{J',M'}^0 \, d\tau \tag{8.202}$$

where J', M', J'', M'' are the rotational and magnetic quantum numbers of the upper and lower states. The Cartesian components of the dipole moment in the x, y and z directions are given by

$$M_x = \bar{M} \sin\theta \cos\varphi, \quad M_y = \bar{M} \sin\theta \sin\varphi, \quad M_z = \bar{M} \cos\theta \tag{8.203}$$

so that the component matrix elements can be written as follows

(a) $\quad R_x(J'', M'', J', M') = \bar{M} \displaystyle\int_0^{2\pi} \int_0^{\pi} \psi_{J'',M''}^{0*} \psi_{J',M'}^0 \sin\theta \cos\varphi \sin\theta d\theta d\varphi$

(b) $\quad R_y(J'', M'', J', M') = \bar{M} \displaystyle\int_0^{2\pi} \int_0^{\pi} \psi_{J'',M''}^{0*} \psi_{J',M'}^0 \sin\theta \sin\varphi \sin\theta d\theta d\varphi$

(c) $\quad R_z(J'', M'', J', M') = \bar{M} \displaystyle\int_0^{2\pi} \int_0^{\pi} \psi_{J'',M''}^{0*} \psi_{J',M'}^0 \cos\theta \sin\theta d\theta d\varphi$

$$\tag{8.204}$$

We recognize immediately that R_x, R_y, R_z differ from zero, that is emission or absorption by the rotator can occur only when the dipole moment $\bar{M}$ is different from zero.

As an example, we will now consider (8.204c) in some detail by substituting the complete time-independent rigid rotator wave function. Using (8.183) we obtain

$$R_z(J'', M'', J', M') = \bar{M}\frac{N_R'' N_R'}{2\pi} \int_0^\pi P_{J''}^{|M''|}(\cos\theta)\cos\theta\, P_{J'}^{|M'|}(\cos\theta)\sin\theta d\theta$$
$$\times \int_0^{2\pi} e^{-i(M''-M')\varphi}d\varphi \tag{8.205}$$

where N_R'' and N_R' are the normalization factors of the two states described by the first integral. The second integral in (8.205) is zero unless $M' = M'' = M$ and the integral is 2π. To evaluate the first integral we employ the recursion formula for the associated Legendre polynomials

$$\cos\theta\, P_J^{|M|}(\cos\theta) = \frac{J+|M|}{2J+1}P_{J-1}^{|M|}(\cos\theta) + \frac{J-|M|+1}{2J+1}P_{J+1}^{|M|}(\cos\theta) \tag{8.206}$$

and obtain the following equation

$$R_z(J'', M'', J', M') = \bar{M}N_R'' N_R'\left(\frac{J'+|M|}{2J'+1}\int_0^\pi P_{J''}^{|M|}P_{J'-1}^{|M|}\sin\theta\, d\theta\right.$$
$$\left. + \frac{J'-|M|+1}{2J'+1}\int_0^\pi P_{J''}^{|M|}P_{J'+1}^{|M|}\sin\theta d\theta\right) \tag{8.207}$$

Now we recognize immediately that the matrix elements $R_z(J'', M, J', M)$ vanish unless

$$J' - J'' = -1, \quad J' - J'' = +1, \quad \text{or} \quad \Delta J = \pm 1 \tag{8.208}$$

and at the same time $M' = M'' = M$ or $\Delta M = 0$. Omitting details, for the remaining directions we find the selection rules $\Delta M = \pm 1$, $\Delta J = \pm 1$ which show that the selection rules for J are the same for the light polarized in each direction.

In order to obtain the line intensity we must calculate the square of the matrix elements. If $J' = J'' + 1$ then the second integral on the right-hand side of (8.207) is zero and we obtain

$$R_z(J'', |M|, J''+1, |M|) = \bar{M}N_R'' N_R'\frac{N_R'}{N_R''}\frac{J''+1+|M|}{2(J''+1)+1}\int_0^\pi P_{J''}^{|M|}P_{J''}^{|M|}\sin\theta d\theta$$
$$= \bar{M}\frac{N_R'}{N_R''}\frac{J''+1+|M|}{2(J''+1)+1} \tag{8.209}$$
$$= \bar{M}\sqrt{\frac{(J''+1-|M|)(J''+1+|M|)}{(2J''+1)(2J''+3)}}$$

For simplicity we replace J'' by J and finally obtain

$$R_z(J, |M|, J+1, |M|) = \bar{M}\sqrt{\frac{(J+1-|M|)(J+1+|M|)}{(2J+1)(2J+3)}} \qquad (8.210)$$

Setting $J' = J'' - 1$, the first integral is zero and we obtain the expression

$$R_z(J, |M|, J-1, |M|) = \bar{M}\sqrt{\frac{(J+|M|)(J-|M|)}{(2J+1)(2J-1)}} \qquad (8.211)$$

By using proper recursion relations R_x and R_y can be found analogously. The results are given, for example, in Penner (1959). The square of the total matrix element may be evaluated by summing over the components and over all allowed values of M'' and M' as stated in

$$\sum_{M',M''} |\mathbf{R}(J'', M'', J', M')|^2 = \begin{cases} \bar{M}^2(J''+1) & \text{for} \quad \Delta J = +1 \\ \bar{M}^2 J'' & \text{for} \quad \Delta J = -1 \end{cases} \qquad (8.212)$$

For the pure rotation spectrum, transitions in absorption always result from $\Delta J = +1$.

Equations (8.201) and (8.212) referring to the harmonic oscillator and the rigid rotator can be superimposed. We consider the case that absorption takes place from the vibrational state v''. Thus for the combined transition (rotator plus oscillator) we may write

$$\sum_{M',M''} |\mathbf{R}(v'', J'', M'', v''+1, J', M')|^2 = M_1^2 \frac{(v''+1)}{2\beta} \begin{cases} (J''+1) & R \text{ branch} \\ J'' & P \text{ branch} \end{cases}$$

$$(8.213)$$

For additional details see, for example, Penner's (1959) description of the matrix elements for the rotational lines belonging to the rotation–vibration bands.

8.6 Influence of thermal distribution of quantum states on line intensities

Now we return to equation (8.113) which requires information on the population of the energy levels. The population N_i of a level having an energy E_i above the ground state of the molecule is given by the *Maxwell–Boltzmann distribution*

$$N_i = \frac{g_i}{Z} N \exp\left(-\frac{E_i}{kT}\right) \quad \text{with} \quad Z = \sum_i g_i \exp\left(-\frac{E_i}{kT}\right) \qquad (8.214)$$

Here, N is the total number of molecules and Z the sum over all states usually called the *partition function*. The quantity g_i is the statistical weight of the energy level E_i. Herzberg (1964a) provides a table showing that for most diatomic molecules at atmospheric temperatures the number of molecules in the first vibrational level is very small compared to the ground state. Hence practically all transitions in absorption observed in the infrared spectrum have $v'' = 0$ as the initial state. For vibrational transitions it is often possible to ignore the ratio N_n/g_n in comparison to N_m/g_m in (8.126).

From (8.181) follows that the number of molecules N_J in the rotational level J of the lowest vibrational state is given by

$$N_J = \frac{2J+1}{Z_R} N \exp\left(-\frac{BJ(J+1)hc}{kT}\right) \tag{8.215}$$

where $2J + 1$ is the statistical weight, $B = h/(8\pi^2 Ic)$ is the so-called *rotational constant* and

$$Z_R = \sum_J (2J+1) \exp\left(-\frac{BJ(J+1)hc}{kT}\right) \tag{8.216}$$

is the *rotational partition function*. The quantity E_i was obtained from (8.181) which was divided by hc so that E_i is expressed in cm^{-1}. Since B is usually quite small, for sufficiently large T, Z_R may be expressed by an integral which can be evaluated analytically. Setting $x = J(J + 1)$ we find the approximate expression

$$Z_R \approx \int_0^\infty \exp\left(-\frac{Bhcx}{kT}\right) dx = \frac{kT}{Bhc} \tag{8.217}$$

resulting for N_J in the simplified expression

$$N_J \approx \frac{NhcB}{kT}(2J+1) \exp\left(-\frac{BJ(J+1)hc}{kT}\right) \tag{8.218}$$

We may now substitute (8.215) into (8.126), remembering that $g_n = 2J'' + 1$, and find for the line intensity the expression

$$S \approx \frac{2\pi^2 vN}{3h\varepsilon_0 c Z_R} \sum_i \sum_k |R(m_i, n_k)|^2 \exp\left(-\frac{BJ''(J''+1)hc}{kT}\right) \tag{8.219}$$

which is identical with the corresponding expression given by Herzberg (1964b). Observing that Z_R is approximately given by (8.217) and by replacing the double

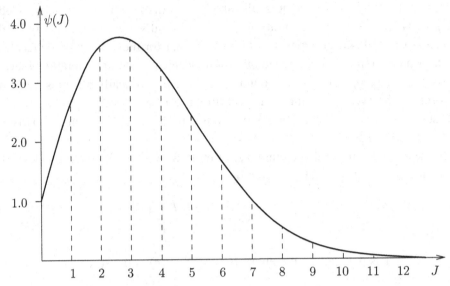

Fig. 8.11 Schematic picture of the thermal distribution of the rotational levels of the HCl molecule in the vibrational ground state, $T = 300$ K. $\psi(J) = (2J + 1)\exp[-BJ(J + 1)hc/(kT)]$. After Herzberg (1964a) *Spectra of Diatomic Molecules.*

sum expression in (8.219) by (8.213) we find that the part of S depending on J'' is given by

$$S_J = \begin{cases} (J + 1)\exp\left(-\dfrac{BJ(J + 1)hc}{kT}\right) & R \text{ branch} \\[3mm] J\exp\left(-\dfrac{BJ(J + 1)hc}{kT}\right) & P \text{ branch} \end{cases} \tag{8.220}$$

where we have replaced J'' by J for brevity. Inspection shows that S_J varies with J almost in the same manner as the population density N_J defined by equation (8.215). For the HCl molecule this variation is depicted in Figure 8.11. Since the factor $(2J + 1)$ varies linearly with J, the number of molecules in the different rotational levels does not from the beginning decrease with the rotational quantum number but goes through a maximum.

8.7 Rotational energy levels of polyatomic molecules

Let us consider the rotation of a polyatomic molecule assuming that it is a rigid structure. By expanding the angular momentum vector (8.12) we find for the vector components J_x, J_y, J_z expressions containing terms of the type $I_x = \sum_i m_i(y_i^2 + z_i^2)$

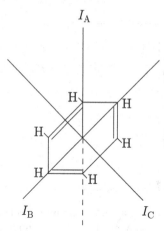

Fig. 8.12 Example of a symmetric top molecule. After Barrow (1962).

and $P_{xy} = \sum_i m_i x_i y_i$. The I_x, I_y, I_z are the moments of inertia about the coordinate axes. The coordinate axes are called *principal axes* for the body at the origin if all the products of inertia P_{xy}, P_{yz}, P_{zx} vanish. If the body (molecule) has symmetry, the direction of one or more of the principal axes going through the center of mass can be found, since axes of symmetry are always principal axes and a plane of symmetry is perpendicular to a principal axis.

The number of rotational quantum numbers which are needed to specify a given rotational state depends on the particular molecular geometry. Only a very brief and incomplete description can be reviewed here. A detailed account is given in *Infrared and Raman Spectra* by Herzberg (1964b).

We will now briefly describe the pure rotation of polyatomic molecules, i.e. we consider non-vibrating molecules in a fixed electronic state. There are four basic types which are distinguished according to their three principal moments of inertia which are denoted by I_A, I_B, I_C. We will give a few atmospheric examples which are taken from Goody (1964a).

(i) Linear molecules (CO_2, N_2O, O_2, N_2, CO): $I_A = 0$, $I_B = I_C \neq 0$
(ii) Symmetric top molecules (no common atmospheric gases): $I_A \neq 0$, $I_B = I_C \neq 0$
(iii) Spherical top molecules (CH_4): $I_A = I_B = I_C$
(iv) Asymmetric top molecules (H_2O, O_3): $I_A \neq I_B \neq I_C$

Of the listed molecules only H_2O has an important pure rotation spectrum. The symmetry property is best demonstrated in case of the benzene molecule which is a symmetric top molecule, see Figure 8.12. The unique moment of inertia of the molecule is usually represented by I_A while the two equal moments of inertia are I_B and I_C.

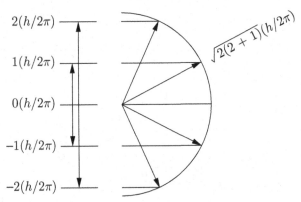

Fig. 8.13 Total and component angular momentum vectors for $J = 2$. The applied field reveals $2J + 1 = 5$ components. Left part: Angular momentum component in direction of an applied field. Right part: angular momentum of rotating molecule. Illustration of the degeneracy $2J + 1$. After Barrow (1962).

8.7.1 Linear molecules

In case of linear molecules and spherical top molecules only one quantum number J is required to describe the rotational state. For linear molecules the solution of Schrödinger's equation for rotation is the same as that for the diatomic molecule. The same selection rules $\Delta J = \pm 1$ are obeyed for dipole transitions. As stated before, unless the molecule possesses a permanent dipole moment no transitions are possible as in case of carbon dioxide which is a symmetrical molecule O–C–O. The N_2O molecule (N–N–O) possesses a permanent dipole moment so that purely rotational transitions are allowed.

As we have previously seen, the rotational energy of the diatomic molecule depends only on the angular momentum quantum number J. The magnitude of the angular momentum itself is given by

$$|\mathbf{J}| = \frac{h}{2\pi}\sqrt{J(J+1)} \tag{8.221}$$

In addition to J, there exists the quantum number $M = 0, \pm 1, \ldots \pm J$. Thus a total of $2J + 1$ wave functions can be constructed. This is called the $2J + 1$ degeneracy of the Jth energy level. If $J = 2$, for example, M assumes the values $-2, -1, 0, 1, 2$ so that five different wave functions can be written down with the help of (8.183). The quantum numbers M enumerate the possible components of the angular momentum $\mathbf{J}$ of the rotating molecule in the direction of an applied (magnetic or electric) field. The resulting angle between $\mathbf{J}$ and the field is not arbitrary, but it is described by the rule that the components in the direction of the applied field are $\pm Mh/2\pi$. For $J = 2$ this is demonstrated in Figure 8.13. The figure implies that the angular momentum vector $\mathbf{J}$ is never in the direction of the applied field since **J** carries out a precessional motion. Thus the maximum component of the angular momentum vector cannot be as large as the vector itself. The reaction of the molecule to an applied field is called *space quantization.*

Since the molecule has an electric dipole moment, the different orientations of **J** relative to the applied field will correspond to different energies. In a magnetic field, all states with different M will have different energies. This observation is called *Zeeman effect.* In case that $J = 2$, energy levels corresponding to $M = 0, \pm 1, \pm 2$ would be observed. In an electric field only those states have different energy which differ in $|M|$. Thus, for $J = 2$ only three different energy levels $M = 0, |\pm 1|, |\pm 2|$ appear. The splitting of the energy levels in the electric field is known as *Stark effect* which often permits precise measurements of the electric dipole moment.

Without going into any details but to demonstrate that the quantum theory describing rotating molecules is very complex, we wish to point out in which way further complications may occur. In addition to the molecular rotation angular momentum, some molecules have angular momentum resulting from the nuclear spin of one or more of their nuclei. A complete description of the rotational states of the molecule must take into account both contributions to the total angular momentum. There exists a nuclear spin quantum number usually denoted by $I = 0, 1/2, 1, 3/2, \ldots$ The spin angular momentum is given by $\sqrt{I(I+1)}h/2\pi$. If there is no interaction (coupling) between the orientation of the nucleus and that of the molecule, the molecule will rotate and leave the spinning nuclei unchanged in orientation. In this case the energy of the given molecular rotational state described in terms of J will not be influenced by the nuclear spin. However, if there is a molecular rotation–nuclear spin coupling the energy of the system will depend on the orientation of the nuclear spin relative to that of the molecular rotation. This coupling will cause a hyperfine splitting of rotational energy values.

8.7.2 Symmetric top molecules

In case of symmetric top molecules, two quantum numbers are needed to specify the rotational states. These are usually denoted by J and K. The axis along which I_A lies is usually called the *figure axis of the molecule.* If $I_A < I_B$ we speak of a prolate symmetric top, and if $I_A > I_B$ an oblate symmetric top. The rotational energy levels for such a molecule are given by

$$E_{J,K} = \frac{h^2}{8\pi^2} \left[\frac{J(J+1)}{I_B} + K^2 \left(\frac{1}{I_A} - \frac{1}{I_B} \right) \right] \tag{8.222}$$

The derivation of this equation can be found, for example, in Pauling and Wilson (1935). The quantum number K has the physical significance that $Kh/2\pi$ is the component along the figure axis of the resultant angular momentum $\sqrt{J(J+1)}h/2\pi$. K takes values between $-J$ and $+J$.

Equation (8.222) verifies that the energy does not depend on the magnetic quantum number $M = 0, \pm1, \pm2, \ldots, \pm J$ and is independent of the sign of K. Therefore, the total degeneracy of the energy level is $2J + 1$ if $K = 0$ and $4J + 2$ if K differs from zero. Most symmetric top molecules that have a dipole moment have this dipole moment in the direction of the figure axis. In such cases the selection rules for the absorption and emission of electromagnetic energy are

$$\begin{aligned} \Delta J &= 0, \pm1, & \Delta K &= 0, & K &\neq 0 \\ \Delta J &= \pm1, & \Delta K &= 0, & K &= 0 \end{aligned} \tag{8.223}$$

For absorption experiments, that part of the selection rules applies which increases the energy of the system, that is

$$\Delta J = +1, \qquad \Delta K = 0 \tag{8.224}$$

8.7.3 Spherical top molecules

The energy levels are given by the same simple formula stated in (8.181). For each value of J there are again $2J + 1$ values of K. Due to the high degree of symmetry, spherical top molecules do not have a permanent dipole moment so that pure rotational spectra do not occur.

8.7.4 Asymmetric top molecules

Theoretical considerations are much more involved than those for linear and symmetric molecules. If we ignore nuclear-spin angular momentum, the total angular momentum is quantized according to the relation (8.221). Unlike the symmetric top molecule, no component of this angular momentum is quantized. No closed general expression is available for any but the lowest few energy values for an asymmetric top molecule. Nevertheless, it is possible to obtain approximate energy values by interpolation involving oblate and prolate symmetric top energy diagrams.

A wealth of information about the approximate formulas of the partition function and the matrix elements can be found in Herzberg's books *Spectra of Diatomic Molecules* (1964a) and *Infrared and Raman Spectra* (1964b) as well as in Penner's book *Quantitative Molecular Spectroscopy and Gas Emissivities* (1959). These books, however, are not introductions to the field of spectroscopy and require much additional reading. A brief treatment for important atmospheric cases is given by Goody (1964a) who assumes some knowledge of spectroscopy. A condensed and quite readable treatment of most of the topics we have introduced in this chapter is given in *Infrared Physics* by Houghton and Smith (1966). Many newer books, too numerous to be mentioned, may serve as guides to get a deeper understanding of the interesting but difficult topics treated in spectroscopy, e.g. *Molecular Spectroscopy* by Barrow (1962), *Molecular Spectroscopy* by Levine (1975), *Molecular Rotation Spectra* by Kroto (1975), *Physikalische Chemie* by Atkins (1993), *Molekülphysik* by Demtröder (2003), and *Lehrbuch der Physikalischen Chemie* by Wedler (2004). Many of the newer books give only very brief outlines of the quantum mechanical derivations or simply state results. For details they often refer to the older textbooks by Pauling and Wilson (1935) and Eyring *et al.* (1965).

8.8 Appendix

8.8.1 The Hamilton function

The Lagrangian function L is defined by

$$L = K(q_k, \dot{q}_k) - V(q_k) \tag{8.225}$$

where K is the kinetic energy and V the potential function. The symbol $q_k (k = 1, 2, \ldots)$ represents the generalized coordinates and $\dot{q}_k$ their time change. Lagrange's equation of motion is defined by

$$\frac{d}{dt}\left(\frac{\partial L}{\partial \dot{q}_k}\right) = \frac{\partial L}{\partial q_k} \tag{8.226}$$

The generalized momenta are

$$p_k = \frac{\partial L}{\partial \dot{q}_k} = \frac{\partial K}{\partial \dot{q}_k} \tag{8.227}$$

Now consider the function

$$H = \sum_k \dot{q}_k p_k - L \tag{8.228}$$

which is a function of the generalized coordinates. Euler's theorem for a homogeneous function f of degree n in the variables $x_1, x_2, \ldots, x_r$ states that

$$\sum_{i=1}^{r} x_i \frac{\partial f}{\partial x_i} = nf \tag{8.229}$$

Application of this theorem to the first term on the right-hand side of (8.228) yields

$$\sum_k \dot{q}_k p_k = \sum_k \dot{q}_k \frac{\partial K}{\partial \dot{q}_k} = 2K \qquad (8.230)$$

which leads to

$$\boxed{H = 2K - K + V = K + V} \qquad (8.231)$$

The function H is known as the *classical Hamilton function*. Detailed treatments of L and H can be found in textbooks on theoretical mechanics, for example in Fowles (1966).

8.8.2 *Macroscopic fields and Maxwell's equations*

In order to describe the absorption and emission of infrared radiation, a number of relations from classical electromagnetic theory will be used which are summarized in this appendix. As before, we will use the mks rationalized units. The electromagnetic state of matter at a given point is described by the following four quantities:

(1) volume density of electric charge ρ;
(2) volume density of electric dipoles, called the *polarization* $\mathbf{P}$;
(3) volume density of magnetic dipoles, called the *magnetization* $\mathbf{M}$;
(4) electric current per unit area, called the *current density* $\mathbf{J}$.

These quantities are considered macroscopically averaged in order to smooth out microscopic variations due to the atomic structure of matter. They are related to the macroscopically averaged electric and magnetic fields $\mathbf{E}$ and $\mathbf{H}$ by *Maxwell's equations*

$$
\begin{aligned}
\nabla \times \mathbf{E} &= -\mu_0 \frac{\partial \mathbf{H}}{\partial t} - \mu_0 \frac{\partial \mathbf{M}}{\partial t} \\
\nabla \times \mathbf{H} &= \varepsilon_0 \frac{\partial \mathbf{E}}{\partial t} + \frac{\partial \mathbf{P}}{\partial t} + \mathbf{J} \\
\nabla \cdot \mathbf{E} &= -\frac{1}{\varepsilon_0}(\nabla \cdot \mathbf{P} - \rho) \\
\nabla \cdot \mathbf{H} &= -\nabla \cdot \mathbf{M}
\end{aligned}
\qquad (8.232)
$$

The constants ε_0 and μ_0 are called the *permittivity* and the *permeability* of the vacuum, respectively. Introducing the abbreviations $\mathbf{D}$ for electric displacement and magnetic induction $\mathbf{B}$

$$\mathbf{D} = \varepsilon_0 \mathbf{E} + \mathbf{P}, \qquad \mathbf{B} = \mu_0(\mathbf{H} + \mathbf{M}) \qquad (8.233)$$

into (8.232) then Maxwell's equations assume the compact form

$$
\begin{array}{ll}
\text{(a)} & \nabla \times \mathbf{E} = -\dfrac{\partial \mathbf{B}}{\partial t} \\[2mm]
\text{(b)} & \nabla \times \mathbf{H} = \dfrac{\partial \mathbf{D}}{\partial t} + \mathbf{J} \\[2mm]
\text{(c)} & \nabla \cdot \mathbf{D} = \rho \\[2mm]
\text{(d)} & \nabla \cdot \mathbf{B} = 0
\end{array}
\tag{8.234}
$$

The response of the conduction electrons to the electric field is given by

$$
\mathbf{J} = \sigma \mathbf{E} \tag{8.235}
$$

where σ is the *electrical conductivity*. Additionally, we introduce the constitutive relation

$$
\mathbf{D} = \varepsilon \mathbf{E} \tag{8.236}
$$

which describes the aggregate response of the bound charges to the electric field and the corresponding magnetic relation

$$
\mathbf{B} = \mu \mathbf{H} \tag{8.237}
$$

Using (8.233) and (8.236) the polarization $\mathbf{P}$ may also be expressed by

$$
\mathbf{P} = (\varepsilon - \varepsilon_0)\mathbf{E} = \chi \varepsilon_0 \mathbf{E} \tag{8.238}
$$

where χ is known as the *electric susceptibility*.

$$
\chi = \frac{\varepsilon}{\varepsilon_0} - 1 \tag{8.239}
$$

This equation is another way of expressing the response of the bound charges to the impressed electric field. For isotropic media such as glass, the susceptibility is a scalar. It is also customary to introduce the so-called *relative permittivity* or *dielectric constant* ε_r and the *relative permeability* μ_r as stated in

$$
\varepsilon_r = \frac{\varepsilon}{\varepsilon_0}, \qquad \mu_r = \frac{\mu}{\mu_0} \tag{8.240}
$$

For non-magnetic media the relative permeability is equal to 1.

Next we are going to separate the electric and magnetic fields $\mathbf{E}$ and $\mathbf{H}$ from the curl relations (8.234a,b). First we take the curl of (8.234a) and the partial derivative of (8.234b) with respect to time. In a region where the volume density of electric charges is zero, we obtain after a few elementary steps the wave equation

$$
\nabla^2 \mathbf{E} - \varepsilon\mu \frac{\partial^2 \mathbf{E}}{\partial t^2} - \mu\sigma \frac{\partial \mathbf{E}}{\partial t} = 0 \tag{8.241}
$$

Analogously we obtain the corresponding wave equation for **H**

$$\nabla^2 \mathbf{H} - \varepsilon\mu\frac{\partial^2 \mathbf{H}}{\partial t^2} - \mu\sigma\frac{\partial \mathbf{H}}{\partial t} = 0 \qquad (8.242)$$

In order to establish two useful operator identities in connection with harmonic waves described by the wave vector **k** and the angular frequency ω, we perform the following operations

$$\frac{\partial}{\partial t}\exp[i(\mathbf{k}\cdot\mathbf{r} - \omega t)] = -i\omega\exp[i(\mathbf{k}\cdot\mathbf{r} - \omega t)]$$

$$\nabla\exp[i(\mathbf{k}\cdot\mathbf{r} - \omega t)] = i\mathbf{k}\exp[i(\mathbf{k}\cdot\mathbf{r} - \omega t)] \qquad (8.243)$$

$$\text{or} \quad \frac{\partial}{\partial t} \to -i\omega, \quad \nabla \to i\mathbf{k}$$

Momentarily we consider a nonconducting medium $\sigma = 0$ and the case $\rho = 0$ so that (8.234) reduces to

$$\nabla \times \mathbf{E} = -\mu\frac{\partial \mathbf{H}}{\partial t}$$

$$\nabla \times \mathbf{H} = \varepsilon\frac{\partial \mathbf{E}}{\partial t} \qquad (8.244)$$

$$\nabla \cdot \mathbf{E} = 0$$

$$\nabla \cdot \mathbf{H} = 0$$

Application of (8.243) to (8.244) leads to Maxwell's equation in the form

$$\mathbf{k} \times \mathbf{E} = \mu\omega\mathbf{H}$$
$$\mathbf{k} \times \mathbf{H} = -\varepsilon\omega\mathbf{E}$$
$$\mathbf{k} \cdot \mathbf{E} = 0 \qquad (8.245)$$
$$\mathbf{k} \cdot \mathbf{H} = 0$$

From the latter set of equations we recognize that the three vectors **E**, **H** and **k** form a mutually orthogonal triad. The electric and the magnetic field vectors are perpendicular to each other, and they are both perpendicular to the wave vector **k** pointing in the direction of propagation. Furthermore, it may be easily seen that the magnitudes of the fields are related by

$$H = \frac{E}{\mu c} = \varepsilon c E \quad \text{since} \quad c = \frac{\omega}{k}, \quad \mathbf{k} = k\mathbf{e}_k \qquad (8.246)$$

Here, $k = 2\pi/\lambda$ is the magnitude of the wave vector, λ is the wavelength, c is the speed of light in the medium and $\mathbf{e}_k$ is the unit vector in the direction of **k**.

The time rate of flow of electromagnetic energy per unit area is given by the *Poynting vector* which is defined as the cross product of the field vectors

$$\mathbf{N} = \mathbf{E} \times \mathbf{H} \tag{8.247}$$

By considering $\mathbf{E}$ and $\mathbf{H}$ as real plane harmonic waves, i.e.

$$\mathbf{E} = \mathbf{E}_0 \cos(\mathbf{k} \cdot \mathbf{r} - \omega t), \qquad \mathbf{H} = \mathbf{H}_0 \cos(\mathbf{k} \cdot \mathbf{r} - \omega t) \tag{8.248}$$

the Poynting vector $\mathbf{N}$ is given by

$$\mathbf{N} = \mathbf{E}_0 \times \mathbf{H}_0 \cos^2(\mathbf{k} \cdot \mathbf{r} - \omega t) \tag{8.249}$$

In the mks-system of units $\mathbf{N}$ is expressed in Watts per square meter.

Since the average value of the cosine squared is $1/2$, we find that the average value of the Poynting vector is given by

$$< \mathbf{N} > = \frac{1}{2}\mathbf{E}_0 \times \mathbf{H}_0 = \frac{E_0^2}{2\mu\omega}\mathbf{k} = \frac{E_0^2}{2\mu c}\mathbf{e}_k = I\mathbf{e}_k \tag{8.250}$$

Hence, it is seen that in isotropic media the direction of $\mathbf{N}$ and $\mathbf{k}$ coincide. For nonisotropic media such as crystals this is not always the case. The magnitude of the averaged Poynting vector is the *intensity I*, also called *irradiance*. The intensity is proportional to the square of E_0. In a conducting medium there is a phase difference between $\mathbf{E}$ and $\mathbf{H}$. Nevertheless, the intensity I is still proportional to E_0^2.

Now we return to the wave equation (8.241). It is easy to verify that the function

$$E = E_0 \exp[i(kz - \omega t)] \tag{8.251}$$

is a solution of (8.241) provided that the following relation is satisfied

$$k^2 = \omega^2 \mu \left(\varepsilon + \frac{i\sigma}{\omega} \right) \tag{8.252}$$

This solution refers to propagation along the z-axis so that z is the distance from a fixed origin.

Finally, we wish to relate (8.252) to the *index of refraction*. For a nonabsorbing medium the refractive index is defined as $\mathcal{N} = c/v$, where c is the speed of light in vacuum. Since for an absorbing medium k is a complex expression, the index of refraction is also a complex quantity and is given by

$$\mathcal{N} = n + i\kappa = \frac{ck}{\omega} \quad \text{where} \quad c = \frac{1}{\sqrt{\varepsilon_0 \mu_0}} \tag{8.253}$$

By separating the real and imaginary parts of the complex refractive index we find from (8.252) and (8.253) the important relations

$$\boxed{n^2 - \kappa^2 = \varepsilon_r \mu_r, \qquad 2n\kappa = \frac{\sigma \mu_r}{\varepsilon_0 \omega}} \tag{8.254}$$

Expressing the wave number k in the solution (8.251) in terms of the complex index of refraction we obtain

$$E = E_0 \exp\left(-\frac{\omega \kappa z}{c}\right) \exp\left[i\omega\left(\frac{nz}{c} - t\right)\right] \tag{8.255}$$

The first exponential factor describes the attenuation along the z-axis from an arbitrary origin, while the second exponential defines the shape of the wave.

By introducing in (8.250) the attenuation along the z-axis we find

$$I = \frac{1}{2\mu c} E_0^2 \exp\left(-\frac{2\omega \kappa z}{c}\right) \tag{8.256}$$

Denoting the initial value of the intensity by $I(z = 0) = I_0$ we obtain *Beer's law*

$$I = I_0 \exp\left(-\frac{2\omega \kappa z}{c}\right) = I_0 \exp(-\alpha z) \tag{8.257}$$

where the absorption coefficient $\alpha = 4\pi \kappa / \lambda$ is expressed as the reciprocal of length.

More complete treatments of the equations of electromagnetic theory are given in standard textbooks on electrodynamics and in related books. Our reference goes to Fowles (1966) and to Houghton and Smith (1966).

8.9 Problems

8.1: In connection with Figure 8.5, show that $y_1 = y_3$.

8.2: Verify equations (8.28) and (8.29).

8.3: Prove equation (8.32).

8.4: Show the validity of equations (8.33).

8.5: Verify equations (8.37), (8.38) and (8.41).

8.6: Show by direct computations that the wave functions for the simple harmonic oscillator are normalized. Use the special cases $n = 1$ and $n = 2$. Show also by direct computations that the wave functions ψ_1 and ψ_2 are orthogonal.

8.7: Show that the selection rules for the harmonic oscillator are given by (8.200).

8.8: Verify that the squares of the matrix elements for the harmonic oscillator are given by (8.201).

8.9: Show the validity of equations (8.210) and (8.211).

9

Light scattering theory for spheres

9.1 Introduction

The evaluation of the radiative transfer equation requires a detailed knowledge of the extinction and scattering properties of atmospheric particles. In most cases we rely on theoretical calculations assuming that the scattering particles have a spherical shape. In order to carry out the calculations we must specify the particle size and the wavelength-dependent *complex index of refraction*. The required calculations are known as *Mie calculations* while the entire theory is called the *Mie theory*. This goes back to Gustav Mie who published the whole theory in 1908. Nowadays a large number of reliable computer programs are available to provide the required information without understanding the theory behind it. This might be sufficient in some cases, but it is more valuable to comprehend the theoretical background which will be presented in the following sections.

The Mie theory of light scattering by homogeneous spheres is based on the formal solution of *Maxwell's equations* using proper boundary conditions. In this text we will follow Stratton's discussion of the *Electromagnetic Theory* (1941). In van De Hulst's (1957) treatise *Light Scattering by Small Particles* many mathematical details of the Mie theory are omitted, but numerous helpful physical explanations are given. In particular he gives a number of approximate formulas for special cases. Due to modern computers the approximate formulas are rarely used since the Mie series solutions can now be evaluated efficiently for practically all situations of interest. A newer more complete treatment of the Mie theory is given, for example, by Bohren and Huffman (1983). We should not overlook the important reference *Principles of Optics* by Born and Wolf (1965) where a detailed treatment of the theory can also be found. While the theory is named after Mie (1908), some of the notations which we will use were introduced by Debye (1909). Special cases of the scattering theory are discussed in various textbooks on electromagnetic theory.

We assume that the student has some familiarity with Bessel functions since they are introduced in most introductory courses on differential equations. The Mie theory makes use of the elements of the so-called *field theory*. In the present case this means that if the scalar function ρ and the vector function **J** denoting the charge density and the distribution of currents are known, we may seek a solution for the vectors **E** and **H** of the electric and the magnetic field by using *Maxwell's equations*. Finally, the entire problem may be reduced to the solution of a standard partial differential equation. A brief and simple but very illuminating mathematical introduction to this type of problem can be found, for example, in Chapter 8 of Buck's *Advanced Calculus* (1965).

A full treatment of the entire Mie theory will now be presented giving all required mathematical details.

9.2 Maxwell's equations

As stated above, the Mie formulas are an exact mathematical solution of Maxwell's equations which contain the basic physics of the problem. From Appendix 8.8.2 of the previous chapter we restate the curl equations (8.234a,b)

$$\nabla \times \mathbf{E} + \frac{\partial \mathbf{B}}{\partial t} = 0, \qquad \nabla \times \mathbf{H} - \frac{\partial \mathbf{D}}{\partial t} - \mathbf{J} = 0 \tag{9.1}$$

The relations involving the vectors **B**, **D**, **E**, **H**, **J** are repeated as

$$\nabla \cdot \mathbf{B} = 0, \qquad \nabla \cdot \mathbf{D} = \rho, \qquad \mathbf{J} = \sigma \mathbf{E}, \qquad \mathbf{D} = \varepsilon \mathbf{E}, \qquad \mathbf{B} = \mu \mathbf{H} \tag{9.2}$$

According to (8.241) and (8.242) the wave equations for **E** and **H** are given by

$$\nabla^2 \mathbf{E} - \varepsilon\mu \frac{\partial^2 \mathbf{E}}{\partial t^2} - \mu\sigma \frac{\partial \mathbf{E}}{\partial t} = 0$$
$$\nabla^2 \mathbf{H} - \varepsilon\mu \frac{\partial^2 \mathbf{H}}{\partial t^2} - \mu\sigma \frac{\partial \mathbf{H}}{\partial t} = 0 \tag{9.3}$$

As usual, we assume that the time dependency of the field vectors is harmonic. Thus we may write for **E** and **H**

$$\mathbf{E} = \tilde{\mathbf{E}} \exp(-i\omega t), \qquad \mathbf{H} = \tilde{\mathbf{H}} \exp(-i\omega t) \tag{9.4}$$

where $\tilde{\mathbf{E}}$ and $\tilde{\mathbf{H}}$ denote the time-independent parts. Substituting (9.4) into (9.3) we find

$$\nabla^2 \mathbf{E} + k^2 \mathbf{E} = 0, \qquad \nabla^2 \mathbf{H} + k^2 \mathbf{H} = 0 \tag{9.5}$$

where we have introduced the square of the wave number

$$k^2 = \omega^2 \left(\varepsilon\mu + \frac{i\mu\sigma}{\omega} \right) \tag{9.6}$$

The following quantities will also be needed in our derivation:

$$
\begin{aligned}
\text{speed of light in vacuum:} \quad & c = \frac{1}{\sqrt{\varepsilon_0\mu_0}} \\
\text{speed of light in a nonconducting medium:} \quad & v = \frac{1}{\sqrt{\varepsilon\mu}} \\
\text{wave number in vacuum:} \quad & k_0 = \frac{\omega}{c} \\
\text{wave number in a nonconducting medium:} \quad & k = \frac{\omega}{v} \\
\text{\textit{complex index of refraction:}} \quad & \mathcal{N} = c\sqrt{\left(\varepsilon\mu + \frac{i\mu\sigma}{\omega} \right)}
\end{aligned}
\tag{9.7}
$$

The latter equation follows from (8.253). The imaginary part of $\mathcal{N}$ indicates that absorption is taking place. From (9.6) and (9.7) one may easily see that

$$k^2 = k_0^2 \mathcal{N}^2 \tag{9.8}$$

From (9.7) it is easily seen that in a dielectric medium with $\sigma = 0$ the wave number reduces to $k = \omega/v$. In a nonabsorbing medium the imaginary part of $\mathcal{N}$ vanishes, that is $\kappa = 0$ in (8.253). The last equation of (9.7) leads to the conclusion that every conductor absorbs electromagnetic waves while every dielectric (insulator) is transparent. Practical examples can be given to contradict this statement. The reason is that not only the refractive index depends on frequency but also the conductivity.

Utilizing (9.8) we may also write (9.5) as

$$\boxed{\nabla^2 \mathbf{E} + k_0^2 \mathcal{N}^2 \mathbf{E} = 0, \qquad \nabla^2 \mathbf{H} + k_0^2 \mathcal{N}^2 \mathbf{H} = 0} \tag{9.9}$$

For these two differential equations we assume the trial solutions

$$\mathbf{E} = \mathbf{E}_0 \exp\left[i(\mathbf{k}\cdot\mathbf{r} - \omega t) \right], \qquad \mathbf{H} = \mathbf{H}_0 \exp\left[i(\mathbf{k}\cdot\mathbf{r} - \omega t) \right] \tag{9.10}$$

where $\mathbf{k}$ and $\mathbf{r}$ are the wave number and position vectors. If $\mathbf{e}_k$ is a unit vector in the direction of wave propagation then the wave number vector is given by

$$\mathbf{k} = k\mathbf{e}_k = k_0 \mathcal{N} \mathbf{e}_k \tag{9.11}$$

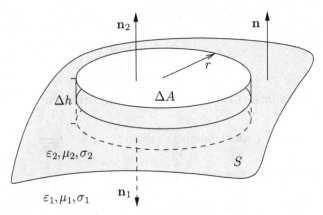

Fig. 9.1 Illustration for the derivation of the normal boundary conditions for **B** and **D**.

9.3 Boundary conditions

The field equations are valid for ordinary points in space in whose neighborhood the field vectors change continuously. Across any surface separating two media sharp changes may occur in the parameters characterizing the media so that the field vectors are expected to exhibit corresponding changes. Let these parameters be denoted by ε_1, μ_1, σ_1 and ε_2, μ_2, σ_2.

For the solution of the wave equations suitable boundary conditions must be supplied. The boundary surface separating the two media is denoted by the symbol S and has the normal unit vector **n**, see Figure 9.1. We imagine a very thin transition layer in which the parameters change rapidly but continuously from their values near S in (1) to their values near S in (2). Within this transition layer as well as in (1) and (2) the field vectors and their first derivatives are assumed to be continuous bounded functions of position and time. Through the transition layer we now construct a small right cylinder of height Δh and cross-section area ΔA. The surface normal unit vectors $\mathbf{n}_1$, $\mathbf{n}_2$ at the bottom and the top of the cylinder have the directions shown in the figure.

Integrating the divergence equations of (9.2) over the volume $\Delta V = \Delta h \Delta A$ of the cylinder and utilizing the divergence theorem of Gauss yields

$$\int_{\Delta V} \nabla \cdot \mathbf{B} \, dV = \oint \mathbf{B} \cdot d\mathbf{A} = 0$$
$$\int_{\Delta V} \nabla \cdot \mathbf{D} \, dV = \oint \mathbf{D} \cdot d\mathbf{A} = q = \int_{\Delta V} \rho \, dV$$

(9.12)

where the integrals over $d\mathbf{A}$ are taken over the closed cylinder surface. For the evaluation of the surface integrals we assume that with vanishing height

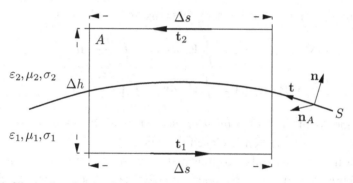

Fig. 9.2 Illustration for the derivation of the tangential boundary conditions for **B** and **D**.

$\Delta h \to 0$ of the cylinder the contribution of the walls may be ignored. Thus we may write

$$\oint \mathbf{B} \cdot d\mathbf{A} = \oint_{\Delta A} \mathbf{B}_1 \cdot \mathbf{n}_1 dA + \oint_{\Delta A} \mathbf{B}_2 \cdot \mathbf{n}_2 dA = 0$$
$$\oint \mathbf{D} \cdot d\mathbf{A} = \oint_{\Delta A} \mathbf{D}_1 \cdot \mathbf{n}_1 dA + \oint_{\Delta A} \mathbf{D}_2 \cdot \mathbf{n}_2 dA = \bar{q}\Delta A$$

(9.13)

Here we have introduced the surface charge density $\bar{q} = q/\Delta A = \rho \Delta h$ which is the charge of the cylinder per unit area. We observe that the total charge q remains constant since it cannot be destroyed. As $\Delta h \to 0$ the volume charge becomes infinite.

Since $\mathbf{n}_2 = -\mathbf{n}_1 = \mathbf{n}$, see Figure 9.1, we have

$$\oint_{\Delta A} (\mathbf{B}_2 - \mathbf{B}_1) \cdot \mathbf{n} dA = 0, \qquad \oint_{\Delta A} (\mathbf{D}_2 - \mathbf{D}_1) \cdot \mathbf{n} dA = \bar{q}\Delta A \qquad (9.14)$$

Assuming that the field vectors are constant within a small area element ΔA we finally obtain the boundary conditions for the normal components of **B** and **D**

$$\boxed{(\mathbf{B}_2 - \mathbf{B}_1) \cdot \mathbf{n} = 0, \qquad (\mathbf{D}_2 - \mathbf{D}_1) \cdot \mathbf{n} = \bar{q}} \qquad (9.15)$$

These equations state that the normal component of **B** across any discontinuity surface is continuous. In contrast, the normal component of **D** experiences an abrupt change whose magnitude is $\bar{q}$.

In order to express the behavior of the tangential components of **B** and **D** at the discontinuity surface S we consider a closed surface A with normal unit vector $\mathbf{n}_A$, see Figure 9.2. The tangential unit vectors along the lower and upper boundary of A are denoted by $\mathbf{t}_1$ and $\mathbf{t}_2$.

Integrating (9.1) over this surface yields together with (9.2)

$$\int_A \nabla \times \mathbf{E} \cdot \mathbf{n}_A dA = \oint \mathbf{E} \cdot ds = -\int_A \frac{\partial \mathbf{B}}{\partial t} \cdot \mathbf{n}_A dA$$
$$\int_A \nabla \times \mathbf{H} \cdot \mathbf{n}_A dA = \oint \mathbf{H} \cdot ds = \int_A \left(\frac{\partial \mathbf{D}}{\partial t} + \mathbf{J}\right) \cdot \mathbf{n}_A dA \tag{9.16}$$

where by means of the *Stokes integral theorem* in each equation the first surface integrals have been rewritten as closed line integrals.

We split the line integrals into four parts and assume that in the limit $\Delta h \to 0$ the contributions from the corresponding sides vanish, see Figure 9.2. Hence we obtain

$$\int_{\Delta s} \mathbf{E}_1 \cdot \mathbf{t}_1 ds + \int_{\Delta s} \mathbf{E}_2 \cdot \mathbf{t}_2 ds = -\int_A \frac{\partial \mathbf{B}}{\partial t} \cdot \mathbf{n}_A dA$$
$$\int_{\Delta s} \mathbf{H}_1 \cdot \mathbf{t}_1 ds + \int_{\Delta s} \mathbf{H}_2 \cdot \mathbf{t}_2 ds = \int_A \left(\frac{\partial \mathbf{D}}{\partial t} + \mathbf{J}\right) \cdot \mathbf{n}_A dA \tag{9.17}$$

For a small surface A we have $\mathbf{t}_2 = -\mathbf{t}_1 = \mathbf{t} = \mathbf{n}_A \times \mathbf{n}$ so that

$$\int_{\Delta s} (\mathbf{E}_2 - \mathbf{E}_1) \cdot \mathbf{n}_A \times \mathbf{n} ds = -\int_A \frac{\partial \mathbf{B}}{\partial t} \cdot \mathbf{n}_A dA$$
$$\int_{\Delta s} (\mathbf{H}_2 - \mathbf{H}_1) \cdot \mathbf{n}_A \times \mathbf{n} ds = \int_A \left(\frac{\partial \mathbf{D}}{\partial t} + \mathbf{J}\right) \cdot \mathbf{n}_A dA \tag{9.18}$$

By assuming that for small values of A all integrands are constant, we obtain from (9.18)

$$(\mathbf{E}_2 - \mathbf{E}_1) \cdot \mathbf{n}_A \times \mathbf{n} = -\frac{\partial \mathbf{B}}{\partial t} \cdot \mathbf{n}_A \Delta h$$
$$(\mathbf{H}_2 - \mathbf{H}_1) \cdot \mathbf{n}_A \times \mathbf{n} = \left(\frac{\partial \mathbf{D}}{\partial t} + \mathbf{J}\right) \cdot \mathbf{n}_A \Delta h \tag{9.19}$$

For finite values of the current $\mathbf{J}$ the right-hand sides of these equations vanish in the limit $\Delta h \to 0$ since the field vectors $\mathbf{B}$ and $\mathbf{D}$ and their derivatives with respect to time remain bounded. Thus, we finally obtain the result that the tangential components of the field vectors are continuous across a discontinuity surface S, i.e.

$$\boxed{\mathbf{n} \times (\mathbf{E}_2 - \mathbf{E}_1) = 0, \qquad \mathbf{n} \times (\mathbf{H}_2 - \mathbf{H}_1) = 0} \tag{9.20}$$

Equations (9.15) and (9.20) describe the boundary conditions which are necessary to solve the differential equations (9.9). We will apply these equations to the scattering of radiation by a sphere (medium 2) which is embedded in the atmosphere (medium 1). The boundary between the two media is the spherical surface of the scatterer. Thus, it will be expedient to state the components of the field vectors in spherical coordinates.

9.4 The solution of the wave equation

9.4.1 Solution of the scalar wave equation in spherical coordinates

The solution of the *scalar form of the wave equation* is instrumental in the construction of the solution of the *vector wave equation*. Given is the scalar form of the wave equation for the scalar quantity $u(x, y, z, t)$

$$\nabla^2 u = \frac{1}{c^2} \frac{\partial^2 u}{\partial t^2} \tag{9.21}$$

For this equation we wish to express the solution in spherical coordinates. Assuming that u can be written in the form

$$u(x, y, z, t) = u^*(x, y, z) \exp(-i\omega t) \tag{9.22}$$

we find

$$\nabla^2 u^* + k^2 u^* = 0 \quad \text{with} \quad k = \frac{\omega}{c} \tag{9.23}$$

and in spherical coordinates

$$\frac{\partial^2 u^*}{\partial r^2} + \frac{2}{r} \frac{\partial u^*}{\partial r} + \frac{1}{r^2 \sin \vartheta} \frac{\partial}{\partial \vartheta} \left(\sin \vartheta \frac{\partial u^*}{\partial \vartheta} \right) + \frac{1}{r^2 \sin^2 \vartheta} \frac{\partial^2 u^*}{\partial \varphi^2} + k^2 u^* = 0 \tag{9.24}$$

Defining the angular part of the Laplacian by

$$\nabla_1^2 u^* = \frac{1}{\sin \vartheta} \frac{\partial}{\partial \vartheta} \left(\sin \vartheta \frac{\partial u^*}{\partial \vartheta} \right) + \frac{1}{\sin^2 \vartheta} \frac{\partial^2 u^*}{\partial \varphi^2} \tag{9.25}$$

the wave equation can be written as

$$\frac{\partial^2 u^*}{\partial r^2} + \frac{2}{r} \frac{\partial u^*}{\partial r} + \frac{1}{r^2} \nabla_1^2 u^* + k^2 u^* = 0 \tag{9.26}$$

We assume that the solution to (9.26) can be written in the form

$$u^*(r, \vartheta, \varphi) = f(r) F(\vartheta, \varphi) \tag{9.27}$$

Substitution of (9.27) into (9.26) yields

$$\frac{r^2}{f} \frac{d^2 f}{dr^2} + \frac{2r}{f} \frac{df}{dr} + k^2 r^2 = -\frac{1}{F} \nabla_1^2 F \tag{9.28}$$

By the method of separation of the variables each side must be equal to a constant C. In order to guarantee unique solutions of these equations we set $C = n(n+1)$ with $n = 0, 1, \ldots$ For details see, for example, Smirnow (1959). Thus we obtain

(a) $$\nabla_1^2 F + CF = 0$$
$$C = n(n+1), \quad n = 0, 1, \ldots \quad (9.29)$$

(b) $$\frac{r^2}{f}\frac{d^2 f}{dr^2} + \frac{2r}{f}\frac{df}{dr} + k^2 r^2 - C = 0$$

The solution to (9.29a) is well-known and is given by

$$F_n(\vartheta, \varphi) = a_n P_n(\cos\vartheta) + \sum_{m=1}^{n} [a_{mn}\cos(m\varphi) + b_{mn}\sin(m\varphi)]\, P_n^m(\cos\vartheta) \quad (9.30)$$

In order to obtain a more suitable form of (9.29b) we introduce the substitutions

$$f(r) = \frac{R_n(r)}{\sqrt{r}}, \qquad \rho = kr \quad (9.31)$$

yielding

$$\frac{d^2 R_n}{d\rho^2} + \frac{1}{\rho}\frac{dR_n}{d\rho} + \left(1 - \frac{(n+1/2)^2}{\rho^2}\right) R_n = 0 \quad (9.32)$$

The quantity R_n is a new variable. Equation (9.32) is known as *Bessel's differential equation* of order $(n + 1/2)$.

Using the notation

$$R_n(\rho) = Z_{n+1/2}(\rho) \quad (9.33)$$

the solution to (9.29b) is given by

$$f_n(r) = \frac{1}{\sqrt{r}} Z_{n+1/2}(kr) \quad (9.34)$$

Employing the latter relation, the solution of the scalar wave equation in spherical coordinates is written as

$$u(\rho, \vartheta, \varphi, t) = \sqrt{\frac{\pi}{2\rho}} Z_{n+1/2}(\rho) F_n(\vartheta, \varphi) \exp(-i\omega t) \quad (9.35)$$

where for convenience we have included the factor $\sqrt{\pi/2k}$. This does not change the solution at all since this factor multiplies the integration constants. By introducing the so-called *spherical Bessel functions*

$$z_n(\rho) = \sqrt{\frac{\pi}{2\rho}} Z_{n+1/2}(\rho) \quad (9.36)$$

the elementary or characteristic wave functions can finally be written as

$$
\boxed{
\begin{aligned}
u^{\mathrm{e}}_{mn}(\rho, \vartheta, \varphi, t) &= \cos(m\varphi) P^m_n(\cos \vartheta) z_n(\rho) \exp(-i\omega t) \\
u_{\mathrm{o},mn}(\rho, \vartheta, \varphi, t) &= \sin(m\varphi) P^m_n(\cos \vartheta) z_n(\rho) \exp(-i\omega t)
\end{aligned}
\quad , \quad m = 0, 1, \ldots, n
}
$$

$$(9.37)$$

Here we have distinguished between the even and odd form of the wave functions u^{e}_{mn} and $u_{\mathrm{o},mn}$ which are denoted according to the even and odd functions $\cos(m\varphi)$ and $\sin(m\varphi)$ occurring in (9.30). Everywhere on the surface of the sphere the elementary wave functions $u^{\mathrm{e}}_{\mathrm{o},mn}$ are finite and single-valued, see also Stratton (1941).

9.4.2 Solution of the vector wave equation in spherical coordinates

Let us reconsider the vector wave equations (9.5). Only if **E** and **H** are resolved in terms of rectangular components we obtain three independent scalar vector wave equations of the form (9.23) for each component of the vectors. In order to solve the vector wave equations (9.5) in spherical coordinates, we make use of the following two theorems.

(a) If ψ satisfies the scalar wave equation (9.23) then the vectors defined by

$$
\mathbf{M}_\psi = \nabla \times (\mathbf{r}\psi), \qquad \mathbf{N}_\psi = \frac{1}{k}\nabla \times \mathbf{M}_\psi
\tag{9.38}
$$

satisfy the vector wave equations

$$
\nabla^2 \mathbf{M}_\psi + k^2 \mathbf{M}_\psi = 0, \qquad \nabla^2 \mathbf{N}_\psi + k^2 \mathbf{N}_\psi = 0
\tag{9.39}
$$

Furthermore, the following relation is valid

$$
\mathbf{M}_\psi = \frac{1}{k}\nabla \times \mathbf{N}_\psi
\tag{9.40}
$$

The quantities **M** and **N** are called *vector wave functions*. The proof of the above statements will be left to the exercises. It will be observed that **M** and **N** are solenoidal vectors, i.e. they satisfy

$$
\nabla \cdot \mathbf{M}_\psi = 0, \qquad \nabla \cdot \mathbf{N}_\psi = 0
\tag{9.41}
$$

(b) If u and v are two solutions of the scalar wave equation, then the vectors $\mathbf{M}_u, \mathbf{M}_v$, $\mathbf{N}_u$ and $\mathbf{N}_v$ represent the derived vector fields which satisfy the vector wave equation. The two vectors **A** and **B** defined by

$$
\mathbf{A} = \mathbf{M}_v - i\mathbf{N}_u, \qquad \mathbf{B} = -\mathcal{N}(\mathbf{M}_u + i\mathbf{N}_v)
\tag{9.42}
$$

satisfy

$$
\nabla \times \mathbf{A} = ik_0 \mathbf{B}, \qquad \nabla \times \mathbf{B} = -ik_0\mathcal{N}^2 \mathbf{A}
\tag{9.43}
$$

In order to evaluate the boundary conditions we must know the components of the **M** and **N** functions. Let us consider the curl of the arbitrary vector **A** which is given by the well-known formula

$$\nabla \times \mathbf{A} = \frac{1}{r^2 \sin \vartheta} \begin{vmatrix} r\mathbf{e}_\vartheta & r\sin\vartheta\,\mathbf{e}_\varphi & \mathbf{e}_r \\ \dfrac{\partial}{\partial \vartheta} & \dfrac{\partial}{\partial \varphi} & \dfrac{\partial}{\partial r} \\ rA_\vartheta & r\sin\vartheta\,A_\varphi & A_r \end{vmatrix} \tag{9.44}$$

Recalling the definition (9.38), we identify the vector **A** by $\psi\mathbf{r} = \psi r\mathbf{e}_r$ where $\mathbf{e}_r$ is the unit vector in direction **r**. Thus we obtain for the components of **M** and **N**

$$M_{\psi,\vartheta} = \frac{1}{\sin\vartheta}\frac{\partial\psi}{\partial\varphi}, \qquad M_{\psi,\varphi} = -\frac{\partial\psi}{\partial\vartheta}, \qquad\qquad M_{\psi,r} = 0$$

$$N_{\psi,\vartheta} = \frac{1}{kr}\frac{\partial^2 r\psi}{\partial r\,\partial\vartheta}, \qquad N_{\psi,\varphi} = \frac{1}{kr\sin\vartheta}\frac{\partial^2 r\psi}{\partial r\,\partial\varphi} \tag{9.45}$$

$$N_{\psi,r} = -\frac{1}{kr\sin\vartheta}\left[\frac{\partial}{\partial\vartheta}\left(\sin\vartheta\frac{\partial\psi}{\partial\vartheta}\right) + \frac{1}{\sin\vartheta}\frac{\partial^2\psi}{\partial\varphi^2}\right]$$

The final expression in (9.45) can be stated in a much simpler form by involving the scalar wave equation in spherical coordinates (9.24) as well as (9.29b). This immediately leads to

$$N_{\psi,r} = kr\psi + \frac{r}{k}\frac{\partial^2\psi}{\partial r^2} + \frac{2}{k}\frac{\partial\psi}{\partial r} = \frac{n(n+1)\psi}{kr} \tag{9.46}$$

These relations will be needed for the evaluation of the boundary conditions.

The **M** and **N** functions obey some very desirable and helpful orthogonality relations which will be derived now. We start the analysis by assuming a harmonic time dependence of the two functions

$$\mathbf{M}_\psi = \mathbf{m}_\psi\exp(-i\omega t), \qquad \mathbf{N}_\psi = \mathbf{n}_\psi\exp(-i\omega t) \tag{9.47}$$

With the help of (9.37) and (9.45) we obtain for the **m** functions

$$\begin{aligned}
\mathbf{m}_{o,mn}^e(r,\vartheta,\varphi) &= \frac{1}{\sin\vartheta}\frac{\partial\psi_{o,mn}^e}{\partial\varphi}\mathbf{e}_\vartheta - \frac{\partial\psi_{o,mn}^e}{\partial\vartheta}\mathbf{e}_\varphi \\[2ex]
&= \begin{pmatrix} -\sin(m\varphi) \\ \cos(m\varphi) \end{pmatrix}\frac{m}{\sin\vartheta}P_n^m(\cos\vartheta)z_n(kr)\mathbf{e}_\vartheta \\[2ex]
&\quad - \begin{pmatrix} \cos(m\varphi) \\ \sin(m\varphi) \end{pmatrix}\frac{dP_n^m}{d\vartheta}z_n(kr)\mathbf{e}_\varphi
\end{aligned} \tag{9.48}$$

Thus it is seen that for $m = 0$ the odd function $\mathbf{m}_{o,m=0,n} = 0$.

To find the orthogonality conditions of the **M** functions, we multiply $\mathbf{m}^e_{0,mn}$ by $\mathbf{m}^e_{0,ml}$. Integrating the product over the unit sphere yields

$$\int_0^{2\pi} \int_0^{\pi} \mathbf{m}^e_{0,mn} \cdot \mathbf{m}^e_{0,ml} \sin \vartheta\, d\vartheta\, d\varphi$$

$$= z_n(kr) z_l(kr) \int_0^{2\pi} \begin{pmatrix} \sin^2(m\varphi) \\ \cos^2(m\varphi) \end{pmatrix} d\varphi \int_0^{\pi} \frac{m^2 P_n^m(\cos\vartheta) P_l^m(\cos\vartheta)}{\sin\vartheta} d\vartheta$$

$$+ z_n(kr) z_l(kr) \int_0^{2\pi} \begin{pmatrix} \cos^2(m\varphi) \\ \sin^2(m\varphi) \end{pmatrix} d\varphi \int_0^{\pi} \frac{d P_n^m}{d\vartheta} \frac{d P_l^m}{d\vartheta} \sin\vartheta\, d\vartheta \qquad (9.49)$$

From textbooks discussing Legendre polynomials we extract the important relation

$$m^2 \int_0^{\pi} \frac{P_n^m(\cos\vartheta) P_l^m(\cos\vartheta)}{\sin\vartheta} d\vartheta + \int_0^{\pi} \frac{d P_n^m}{d\vartheta} \frac{d P_l^m}{d\vartheta} \sin\vartheta\, d\vartheta$$

$$= \frac{2n(n+1)}{2n+1} \frac{(n+m)!}{(n-m)!} \delta_{n,l} \qquad (9.50)$$

By recalling the well-known orthogonality relations of the trigonometric functions

$$\frac{1}{\pi} \int_0^{2\pi} \cos(mx)\cos(kx)dx = (1 + \delta_{0,m})\delta_{m,k}$$

$$\frac{1}{\pi} \int_0^{2\pi} \sin(mx)\sin(kx)dx = (1 - \delta_{0,m})\delta_{m,k} \qquad (9.51)$$

the orthogonality relation of the **m** functions can finally be stated as

$$\boxed{\begin{aligned} & \int_0^{2\pi} \int_0^{\pi} \mathbf{m}^e_{0,mn} \cdot \mathbf{m}^e_{0,mn} \sin\vartheta\, d\vartheta\, d\varphi \\ & = \pi \begin{pmatrix} (1+\delta_{0,m}) \\ (1-\delta_{0,m}) \end{pmatrix} \frac{2n(n+1)}{2n+1} \frac{(n+m)!}{(n-m)!} [z_n(kr)]^2 \end{aligned}} \qquad (9.52)$$

Since the **m** functions differ from the **M** functions only by the time factor $\exp(-i\omega t)$, we have obtained the desired result.

Utilizing (9.45) and (9.46) we obtain for the **n** functions

$$\mathbf{n}^e_{0,mn} = \frac{1}{kr} \frac{\partial^2}{\partial r \partial \vartheta}\left(r\psi^e_{0,mn}\right)\mathbf{e}_\vartheta + \frac{1}{kr\sin\vartheta} \frac{\partial^2}{\partial r \partial \varphi}\left(r\psi^e_{0,mn}\right)\mathbf{e}_\varphi + \frac{n(n+1)}{kr}\psi^e_{0,mn}\mathbf{e}_r \qquad (9.53)$$

Carrying out the differentiations we find after a few easy steps

$$
\mathbf{n}^{\mathrm{e}}_{\mathrm{o},mn} = \frac{1}{kr}\frac{\partial}{\partial r}[rz_n(kr)]\frac{d\,P^m_n}{d\vartheta}\begin{pmatrix}\cos(m\varphi)\\ \sin(m\varphi)\end{pmatrix}\mathbf{e}_\vartheta
$$

$$
\times\begin{pmatrix}-\sin(m\varphi)\\ +\cos(m\varphi)\end{pmatrix}\frac{m}{kr\sin\vartheta}\frac{\partial}{\partial r}[rz_n(kr)]P^m_n(\cos\vartheta)\mathbf{e}_\varphi
$$

$$
+\frac{n(n+1)}{kr}z_n(kr)P^m_n(\cos\vartheta)\begin{pmatrix}\cos(m\varphi)\\ \sin(m\varphi)\end{pmatrix}\mathbf{e}_r \qquad (9.54)
$$

Again we observe that for $m = 0$ the odd function $\mathbf{n}_{\mathrm{o},m=0,n} = 0$.

The orthogonality relations for the $\mathbf{n}$ functions can be obtained analogously to the $\mathbf{m}$ functions. After a few easy steps we find

$$
\int_0^{2\pi}\int_0^\pi \mathbf{n}^{\mathrm{e}}_{\mathrm{o},mn}\cdot\mathbf{n}^{\mathrm{e}}_{\mathrm{o},ml}\sin\vartheta\,d\vartheta\,d\varphi
$$

$$
= \frac{1}{(kr)^2}\frac{\partial}{\partial r}[rz_n(kr)]\frac{\partial}{\partial r}[rz_l(kr)]\int_0^\pi\frac{d\,P^m_n}{d\vartheta}\frac{d\,P^m_l}{d\vartheta}\sin\vartheta\,d\vartheta\int_0^{2\pi}\begin{pmatrix}\cos^2(m\varphi)\\ \sin^2(m\varphi)\end{pmatrix}d\varphi
$$

$$
+\frac{m^2}{(kr)^2}\frac{\partial}{\partial r}[rz_n(kr)]\frac{\partial}{\partial r}[rz_l(kr)]\int_0^\pi\frac{P^m_n(\cos\vartheta)P^m_l(\cos\vartheta)}{\sin\vartheta}\,d\vartheta
$$

$$
\times\int_0^{2\pi}\begin{pmatrix}\sin^2(m\varphi)\\ \cos^2(m\varphi)\end{pmatrix}d\varphi + \frac{n(n+1)}{kr}\frac{l(l+1)}{kr}z_n(kr)z_l(kr)
$$

$$
\times\int_0^\pi P^m_n(\cos\vartheta)P^m_l(\cos\vartheta)\sin\vartheta\,d\vartheta\int_0^{2\pi}\begin{pmatrix}\cos^2(m\varphi)\\ \sin^2(m\varphi)\end{pmatrix}d\varphi \qquad (9.55)
$$

In order to evaluate the spherical Bessel functions we make use of the following recurrence relations

$$
z_{n-1}(r) + z_{n+1}(r) = \frac{2n+1}{r}z_n(r),
$$

$$
\frac{dz_n(r)}{dr} = \frac{1}{2n+1}[nz_{n-1}(r) - (n+1)z_{n+1}(r)] \qquad (9.56)
$$

so that we finally obtain the orthogonality relations for the $\mathbf{n}$ functions in the form

$$
\int_0^{2\pi} \int_0^{\pi} \mathbf{n}_{o,mn}^e \cdot \mathbf{n}_{o,mn}^e \sin \vartheta \, d\vartheta \, d\varphi = \pi \begin{pmatrix} (1 + \delta_{0,m}) \\ (1 - \delta_{0,m}) \end{pmatrix} \frac{2n(n+1)}{(2n+1)^2} \frac{(n+m)!}{(n-m)!}
$$
$$
\times \{(n+1) \, [z_{n-1}(kr)]^2 + n \, [z_{n+1}(kr)]^2\}
$$

(9.57)

The orthogonality relations involving mixed products of the $\mathbf{m}$ and $\mathbf{n}$ functions may be obtained in the same way as described above. A detailed derivation of the corresponding relations is given by Stratton (1941). For completeness we list the additional orthogonality relations

$$
\int_0^{2\pi} \int_0^{\pi} \mathbf{m}_{o,mn}^e \cdot \mathbf{n}_{o,ml}^e \sin \vartheta \, d\vartheta \, d\varphi = 0
$$
$$
\int_0^{2\pi} \int_0^{\pi} \mathbf{m}_{o,mn}^e \cdot \mathbf{n}_{e,ml}^o \sin \vartheta \, d\vartheta \, d\varphi = 0
$$
$$
\int_0^{2\pi} \int_0^{\pi} \mathbf{m}_{o,mn}^e \cdot \mathbf{m}_{e,ml}^o \sin \vartheta \, d\vartheta \, d\varphi = 0
$$
$$
\int_0^{2\pi} \int_0^{\pi} \mathbf{n}_{o,mn}^e \cdot \mathbf{n}_{e,ml}^o \sin \vartheta \, d\vartheta \, d\varphi = 0
$$

(9.58)

9.5 Mie's scattering problem

We are now prepared to discuss the actual scattering problem. Given is a spherical particle of radius a which is embedded in a vacuum. This particle is illuminated by a plane electromagnetic wave of wavelength λ. We assume that the wave is linearly polarized and propagating in the positive z-direction. It is traditional to designate the direction of the electric vector as the direction of polarization.[1] The task at hand is to determine the scattered electromagnetic field.

9.5.1 The incoming wave

The center of the particle is located at the origin of the coordinate system and the incoming electric vector $\mathbf{E}^i$ is directed along the x-axis as shown in Figure 9.3.

---

[1] A detailed treatment of the effects of polarization on radiative transfer will be given in the next chapter.

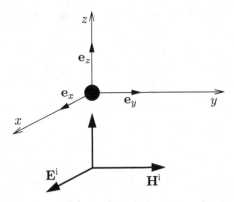

Fig. 9.3 Mie's scattering problem: the incoming wave.

Again assuming a harmonic time dependency, the field vectors can be described by

$$\mathbf{E}^{\mathrm{i}} = \mathbf{e}_x E_0^{\mathrm{i}} \exp\left[i(k_0 z - \omega t)\right], \qquad \mathbf{H}^{\mathrm{i}} = \mathbf{e}_y H_0^{\mathrm{i}} \exp\left[i(k_0 z - \omega t)\right] \qquad (9.59)$$

whereby $(\mathbf{e}_x, \mathbf{e}_y)$ are the unit vectors in (x, y)-direction. $\mathbf{E}^{\mathrm{i}}$ and $\mathbf{H}^{\mathrm{i}}$ satisfy the vector wave equations

$$\nabla^2 \mathbf{E}^{\mathrm{i}} - \frac{1}{c^2}\frac{\partial^2 \mathbf{E}^{\mathrm{i}}}{\partial t^2} = 0, \qquad \nabla^2 \mathbf{H}^{\mathrm{i}} - \frac{1}{c^2}\frac{\partial^2 \mathbf{H}^{\mathrm{i}}}{\partial t^2} = 0 \qquad (9.60)$$

We will now attempt to express $\mathbf{E}^{\mathrm{i}}$ and $\mathbf{H}^{\mathrm{i}}$ by means of the vectors $\mathbf{M}$ and $\mathbf{N}$. Substituting (9.10) into (9.1) yields

$$\nabla \times \mathbf{E} = i\omega\mathbf{B}, \qquad \nabla \times \mathbf{B} = -i\omega\frac{\mathcal{N}^2}{c^2}\mathbf{E} \qquad (9.61)$$

where (9.2) and the defining expression of the *complex index of refraction* $\mathcal{N}$ (9.7) have been used.

For brevity we introduce the vector $\mathbf{B}^*$ by means of

$$\mathbf{B}^* = c\mathbf{B} = \sqrt{\frac{\mu_0}{\varepsilon_0}}\mathbf{H} \qquad (9.62)$$

so that (9.61) assumes the form

$$\nabla \times \mathbf{E} = ik_0\mathbf{B}^*, \qquad \nabla \times \mathbf{B}^* = -ik_0\mathcal{N}^2\mathbf{E} \qquad (9.63)$$

with $k_0 = \omega/c$. Furthermore, according to (9.9), $\mathbf{E}$ and $\mathbf{B}^*$ fulfill the vector wave equations

$$\nabla^2\mathbf{E} + k_0^2\mathcal{N}^2\mathbf{E} = 0, \qquad \nabla^2\mathbf{B}^* + k_0^2\mathcal{N}^2\mathbf{B}^* = 0 \qquad (9.64)$$

By comparing (9.63) with (9.43), utilizing (9.42) and (9.62), it is seen that the field vectors of the incoming wave may be written as

$$\mathbf{E}^i = E_0^i(\mathbf{M}_v - i\mathbf{N}_u), \qquad \mathbf{H}^i = -H_0^i(\mathbf{M}_u + i\mathbf{N}_v) \quad \text{with} \quad H_0^i = \sqrt{\frac{\varepsilon_0}{\mu_0}} E_0^i$$

(9.65)

In order to apply the boundary conditions to the scattering sphere, we introduce the spherical coordinates (ϑ, φ, r) by means of the well-known transformation equations

$$x = r \sin \vartheta \cos \varphi, \qquad y = r \sin \vartheta \sin \varphi, \qquad z = r \cos \vartheta \qquad (9.66)$$

The relations between the unit vectors $(\mathbf{e}_x, \mathbf{e}_y, \mathbf{e}_z)$ of the Cartesian system and $(\mathbf{e}_\vartheta, \mathbf{e}_\varphi, \mathbf{e}_r)$ of the spherical coordinate system are

$$
\begin{aligned}
\mathbf{e}_\vartheta &= \cos \vartheta \cos \varphi \mathbf{e}_x + \cos \vartheta \sin \varphi \mathbf{e}_y - \sin \vartheta \mathbf{e}_z \\
\mathbf{e}_\varphi &= -\sin \varphi \mathbf{e}_x + \cos \varphi \mathbf{e}_y \\
\mathbf{e}_r &= \sin \vartheta \cos \varphi \mathbf{e}_x + \sin \vartheta \sin \varphi \mathbf{e}_y + \cos \vartheta \mathbf{e}_z \\
\mathbf{e}_x &= \sin \vartheta \cos \varphi \mathbf{e}_r + \cos \vartheta \cos \varphi \mathbf{e}_\vartheta - \sin \varphi \mathbf{e}_\varphi \\
\mathbf{e}_y &= \sin \vartheta \sin \varphi \mathbf{e}_r + \cos \vartheta \sin \varphi \mathbf{e}_\vartheta + \cos \varphi \mathbf{e}_\varphi \\
\mathbf{e}_z &= \cos \vartheta \mathbf{e}_r - \sin \vartheta \mathbf{e}_\vartheta
\end{aligned}
$$

(9.67)

A detailed derivation of these relations may be found, for instance, in Zdunkowski and Bott (2003). Replacing in (9.59) $\mathbf{e}_x$ and $\mathbf{e}_y$ by the corresponding expressions in (9.67) yields

(a) $\mathbf{E}^i = E_0^i \exp\left[i(k_0 r \cos \vartheta - \omega t)\right](\sin \vartheta \cos \varphi \mathbf{e}_r + \cos \vartheta \cos \varphi \mathbf{e}_\vartheta - \sin \varphi \mathbf{e}_\varphi)$

(b) $\mathbf{H}^i = H_0^i \exp\left[i(k_0 r \cos \vartheta - \omega t)\right](\sin \vartheta \sin \varphi \mathbf{e}_r + \cos \vartheta \sin \varphi \mathbf{e}_\vartheta + \cos \varphi \mathbf{e}_\varphi)$

(9.68)

The representation of the field vectors as stated in (9.65) and (9.68) shall now be harmonized by writing series expressions for $\mathbf{E}^i$ and $\mathbf{H}^i$ in terms of still unknown expansions coefficients A_{mn}, B_{mn}, C_{mn} and D_{mn}. Recalling (9.47) we may write

$$\mathbf{E}^i = E_0^i \sum_{n=0}^{\infty} \sum_{m=0}^{n} (A_{mn}\mathbf{m}_{mn} + B_{mn}\mathbf{n}_{mn}) \exp(-i\omega t)$$

$$\mathbf{H}^i = H_0^i \sum_{n=0}^{\infty} \sum_{m=0}^{n} (C_{mn}\mathbf{m}_{mn} + D_{mn}\mathbf{n}_{mn}) \exp(-i\omega t)$$

(9.69)

Comparison of coefficients of the terms involving the $\cos \varphi$ and $\sin \varphi$ functions in (9.68), in view of the defining equations (9.48) and (9.54) for $\mathbf{m}$ and $\mathbf{n}$, shows that unless $m = 1$ the coefficients A_{mn}, B_{mn}, C_{mn} and D_{mn} vanish. For this reason the sums over m may be evaluated in (9.69) yielding

(a) $\quad \mathbf{E}^i = E_0^i \sum_{n=0}^{\infty} \left(A_n \mathbf{m}_{o,1n} + B_n \mathbf{n}_{1n}^e \right) \exp(-i\omega t)$

(b) $\quad \mathbf{H}^i = H_0^i \sum_{n=0}^{\infty} \left(C_n \mathbf{m}_{1n}^e + D_n \mathbf{n}_{o,1n} \right) \exp(-i\omega t)$

$$(9.70)$$

with $A_{1n} = A_n$, $B_{1n} = B_n$, $C_{1n} = C_n$ and $D_{1n} = D_n$. The reason that either the odd or the even parts of the $\mathbf{m}$ and $\mathbf{n}$ functions occur in (9.70) results from the comparison of the $\cos \varphi$ and the $\sin \varphi$ terms appearing in (9.68) with (9.48) and (9.54).

In order to determine the expansion coefficients A_n, B_n, C_n and D_n we make use of the orthogonality relations (9.52) and (9.57) of the $\mathbf{m}$ and $\mathbf{n}$ functions. We will demonstrate how to proceed by determining A_n. The remaining coefficients are found analogously.

First we set (9.68a) equal to (9.70a). Then we multiply both sides by the function $\mathbf{m}_{o,1l}$ and integrate the resulting expression over the unit sphere. Thus we obtain

$$\int_0^{2\pi} \int_0^{\pi} \exp(ik_0 r \cos \vartheta)(\sin \vartheta \cos \varphi \mathbf{e}_r + \cos \vartheta \cos \varphi \mathbf{e}_\vartheta - \sin \varphi \mathbf{e}_\varphi) \cdot \mathbf{m}_{o,1l} \sin \vartheta \, d\vartheta \, d\varphi$$

$$= \int_0^{2\pi} \int_0^{\pi} A_n \mathbf{m}_{o,1n} \cdot \mathbf{m}_{o,1l} \sin \vartheta \, d\vartheta \, d\varphi \qquad (9.71)$$

Due to the orthogonality relations of the trigonometric functions, see (9.51), and by using (9.52) we first find

$$\pi z_n(k_0 r) \int_0^{\pi} \exp(ik_0 r \cos \vartheta) \left(\frac{\cos \vartheta}{\sin \vartheta} P_n^1(\cos \vartheta) + \frac{dP_n^1}{d\vartheta} \right) \sin \vartheta \, d\vartheta$$

$$= A_n z_n^2(k_0 r) 2\pi \frac{n(n+1)}{2n+1} \frac{(n+1)!}{(n-1)!} \qquad (9.72)$$

Recalling the relationship between the associated and the ordinary Legendre polynomials

$$P_n^m(x) = (1 - x^2)^{m/2} \frac{d^m}{dx^m} P_n(x) \qquad (9.73)$$

we may replace the expression in parenthesis on the left-hand side of (9.72) by

$$\frac{\cos \vartheta}{\sin \vartheta} P_n^1(\cos \vartheta) + \frac{dP_n^1}{d\vartheta} = -(1 - x^2) \frac{d^2 P_n}{dx^2} + 2x \frac{dP_n}{dx}$$

$$= n(n+1)P_n(x) \quad \text{with} \quad x = \cos \vartheta$$

$$(9.74)$$

where we have made use of *Legendre's differential equation*

$$(1 - x^2)\frac{d^2 P_n}{dx^2} - 2x\frac{d}{dx}P_n + n(n+1)P_n(x) = 0 \qquad (9.75)$$

Thus we obtain

$$A_n = \frac{1}{z_n(k_0 r)}\frac{2n+1}{2n(n+1)}\int_0^\pi \exp(ik_0 r \cos\vartheta)P_n(\cos\vartheta)\sin\vartheta\,d\vartheta \qquad (9.76)$$

which contains the unspecified spherical Bessel function $z_n(k_0 r)$. We choose $z_n = j_n$ which is defined by means of

$$j_n(k_0 r) = \frac{i^{-n}}{2}\int_0^\pi \exp(ik_0 r \cos\vartheta)P_n(\cos\vartheta)\sin\vartheta\,d\vartheta \qquad (9.77)$$

The function j_n is known as the *spherical Bessel function of the first kind* which is finite at the origin. This spherical Bessel function is related to the ordinary Bessel function of the first kind by

$$j_n(k_0 r) = \sqrt{\frac{\pi}{2k_0 r}}J_{n+1/2}(k_0 r) \qquad (9.78)$$

Introducing (9.77) into (9.76) gives the final form of the expansion coefficient A_n

$$\boxed{A_n = i^n\frac{2n+1}{n(n+1)}, \qquad B_n = -iA_n, \qquad C_n = -A_n, \qquad D_n = B_n} \qquad (9.79)$$

The remaining expansion coefficients are also stated as part of this equation. Thus (9.70) can be rewritten as

$$\boxed{\begin{aligned} \mathbf{E}^i &= E_0^i \sum_{n=0}^\infty i^n \frac{2n+1}{n(n+1)}\left(\mathbf{m}_{o,1n} - i\mathbf{n}_{1n}^e\right)\exp(-i\omega t) \\ \mathbf{H}^i &= -H_0^i \sum_{n=0}^\infty i^n \frac{2n+1}{n(n+1)}\left(\mathbf{m}_{1n}^e + i\mathbf{n}_{o,1n}\right)\exp(-i\omega t) \end{aligned}} \qquad (9.80)$$

Due to the choice $z_n = j_n$ the solutions (u_n, v_n) of the scalar wave equation appearing in the $\mathbf{M}$ and $\mathbf{N}$ functions for the incoming wave can be written down immediately. From (9.37), including the factor $i^n(2n+1)/n(n+1)$ in the definition of the u_n and v_n functions, we obtain

$$\begin{aligned} u_n^i(r, \vartheta, \varphi, t) &= i^n\frac{2n+1}{n(n+1)}\cos\varphi\, P_n^1(\cos\vartheta)j_n(k_0 r)\exp(-i\omega t) \\ v_n^i(r, \vartheta, \varphi, t) &= i^n\frac{2n+1}{n(n+1)}\sin\varphi\, P_n^1(\cos\vartheta)j_n(k_0 r)\exp(-i\omega t) \end{aligned} \qquad (9.81)$$

In view of equations (9.47) we may thus write the series expressions for the field vectors of the incoming wave as

$$\mathbf{E}^i = E_0^i \sum_{n=0}^{\infty} \left(\mathbf{M}_{v_n^i} - i\mathbf{N}_{u_n^i}\right), \quad \mathbf{H}^i = -H_0^i \sum_{n=0}^{\infty} \left(\mathbf{M}_{u_n^i} + i\mathbf{N}_{v_n^i}\right) \quad \text{with} \quad H_0^i = \sqrt{\frac{\varepsilon_0}{\mu_0}} E_0^i$$

$$(9.82)$$

9.5.2 The scattered and the interior waves

A periodic wave which is incident on the particle gives rise to a forced oscillation of free and bound charges synchronous with the applied field. These motions of the charge set up a secondary field both inside and outside of the particle. The resultant field at any point is the vector sum of the primary and the secondary fields. After the transient oscillations are damped out a steady-state situation will occur which will now be investigated.

As the incident wave interacts with the particle, the induced secondary field must be constructed in two parts. The interior part of the sphere we will call transmitted (t) while the other part, denoted by (s), refers to scattering. In analogy to the structure of the incident field vectors we now use the forms

$$\mathbf{E}^{s,t} = E_0^i \sum_{n=0}^{\infty} \left(\mathbf{M}_{v_n^{s,t}} - i\mathbf{N}_{u_n^{s,t}}\right), \qquad \mathbf{H}^{s,t} = -H_0^{s,t} \sum_{n=0}^{\infty} \left(\mathbf{M}_{u_n^{s,t}} + i\mathbf{N}_{v_n^{s,t}}\right)$$

$$(9.83)$$

$$\text{with} \quad H_0^s = \frac{E_0^i}{\mu_0 c} = H_0^i, \qquad H_0^t = \frac{N E_0^i}{\mu c}$$

In contrast to the incoming wave, for the wave functions u_n^s and v_n^s of the scattered field we select $z_n = h_n$ where h_n are the *spherical Hankel functions of the first kind*. The reason for this particular choice is the asymptotic behavior of the spherical Hankel functions. For large values of the argument we have

$$h_n^1(k_0 r) \sim \frac{\exp(i k_0 r)}{k_0 r}(-i)^{n+1}$$

$$(9.84)$$

If this expression is multiplied by the factor $\exp(-i\omega t)$ it represents an outgoing spherical wave (of amplitude 1), as required for the scattered wave. Thus, analogously to (9.81) we may write

$$u_n^s(r, \vartheta, \varphi, t) = i^n \frac{2n+1}{n(n+1)} a_n^s \cos\varphi \, P_n^1(\cos\vartheta) h_n^1(k_0 r) \exp(-i\omega t)$$

$$v_n^s(r, \vartheta, \varphi, t) = i^n \frac{2n+1}{n(n+1)} b_n^s \sin\varphi \, P_n^1(\cos\vartheta) h_n^1(k_0 r) \exp(-i\omega t)$$

$$(9.85)$$

For the inside wave we choose the function $j_n(\mathcal{N}k_0 r)$ because the refractive index is finite and the spherical Bessel function is finite at the origin. This results in

$$u_n^t(r, \vartheta, \varphi, t) = i^n \frac{2n+1}{n(n+1)} a_n^t \cos\varphi P_n^1(\cos\vartheta) j_n(\mathcal{N}k_0 r)\exp(-i\omega t)$$

$$v_n^t(r, \vartheta, \varphi, t) = i^n \frac{2n+1}{n(n+1)} b_n^t \sin\varphi P_n^1(\cos\vartheta) j_n(\mathcal{N}k_0 r)\exp(-i\omega t) \tag{9.86}$$

with $0 \le r \le a$. For the calculation of the unknown coefficients $a_n^{s,t}$ and $b_n^{s,t}$ we apply the boundary conditions at the spherical surface which have been derived in Section 9.3. Replacing in (9.20) the unit normal vector $\mathbf{n}$ by $\mathbf{e}_r$ we obtain

$$\mathbf{e}_r \times (\mathbf{E}^i + \mathbf{E}^s) = \mathbf{e}_r \times \mathbf{E}^t, \qquad \mathbf{e}_r \times (\mathbf{H}^i + \mathbf{H}^s) = \mathbf{e}_r \times \mathbf{H}^t \tag{9.87}$$

The boundary conditions at the particle's surface $r = a$ can now be written down for the various components of the $\mathbf{M}$ and $\mathbf{N}$ functions

$$\begin{aligned}
\text{(a)} &\quad \left(M_{v_n^i} - iN_{u_n^i}\right)_\vartheta + \left(M_{v_n^s} - iN_{u_n^s}\right)_\vartheta = \left(M_{v_n^t} - iN_{u_n^t}\right)_\vartheta \\
\text{(b)} &\quad \left(M_{v_n^i} - iN_{u_n^i}\right)_\varphi + \left(M_{v_n^s} - iN_{u_n^s}\right)_\varphi = \left(M_{v_n^t} - iN_{u_n^t}\right)_\varphi \\
\text{(c)} &\quad \left(M_{u_n^i} + iN_{v_n^i}\right)_\vartheta + \left(M_{u_n^s} + iN_{v_n^s}\right)_\vartheta = \frac{\mathcal{N}\mu_0}{\mu}\left(M_{u_n^t} + iN_{v_n^t}\right)_\vartheta \\
\text{(d)} &\quad \left(M_{u_n^i} + iN_{v_n^i}\right)_\varphi + \left(M_{u_n^s} + iN_{v_n^s}\right)_\varphi = \frac{\mathcal{N}\mu_0}{\mu}\left(M_{u_n^t} + iN_{v_n^t}\right)_\varphi
\end{aligned} \tag{9.88}$$

Let us examine in detail equation (9.88a). Using (9.45) we obtain at $r = a$

$$\frac{1}{\sin\vartheta}\frac{\partial}{\partial\varphi}\left(v_n^i + v_n^s - v_n^t\right) - \frac{i}{k_0 r}\frac{\partial^2}{\partial r\partial\vartheta}\left(ru_n^i + ru_n^s - \frac{ru_n^t}{\mathcal{N}}\right) = 0 \tag{9.89}$$

Employing the required equations (9.81), (9.85) and (9.86) we find at $r = a$

$$\mathcal{N}a_n^s[\rho_0 h_n^1(\rho_0)]' = a_n^t[\rho j_n(\rho)]' - \mathcal{N}[\rho_0 j_n(\rho_0)]'$$

$$b_n^s h_n^1(\rho_0) = b_n^t j_n(\rho) - j_n(\rho_0) \quad \text{with}$$

$$[\rho_0 j_n(\rho_0)]' = \frac{d}{d\rho_0}[\rho_0 j_n(\rho_0)]\bigg|_{r=a}, \qquad [\rho_0 h_n^1(\rho_0)]' = \frac{d}{d\rho_0}[\rho_0 h_n^1(\rho_0)]\bigg|_{r=a}$$

$$[\rho j_n(\rho)]' = \frac{d}{d\rho}[\rho j_n(\rho)]\bigg|_{r=a}, \qquad \rho_0 = k_0 r, \quad \rho = \mathcal{N}\rho_0 \tag{9.90}$$

Had we used the azimuthal components we would have obtained the same result. With the help of (9.88c,d) we can obtain two additional relations for the coefficients

a_n and b_n. Altogether we find

$$a_n^{\rm t} \left[\rho j_n(\rho)\right]' - \mathcal{N} a_n^{\rm s} \left[\rho_0 h_n^1(\rho_0)\right]' = \mathcal{N} \left[\rho_0 j_n(\rho_0)\right]'$$

$$\mu_0 \mathcal{N} a_n^{\rm t} j_n(\rho) - \mu a_n^{\rm s} h_n^1(\rho_0) = \mu j_n(\rho_0)$$

$$\mu_0 b_n^{\rm t} \left[\rho j_n(\rho)\right]' - \mu b_n^{\rm s} \left[\rho_0 h_n^1(\rho_0)\right]' = \mu \left[\rho_0 j_n(\rho_0)\right]' \tag{9.91}$$

$$b_n^{\rm t} j_n(\rho) - b_n^{\rm s} h_n^1(\rho_0) = j_n(\rho_0)$$

from which the coefficients a_n and b_n of the scattered wave can be calculated as

$$
\boxed{
\begin{aligned}
a_n^{\rm s} &= -\frac{\mu j_n(\rho_0) \left[\rho j_n(\rho)\right]' - \mathcal{N}^2 \mu_0 j_n(\rho) \left[\rho_0 j_n(\rho_0)\right]'}{\mu h_n^1(\rho_0) \left[\rho j_n(\rho)\right]' - \mathcal{N}^2 \mu_0 j_n(\rho) \left[\rho_0 h_n^1(\rho_0)\right]'} \\[2mm]
b_n^{\rm s} &= -\frac{\mu_0 j_n(\rho_0) \left[\rho j_n(\rho)\right]' - \mu j_n(\rho) \left[\rho_0 j_n(\rho_0)\right]'}{\mu_0 h_n^1(\rho_0) \left[\rho j_n(\rho)\right]' - \mu j_n(\rho) \left[\rho_0 h_n^1(\rho_0)\right]'}
\end{aligned}
}
\tag{9.92}
$$

The coefficients $a_n^{\rm s}$ and $b_n^{\rm s}$ are of paramount importance for the Mie computations.

The evaluation of the Mie expressions in some cases may pose numerical difficulties. In order to avoid these see, for example, Deirmendjian (1969). It is possible to state (9.92) in a somewhat simplified form by introducing the so-called *Riccati–Bessel functions* which differ from the spherical Bessel functions. For details see van de Hulst (1957). Moreover, equations (9.92) simplify by setting $\mu = \mu_0$ which is permissible if the scattering particle is non-magnetic.

The *transmission coefficients* have not been stated explicitly. They are also of great importance for the calculation of the electromagnetic energy of a scattering sphere. This is, for instance, necessary if photochemical reactions within water drops are calculated. A detailed analysis of this topic is given by Bott and Zdunkowski (1987).

We will now return to equation (9.85) and replace the spherical Hankel function by its asymptotic expression (9.84). Thus we find for the wave functions of the scattered wave

$$u_n^{\rm s}(r, \vartheta, \varphi, t) = -i \frac{2n+1}{n(n+1)} a_n^{\rm s} \frac{\cos\varphi}{k_0 r} P_n^1(\cos\vartheta) \exp\left[i(k_0 r - \omega t)\right]$$

$$v_n^{\rm s}(r, \vartheta, \varphi, t) = -i \frac{2n+1}{n(n+1)} b_n^{\rm s} \frac{\sin\varphi}{k_0 r} P_n^1(\cos\vartheta) \exp\left[i(k_0 r - \omega t)\right] \tag{9.93}$$

The following abbreviations will be introduced

$$\pi_n(\cos\vartheta) = \frac{P_n^1(\cos\vartheta)}{\sin\vartheta} = \frac{dP_n}{d\cos\vartheta}$$

$$\tau_n(\cos\vartheta) = \frac{dP_n^1}{d\vartheta} = \cos\vartheta\, \pi_n(\cos\vartheta) - \sin^2\vartheta \frac{d\pi_n}{d\cos\vartheta} \tag{9.94}$$

where we have used (9.74). First let us treat the components of the field vectors of the scattered wave. According to (9.83) they are given by

$$E_\vartheta^s = E_0^i \sum_{n=0}^\infty \left(M_{v_n^s} - i N_{u_n^s}\right)_\vartheta, \qquad E_\varphi^s = E_0^i \sum_{n=0}^\infty \left(M_{v_n^s} - i N_{u_n^s}\right)_\varphi, \qquad E_r^s = -i E_0^i \sum_{n=0}^\infty \left(N_{u_n^s}\right)_r$$

$$H_\vartheta^s = -H_0^i \sum_{n=0}^\infty \left(M_{u_n^s} + i N_{v_n^s}\right)_\vartheta, \qquad H_\varphi^s = -H_0^i \sum_{n=0}^\infty \left(M_{u_n^s} + i N_{v_n^s}\right)_\varphi, \qquad H_r^s = -H_0^i \sum_{n=0}^\infty i \left(N_{v_n^s}\right)_r$$

$$(9.95)$$

Employing (9.45) and (9.46) we may write for the horizontal components

$$E_\vartheta^s = E_0^i \sum_{n=0}^\infty \left(\frac{1}{\sin\vartheta} \frac{\partial v_n^s}{\partial\varphi} - \frac{i}{k_0 r} \frac{\partial^2 (r u_n^s)}{\partial r \partial\vartheta}\right)$$

$$E_\varphi^s = E_0^i \sum_{n=0}^\infty \left(-\frac{\partial v_n^s}{\partial\vartheta} - \frac{i}{k_0 r \sin\vartheta} \frac{\partial^2 (r u_n^s)}{\partial r \partial\varphi}\right)$$

$$H_\vartheta^s = -H_0^i \sum_{n=0}^\infty \left(\frac{1}{\sin\vartheta} \frac{\partial u_n^s}{\partial\varphi} + \frac{i}{k_0 r} \frac{\partial^2 (r v_n^s)}{\partial r \partial\vartheta}\right)$$

$$H_\varphi^s = -H_0^i \sum_{n=0}^\infty \left(-\frac{\partial u_n^s}{\partial\vartheta} + \frac{i}{k_0 r \sin\vartheta} \frac{\partial^2 (r v_n^s)}{\partial r \partial\varphi}\right)$$

$$(9.96)$$

The r-component of the scattered field is proportional to r^{-2} so that it may be ignored at large distances from the scattering center.

Inserting (9.93) into (9.96) and using the definitions (9.94) we find

$$E_\vartheta^s = -E_0^i \cos\varphi \sum_{n=1}^\infty \frac{2n+1}{n(n+1)} \left[a_n^s \tau_n(\cos\vartheta) + b_n^s \pi_n(\cos\vartheta)\right] \frac{i}{k_0 r} \exp\left[i(k_0 r - \omega t)\right]$$

$$E_\varphi^s = E_0^i \sin\varphi \sum_{n=1}^\infty \frac{2n+1}{n(n+1)} \left[a_n^s \pi_n(\cos\vartheta) + b_n^s \tau_n(\cos\vartheta)\right] \frac{i}{k_0 r} \exp\left[i(k_0 r - \omega t)\right]$$

$$H_\vartheta^s = -H_0^i \sin\varphi \sum_{n=1}^\infty \frac{2n+1}{n(n+1)} \left[a_n^s \pi_n(\cos\vartheta) + b_n^s \tau_n(\cos\vartheta)\right] \frac{i}{k_0 r} \exp\left[i(k_0 r - \omega t)\right]$$

$$H_\varphi^s = -H_0^i \cos\varphi \sum_{n=1}^\infty \frac{2n+1}{n(n+1)} \left[a_n^s \tau_n(\cos\vartheta) + b_n^s \pi_n(\cos\vartheta)\right] \frac{i}{k_0 r} \exp\left[i(k_0 r - \omega t)\right]$$

$$(9.97)$$

Inspection of these equations shows that the following relations hold

$$H_\vartheta^s = -\sqrt{\frac{\varepsilon_0}{\mu_0}} E_\varphi^s, \qquad H_\varphi^s = \sqrt{\frac{\varepsilon_0}{\mu_0}} E_\vartheta^s \qquad (9.98)$$

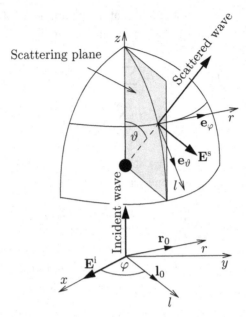

Fig. 9.4 The (l, r, z)-coordinate system and the scattering plane.

since $H_0^i = \sqrt{\varepsilon_0/\mu_0}E_0^i$, see (9.82). For this reason, in the following it is sufficient to discuss the electric field only.

In order to treat scattering of electromagnetic waves on spherical particles it is advantageous to introduce the so-called (l, r)-system. For the incoming wave the new system is obtained by rotating the original (x, y, z)-system about the z-axis in counterclockwise direction by the angle φ. For the scattered wave the directions of the l- and r-axes agree with the directions ϑ and φ of the spherical coordinate system. Thus the l-axes are located in the plane defined by the directions of the incoming and the scattered wave. This particular plane is called the *scattering plane*. Figure 9.4 depicts the directions of the l- and r-axes as well as the scattering plane. The labels l and r are taken from the last letters of the words 'parallel' and 'perpendicular' describing the directions of the components of the electric field vectors with respect to the scattering plane.

If $\mathbf{l}_0$ and $\mathbf{r}_0$ are unit vectors along the l and r axis of the incident wave, according to Figure 9.4 we may express $\mathbf{E}^i$ as

$$\mathbf{E}^i = E_l^i\mathbf{l}_0 + E_r^i\mathbf{r}_0 = E_0^i \cos\varphi \exp\left[i(k_0z - \omega t)\right]\mathbf{l}_0 - E_0^i \sin\varphi \exp\left[i(k_0z - \omega t)\right]\mathbf{r}_0$$

$$(9.99)$$

The scattered electric vector can be written as

$$\mathbf{E}^s = E_\vartheta^s \mathbf{e}_\vartheta + E_\varphi^s \mathbf{e}_\varphi = E_l^s \mathbf{e}_\vartheta + E_r^s \mathbf{e}_\varphi \quad \text{since} \quad E_\vartheta^s = E_l^s, \quad E_\varphi^s = E_r^s \quad (9.100)$$

In order to have a shorthand notation we introduce the amplitude functions S_1 and S_2 by

$$S_1(\cos\vartheta) = \sum_{n=1}^\infty \frac{2n+1}{n(n+1)} \left[a_n^s \pi_n(\cos\vartheta) + b_n^s \tau_n(\cos\vartheta) \right]$$

$$S_2(\cos\vartheta) = \sum_{n=1}^\infty \frac{2n+1}{n(n+1)} \left[a_n^s \tau_n(\cos\vartheta) + b_n^s \pi_n(\cos\vartheta) \right] \quad (9.101)$$

By using (9.19), and (9.100) and (9.101) the first two equations of (9.96) can be written in matrix notation as

$$\begin{pmatrix} E_l^s \\ E_r^s \end{pmatrix} = \frac{\exp\left[ik_0(r-z)\right]}{ik_0 r} \begin{pmatrix} S_2(\cos\vartheta) & 0 \\ 0 & S_1(\cos\vartheta) \end{pmatrix} \begin{pmatrix} E_l^i \\ E_r^i \end{pmatrix} \quad (9.102)$$

In Chapter 1 the symbol Θ was introduced for the scattering angle, see Figure 1.17. In the spherical coordinate system used to derive (9.102) the scattering angle corresponds to the polar angle ϑ. In order to obtain consistency with the earlier notation, from now on we will again use the symbol Θ for the scattering angle.

9.5.3 Rayleigh scattering

Rayleigh (1871) has developed a scattering theory for particles which are small in comparison with the wavelength. With the help of the formulas which he derived he was able to explain the blue color of the sky. Since the Mie theory is valid for spherical particles of any size, it is possible to derive the Rayleigh formulas as special cases of the Mie equations rather than following Rayleigh's original work. Thus, in the following we will assume that the radius a of the scattering particle (e.g. an air molecule) is small compared to the wavelength λ of the incident electromagnetic wave.

Before we begin with the simplification of the Mie equations we need to state a number of formulas. Employing the definition of the spherical Bessel functions as stated in (9.36), with the help of the sum formula for the ordinary Bessel function of the first kind, (see e.g. Abramowitz and Stegun, 1972) we have

$$j_n(x) = \frac{\sqrt{\pi}}{2} \left(\frac{x}{2}\right)^n \sum_{k=0}^\infty \frac{\left(-\frac{1}{4}x^2\right)^k}{k!\,\Gamma(n+3/2+k)} \quad (9.103)$$

From (9.56) we first find

$$[xj_n(x)]' = xj_{n-1}(x) - nj_n(x) \tag{9.104}$$

and then the sum representation

$$[xj_n(x)]' = \sqrt{\pi} \left(\frac{x}{2}\right)^n \sum_{k=0}^{\infty} \frac{\left(-\frac{1}{4}x^2\right)^k}{k!} \left(\frac{1}{\Gamma(n+1/2+k)} - \frac{n}{2\Gamma(n+3/2+k)}\right) \tag{9.105}$$

which will be used to treat some special cases as required for the Rayleigh theory. We also need the definition of the spherical Hankel function of the first kind which is given by

$$h_n^1(x) = j_n(x) - i(-1)^n j_{-(n+1)}(x) \tag{9.106}$$

The sum representation follows directly

$$
h_n^1(x) = \frac{\sqrt{\pi}}{2} \left(\frac{x}{2}\right)^n \sum_{k=0}^{\infty} \frac{\left(-\frac{1}{4}x^2\right)^k}{k!\Gamma(n+3/2+k)}
$$
$$
- \frac{i\sqrt{\pi}}{2}(-1)^n \left(\frac{x}{2}\right)^{-(n+1)} \sum_{k=0}^{\infty} \frac{\left(-\frac{1}{4}x^2\right)^k}{k!\Gamma(-n+1/2+k)} \tag{9.107}
$$

from which we obtain the derivative formula

$$
[xh_n^1(x)]' = \sqrt{\pi} \left(\frac{x}{2}\right)^n \sum_{k=0}^{\infty} \frac{\left(-\frac{1}{4}x^2\right)^k}{k!} \left(\frac{1}{\Gamma(n+1/2+k)} - \frac{n}{2\Gamma(n+3/2+k)}\right)
$$
$$
- i(-1)^n \sqrt{\pi} \left(\frac{x}{2}\right)^{-(n+1)} \sum_{k=0}^{\infty} \frac{\left(-\frac{1}{4}x^2\right)^k}{k!}
$$
$$
\times \left(\frac{1}{\Gamma(-n-1/2+k)} - \frac{n+1}{2\Gamma(-n+1/2+k)}\right) \tag{9.108}
$$

Now we are ready to obtain the simplified formulas which will be needed soon.

The arguments of the Bessel functions are of the form $x = ka$ where, as usual, k is the wave number and a is the radius of the scattering particle. If $a \ll \lambda$ then the argument approaches zero. Thus it is sufficient to discontinue the various sums after the first term. Evaluation of (9.103) for small arguments yields

$$j_n(x) \approx \frac{\sqrt{\pi}}{2} \left(\frac{x}{2}\right)^n \frac{1}{\Gamma(n+3/2)} \tag{9.109}$$

Similarly, we obtain from the derivative formula (9.105)

$$[xj_n(x)]' \approx \sqrt{\pi} \left(\frac{x}{2}\right)^n \left(\frac{1}{\Gamma(n+1/2)} - \frac{n}{2\Gamma(n+3/2)}\right) \qquad (9.110)$$

while (9.107) and (9.108) result in

$$h_n^1(x) \approx \frac{\sqrt{\pi}}{2} \left(\frac{x}{2}\right)^n \frac{1}{\Gamma(n+3/2)} - \frac{i\sqrt{\pi}}{2}(-1)^n \left(\frac{x}{2}\right)^{-(n+1)} \frac{1}{\Gamma(-n+1/2)}$$

$$(9.111)$$

and

$$\left[xh_n^1(x)\right]' = \sqrt{\pi} \left(\frac{x}{2}\right)^n \left(\frac{1}{\Gamma(n+1/2)} - \frac{n}{2\Gamma(n+3/2)}\right)$$

$$- i(-1)^n \sqrt{\pi} \left(\frac{x}{2}\right)^{-(n+1)} \left(\frac{1}{\Gamma(-n-1/2)} - \frac{n+1}{2\Gamma(-n+1/2)}\right)$$

$$(9.112)$$

We will now investigate the coefficients a_n^s and b_n^s as given by (9.92) and find out in which way these simplify if we apply the above approximation formulas. The numerators of both coefficients contain powers of $(x/2)^n$. Moreover, the real part of the denominator also contains powers of $(x/2)^n$ while the imaginary part is a function of x^{-1}. Thus for very small arguments of the functions and for $n > 1$ we may set approximately

$$a_n^s \approx 0, \qquad b_n^s \approx 0 \quad \text{for} \quad n > 1 \qquad (9.113)$$

implying that in (9.101) the S_1 and S_2 series may be discontinued after the first term. Now let us find explicit expressions for a_1^s and b_1^s. Inspection of the definitions of the spherical Bessel functions shows that these contain the gamma functions. By using the formulas

$$\Gamma(1/2) = \sqrt{\pi}, \qquad \Gamma(x+1) = x\Gamma(x) \qquad (9.114)$$

we find from (9.109) and (9.110) for small values of the argument

$$j_1(x) \approx \frac{x}{3}, \qquad [xj_1(x)]' \approx \frac{2x}{3} \qquad (9.115)$$

Likewise, from (9.111) and (9.112) we obtain

$$h_1(x) \approx \frac{x}{3} - \frac{i}{x^2}, \qquad [xh_1(x)]' \approx \frac{2x}{3} + \frac{i}{x^2} \qquad (9.116)$$

Substitution of these expressions into (9.92) yields the approximate expressions for a_1^s

$$a_1^s \approx \frac{\frac{2\rho_0}{3}\left(\mathcal{N}^3 - \mathcal{N}\right)}{2\left(\frac{\rho_0}{3} - \frac{i}{\rho_0^2}\right)\mathcal{N} - \left(\frac{2\rho_0}{3} + \frac{i}{\rho_0^2}\right)\mathcal{N}^3}$$

$$= -\frac{\rho_0^3\left(\mathcal{N}^2 - 1\right)\left[\rho_0^3\left(\mathcal{N}^2 - 1\right) - \frac{3}{2}i\left(\mathcal{N}^2 + 2\right)\right]}{\rho_0^6(\mathcal{N}^2 - 1)^2 + \frac{9}{4}(\mathcal{N}^2 + 2)^2}$$

(9.117)

Finally, by ignoring in (9.117) the sixth power of ρ_0 gives

$$\boxed{a_1^s \approx \frac{2}{3}i\rho_0^3 \frac{\mathcal{N}^2 - 1}{\mathcal{N}^2 + 2}}$$

(9.118)

By the same procedure we find that the coefficient b_1^s vanishes, that is

$$\boxed{b_1^s \approx 0}$$

(9.119)

Thus for Rayleigh scattering we obtain from (9.101) the amplitude functions

$$S_1(\cos\Theta) \approx i\rho_0^3 \frac{\mathcal{N}^2 - 1}{\mathcal{N}^2 + 2}\pi_1(\cos\Theta), \qquad S_2(\cos\Theta) \approx i\rho_0^3 \frac{\mathcal{N}^2 - 1}{\mathcal{N}^2 + 2}\tau_1(\cos\Theta)$$

(9.120)

From (9.94) we get for $n = 1$

$$\pi_1(\cos\Theta) = 1, \qquad \tau_1(\cos\Theta) = \cos\Theta$$

(9.121)

so that the amplitude functions assume the simple forms

$$S_1(\cos\Theta) \approx i\rho_0^3 \frac{\mathcal{N}^2 - 1}{\mathcal{N}^2 + 2}, \qquad S_2(\cos\Theta) \approx i\rho_0^3 \frac{\mathcal{N}^2 - 1}{\mathcal{N}^2 + 2}\cos\Theta$$

(9.122)

For the $\Theta = 0$ we observe that $S_1 = S_2$.

Utilizing the above approximations, for Rayleigh scattering the relationship (9.102) between the components of the incoming and the scattered electric vector with reference to the scattering plane are given by

$$\boxed{\begin{pmatrix} E_l^s \\ E_r^s \end{pmatrix} = \frac{\rho_0^3}{k_0 r} \frac{\mathcal{N}^2 - 1}{\mathcal{N}^2 + 2}\exp\left[ik_0(r - z)\right] \begin{pmatrix} \cos\Theta & 0 \\ 0 & 1 \end{pmatrix}\begin{pmatrix} E_l^i \\ E_r^i \end{pmatrix}}$$

(9.123)

Since $\rho_0 = k_0 a = 2\pi a/\lambda$ it is seen that for Rayleigh scattering the electric field vector of the scattered wave is proportional to λ^{-2} so that the radiance of the scattered light is proportional to λ^{-4}. Thus in the atmosphere scattering of the visible sunlight on air molecules is most efficient for the shorter wavelengths, i.e.

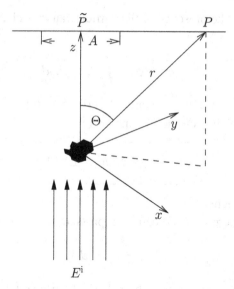

Fig. 9.5 Hypothetical experiment to define the extinction coefficient.

for blue light. This explains why the sky is blue, or in other words, why the sun appears red during dawn and dusk. In the next section we are going to discuss the scattering properties in a little more detail.

9.6 Material characteristics and derived directional quantities

9.6.1 Extinction, scattering and absorption coefficients

Let us consider a single fixed particle of arbitrary shape and composition being illuminated by an electromagnetic wave as shown in Figure 9.5. The origin of the (x, y, z)-coordinate system is somewhere within the particle. One part of the energy which is incident on the particle is scattered and, in general, another part is being absorbed. At a point $\tilde{P}$ which is located at a distance $r = z$ from the origin, the radiative flux ϕ incident on the surface A is given by

$$\phi = \int_A E \, dA \tag{9.124}$$

Now we remove the particle so that none of the incident light is lost. Then at point $\tilde{P}$ the observed flux is given by

$$\phi^i = \int_A E^i \, dA \tag{9.125}$$

The extinction due to the presence of the particle causes a change ϕ^e in the radiant flux as given by

$$\phi^e = \phi^i - \phi = \int_A (E^i - E)dA \qquad (9.126)$$

The change of the radiant flux can be described with the help of the *extinction cross-section* C_{ext} which is defined by means of

$$C_{ext} = \frac{1}{E^i} \int_A (E^i - E)dA \qquad (9.127)$$

Thus C_{ext} has the dimension (m²).

Similarly, the scattered flux ϕ^s can be expressed by means of

$$\phi^s = \oint_{4\pi r^2} E^s dA \qquad (9.128)$$

where the integration extends over the spherical surface defined by the radius r. Analogously to (9.127) we define the *scattering cross-section* by means of

$$C_{sca} = \frac{1}{E^i} \oint_{4\pi r^2} E^s dA \qquad (9.129)$$

Finally, we need to define the *absorption cross-section*. Since the extinction may be treated as a combination of scattering and absorption, we may introduce

$$C_{abs} = \frac{1}{E^i} \oint_{4\pi r^2} (E^i - E^s - E)dA \qquad (9.130)$$

as the defining equation for C_{abs} so that

$$C_{abs} + C_{sca} = C_{ext} \qquad (9.131)$$

Hence in the absence of absorption $C_{ext} = C_{sca}$.

If we divide the quantities C_{ext}, C_{sca} and C_{abs} by the geometric cross-section G of the particle, we obtain the dimensionless *efficiency factors for absorption, scattering* and *extinction*

$$Q_{abs} = \frac{C_{abs}}{G}, \qquad Q_{sca} = \frac{C_{sca}}{G}, \qquad Q_{ext} = \frac{C_{ext}}{G} \qquad (9.132)$$

In case of the Mie theory which was developed for spherical particles, we use $G = \pi a^2$ since the cross-section of a sphere with radius a is a circle having the same radius.

In the following analysis we will assume that the particles have spherical shape. Reference to equation (9.102) shows that each component of the electric vector can

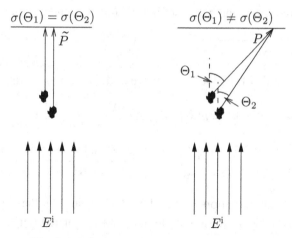

Fig. 9.6 Influence of the origin on the phase.

be expressed in the form

$$u^s = S(\Theta) \frac{\exp[ik_0(r - z)]}{ik_0 r} u^i \tag{9.133}$$

Here, u^s and u^i are the amplitudes of the scattered and the incident wave and $S(\Theta)$ is the amplitude function. The amplitude function is in general complex and can also be written in the form

$$S(\Theta) = s(\Theta) \exp[i\sigma(\Theta)] \tag{9.134}$$

where σ is the phase of the scattered wave. The quantity s is positive and σ is real.

In general $s(\Theta)$ is independent of the choice of the origin while $\sigma(\Theta)$ depends on this choice. The only exception is the case with $\Theta = 0$ since a displacement of the origin does not change the scattering angle. Figure 9.6 illustrates the situation with $\Theta = 0$ (left panel) and $\Theta \neq 0$ (right panel).

At every point outside the particle the electromagnetic wave can be separated into two parts representing the scattered wave in direction P and the incoming wave. In order to determine the extinction caused by the presence of the particle we need to know the amplitude of the wave at point $\tilde{P}$ which is located at a distance $r = z$ from the scattering center where the scattering angle is zero, see Figure 9.5. Since in this case the directions of the incoming plane wave and the scattered spherical wave are the same it is impossible to distinguish between the two waves. In order to make a distinction possible one chooses a point which is very close to $\tilde{P}$ so that the scattering angle for all practical purposes is still zero. For the coordinates of this point we have $z \gg x$ and $z \gg y$ so that we may use the approximation

$$r = \sqrt{x^2 + y^2 + z^2} \approx z + \frac{x^2 + y^2}{2z} \approx z + \frac{x^2 + y^2}{2r} \qquad (9.135)$$

Adding the amplitude of the incoming wave and the scattered wave we may write

$$u = u^i + u^s = u^i \left(1 + S(0)\frac{\exp\left[ik_0(r - z)\right]}{ik_0 r} \right)$$
$$\approx u^i \left(1 + S(0)\frac{\exp\left[ik_0(x^2 + y^2)/2r\right]}{ik_0 r} \right) \qquad (9.136)$$

The radiance of the radiation is equal to the square of the amplitude. Since no integration over the scattering angle needs to be carried out to obtain the flux density E we have

$$E = E^i \left(1 + S(0)\frac{\exp\left[ik_0(x^2 + y^2)/2r\right]}{ik_0 r} \right) \left(1 + S(0)\frac{\exp\left[ik_0(x^2 + y^2)/2r\right]}{ik_0 r} \right)^*$$
$$= E^i(1 + a + ib)(1 + a - ib) = E^i(1 + a^2 + 2a + b^2) \quad \text{with}$$
$$a = \Re \left(S(0)\frac{\exp\left[ik_0(x^2 + y^2)/2r\right]}{ik_0 r} \right), \quad b = \Im \left(S(0)\frac{\exp\left[ik_0(x^2 + y^2)/2r\right]}{ik_0 r} \right)$$

$$\qquad (9.137)$$

Since a^2 and b^2 are proportional to r^{-2} we ignore these two terms for large r. Hence we obtain for the total flux density

$$E = E^i \left[1 + 2\Re \left(S(0)\frac{\exp\left[ik_0(x^2 + y^2)/2r\right]}{ik_0 r} \right) \right] \qquad (9.138)$$

from which we find the radiative flux with respect to the surface A as

$$\phi = \int_A E \, dA = \int_A E^i \, dA + \int_A E^i 2\Re \left(S(0)\frac{\exp\left[ik_0(x^2 + y^2)/2r\right]}{ik_0 r} \right) dA \quad (9.139)$$

By comparing (9.127) with (9.139) we obtain for the extinction coefficient

$$C_{\text{ext}} = -\int_A 2\Re \left(S(0)\frac{\exp\left[ik_0(x^2 + y^2)/2r\right]}{ik_0 r} \right) dA \qquad (9.140)$$

which is an useless expression unless we are able to carry out the integration. To accomplish this, first of all we use the substitutions

$$\alpha^2 = \frac{k_0 x^2}{2r}, \quad \beta^2 = \frac{k_0 y^2}{2r} \implies dA = dx \, dy = \frac{2r}{k_0} d\alpha d\beta \qquad (9.141)$$

After a little algebra we obtain

$$
\begin{aligned}
C_{\text{ext}} = &-\int_{\alpha_1}^{\alpha_2}\int_{\beta_1}^{\beta_2}\frac{4}{k_0^2}\Re[S(0)(\cos\alpha^2\sin\beta^2 + \cos\beta^2\sin\alpha^2)]d\beta d\alpha \\
&-\int_{\alpha_1}^{\alpha_2}\int_{\beta_1}^{\beta_2}\frac{4}{k_0^2}\Im[S(0)(\cos\alpha^2\cos\beta^2 - \sin\beta^2\sin\alpha^2)]d\beta d\alpha
\end{aligned}
\tag{9.142}
$$

The integrals in (9.142) can be evaluated if the limits of the double integral are extended to infinity. This procedure is legal since only points $x \ll z$ and $y \ll z$ effectively contribute to the result. Observing

$$
\int_{-\infty}^{\infty}\sin x^2 dx = \int_{-\infty}^{\infty}\cos x^2 dx = \sqrt{\frac{\pi}{2}}
\tag{9.143}
$$

we obtain the final result for the extinction cross-section

$$
\boxed{C_{\text{ext}} = -\frac{4\pi}{k_0^2}\Re[S(0)]}
\tag{9.144}
$$

The radiance of the light wave components I_l^s and I_r^s are obtained by squaring the corresponding components of the electric vector. Thus we find from (9.102)

$$
I_l^s = \frac{1}{k_0^2 r^2}S_2(\Theta)S_2^*(\Theta)I_l^i, \qquad I_r^s = \frac{1}{k_0^2 r^2}S_1(\Theta)S_1^*(\Theta)I_r^i
\tag{9.145}
$$

The total scattered light intensity is given by

$$
I^s = I_l^s + I_r^s = \frac{1}{2k_0^2 r^2}[S_1(\Theta)S_1^*(\Theta) + S_2(\Theta)S_2^*(\Theta)]I^i
\tag{9.146}
$$

with $I_l^i = I_r^i = I^i/2$ which is true for natural light. In the general case where the light is polarized, the zeros in the matrix on the right-hand side of (9.102) are replaced by other quantities. In which way polarized light must be treated in radiative transfer will be discussed in the following chapter.

We define the *Mie intensity functions* by means of

$$
\boxed{i_1(\cos\Theta) = S_1(\cos\Theta)S_1^*(\cos\Theta), \qquad i_2(\cos\Theta) = S_2(\cos\Theta)S_2^*(\cos\Theta)}
\tag{9.147}
$$

so that I^s can be written as

$$
I^s = \frac{i_1(\cos\Theta) + i_2(\cos\Theta)}{2k_0^2 r^2}I^i
\tag{9.148}
$$

The functions i_1 and i_2 refer to light vibrating perpendicularly and parallel to the scattering plane. For a given scattering angle Θ, the flux densities can be stated as

$$
E^s = \frac{i_1(\cos\Theta) + i_2(\cos\Theta)}{2k_0^2 r^2}E^i
\tag{9.149}
$$

Hence, with reference to equation (9.129) the scattering cross-section is given by

$$C_{sca} = \int_0^{2\pi} \int_{-1}^1 \frac{i_1(x) + i_2(x)}{2k_0^2} dx\, d\varphi \quad \text{with} \quad x = \cos\Theta \tag{9.150}$$

In this equation we substitute for i_1 and i_2 the definitions S_1 and S_2 as given by (9.101) and find

$$C_{sca} = \frac{\pi}{k_0^2} \int_{-1}^1 \sum_{n=1}^{\infty} \sum_{m=1}^{\infty} c_n c_m \left(a_n^s a_m^{s*} + b_n^s b_m^{s*}\right) [\pi_n(x)\pi_m(x) + \tau_n(x)\tau_m(x)]\, dx$$

$$+ \frac{\pi}{k_0^2} \int_{-1}^1 \sum_{n=1}^{\infty} \sum_{m=1}^{\infty} c_n c_m \left(a_n^s b_m^{s*} + b_n^s a_m^{s*}\right) [\pi_n(x)\tau_m(x) + \tau_n(x)\pi_m(x)]\, dx \tag{9.151}$$

with $c_i = (2i + 1)/[i(i + 1)]$. According to (9.94) the functions $\pi_n(x)$ and $\tau_n(x)$ may be written as

$$\pi_n(x) = \frac{P_n^1(x)}{\sqrt{1 - x^2}}, \quad \tau_n(x) = -\sqrt{1 - x^2}\frac{dP_n^1}{dx} \tag{9.152}$$

Recalling the orthogonality relation (9.50) we find the condition

$$\int_{-1}^1 [\pi_n(x)\pi_m(x) + \tau_n(x)\tau_m(x)]\, dx$$

$$= \int_{-1}^1 \left[\frac{1}{1 - x^2}P_n^1(x)P_m^1(x) + (1 - x^2)\frac{dP_n^1}{dx}\frac{dP_m^1}{dx}\right] dx = \frac{2n^2(n + 1)^2}{2n + 1}\delta_{n,m} \tag{9.153}$$

Furthermore, we easily obtain the orthogonality condition

$$\int_{-1}^1 [\pi_n(x)\tau_m(x) + \tau_n(x)\pi_m(x)]\, dx = -\int_{-1}^1 \frac{d}{dx}[P_n^1(x)P_m^1(x)]\, dx = 0 \tag{9.154}$$

so that the scattering cross-section C_{sca} can finally be written as

$$\boxed{C_{sca} = \frac{2\pi}{k_0^2}\sum_{n=1}^{\infty}(2n + 1)\left(a_n^s a_n^{s*} + b_n^s b_n^{s*}\right)} \tag{9.155}$$

Thus C_{sca} can be expressed entirely by the coefficients a_n^s and b_n^s.

Next we will show that C_{ext} can also be formulated in terms of these coefficients. From textbooks on spherical harmonics and from mathematical tables, such as

Jahnke and Emde (1945), we may extract the expression

$$P_n^m(x) = \frac{(n+m)!}{2^m m!(n-m)!}(1-x^2)^{m/2}\left(1 - \frac{(n-m)(n+m+1)}{m+1}\frac{1-x}{2} + \cdots\right)$$

$$(9.156)$$

for the associated Legendre polynomials. For the scattering angle $\Theta = 0$ so that $x = 1$ we find for $m = 1$ the following simple expressions for the π_n and τ_n functions

$$\pi_n(x)\big|_{x=1} = \frac{n(n+1)}{2}, \qquad \tau_n(x)\big|_{x=1} = \frac{n(n+1)}{2} \qquad (9.157)$$

Thus for the function $S(0)$ appearing in (9.144) we obtain

$$S(0) = S_1(0) = S_2(0) = \sum_{n=1}^{\infty} \frac{2n+1}{2}(a_n^s + b_n^s) \qquad (9.158)$$

yielding for C_{ext}

$$\boxed{C_{\text{ext}} = -\frac{2\pi}{k_0^2}\Re\left(\sum_{n=1}^{\infty}(2n+1)(a_n^s + b_n^s)\right)} \qquad (9.159)$$

We will now briefly discuss Rayleigh scattering which we have treated as a special case of Mie scattering. Without going into details, there is a problem if we try to evaluate the Rayleigh extinction coefficient for a real index of refraction by using equation (9.144). In this case we obtain $C_{\text{ext}} = 0$ which cannot be correct. Van de Hulst (1957) and Goody (1964a) trace this problem back to the Rayleigh theory where radiation reaction on the oscillating dipole is neglected. Because of this the phase of the scattered wave is incorrect. Due to our simplified treatment of a_n^s and b_n^s we have introduced the same type of problem.

We overcome this problem by recalling that in the absence of absorption the extinction cross-section is equal to the scattering cross-section. Thus we find the scattering cross-section by substituting (9.118) and (9.119) into (9.155). A brief derivation shows that the scattering cross-section for a dielectric sphere of radius a can then be expressed by

$$C_{\text{sca}} = \frac{8\pi}{3k_0^2}\frac{(N^2-1)^2}{(N^2+2)^2}\rho_0^6 \qquad (9.160)$$

This equation was derived for incident linearly polarized light, but it is also valid for incident natural light. The reason for this is that unpolarized light can be treated as two beams of incoherent light of equal intensity. We have used this fact already when deriving equation (9.155).

We will close this section with a few remarks stating that the Rayleigh theory can be improved. Since molecules are not perfect dielectric spheres a correction term must be included in equation (9.160) which is known as a *depolarization correction*. Practically all tabulations of the scattering coefficient include this factor showing that this increases the extinction coefficient by about 7%. A condensed derivation of the correction factor is given by Goody (1964a). Finally, we would like to remark that the inverse power law for the scattering coefficient is not exactly a fourth power law since the index of refraction is wavelength dependent and slightly decreasing with increasing wavelength. If this fact is incorporated into the theory then within the visible spectrum the scattering coefficient is proportional to $\lambda^{-4.08}$.

9.6.2 The scattering function and the scattering phase function

At this point we wish to discuss the scattering of unpolarized light by a spherical particle contained in a small volume element ΔV. In analogy to the scattering function $S(\mu, \varphi, \mu', \varphi')$ introduced in terms of the principles of invariance (see Chapter 3), we now wish to define the *scattering function* $\tilde{P}(\cos\Theta)$ in the form needed to evaluate the RTE. This type of scattering function is defined as the ratio of the scattered radiant flux $d\phi^s$ per solid angle element $d\Omega$ and volume element ΔV to the incoming radiant flux density dE^i, that is

$$\tilde{P}(\cos\Theta) = \frac{d\phi^s(\cos\Theta)}{\Delta V d\Omega^s dE^i} = \frac{r^2 dE^s(\cos\Theta)}{\Delta V dE^i} \qquad (9.161)$$

The second expression follows from $d\Omega = dA/r^2$ and $dE^s = d\phi^s/dA$. The units of $\tilde{P}(\cos\Theta)$ are ($m^{-1}\,sr^{-1}$). Utilizing (9.149) we obtain immediately

$$\tilde{P}(\cos\Theta) = \frac{i_1(\cos\Theta) + i_2(\cos\Theta)}{2k_0^2 \Delta V} \qquad (9.162)$$

The scattering function agrees with the differential scattering coefficient introduced in Section 1.6.2, see (1.42). Thus the ordinary *scattering coefficient*, defined in (1.44), may also be written as

$$k_{sca} = \oint_{4\pi} \tilde{P}(\cos\Theta) d\Omega \qquad (9.163)$$

Analogously to (1.45) we obtain the *scattering phase function* $\mathcal{P}(\cos\Theta)$ as

$$\mathcal{P}(\cos\Theta) = \frac{4\pi}{k_{sca}} \tilde{P}(\cos\Theta) \qquad (9.164)$$

Hence we directly see that $\mathcal{P}(\cos\Theta)$ is normalized, that is

$$\oint_{4\pi} \mathcal{P}(\cos\Theta) d\Omega = 4\pi \qquad (9.165)$$

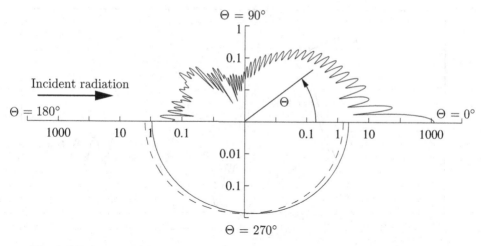

Fig. 9.7 Polar plot of $\mathcal{P}(\cos \Theta)$ for $\mathcal{N} = 1.33 + 0i$ and three different size parameters. Dashed curve: $x = 0.01$, lower solid curve: $x = 1$, upper solid curve: $x = 50$.

Finally, it is a simple task to show that for Rayleigh scattering the phase function reduces to

$$\mathcal{P}_{\text{Rayleigh}}(\cos \Theta) = \frac{3}{4}(1 + \cos^2 \Theta) \qquad (9.166)$$

The proof of this equation will be left to the exercises. Application of (9.166) to (6.25) shows that in case of Rayleigh scattering the *backscattering coefficient* is given by 0.5.

9.7 Selected results from Mie theory

In this section we will present various results which follow from the Mie theory. Figure 9.7 depicts a polar plot of the Mie scattering phase function $\mathcal{P}(\cos \Theta)$ for a nonabsorbing sphere with refractive index $\mathcal{N} = 1.33 + 0i$. The figure shows three curves for the size parameter $x = 2\pi r/\lambda$ selected as 0.01, 1 and 50. The value of $\mathcal{P}(\cos \Theta)$ is given by the radial distance between a point on the curve and the origin of the polar plot. The Mie phase function has rotational symmetry with respect to the axis of incidence. Thus, for better comparison all three curves have been drawn in a single plot, i.e. for each size parameter the full phase function is obtained by reflecting the corresponding curve on the axis of incidence.

As can be seen from the figure, for very small size parameters of $x = 0.01$ (dashed curve) the phase function is symmetric with respect to the ordinate denoting the scattering directions $\Theta = 90°$ and $\Theta = 270°$. For such a small size parameter, $\mathcal{P}(\cos \Theta)$ approaches the Rayleigh phase function (9.166). For $x = 1$ this symmetry

Light scattering theory for spheres

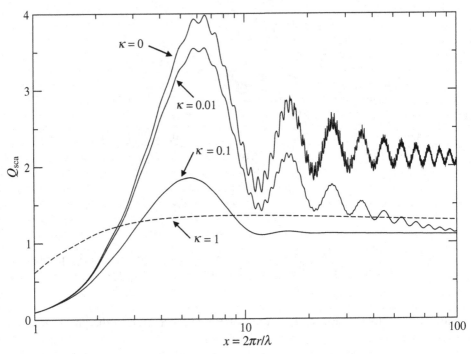

Fig. 9.8 Efficiency factor for scattering for different values of κ.

of the phase function with respect to the ordinate is approximate only with a small enhancement in the forward direction. Nevertheless the curve is still rather smooth. For $x = 50$ the phase function is very asymmetric with respect to the ordinate showing a very strong peak in the forward direction. Note the logarithmic scale of the axes of the polar plot. Moreover, the curve is characterized by numerous ripples indicating the complex scattering behavior of particles with large size parameters. For visible wavelengths $x = 1$ is representative for a small particle having a radius of approximately 0.1 μm, while $x = 50$ is typical for a cloud droplet with radius 5 μm.

In order to calculate the wavelength-dependent efficiency factors Q_{abs}, Q_{sca} and Q_{ext} as defined in (9.132), we must specify the radius r of the scattering particle and the wavelength λ. These two physically significant quantities enter the Mie scattering coefficients a_n^s and b_n^s in terms of the size parameters ρ_0 and ρ, see (9.90) and (9.92). For the evaluation of (9.92), the wavelength-dependent complex index of refraction $\mathcal{N}$, defined in (8.253), is also needed. This quantity can be extracted from suitable tables. As soon as the coefficients a_n^s, b_n^s are available, the scattering and extinction cross-sections C_{sca}, C_{ext} can be calculated from (9.155) and (9.159), whereas the absorption cross-section follows from (9.131).

As a first example, let us consider Figures 9.8–9.10 which show the distribution of Q_{sca}, Q_{abs} and Q_{ext} as a function of the size parameter $x = \rho_0 = 2\pi r/\lambda$ for a fixed

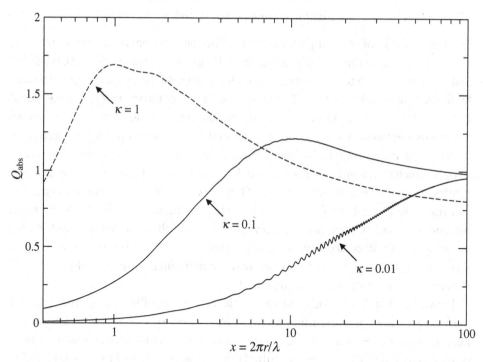

Fig. 9.9 Efficiency factor for absorption for different values of κ.

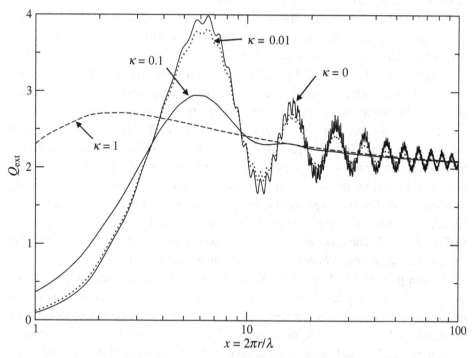

Fig. 9.10 Efficiency factor for extinction for different values of κ.

real part $n = 1.33$ of the complex index of refraction. This value is representative of pure water in the visible part of the spectrum. By assuming values of $\kappa = 0, 0.01, 0.1$ and $\kappa = 1.0$, we wish to investigate in which way the efficiency factors are modified by nonzero absorption indices. For these κ-values four curves of Q_{sca} are depicted in Figure 9.8. In the right part of the figure, the distribution Q_{sca} not only shows a wave-like behavior with decreasing amplitudes but also many ripples interfering with the distribution curve. The major maxima and minima result from interference of light which is transmitted and diffracted by the spherical particle. The superposed ripples are not numerical inaccuracies. They are due to edge rays which are grazing and traveling around the sphere thereby emitting small amounts of energy in all directions. These ripples, however, are not of major physical concern and importance to us. As will be seen, even small κ-values, e.g. $\kappa = 0.01$, will almost remove this fine structure. For $\kappa = 0.1$ only one major maximum is observed while for $\kappa = 1$ the entire wave structure has disappeared.

Figure 9.9 displays the distributions of the absorption efficiency factors Q_{abs} for the values of the absorption index κ used in the previous figure. If no absorption takes place, i.e. $\kappa = 0$, we must have $Q_{abs} = 0$. It is seen that with increasing values of κ the maxima of the curves are shifted toward lower values of the size parameter. Ripple patterns are barely visible, only for $\kappa = 0.01$ they can be identified for values of the size parameter of approximately 5–50.

Figure 9.10 depicts the distributions of the extinction efficiency factors Q_{ext}. For $\kappa = 0$, of course, the efficiency factor for scattering and extinction are identical. Of particular interest is the asymptotic behavior of Q_{ext} for very large values of x where Q_{ext} approaches the value of 2 for all κ. At first it is surprising that the extinction cross-section is twice as large as the geometrical cross-section of the spherical particle. This apparently contradicts the observations, but the effect is real. A part of the light is scattered in the forward direction and cannot be distinguished from the incoming light. Since the particle is large, the so-called *extinction paradox* can be explained in terms of geometric optics. A very minute part of the incoming light traverses the sphere in the direction of the scattering angle zero. The remaining light intercepted by the large particle suffers a change in direction by reflection and refraction and is, therefore, scattered out of the forward beam. This explains one half of $Q_{ext} = 2$. The other half is the radiation which is diffracted by the 'edge' of the sphere. According to *Babinet's principle* an opaque circular disk forms the same diffraction pattern as a hole of the same radius in an opaque screen. *Fraunhofer diffraction theory* (incident and diffracted wave are essentially plane) shows that all rays passing through the hole, except for the axial ray, are deviated or scattered. Most of the edge-diffracted light is contained within the maximum of the diffraction pattern centered around the forward direction whose angular width is defined by the

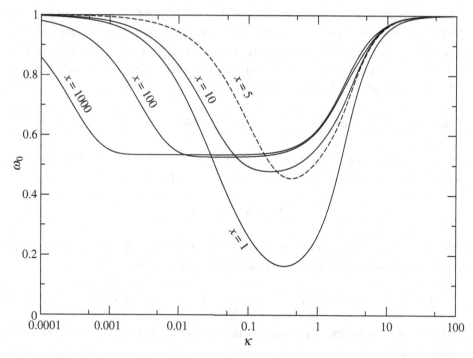

Fig. 9.11 Single scattering albedo as a function of the absorption index for different values of the size parameter.

first minimum of the diffraction pattern. For a simple yet illuminating discussion of the Fraunhofer diffraction pattern of a single slit see, for example, Fowles (1967). Discussions of the extinction paradox can be found in Van de Hulst (1957), Johnson (1960), Goody (1964a), Houghton (1985) and elsewhere.

As another example for the results of the Mie theory Figure 9.11 shows the single scattering albedo ω_0 as a function of κ for different values of the size parameter x. It is not surprising that for a given x with increasing κ, i.e. with increasing absorption, the single scattering albedo decreases strongly. However, instead of going to zero ω_0 decreases to a minimum value and then rises steadily to $\omega_0 = 1$ which corresponds to *perfect scattering*. To explain this effect in simple terms seems difficult, but it is the result of exact Mie computations. In order to at least qualitatively appreciate this result, let us consider a plane wave which is incident on the boundary of a medium having a complex index of refraction as defined by (8.253). For normal incidence a simple reflection formula exists as is shown, for example, in Fowles (1967). For metals the absorption index κ is large, resulting in a high value of the reflectance which approaches unity as κ becomes infinite. This compares with $\omega_0 = 1$ when perfect scattering takes place.

In order to model the attenuation caused by a particle population as observed in a cloud, we need to introduce a particle distribution function $n(r)$. Such a function may have various mathematical forms. Here we will model a cloud droplet population with the help of the rather versatile standard modified gamma distribution, a type of function already used by Deirmendjian (1969). It is expressed as

$$n(r) = Cr^{(1-3b)/b} \exp\left(-\frac{r}{ab}\right), \qquad C = N(ab)^{(2b-1)/b}\Gamma\left(\frac{1-2b}{b}\right) \qquad (9.167)$$

where C is a normalization constant and N (cm^{-3}) the total number of cloud particles per unit volume of air which is related to $n(r)$ by

$$N = \int_0^\infty n(r)dr \qquad (9.168)$$

It can be shown that the parameters a and b are equal to the *effective particle radius* and the dimensionless *effective variance* of the size distribution, i.e.

$$a = \frac{1}{G}\int_0^\infty r\pi r^2 n(r)dr, \qquad b = \frac{1}{Ga^2}\int_0^\infty (r-a)^2\pi r^2 n(r)dr \qquad (9.169)$$

G is the total geometric cross-section of the entire particle population per unit volume, that is

$$G = \int_0^\infty \pi r^2 n(r)dr \qquad (9.170)$$

In the exercises it will be shown that the integrations of (9.169) indeed result in the effective radius and the effective variance. As will be seen, the units of the distribution function $n(r)$ are (cm^{-3} cm^{-1}) which is number of particles per cm^3 and per radius interval measured in cm. Of course, we could have used the length unit m, but it is customary to use the smaller unit cm. By adjusting the length parameter a (cm) and the dimensionless parameter b, equation (9.167) may also be applied to describe an aerosol particle spectrum.

For $b = 0.01, 0.1, 0.2$ the gamma function appearing in (9.167) assumes the values $97!, 7!, 2!$. To get an impression of the resulting shape of the size distribution, the normalized form ($N = 1$) cm^{-3} is plotted in Figure 9.12 with $f(r) = n(r, N = 1)$. Six different distributions are shown, three curves for each $a = 1$ and $a = 10$. Two curves refer to $b = 0.01$ (solid), two for $b = 0.1$ (dotted) and finally two curves for $b = 0.2$ (dashed). Inspection of the figure shows that the widths of the particle distribution curves increase with increasing values of b while the heights of the curves decrease with increasing a.

An application of the cloud droplet distribution function is shown in Figure 9.13 displaying the scattering efficieny factor Q_{sca} as a function of the size parameter $x = 2\pi a/\lambda$. Here a, as defined in (9.169) is the effective radius of the standard

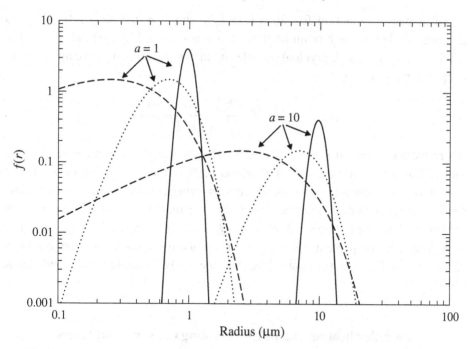

Fig. 9.12 Particle distribution functions for different values of the parameters a and b.

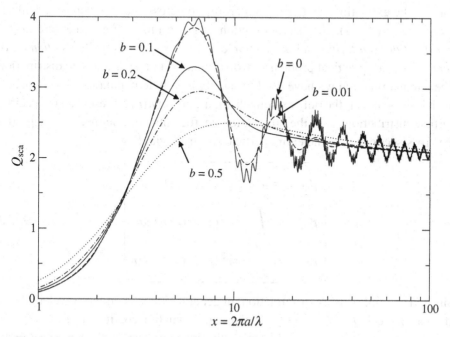

Fig. 9.13 Scattering efficiency factor Q_{sca} for various b and the standard modified gamma distribution function.

modified gamma function for $\mathcal{N} = 1.33 + 0i$ and $b = 0, 0.01, 0.1, 0.2, 0.5$. To produce this figure, each point on the curve was obtained by first calculating the efficiency factor for an individual particle and then integrating over the number size distribution according to

$$Q_{\text{sca}}(a, b) = \frac{\int_0^\infty Q_{\text{sca}}(r)n(r, a, b)dr}{\int_0^\infty n(r, a, b)dr} \tag{9.171}$$

where we have written $n(r, a, b)$ to explicitly show that the particle distribution function depends on the parameters a and b. A further inspection of the figure reveals that with increasing b and increasing values of the size parameter x the ripples, often called the resonances, disappear very quickly even for the rather small value $b = 0.01$. Again the limiting value $Q_{\text{sca}} = 2$ is obtained for large values of x. Note that in the present case $Q_{\text{ext}} = Q_{\text{sca}}$ since the absorption index has been chosen to $\kappa = 0$. More informations of this type can be found in Hansen and Travis (1974).

9.8 Solar heating and infrared cooling rates in cloud layers

At this point we have some idea about the behavior of the important efficiency factors which are needed to evaluate the radiative transfer equation. For a single particle the extinction, scattering and absorption cross-sections can be calculated according to (9.132). If the cross-sections refer to unit volume, they are known as *extinction, scattering* and *absorption coefficients* or collectively as *attenuation coefficients*. All practical problems involve some sort of a particle size distribution. The gamma function we have used to specify a particle population is very useful, but it is by no means the only existing distribution. Let $n(r)$ represent a nonspecified particle distribution, then the attenuation coefficients are obtained by integrating the cross-sections over the entire particle spectrum yielding

$$
\begin{aligned}
\beta_{\text{ext},\lambda} &= \int_0^\infty \pi r^2 Q_{\text{ext},\lambda}(r)n(r)dr \\
\beta_{\text{sca},\lambda} &= \int_0^\infty \pi r^2 Q_{\text{sca},\lambda}(r)n(r)dr \\
\beta_{\text{abs},\lambda} &= \int_0^\infty \pi r^2 Q_{\text{abs},\lambda}(r)n(r)dr
\end{aligned} \tag{9.172}
$$

Since the complex index of refraction, in general, depends on the wavelength, the efficieny factors $Q_{\text{ext}}, Q_{\text{sca}}, Q_{\text{abs}}$ and thus the attenuation coefficients $\beta_{\text{ext}}, \beta_{\text{sca}}, \beta_{\text{abs}}$ are also wavelength dependent. Application of the wavelength dependent attenuation coefficients, in connection with the RTE, results in monochromatic flux

Table 9.1 *Solar heating rates at cloud top and cloud base for two liquid water contents w_1. Cloud base height is 3000 m or 500 m, cloud thickness is 500 m, $A_g = 0$*

Cloud base at 3000 m	$w_1 = 0.1\,\mathrm{g\,m^{-3}}$	$w_1 = 0.2\,\mathrm{g\,m^{-3}}$
Cloud top	1.6 K h^{-1}	2.4 K h^{-1}
Cloud base	0.15 K h^{-1}	0.15 K h^{-1}
Cloud base at 500 m		
Cloud top	0.9 K h^{-1}	1.4 K h^{-1}
Cloud base	0.1 K h^{-1}	0.15 K h^{-1}

densities. To complete a particular problem, an integration over the spectrum must be carried out.

Now we will briefly consider solar heating and infrared cooling rates of cloud layers. It is very difficult to give representative results since radiative temperature changes depend on many parameters. To calculate solar heating rates, we must specify the position of the Sun, the heights of cloud base and cloud top, wavelength-dependent optical parameters, the ground albedo and the cloud temperature which determines the cloud saturation water vapor content. Moreover, the liquid water content and the droplet distribution function must be known. Additional parameters are involved. With the exception of the Sun's position, infrared cooling rates depend on the same quantities. In order to be brief, omitting details, we will simply state a few results. The calculation methods were described previously.

In all cases we assume that the clouds are stratified, that the solar zenith angle $\theta_0 = 30°$ is fixed and that the cloud has a thickness of 500 m. The base of the cloud is placed at 3000 m or 500 m while the ground albedo is assumed to be 0 or 40%. The calculations are based on the Best (1951) droplet size distribution which is similar to (9.167). Both distributions contain exponential factors with negative arguments, and both involve the radius r raised to a negative power. From Welch *et al.* (1976) we have extracted the information stated in Table 9.1. The table shows that for high values of w_1 the heating rates at cloud top are substantially larger than for $w_1 = 0.1\,\mathrm{g\,m^{-3}}$. At the cloud base both heating rates are nearly identical. The reason is that the solar beam arriving at the bottom of either cloud is nearly exhausted.

Suppose that the ground albedo is raised to the rather high value of 0.4. In case of the smaller liquid water content $w_1 = 0.1\,\mathrm{g\,m^{-3}}$ and cloud base height of 500 m, the heating rate at cloud base is increased to 0.20 K h^{-1}. In the upper half of the cloud no distinction of the heating rates for $A_g = 0$ and $A_g = 0.4$ is found. The differences in heating rates for $A_g = 0$ at the cloud tops but different cloud base heights is explained easily. At the cloud top of the lower cloud, the solar flux density of the parallel solar radiation is substantially smaller than at the higher cloud top so that less energy is available for solar heating. This difference results from the water vapor absorption along the atmospheric path from 3500 to 1000 m.

Next we will give a few results on infrared cooling. Of some interest are the cooling rates which are calculated as part of a radiation fog prediction model. Since the model contains detailed microphysics, the droplet distribution is not prescribed but actually calculated as a time-dependent table. During the time period which we briefly consider, the liquid water content in the upper part of a 30 m deep ground fog varies considerably resulting in varying cooling rates. For a particular situation during a period of about 30 min, maximum cooling rates of 3–4 K h^{-1} are calculated in the upper part of the fog. The liquid water content at the points of maximum cooling was about 0.1 g m^{-3}. In general, in a developing fog cooling rates vary considerably with height and time. Detailed investigations of fog modeling show the importance of reliable radiative transport models. Some details regarding the influence of radiative cooling on the development of radiation fog are given by Bott *et al.* (1990) and Siebert *et al.* (1992).

Fu *et al.* (1997) have calculated cooling rates of various clouds. For a low cloud (cloud base height at 1 km, cloud thickness 1 km) and a liquid water content $w_l = 0.22$ g m^{-3} they find a cloud top cooling of almost 1.5 K h^{-1} and a small amount of cloud-base heating. For $w_l = 0.28$ g m^{-3} they calculate a cloud-top cooling of 2.5 K h^{-1} for a middle high cloud (cloud base height at 4 km, cloud thickness 1 km) and a cloud-base heating of 0.7 K h^{-1}. The heating of the cloud base is caused by the trapping of long-wave energy emitted by the considerably warmer surface. The results were calculated by assuming mid-latitude summer conditions.

By utilizing a detailed spectral cloud microphysics approach, Bott *et al.* (1996) have investigated the role of atmospheric radiative transfer for the evolution of the cloud-topped marine boundary layer. Bott (1997) has studied the impact of aerosol particles on the radiative forcing of stratiform clouds. He showed that the radiative forcing of the clouds is strongly affected by the physico-chemical properties of the aerosol particles yielding different reflectivities and absorption characteristics of the clouds.

9.9 Problems

9.1: As a review: Use the curl equations (9.1) to obtain the wave equations.

9.2: Show that equation (9.29b) can be transformed to (9.32).

9.3: The auxiliary vector $\mathbf{M}_\psi$ is defined in (9.38). Show that this vector satisfies the wave equation (9.39).

9.4: Show that the relations (9.43) are valid.

9.5: Show in detail that $\mathbf{N}_{\psi,r}$ as defined in (9.45) can be written as (9.46).

9.6: Use the definition of $\mathbf{N}_\psi$ in (9.38) to show that $\mathbf{M}_\psi$ can be written as (9.40).

9.7: Starting with the proper boundary conditions, find by detailed mathematical operations the second, the third and the fourth equation of (9.91).

9.8: Verify that $B_n = -i\,A_n$, $C_n = -A_n$ and $B_n = D_n$, see (9.79).

9.9: Show that (9.142) follows from (9.140).

9.10: Verify (9.169).

9.11: Verify equation (9.166).

10

Effects of polarization in radiative transfer

So far we have considered radiative transfer in the atmosphere by means of the scalar form of the radiative transfer equation. Since light is a vector quantity the formulation of the radiative transfer equation was only approximate but sufficiently accurate for most practical purposes in connection with atmospheric energetics. A complete description of the radiation field, however, must include the state of polarization since scattering, in general, produces polarized light. In this chapter we will derive some mathematical statements about the *polarization ellipse. Linear* and *circular polarization* follow as special cases.

At this point we wish to refer the reader to the interesting article by Hovenier and van der Mee (1983) about the fundamental relationships relevant to the transfer of polarized light in a scattering atmosphere. Here we can only give an introductory discussion which may serve as a basis to tackle the more advanced papers.

10.1 Description of elliptic, linear and circular polarization

Let us consider the propagation of a plane time harmonic wave in a Cartesian coordinate system. The components of the electric (magnetic) vector are of the form $\Re[a \exp(-i(\tau + \delta))] = a \cos(\tau + \delta)$ where a and δ are the amplitude and the phase angle, respectively. The quantity τ, defined by

$$\tau = \omega t - \mathbf{k} \cdot \mathbf{r} \tag{10.1}$$

is the variable part of the phase factor and $\mathbf{k}$, as before, represents the propagation vector of the electromagnetic wave. To be specific, let us assume that the wave propagates along the positive z-axis. Then the electric vector can be written as

$$\mathbf{E} = a_1 \Re[\exp(-i(\tau + \delta_1))]\, \mathbf{i} + a_2 \Re[\exp(-i(\tau + \delta_2))]\, \mathbf{j} = E_x \mathbf{i} + E_y \mathbf{j}$$

with $\quad E_x = a_1 \cos(\tau + \delta_1), \quad E_y = a_2 \cos(\tau + \delta_2), \quad E_z = 0$

$$\tag{10.2}$$

where δ_1 and δ_2 are constant phase angles. $E_z = 0$ since the field is transversal. The point with coordinates (E_x, E_y) describes a certain curve in space which will now be investigated in some detail.

In order to eliminate τ in (10.2) we first expand the cosine functions yielding

$$\text{(a)} \quad \frac{E_x}{a_1} = \cos \tau \cos \delta_1 - \sin \tau \sin \delta_1$$

$$\text{(b)} \quad \frac{E_y}{a_2} = \cos \tau \cos \delta_2 - \sin \tau \sin \delta_2 \tag{10.3}$$

Next we multiply (10.3a) by $\sin \delta_2$ and (10.3b) by $\sin \delta_1$ and subtract the resulting equations from each other. After some simple manipulations we obtain

$$\frac{E_x}{a_1} \sin \delta_2 - \frac{E_y}{a_2} \sin \delta_1 = \cos \tau \sin(\delta_2 - \delta_1) \tag{10.4}$$

Multiplication of (10.3a) by $\cos \delta_2$ and (10.3b) by $\cos \delta_1$ and subtraction of the resulting equations gives

$$\frac{E_x}{a_1} \cos \delta_2 - \frac{E_y}{a_2} \cos \delta_1 = \sin \tau \sin(\delta_2 - \delta_1) \tag{10.5}$$

Squaring and adding (10.4) and (10.5) results in

$$\left(\frac{E_x}{a_1}\right)^2 + \left(\frac{E_y}{a_2}\right)^2 - \frac{2E_x E_y}{a_1 a_2} \cos \delta = \sin^2 \delta \quad \text{with} \quad \delta = \delta_2 - \delta_1 = const \tag{10.6}$$

which is independent of τ, i.e. independent of the space coordinate z and of time t.

Now we consider the general equation of a conic

$$Ax^2 + Bxy + Cy^2 + Dx + Ey + F = 0 \tag{10.7}$$

The conditions to represent an ellipse, parabola or hyperbola are

$$B^2 - 4AC \begin{cases} < 0 & \text{ellipse} \\ = 0 & \text{parabola} \\ > 0 & \text{hyperbola} \end{cases} \tag{10.8}$$

Comparison of (10.7) and (10.6) with

$$\frac{4(\cos^2 \delta - 1)}{a_1^2 a_2^2} \leq 0 \tag{10.9}$$

shows that (10.6) is the equation of an ellipse. Hence the endpoint of the electric field vector traces out an ellipse as shown in Figure 10.1. Therefore, the light is elliptically polarized. This ellipse is inscribed into a rectangle whose sides are parallel to the coordinate axes. The sides of the rectangle have the lengths $2a_1$ and $2a_2$. The ellipse touches the rectangle at the four points P_i, $i = 1, \ldots, 4$. From

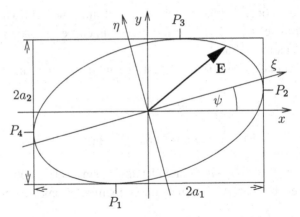

Fig. 10.1 Vibrational ellipse for the electric field vector.

(10.6) and Figure 10.1 one may easily verify that at these points the values of (E_x, E_y) are given by

$$P_1: \quad (E_x, E_y) = (-a_1 \cos \delta, -a_2), \quad P_2: \quad (E_x, E_y) = (a_1, a_2 \cos \delta)$$
$$P_3: \quad (E_x, E_y) = (a_1 \cos \delta, a_2), \quad P_4: \quad (E_x, E_y) = (-a_1, -a_2 \cos \delta)$$
$$(10.10)$$

Figure 10.1 shows that the axes of the ellipse are not in the x- and y-directions. Thus, it seems expedient to rotate the (x, y)-cooordinate system in the counterclockwise direction by the angle ψ so that the directions ξ and η of the new coordinate system are along the semi-axes of the ellipse. The components (E_x, E_y) and (E_ξ, E_η) are related by the well-known transformation rule

$$\begin{pmatrix} E_\xi \\ E_\eta \end{pmatrix} = \begin{pmatrix} \cos \psi & \sin \psi \\ -\sin \psi & \cos \psi \end{pmatrix} \begin{pmatrix} E_x \\ E_y \end{pmatrix} \tag{10.11}$$

With respect to the (ξ, η)-system the equation of the ellipse assumes the normal form

$$\frac{E_\xi^2}{a^2} + \frac{E_\eta^2}{b^2} = 1 \tag{10.12}$$

where a and b are the lengths of the semi-axes with $a > b$. In the parametric form we may write

$$E_\xi = a \cos (\tau + \delta_0), \quad E_\eta = \pm b \sin (\tau + \delta_0) \quad \text{with} \quad \delta_0 = const \tag{10.13}$$

The chosen sign of the second equation specifies one of the two possibilities in which way the endpoint of the electric vector describes the ellipse.

We will now determine the still unknown quantities a and b appearing in (10.12). Substituting (10.2) and (10.13) into (10.11) we obtain

$$
\begin{aligned}
a\cos(\tau + \delta_0) &= a_1\cos(\tau + \delta_1)\cos\psi + a_2\cos(\tau + \delta_2)\sin\psi \\
\pm b\sin(\tau + \delta_0) &= -a_1\cos(\tau + \delta_1)\sin\psi + a_2\cos(\tau + \delta_2)\cos\psi
\end{aligned}
\qquad (10.14)
$$

Developing (10.14) with the help of the addition theorems for the trigonometric functions and equating the coefficients of $\cos\tau$ and $\sin\tau$ we find

$$
\begin{aligned}
a\cos\delta_0 &= a_1\cos\delta_1\cos\psi + a_2\cos\delta_2\sin\psi \\
a\sin\delta_0 &= a_1\sin\delta_1\cos\psi + a_2\sin\delta_2\sin\psi \\
\pm b\cos\delta_0 &= a_1\sin\delta_1\sin\psi - a_2\sin\delta_2\cos\psi \\
\pm b\sin\delta_0 &= -a_1\cos\delta_1\sin\psi + a_2\cos\delta_2\cos\psi
\end{aligned}
\qquad (10.15)
$$

Omitting further details of the calculations, from these equations the following relations may be easily derived

(a) $a^2 + b^2 = a_1^2 + a_2^2$

(b) $\pm ab = a_1 a_2 \sin\delta$ (10.16)

(c) $(a_1^2 - a_2^2)\sin(2\psi) = 2a_1 a_2 \cos\delta \cos(2\psi)$

It will be convenient for the following discussion to introduce the auxiliary angles α and β by means of

(a) $\tan\alpha = \dfrac{a_2}{a_1}, \quad 0 \le \alpha \le \dfrac{\pi}{2}$

(10.17)

(b) $\tan\beta = \pm\dfrac{b}{a}, \quad -\dfrac{\pi}{4} \le \beta \le \dfrac{\pi}{4}$

Thus, the numerical value of $\tan\beta$ represents the ratio of the axes of the ellipse, called the ellipticity, while the sign of β distinguishes the sense in which the ellipse may be described. We will soon see that β may also be expressed in terms of the angles α and δ.

Substituting (10.17a) into (10.16c) first gives

$$
\tan(2\psi) = \frac{2a_1 a_2 \cos\delta}{a_1^2 - a_2^2} = \frac{2\tan\alpha\cos\delta}{1 - \tan^2\alpha} = \tan(2\alpha)\cos\delta
\qquad (10.18)
$$

Combining the well-known trigonometric identity $\sin(2\beta) = 2\tan\beta/(1 + \tan^2\beta)$ with (10.17b) and (10.16a,b) yields

$$
\sin(2\beta) = \frac{2\tan\beta}{1 + \tan^2\beta} = \pm\frac{2ab}{a^2 + b^2} = \frac{2a_1 a_2}{a_1^2 + a_2^2}\sin\delta
\qquad (10.19)
$$

Similarly, from $\sin(2\alpha) = 2\tan\alpha/(1 + \tan^2\alpha)$ and (10.17a) we have

$$
\sin(2\alpha) = \frac{2a_1 a_2}{a_1^2 + a_2^2}
\qquad (10.20)
$$

so that we finally obtain

$$\sin(2\beta) = \sin(2\alpha) \sin \delta \qquad (10.21)$$

We summarize our results: if a plane electromagnetic wave moving in the positive z-direction is described by means of the quantities (a_1, a_2, δ) it can also be expressed in terms of the quantities (a, b, ψ). Conversely, if (a, b, ψ) are given it is easy to find the amplitudes (a_1, a_2) and the phase difference δ of the wave.

Now we are going to introduce some useful terminology and designate the direction of the electric field as the direction of polarization. In connection with equation (10.13) we stated that the $\pm$ sign determines the two possible senses in which the endpoint of the electric vector may describe the ellipse. Thus we distinguish two types of polarization. If the electric field vector appears to be rotating clockwise at angular speed ω as viewed by an observer toward whom the wave is moving (i.e. looking back at the source or looking against the direction of propagation) we call the polarization right-handed. In the opposite case the polarization is called left-handed.

Substituting the phase difference $\delta = \delta_2 - \delta_1$ and $\tau' = \tau + \delta_1$ into (10.2) we may write

$$E_x = a_1 \cos \tau', \quad E_y = a_2 \cos(\tau' + \delta) = a_2 \left(\cos \tau' \cos \delta - \sin \tau' \sin \delta \right) \quad (10.22)$$

To give an example, we choose two different times t_0 and $t_1 > t_0$ in such a way that for a fixed z-value $\tau'(t_0) = 0$ and $\tau'(t_1) = \pi/2$. Evaluating (10.22) at t_0 and t_1 then gives

$$
\begin{aligned}
E_x(t_0) &= a_1, & E_y(t_0) &= a_2 \cos \delta \\
E_x(t_1) &= 0, & E_y(t_1) &= -a_2 \sin \delta
\end{aligned}
\qquad (10.23)
$$

From this equation we may easily see that the polarization is right-handed if $0 \le \delta \le \pi$ whereas it is left-handed if $\pi \le \delta \le 2\pi$. Unfortunately, the terminology we have used is not of universal usage. Some authors employ the opposite convention so that some care must be exercised when consulting other textbooks.

We conclude this section by investigating two special cases of polarization.

10.1.1 Linear polarization

If the phase difference is $\delta = \delta_2 - \delta_1 = m\pi$, $m = 0, \pm 1, \pm 2, \ldots$, we obtain from (10.22)

$$\frac{E_y}{E_x} = \frac{a_2 \cos(\tau' + m\pi)}{a_1 \cos \tau'} = \frac{a_2}{a_1} \cos m\pi = \frac{a_2}{a_1}(-1)^m \qquad (10.24)$$

Thus, the ellipse has degenerated to a straight line with slope $(a_2/a_1)(-1)^m$ passing through the origin.

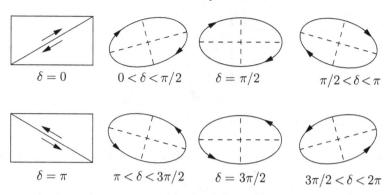

$\delta = 0$ $0 < \delta < \pi/2$ $\delta = \pi/2$ $\pi/2 < \delta < \pi$

$\delta = \pi$ $\pi < \delta < 3\pi/2$ $\delta = 3\pi/2$ $3\pi/2 < \delta < 2\pi$

Fig. 10.2 Various forms of elliptical polarization. Upper panel: right-handed polarization, lower panel: left-handed polarization. After Born and Wolf (1965).

10.1.2 Circular polarization

For $\delta = \pm\pi/2 + 2m\pi$, $m = 0, \pm1, \pm2, \ldots$ and $a_1 = a_2 = a$ (10.22) results in the parametric equation of a circle, that is

(a) $\delta = \dfrac{\pi}{2} + 2m\pi$: $E_x = a \cos \tau'$, $E_y = a \cos(\tau' + \delta) = -a \sin(\tau')$

(b) $\delta = -\dfrac{\pi}{2} + 2m\pi$: $E_x = a \cos \tau'$, $E_y = a \cos(\tau' + \delta) = a \sin(\tau')$

$$(10.25)$$

Thus, the **E** vector is moving on a circle with radius a and the polarization is right-handed in (10.25a) and left-handed in (10.25b). Figure 10.2 depicts different elliptic polarization possibilities and the corresponding phase differences.

10.2 The Stokes parameters

Again we consider a monochromatic plane electromagnetic wave which is moving in the positive z-direction. As already mentioned the wave is completely determined in terms of the parameters (a_1, a_2, δ) or equivalently as function of (a, b, ψ). Stokes (1852) introduced the following four parameters to describe the electromagnetic wave

$$
\boxed{
\begin{array}{ll}
S_0 = a_1^2 + a_2^2, & S_1 = a_1^2 - a_2^2 \\
S_2 = 2a_1a_2 \cos \delta, & S_3 = 2a_1a_2 \sin \delta
\end{array}
}
$$

$$(10.26)$$

Evidently, S_0 is proportional to the intensity of the light.[1] From (10.26) follows

$$S_0^2 = S_1^2 + S_2^2 + S_3^2 \qquad (10.27)$$

---

[1] Here and elsewhere, the expression 'intensity' is used quite loosely. In reality S_0 may be a flux density.

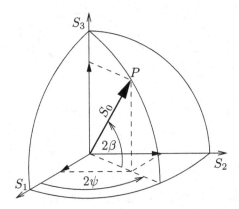

Fig. 10.3 The Poincaré sphere.

so that, analogously to the three parameters (a_1, a_2, δ) or (a, b, ψ), only three of the Stokes parameters are independent. The radiation is said to be completely polarized if (10.27) is valid.

The Stokes parameters can also be expressed with the help of the variables (a, b, ψ)

$$
\begin{aligned}
&S_0 = a^2 + b^2, &&S_1 = S_0 \cos(2\beta) \cos(2\psi) \\
&S_2 = S_0 \cos(2\beta) \sin(2\psi), &&S_3 = S_0 \sin(2\beta)
\end{aligned}
\tag{10.28}
$$

The proof that (10.26) can be transformed to (10.28) is straightforward and will be left to the exercises.

With the help of the Stokes parameters we will now present a graphical description of polarization. Inspection of (10.27) and (10.28) reveals that between the Stokes parameters (S_1, S_2, S_3) and the quantities $(2\psi, 2\beta, S_0)$ the same transformations are valid as between the Cartesian coordinates (x, y, z) and the coordinates (ϑ, φ, r) of a spherical coordinate system. Hence, if we replace the (x, y, z)-coordinate system by an (S_1, S_2, S_3)-system, a point $P(S_1, S_2, S_3)$ on the surface of a sphere may also be expressed by means of $P(2\psi, 2\beta, S_0)$. This type of a sphere is known as the *Poincaré sphere*.

Figure 10.3, depicting Poincaré's sphere, can be used to display various types of polarization. From (10.21) we see that for $0 < \alpha < \pi/2$ and $\sin \delta > 0$ denoting right-handed polarization we obtain $\sin(2\beta) > 0$. From (10.26) and (10.28) we conclude that in this case $S_3 > 0$, i.e. the point P is located in the upper half of the Poincaré sphere. If the light is left-handed polarized we have $S_3 < 0$ so that P is located below the equatorial plane. Linear polarization, expressed by $\delta = m\pi$, $m = 0, \pm 1, \ldots$, yields $\sin \delta = 0$, $\sin(2\beta) = 0$ and, therefore, $S_3 = 0$, i.e. P is located within the equatorial plane. Finally, from (10.26) we see that in case of a circularly polarized electromagnetic wave with $\delta = \pm \pi/2 + 2m\pi$, $m = 0$, $\pm 1, \pm 2, \ldots$ and $a_1 = a_2 = a$ the parameters S_1 and S_2 vanish so that, according to (10.28), $\cos(2\beta) = 0$. This in turn means that P is on the north pole or on the south pole of Poincare's sphere, denoting right-handed and left-handed circular polarization, respectively.

Partially polarized light is a mixture of natural and polarized light. If E_p and E_u represent the constituent flux densities of the polarized and the unpolarized light, then the degree of polarization D_p is defined by

$$D_p = \frac{E_p}{E_p + E_u} \tag{10.29}$$

As can be shown, for partially polarized light the degree of polarization can also be expressed in terms of the Stokes parameters by means of

$$D_p = \frac{\left(S_1^2 + S_2^2 + S_3^2\right)^{1/2}}{S_0} \tag{10.30}$$

For completely polarized light we will now describe the state of polarization in terms of the *Stokes vector* consisting of the four components (S_0, S_1, S_2, S_3). For a concise description we introduce the following notation. Linearly also called plane polarized light is said to be in the *P-state* while right and left circularly polarized light are in the *R-state* and *L-state*. Analogously, elliptically polarized light is in the *E-state*.

Often the Stokes parameters are expressed in normalized form by dividing all elements by S_0. For completely polarized light a table of Stokes vectors may be computed. We will briefly show how this can be done for linearly polarized light. If a_1 and a_2 label the amplitudes of the horizontal and the vertical component of the electromagnetic wave, we have a horizontally polarized wave if $a_1 > 0$ and $a_2 = 0$. Utilizing (10.26) we see that the normalized Stokes vector is now given by

$$\frac{1}{S_0} \begin{pmatrix} S_0 \\ S_1 \\ S_2 \\ S_3 \end{pmatrix} = \begin{pmatrix} 1 \\ 1 \\ 0 \\ 0 \end{pmatrix} \tag{10.31}$$

In this case the light is said to be in the horizontal P-state.

Analogously to (10.31) other polarization states can be identified. Table 10.1 summarizes some polarization states which may be easily obtained by applying the appropriate values of the variables (a_1, a_2, δ) or (a, b, ψ) to (10.26) or (10.28). More complete tables may be found in Shurcliffe (1966) or Deirmendjian (1969).

Table 10.1 *Normalized Stokes vectors for various types of polarized light.
Upper panel: linear polarization. Lower panel: circular and general
elliptical polarization*

horizontal P-state	vertical P-state	45° P-state	−45° P-state
$\begin{pmatrix} 1 \\ 1 \\ 0 \\ 0 \end{pmatrix}$	$\begin{pmatrix} 1 \\ -1 \\ 0 \\ 0 \end{pmatrix}$	$\begin{pmatrix} 1 \\ 0 \\ 1 \\ 0 \end{pmatrix}$	$\begin{pmatrix} 1 \\ 0 \\ -1 \\ 0 \end{pmatrix}$

R-state	L-state	E-state
$\begin{pmatrix} 1 \\ 0 \\ 0 \\ 1 \end{pmatrix}$	$\begin{pmatrix} 1 \\ 0 \\ 0 \\ -1 \end{pmatrix}$	$\begin{pmatrix} 1 \\ \cos(2\beta)\cos(2\psi) \\ \cos(2\beta)\sin(2\psi) \\ \sin(2\beta) \end{pmatrix}$

The previous section largely followed the standard textbook by Born and Wolf (1965) where additional information may be found. We have also made use of the discussions on polarization given in the textbooks by Shurcliffe (1966), Fowles (1966) and Hecht (1987).

10.3 The scattering matrix

10.3.1 Representation of the electric vector in the scattering plane

Let us consider an electromagnetic wave of wavelength λ which is scattered by a spherical particle. In order to describe the scattering process various coordinate systems will now be introduced as shown in Figure 10.4. The basic Cartesian (x, y, z)-coordinate system is fixed in space with origin in the center of the scattering sphere. By drawing meridians from the z-axis to the x-axis and to the y-axis we construct a sphere whose equatorial plane coincides with the (x, y)-plane. The incident wave propagating in the direction $\mathbf{k}^i$ is defined in terms of the Cartesian coordinate system (x^i, y^i, z^i) with unit vectors $(\mathbf{i}^i, \mathbf{j}^i, \mathbf{k}^i)$. Similarly, the scattered wave moving in direction $\mathbf{k}^s$ is defined with respect to the Cartesian coordinate system (x^s, y^s, z^s) with unit vectors $(\mathbf{i}^s, \mathbf{j}^s, \mathbf{k}^s)$. The *scattering plane* is formed by the straight lines in directions $\mathbf{k}^i$ and $\mathbf{k}^s$. The intersection of the scattering plane with the sphere defines the scattering angle Θ. The straight lines pointing in the directions $\mathbf{k}^i$ and $\mathbf{k}^s$ together with the z-axis form two vertical planes whose intersections with

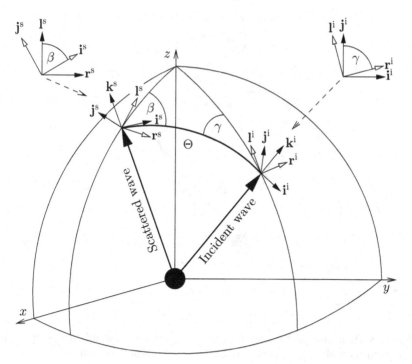

Fig. 10.4 Definition of various coordinate systems needed to describe the scattering on a sphere, see also text.

the sphere define two meridians. The unit vectors $\mathbf{l}^i$ and $\mathbf{l}^s$ are tangential to the meridians while the unit vectors $\mathbf{r}^i$ and $\mathbf{r}^s$ are perpendicular to the corresponding vertical planes. The two systems with the unit vectors $(\mathbf{i}^i, \mathbf{j}^i, \mathbf{k}^i)$ and $(\mathbf{i}^s, \mathbf{j}^s, \mathbf{k}^s)$ are chosen in such a way that $\mathbf{i}^i$ and $\mathbf{i}^s$ are parallel to the scattering plane while $\mathbf{j}^i$ and $\mathbf{j}^s$ are perpendicular to this plane.

We are now ready to apply the results of the Mie theory. In the following the subscripts i and j refer to the components of the incident electric vector along the unit vectors $\mathbf{i}^i$ and $\mathbf{j}^i$. According to (10.2) the incoming electric vector may be expressed by

$$
\begin{aligned}
\mathbf{E}^i &= \left[a_i^i \exp(-i\delta_1)\mathbf{i}^i + a_j^i \exp(-i\delta_2)\mathbf{j}^i \right] \exp\left[i(k_0 z - \omega t) \right] \\
&= \left(E_i^i \mathbf{i}^i + E_j^i \mathbf{j}^i \right) \exp\left[i(k_0 z - \omega t) \right]
\end{aligned}
\tag{10.32}
$$

where a_i^i and a_j^i are real amplitudes. With the help of (9.102) we obtain

$$
\begin{pmatrix} E_i^s \\ E_j^s \end{pmatrix} = \frac{\exp[ik_0(z^s - z^i)]}{ik_0 z^s} \begin{pmatrix} S_2(\cos\Theta) & 0 \\ 0 & S_1(\cos\Theta) \end{pmatrix} \begin{pmatrix} E_i^i \\ E_j^i \end{pmatrix}
\tag{10.33}
$$

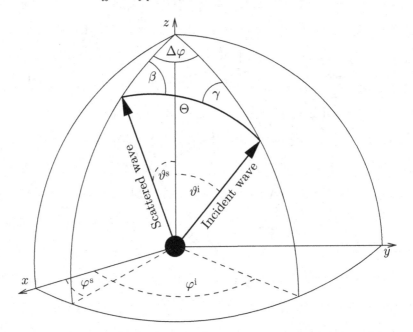

Fig. 10.5 Definition of the variables $(\vartheta^i, \vartheta^s, \Delta\varphi)$ for the formulation of the scattering matrix with respect to the fixed (x, y, z)-coordinate system.

The amplitude functions $S_1(\cos\Theta)$ and $S_2(\cos\Theta)$ are defined in (9.101). Hence the scattered electric vector depends only on the scattering angle Θ. The advantage of this representation is the rotational symmetry of the scattered light with respect to the scattering plane. The disadvantage is that the scattering plane is not fixed in space, but it is continually changing its position. This implies that the corresponding coordinate systems describing the incident and the scattered light are not fixed in space either. In order to avoid this disadvantage we will transform (10.33) in such a way that only variables appear which can be expressed with respect to the fixed (x, y, z)-coordinate system. The required variables are $(\vartheta^i, \vartheta^s)$ and $\Delta\varphi = \varphi^i - \varphi^s$ as shown in Figure 10.5.

With the help of Figure 10.4 we easily find the transformations

$$
\begin{pmatrix} E_i^i \\ E_j^i \end{pmatrix} = \begin{pmatrix} -\cos\gamma & \sin\gamma \\ \sin\gamma & \cos\gamma \end{pmatrix} \begin{pmatrix} E_l^i \\ E_r^i \end{pmatrix}
$$
$$
\begin{pmatrix} E_l^s \\ E_r^s \end{pmatrix} = \begin{pmatrix} \cos\beta & \sin\beta \\ \sin\beta & -\cos\beta \end{pmatrix} \begin{pmatrix} E_i^s \\ E_j^s \end{pmatrix}
$$

(10.34)

Substituting these relations into (10.33) yields

$$
\begin{pmatrix} E_l^s \\ E_r^s \end{pmatrix} = \frac{\exp[ik_0(z^s - z^i)]}{ik_0 z^s} \begin{pmatrix} \cos\beta & \sin\beta \\ \sin\beta & -\cos\beta \end{pmatrix} \begin{pmatrix} S_2(\cos\Theta) & 0 \\ 0 & S_1(\cos\Theta) \end{pmatrix}
$$

$$
\times \begin{pmatrix} -\cos\gamma & \sin\gamma \\ \sin\gamma & \cos\gamma \end{pmatrix} \begin{pmatrix} E_l^i \\ E_r^i \end{pmatrix} \tag{10.35}
$$

$$
= \frac{\exp[ik_0(z^s - z^i)]}{ik_0 z^s} \begin{pmatrix} S_{11} & S_{12} \\ S_{21} & S_{22} \end{pmatrix} \begin{pmatrix} E_l^i \\ E_r^i \end{pmatrix}
$$

where the following abbreviations have been introduced

$$
\begin{aligned}
S_{11} &= -\cos\beta\cos\gamma\, S_2(\cos\Theta) + \sin\beta\sin\gamma\, S_1(\cos\Theta) \\
S_{12} &= \cos\beta\sin\gamma\, S_2(\cos\Theta) + \sin\beta\cos\gamma\, S_1(\cos\Theta) \\
S_{21} &= -\sin\beta\cos\gamma\, S_2(\cos\Theta) - \cos\beta\sin\gamma\, S_1(\cos\Theta) \\
S_{22} &= \sin\beta\sin\gamma\, S_2(\cos\Theta) - \cos\beta\cos\gamma\, S_1(\cos\Theta)
\end{aligned} \tag{10.36}
$$

The spherical triangle with the sides $(\Theta, \vartheta^i, \vartheta^s)$ will now be used to determine the relationship between the scattering angle and the angles $(\vartheta^i, \vartheta^s, \Delta\varphi)$, see Figure 10.5. Spherical trigonometry provides the law of cosines for the sides

$$
\begin{aligned}
&\text{(a)} \quad \cos\Theta = \cos\vartheta^i \cos\vartheta^s + \sin\vartheta^i \sin\vartheta^s \cos\Delta\varphi \\
&\text{(b)} \quad \cos\vartheta^s = \cos\Theta \cos\vartheta^i + \sin\vartheta^i \sin\Theta \cos\gamma \\
&\text{(c)} \quad \cos\vartheta^i = \cos\Theta \cos\vartheta^s + \sin\vartheta^s \sin\Theta \cos\beta
\end{aligned} \tag{10.37}
$$

the law of sines

$$
\frac{\sin\Delta\varphi}{\sin\Theta} = \frac{\sin\gamma}{\sin\vartheta^s} = \frac{\sin\beta}{\sin\vartheta^i} \tag{10.38}
$$

and the law of cosines of the angles

$$
\begin{aligned}
\cos\Delta\varphi &= -\cos\beta\cos\gamma + \sin\beta\sin\gamma\cos\Theta \\
\cos\beta &= -\cos\gamma\cos\Delta\varphi + \sin\gamma\sin\Delta\varphi\cos\vartheta^i \\
\cos\gamma &= -\cos\beta\cos\Delta\varphi + \sin\beta\sin\Delta\varphi\cos\vartheta^s
\end{aligned} \tag{10.39}
$$

(10.37a) is a fundamental equation for $\cos\Theta$ and has already been derived earlier, see (2.65). Substituting (10.38) into (10.37b,c) yields

$$
\begin{aligned}
\cos\vartheta^i \sin\Delta\varphi &= \cos\vartheta^s \sin\Delta\varphi\cos\Theta + \cos\beta\sin\gamma\sin^2\Theta \\
\cos\vartheta^s \sin\Delta\varphi &= \cos\vartheta^i \sin\Delta\varphi\cos\Theta + \sin\beta\cos\gamma\sin^2\Theta
\end{aligned} \tag{10.40}
$$

which may be reformulated as

$$\cos \vartheta^i \sin \Delta\varphi = \cos \beta \sin \gamma + \sin \beta \cos \gamma \cos \Theta$$
$$\cos \vartheta^s \sin \Delta\varphi = \cos \beta \sin \gamma \cos \Theta + \sin \beta \cos \gamma \tag{10.41}$$

Introducing the abbreviation

$$\cos \chi = \cos \vartheta^i \cos \vartheta^s \cos \Delta\varphi + \sin \vartheta^i \sin \vartheta^s \tag{10.42}$$

after a few easy manipulations we find the following relations

$$\cos \beta \cos \gamma = \frac{\cos \Theta \cos \chi - \cos \Delta\varphi}{\sin^2 \Theta},$$

$$\cos \beta \sin \gamma = \frac{(\cos \vartheta^i - \cos \vartheta^s \cos \Theta) \sin \Delta\varphi}{\sin^2 \Theta}$$

$$\sin \beta \sin \gamma = \frac{\cos \chi - \cos \Theta \cos \Delta\varphi}{\sin^2 \Theta}, \tag{10.43}$$

$$\sin \beta \cos \gamma = \frac{(\cos \vartheta^s - \cos \vartheta^i \cos \Theta) \sin \Delta\varphi}{\sin^2 \Theta}$$

Thus, the matrix coefficients of (10.36) assume the form

$$\boxed{\begin{aligned} S_{11} &= T_1 \cos \Delta\varphi + T_2 \cos \chi, & S_{12} &= (T_1 \cos \vartheta^i + T_2 \cos \vartheta^s) \sin \Delta\varphi \\ S_{21} &= -(T_1 \cos \vartheta^s + T_2 \cos \vartheta^i) \sin \Delta\varphi, & S_{22} &= T_1 \cos \chi + T_2 \cos \Delta\varphi \end{aligned}}$$

$$\tag{10.44}$$

where we have used the abbreviations

$$T_1(\vartheta^i, \vartheta^s, \Delta\varphi) = \frac{S_2(\cos \Theta) - \cos \Theta S_1(\cos \Theta)}{\sin^2 \Theta}$$

$$T_2(\vartheta^i, \vartheta^s, \Delta\varphi) = \frac{S_1(\cos \Theta) - \cos \Theta S_2(\cos \Theta)}{\sin^2 \Theta} \tag{10.45}$$

Detailed calculation steps will be left to the exercises.

Equation (10.44), together with (10.37a), (10.42) and (10.45), shows that the matrix elements S_{ij} are functions of the variables $(\vartheta^i, \vartheta^s, \Delta\varphi)$ only so that now the scattering process is completely described in terms of quantities which are known in the fixed (x, y, z)-coordinate system, see Figure 10.5.

From (10.45) we see that for $\Theta = 0$ or $\Theta = \pi$ the denominators of T_1 and T_2 vanish. Therefore, we must show that in these cases the values of both quantities are non-singular and well-defined. From the definitions (9.94) and (9.101) it can be seen that

$$\tau_n(\cos\Theta)\big|_{\Theta=0} = \pi_n(\cos\Theta)\big|_{\Theta=0}, \quad \tau_n(\cos\Theta)\big|_{\Theta=\pi} = -\pi_n(\cos\Theta)\big|_{\Theta=\pi} \implies$$
$$S_1(\cos\Theta)\big|_{\Theta=0} = S_2(\cos\Theta)\big|_{\Theta=0}, \quad S_1(\cos\Theta)\big|_{\Theta=\pi} = -S_2(\cos\Theta)\big|_{\Theta=\pi}$$

$$(10.46)$$

Thus in the limits $\Theta \to 0$ and $\Theta \to \pi$ we easily find from (10.45)

$$T_1\big|_{\Theta=0} = T_2\big|_{\Theta=0} = \frac{1}{2}S_1\big|_{\Theta=0}, \quad T_1\big|_{\Theta=\pi} = -T_2\big|_{\Theta=\pi} = -\frac{1}{2}S_1\big|_{\Theta=\pi} \quad (10.47)$$

verifying that T_1 and T_2 are nonsingular.

10.3.2 Transformation of the Stokes vector

For the consideration of polarization effects in the RTE we need a transformation law for the Stokes parameters in analogy to the transformation law (10.35) for the electric vector. Equation (10.26) introduced the Stokes parameters in terms of the amplitudes a_1, a_2 and the phase angle δ. Instead of using the symbols (S_0, S_1, S_2, S_3), in honor of Stokes, it is customary to use the symbols (I, Q, U, V), defined by

$$I = E_i E_i^* + E_j E_j^*, \quad Q = E_i E_i^* - E_j E_j^*$$
$$U = E_i E_j^* + E_j E_i^*, \quad V = -i(E_i E_j^* - E_j E_i^*)$$

$$(10.48)$$

whereby the electric vector components are given by

$$E_i = a_i \exp(-i\delta_i) \exp[i(k_0 z - \omega t)], \quad E_j = a_j \exp(-i\delta_j) \exp[i(k_0 z - \omega t)]$$

$$(10.49)$$

As before, the subscripts i and j denote the directions parallel and perpendicular to the scattering plane, see Figure 10.4. Substituting (10.49) into (10.48) yields the Stokes parameters as

$$I = a_i^2 + a_j^2, \quad Q = a_i^2 - a_j^2$$
$$U = 2a_i a_j \cos\delta, \quad V = 2a_i a_j \sin\delta \quad \text{with} \quad \delta = \delta_j - \delta_i$$

$$(10.50)$$

Of course, (10.26), (10.48) and (10.50) are equivalent. Since the intensity components of a radiation source can be measured directly, it seems practical to introduce the so-called *modified Stokes parameters*

$$I_i = CE_i E_i^* = Ca_i^2, \qquad I_j = CE_j E_j^* = Ca_j^2$$
$$U = 2Ca_i a_j \cos\delta = 2C\Re(E_i E_j^*), \quad V = 2Ca_i a_j \sin\delta = 2C\Im(E_i E_j^*)$$

$$(10.51)$$

Both sets (10.50) and (10.51) give a complete description of polarization of an electromagnetic plane wave since $I = I_i + I_j$ and $Q = I_i - I_j$. For dimensional reasons, in (10.51) we have included a proportionality factor C to indicate that the Stokes parameters carry the dimension of intensity. The factor C depends on the system of electromagnetic units. As already mentioned, the expression 'intensity' is used quite loosely and, in reality, could be a flux density which is expressed in units of energy per unit time and unit area.

Next we are going to carry out the linear transformation

$$
\begin{pmatrix} I_l^s \\ I_r^s \\ U^s \\ V^s \end{pmatrix} = \frac{\mathbb{A}}{(k_0 r)^2} \begin{pmatrix} I_l^i \\ I_r^i \\ U^i \\ V^i \end{pmatrix}
\tag{10.52}
$$

where, as before, the superscripts i and s denote the incoming and the scattered light and the subscripts l and r refer to parallel and perpendicular to the meridional planes. Furthermore, in the denominator we have replaced z^s by the radius r of the scattering sphere. The constituents required to find the 4×4 matrix $\mathbb{A}$ are provided by (10.35). Analogously to (9.161) we call the term

$$
\boxed{\tilde{\mathbb{P}}(\cos \Theta) = \frac{\mathbb{A}}{k_0^2 \Delta V}}
\tag{10.53}
$$

the *scattering matrix* for the intensities whereby ΔV is a small volume element around the scattering sphere.

Before proceeding, we wish to recall a few helpful formulas from complex variable theory. Let (z_1, z_2) represent complex variables. Then the following relations hold

$$
\Re(z_1 z_2) = \Re(z_1)\Re(z_2) - \Im(z_1)\Im(z_2), \quad \Im(z_1 z_2) = \Re(z_1)\Im(z_2) + \Im(z_1)\Re(z_2)
$$
$$
\Re(z_1 z_2^*) = \Re(z_1^* z_2), \quad\quad\quad\quad \Im(z_1 z_2^*) = -\Im(z_1^* z_2)
\tag{10.54}
$$

Utilizing these relations, it is straightforward to show that $\mathbb{A}$ may be written as

$$
\mathbb{A} = \begin{pmatrix}
S_{11}S_{11}^* & S_{12}S_{12}^* & \Re(S_{11}S_{12}^*) & -\Im(S_{11}S_{12}^*) \\
S_{21}S_{21}^* & S_{22}S_{22}^* & \Re(S_{21}S_{22}^*) & -\Im(S_{21}S_{22}^*) \\
2\Re(S_{11}S_{21}^*) & 2\Re(S_{12}S_{22}^*) & \Re(S_{11}S_{22}^* + S_{12}S_{21}^*) & -\Im(S_{11}S_{22}^* - S_{12}S_{21}^*) \\
2\Im(S_{11}S_{21}^*) & 2\Im(S_{12}S_{22}^*) & \Im(S_{11}S_{22}^* + S_{12}S_{21}^*) & \Re(S_{11}S_{22}^* - S_{12}S_{21}^*)
\end{pmatrix}
\tag{10.55}
$$

This form of the transformation matrix for the Stokes vectors was first derived by Sekera (1956). The detailed derivation of (10.55) will be left as a problem of the exercises.

We will conclude this section with a few additional remarks about the Stokes parameters. In the previous sections we have considered strictly monochromatic waves. We have shown that every wave of the type (10.2) is elliptically polarized. This implies that with increasing time the endpoint of the electric (and also of the magnetic) vector at each point in space periodically describes the circumference of an ellipse. In special cases the ellipse degenerates into a straight line or into a circle. Moreover, the amplitudes a_1 and a_2 and the difference in the phase angles δ are constants, i.e. they are independent of time. However, no perfectly monochromatic radiation exists. Even in the best so-called *monochromatic sources* there is always some finite frequency spread centered about a mean frequency. Let us briefly consider a hypothetical quasi-monochromatic light source having the following property: the oscillation and the subsequent field varies sinusoidally for a certain time and then changes phase abruptly. This time is known as the *coherence time*. The sequence keeps repeating indefinitely, and the phase change after each coherence time occurs randomly. This type of a field may be regarded as an approximation to that of a radiating atom. The abrupt changes of phase may result from collisions.

To continue the discussion on Stokes parameters we consider the complex amplitudes $a_i \exp(-i\delta_i)$ and $a_j \exp(-i\delta_j)$, which are no longer constants but functions of time, i.e. $a_i = a_i(t), a_j = a_j(t), \delta_i = \delta_i(t)$ and $\delta_j = \delta_j(t)$. Over time intervals of the order of the period of the oscillation $2\pi/\omega$ they vary slowly. However, for a time interval large in comparison with the period, the amplitudes fluctuate in some way, perhaps independently or perhaps with some correlation. If the complex amplitudes are completely uncorrelated, the light is *natural* or *unpolarized*. In this case, over sufficiently long periods of time, vibration ellipses of all shapes, handedness and orientation will have been traced out so that there exists no preferred polarization ellipse. In contrast, if $a_i \exp(-i\delta_i)$ and $a_j \exp(-i\delta_j)$ are completely correlated, the light is called *polarized*. This definition includes strictly *monochromatic light*, but it is somewhat more general: $(a_i, a_j, \delta_i, \delta_j)$ may separately fluctuate provided that the ratio a_i/a_j of the real amplitudes and the phase difference $\delta = \delta_j - \delta_i$ are independent of time. If $a_i \exp(-i\delta_i)$ and $a_j \exp(-i\delta_j)$ are partially correlated, the light is said to be *partially polarized*. Ignoring some statistical fluctuations, such a partially polarized beam is characterized by a preference in handedness, or ellipticity, or azimuth, which is the angle between the major axis of the ellipse and an arbitrary reference direction.

The Stokes parameters of a quasi-monochromatic beam (omitting again the constant C) are defined by

$$I = \langle E_i E_i^* \rangle + \langle E_j E_j^* \rangle, \qquad Q = \langle E_i E_i^* \rangle - \langle E_j E_j^* \rangle$$
$$U = 2\Re \left(\langle E_i E_j^* \rangle \right), \qquad V = 2\Im \left(\langle E_i E_j^* \rangle \right) \tag{10.56}$$

where the symbols $\langle \ldots \rangle$ refer to a time average over an interval which is long in comparison with the period of the vibration. It can be shown that in this case the inequality

$$Q^2 + U^2 + V^2 \leq I^2 \tag{10.57}$$

must hold. When this situation occurs we speak of *partial polarization*. As we already know from (10.27), the equality sign refers to completely polarized light.

Let us briefly look at the Stokes parameters from the experimental point of view. Since the fluctuations are extremely swift in comparison with the duration of any measurement, it is possible to measure only mean amplitudes and mean phases so that all measured Stokes parameters refer to mean quantities.

We will now consider the Stokes vector of natural or unpolarized light such as the *Planckian black body radiation* and the parallel (unscattered) solar radiation. Due to the randomness of the fluctuations, experimental techniques fail to measure any differences between the phase angles and the intensity components. Thus the intensity in any direction in the transverse plane is the same so that the Stokes parameters $Q = U = V = 0$. This is a necessary and sufficient condition for light to be natural. That Q is zero follows directly from (10.56). U and V vanish because $\cos \delta$ and $\sin \delta$ average to zero independently of the amplitudes. Thus the Stokes vector for natural light is $(I, 0, 0, 0)$, that is the intensity I by itself is sufficient to specify unpolarized radiation.

It can be shown, see for example Chandrasekhar (1960) and Born and Wolf (1965), that in all cases partially polarized light of intensity I may be regarded to consist of one part of completely (elliptically) polarized light I_{pol} and another part of completely unpolarized light $I - I_{pol}$. These parts are independent of each other, and this representation is unique. The Stokes vector for the completely polarized part of the radiation is specified by (I_{pol}, Q, U, V) with $I_{pol} = (Q^2 + U^2 + V^2)^{1/2}$ while the unpolarized part of the radiation is described by the Stokes vector $(I - I_{pol}, 0, 0, 0)$.

Let us consider two or more quasi-monochromatic beams, traveling in the same direction, which are superposed incoherently. This means that there is no fixed permanent relationship among the phases of the separate beams. The total irradiance is the sum of the irradiances of the individual beam. Because of the definition of the Stokes parameters they are additive if a collection of incoherent sources is involved.

We have previously shown that for monochromatic light the Stokes vectors of the scattered and the incoming light are related by a transformation matrix of the type (10.55). Each one of the 16 elements of this matrix is a real number. If the light is partially polarized, which is the case for most physical atmospheric situations, the same transformation matrix is still valid. The reason for this is that the Stokes parameters for monochromatic and quasi-monochromatic light have the same mathematical form.

The previous sections are based on the discussions presented by Bohren and Huffman (1983), Born and Wolf (1965), Chandrasekhar (1960) and van de Hulst (1957) where additional information may be found.

10.4 The vector form of the radiative transfer equation

Let us now return to the scalar form of the RTE for a horizontally homogeneous atmosphere

$$\mu \frac{d}{d\tau} I(\tau, \mu, \varphi) = I(\tau, \mu, \varphi) - J(\tau, \mu, \varphi) \tag{10.58}$$

where the *source function* J is given by

$$J(\tau, \mu, \varphi) = \frac{\omega_0}{4\pi} \int_0^{2\pi} \int_{-1}^{1} \mathcal{P}(\cos\Theta) I(\tau, \mu', \varphi') d\mu' d\varphi'$$
$$+ \frac{\omega_0}{4\pi} \mathcal{P}(\cos\Theta_0) S_0 \exp\left(-\frac{\tau}{\mu_0}\right) + (1 - \omega_0) B(\tau) \tag{10.59}$$

see (2.53). In order to treat polarization effects in radiative transfer, the scalar quantities I and J must be replaced by the vectors $\mathbf{I}$ and $\mathbf{J}$ as defined by

$$\mathbf{I} = \begin{pmatrix} I_l \\ I_r \\ U \\ V \end{pmatrix}, \qquad \mathbf{J} = \begin{pmatrix} J_l \\ J_r \\ J_U \\ J_V \end{pmatrix} \tag{10.60}$$

This yields the *vector form of the RTE*

$$\boxed{\mu \frac{d}{d\tau} \mathbf{I}(\tau, \mu, \varphi) = \mathbf{I}(\tau, \mu, \varphi) - \mathbf{J}(\tau, \mu, \varphi)} \tag{10.61}$$

The scalar form of the source function (10.59) will now be generalized to the vector form. *Planckian emission* $B(\tau)$ and the direct solar radiation S_0 must be classified as natural light so that the corresponding *Stokes vectors* can be written as

$$\mathbf{B}(\tau) = \frac{1}{2} B(\tau) \begin{pmatrix} 1 \\ 1 \\ 0 \\ 0 \end{pmatrix}, \qquad \mathbf{S}_0 = \frac{1}{2} S_0 \begin{pmatrix} 1 \\ 1 \\ 0 \\ 0 \end{pmatrix} \tag{10.62}$$

Utilizing these expressions the generalization of the scalar source function to the vector form is

$$
\begin{aligned}
\mathbf{J}(\tau, \mu, \varphi) = {} & \frac{\omega_0}{4\pi} \int_0^{2\pi} \int_{-1}^{1} \mathbb{P}(\cos\Theta) \cdot \mathbf{I}(\tau, \mu', \varphi') d\mu' d\varphi' \\
& + \frac{\omega_0}{4\pi} \mathbb{P}(\cos\Theta_0) \cdot \mathbf{S}_0 \exp\left(-\frac{\tau}{\mu_0}\right) + (1 - \omega_0)\mathbf{B}(\tau)
\end{aligned}
\tag{10.63}
$$

The *phase matrix* $\mathbb{P}(\cos\Theta)$ and the scattering matrix $\tilde{\mathbb{P}}(\cos\Theta)$ are related by

$$
\mathbb{P}(\cos\Theta) = \frac{4\pi}{k_{sca}} \tilde{\mathbb{P}} \quad \text{with} \quad \tilde{\mathbb{P}} = \frac{N}{k_0^2 \Delta V} \mathbb{A}
\tag{10.64}
$$

The quantity N refers to the number of identical scattering particles in the scattering volume ΔV. Obviously, in (10.53) we have $N = 1$ expressing the scattering by a single particle. If ΔV does not contain identical particles, then for each scattering angle Θ the elements in (10.55) must be integrated over the particle size distribution.

Finally, we will list a few authors who have presented radiation calculations in case of multiple scattering including polarization. The list could be easily extended. Chandrasekhar (1960) was the first to show how to accurately compute the intensity and polarization of radiation in case of multiple *Rayleigh scattering*. The work was extended notably by Sekera and co-workers so that some extensive tables of numerical results are now available, see, for example, Coulson *et al.* (1960), Sekera and Kahle (1966).

Herman *et al.* (1971) belonged to the first group of investigators to apply (10.55) in order to study the influence of atmospheric aerosols on scattered sunlight. They used a numerical scheme applicable to small and moderate optical thicknesses. Later a modified version of this scheme was used by others to solve the vector form of the RTE. However, due to the occurrence of large optical depths of clouds, the method cannot be applied to study radiative transfer in the cloudy atmosphere.

In Chapter 4 we have shown how to apply the adding–doubling method (MOM) in case of the scalar form of the radiative transfer equation. The adding–doubling procedure, originally introduced by van de Hulst (1963), did not consider polarization effects. However, Hansen (1971) generalized this method to include polarization. He showed that the MOM is capable of handling strongly anisotropic phase matrices. For selected wavelengths in the near infrared he found that in case of planetary clouds polarization is more sensitive than the intensity to changes of cloud microstructure such as the particle size distribution. He concluded that polarization measurements are potentially a valuable tool for cloud identification and for microphysical studies. Moreover, his case studies revealed that the radiance computed with the exact theory which includes polarization differs by 1% or less from the results obtained with the help of the scalar theory where polarization is ignored. Thus he concluded that for energetic studies polarization effects can be neglected.

In a related paper Hovenier (1971) also showed how to generalize the adding–doubling method by including polarization. He confirmed Hansen's conclusion that generally polarization effects can be neglected if the radiative intensity (radiance) is of interest only. The adding–doubling method for multiple scattering calculations of polarized light was also treated by de Haan *et al.* (1987). To evaluate numerically the combinations of an integration and a matrix multiplication, as they occur in the adding method, they introduced the *concept of a supermatrix*. Using supermatrices, such combinations are treated as a single matrix product thus simplifying the computational procedures.

The research on this topic is going on. A fairly complete list of references can be found in Liou's (2002) book *An Introduction to Atmospheric Radiation*.

10.5 Problems

10.1: The electric vector of a certain wave is given by

$$\mathbf{E} = \mathbf{i}E_0 \cos(\omega t - kz + \pi/2) + \mathbf{j}E_0 \cos(\omega t - kz)$$

Discuss the state of polarization.

10.2: Consider a linearly polarized harmonic wave of amplitude E_0. Assume that the wave is propagating along a straight line in the (x, y)-plane which is the plane of vibration. For a straight line which is inclined $45°$ to the x-axis find the electric vector.

10.3: Consider the superposition of an R-state and an L-state. Assume equal amplitudes of the constituent waves. Show that the resulting wave is a P-state.

10.4: Suppose that a wave is described by

$$\mathbf{E}_x = \mathbf{i}a_x \cos(kz - \omega t), \qquad \mathbf{E}_y = \mathbf{j}a_y \cos(kz - \omega t + \delta)$$

(a) Discuss the state of polarization for $\delta = \pi/2$.
(b) Sketch the vibrational figures for $\delta = n\pi/4, \ n = 0, 1, \ldots, 7$.

10.5: Suppose that the components of the electric vector are written in the form

$$E_x = E_{x,0} \cos(kz - \omega t), \qquad E_y = E_{y,0} \cos(kz - \omega t + \delta)$$

By performing the required operations, show that again we obtain the form (10.6).

10.6: Verify that the coordinates P_3 and P_4 of equation (10.10) are stated correctly.

10.7: Show that in (10.28) the quantities S_1 and S_2 are correct statements.

10.8: Show that the Stokes vector in Table 10.1 for $45°$ is correctly stated.

10.9: Show that the transformation matrix in (10.35) is made up of the elements listed in (10.44). Follow the derivation in the text and fill in the missing steps.

10.10: For spherical particles the electric vector of the incoming and the scattered wave in terms of the amplitude scattering matrix A can be written as

$$\begin{pmatrix} E_i^s \\ E_j^s \end{pmatrix} = \begin{pmatrix} A_2 & 0 \\ 0 & A_1 \end{pmatrix} \begin{pmatrix} E_i^i \\ E_j^i \end{pmatrix}$$

where E_i and E_j are given by (10.32). Find the corresponding transformation matrix F for the incoming and the scattered Stokes vectors, i.e.

$$\begin{pmatrix} I_i^s \\ I_j^s \\ U^s \\ V^s \end{pmatrix} = F \begin{pmatrix} I_i^i \\ I_j^i \\ U^i \\ V^i \end{pmatrix}$$

Specify each element of the matrix. The transformation to the vertical plane is not required for this problem.

10.11: Verify by a direct calculation that the elements of the matrix (10.55) are correctly stated.[2]

---

[2] In setting up the first four problems of Chapter 10, we acknowledge the assistance of the textbook *Optics* (Hecht, 1987)

11

Remote sensing applications of radiative transfer

11.1 Introduction

In this chapter we will deal with the application of radiative transfer theory to atmospheric *remote sensing*. Remote sensing means that measurements are performed at a large distance from the physical object or medium under consideration with the purpose of retrieving its physical properties. In many cases the carrier of the physical information is electromagnetic waves. Nevertheless it is possible to observe the atmosphere, the soil, or the ocean by means of sound waves. *Sodar* (sound detection and ranging) and *sonar* (sound navigation and ranging) are techniques that employ acoustic waves.

Two basic methods known as *active* and *passive remote sensing* can be applied. 'Active' means that the source of the waves is man-made; for example, a laser transmitter can be used to emit light pulses which propagate through the medium under consideration. The laser light is scattered by air molecules, or it is scattered and partly absorbed by aerosol and cloud particles. The scattered laser light is then collected by a detector telescope. The amount and the amplitude of the detected pulses can then be used as a measure for transmission losses. This particular technique is called *lidar* (light detection and ranging). Another widely employed active technique is *radar* (radiation detection and ranging) where antennas emitting microwave radiation are being used.

In contrast to the active techniques, passive remote sensing makes use of natural radiation sources. The observation of sunlight propagating through the atmosphere, being reflected by the Earth's surface, then traveling upward to finally enter a radiometer aboard a satellite, constitutes one of several other important methods to investigate the physical and chemical properties of the Earth–atmosphere system. Instead of solar radiation one can also exploit the long-wave infrared emission of both the surface and the atmosphere. Another passive technique, which is based on the *microwave emission* by atmospheric constituents, provides the beneficial feature that light is transmitted through clouds. This feature is due to the fact that the size of cloud particles is very small in comparison to the observed wavelengths.

Depending on the location of the source and/or the detecting instrument, one further distinguishes between *ground-based, airborne* or *spaceborne remote sensing*. We will mainly focus on spaceborne remote sensing, i.e. to cases where an instrument flown on a satellite is used for remote sensing.

In the following we will restrict the discussion to passive remote sensing. Clearly, in a single chapter we cannot present an exhaustive and detailed treatment of this subject. The main purpose of the following sections is to demonstrate how radiative transfer theory can be used as a sound physical and mathematical basis to retrieve, for example, the atmospheric temperature profile, or to make use of the forward–adjoint-based perturbation theory to retrieve the atmospheric ozone profile.

In the *retrieval process* we have to distinguish two different steps, the *forward problem* and the *inverse problem*. The easier and more straightforward task is the forward problem in which the RTE is used to simulate the radiation field at the detector's location. This task requires as input all important geophysical and optical parameters of the Earth–atmosphere system. If we assume that a measurable set of such parameters is available, the only work to be accomplished is the computation of the radiation field. In contrast to this, the inverse problem attempts to find the inverse relationship. The task is to derive from the detected radiation field the physical atmospheric properties which are relevant for the radiative transfer.

Since the radiation field at the satellite's position depends in a complex and generally nonlinear way on the parameters to be retrieved (total gas columns, vertical profiles of gas concentrations, extinction properties of aerosol and cloud particles, temperature and pressure profile, etc.), the inverse problem is much more difficult to solve than the forward problem. As we will see later, this difficulty is intimately related to the so-called *ill-posedness of the inverse problem*. An ill-posed problem may, for example, imply that there are far less independent measurements available than the number of unknowns characterizing the problem. Therefore, the difficulty is to properly add additional information that enables us to establish an approximate inverse relationship between the unknown quantities and the radiation measurements. The situation is similar to the inversion of a matrix which is singular or at least close to be singular. The inversion of the matrix is either impossible, or the solution strongly depends on the accuracy of the matrix elements. Thus is becomes clear that the additional information mentioned above acts as a regularization of the problem.

Figure 11.1 illustrates the connection between the forward and the inverse problem. One also speaks of setting up the forward model $\mathbf{y} = F(\mathbf{x})$ and the corresponding inverse model $\mathbf{x} = F^{-1}(\mathbf{y})$, where $\mathbf{y}$ designates the measurement vector and $\mathbf{x}$ is the state vector of the atmosphere to be retrieved.

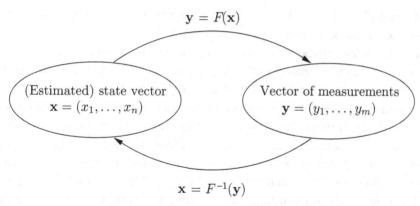

Fig. 11.1 Forward and inverse problem in remote sensing applications. Note that the solution to the inverse problem in general provides only an estimate for the true state of the atmosphere.

For the forward problem the function F can be seen as the RTE. It should be kept in mind, however, that this is only true for an ideal instrument. In practice, F has to consider the instrumental properties, such as *slit function, response function, field of view* and *noise*. This means that the forward model has to simulate the *instrument signal* by performing a convolution of the radiance spectrum as seen by an ideal instrument with the various instrument characterizing functions.

While $\mathbf{y} = (y_1, y_2, \ldots, y_m)$ is the radiance at the instrument's location for a set of m different wavelengths, i.e. a radiance spectrum, the vector $\mathbf{x} = (x_1, x_2, \ldots, x_n)$ could be temperature values $T(z_1), T(z_2), \ldots, T(z_n)$ at n altitudes $z_1, z_2, \ldots, z_n$.

The determination of a medium's state based on measured spectra is not limited to atmospheric applications. Similar problems arise in other disciplines, e.g. the determination of the composition of the Earth's interior by exploiting seismic waves, the derivation of the properties of single stars and galaxies on the basis of radio waves, infrared or gamma ray observations, or in medicine using nuclear spin computer tomography, techniques of nuclear medicine or ultrasound. In all these applications the derivation of the target's properties and/or composition is called an inverse problem.

In general it is not possible to exactly reconstruct the state of the target under investigation. One reason for this is the fact that any measurement contains to some degree noise signals. Thus the relation $\mathbf{y} = F(\mathbf{x})$ is only approximately fulfilled. Likewise the measurement apparatus may possess certain systematic inaccuracies which lead to a distorted observation, and the forward model F renders only an approximate solution to the real problem. Finally, one has to keep in mind that with a discrete set of observations, that is with a limited number of observations, one cannot reconstruct field properties of the medium which depend on certain geophysical parameters in a continuous manner.

In principle, the number n of unknown parameters must be smaller than the number m of independent measurements, $n \leq m$. In the ideal case that each individual measurement provides an independent piece of information, we would be able to find a unique solution for the n required parameters. In practice, however, it is not an easy task to determine the information content of a set of measurements. To give an example, the observation of the nadir radiance from space in the wavelength region 290 to 330 nm allows one not only to determine the total column amount of ozone below the sub-satellite point, it is also possible to 'sound' the atmosphere as a function of increasing distance from the platform if the wavelength is scanned from smaller to larger wavelengths. This is due to the fact that ozone molecules absorb solar radiation very strongly at short wavelengths, i.e. photons entering the atmosphere are not able to pass the ozone layer, with maximum concentration near 20 to 25 km altitude at the short-wave end of this particular wavelength region. On the other hand, for gradually longer wavelengths the chances will increase that the photons will reach a greater depth (lower altitude) before absorption occurs. This particular example illustrates the basic principle of passive (or active) remote sensing: the spectral absorption or emission characteristics in combination with the monotonously increasing path length provide a direct link between altitude, absorber amount and magnitude of the observed radiance.

In the following section we will discuss different topics which are necessary to understand the principles of remote sensing from satellites. Section 11.2 provides some insight into solar–terrestrial relations. The physical principles of remote sensing based on the extinction of solar radiation and in the long-wave spectral region will be described in Section 11.3. In Section 11.4 the inversion of the atmospheric temperature profile will be analyzed by means of various classical methods. The final Section 11.5 will make use of the radiative perturbation technique introduced in Chapter 6 to provide an efficient and accurate algorithm for determining the so-called *weighting function* which is of key importance to any physically mathematically based inversion technique.

11.2 Remote sensing based on short- and long-wave radiation

At a wavelength of about 4 μm the spectrum can be separated into the short-wave region ($\lambda < 3.5$ μm), where the Sun is the radiation source, and the long-wave region ($\lambda > 3.5$ μm), where the thermal emission of the Earth itself and the Earth's atmosphere are the sources of radiation. A particularly important wavelength region is the UV/VIS (UV: ultraviolet, VIS: visible) spectral region in which scattering of solar radiation by air molecules, aerosol and cloud particles plays an important role. An additional feature of the solar radiation can be exploited, namely that both the diffuse as well as the direct solar radiation can be separately used for remote sensing.

The different radiation sources lead to different remote sensing algorithms, each taking advantage of a particular feature of the radiative transfer process. One class of methods is based on the extinction process, while a second class has its focus on the scattering process for short-wave radiation. In the long-wave spectral region the thermal emission as function of the temperature field is in the center of the investigation. The different methods can also be based on observations from the ground (ground based), from instruments flown on aircrafts (airborne) or from instruments aboard rockets or satellites (spaceborne). Regarding satellites, an important observation geometry is the *nadir-looking mode*, i.e. the instrument looks at the surface of the Earth in the nadir direction. *Limb sounding* means that the viewing path as seen from the instrument represents a path which is tangential to the Earth and which, in certain situations, avoids the influence by the Earth's surface. For nadir-looking instruments, or instruments covering a wide range of viewing angles, the contributions of the Earth's surface play a major role as a function of wavelength. These contributions are further modified by the radiative properties of the intervening atmospheric layers. Thus a more or less uniformly distributed (with respect to wavelength) surface contribution will be modified in a manner that it carries the spectrally highly variable absorption and emission features of individual atmospheric trace gases.

11.2.1 Methods based on the extinction of solar radiation

If remote sensing is based on the measurement of short-wave radiation, the following assumptions can be made.

(i) The thermal emission source is neglected in the RTE, i.e. $J_\nu^e = 0$.
(ii) The instrument detects the direct solar radiation only in a very narrow angular solid angle interval centered around the solar disk. This narrow field of view has the advantage that contributions due to single or multiple scattering processes of solar photons can safely be neglected. Thus we can also omit the single and multiple scattering terms in the RTE.

Introducing these simplifications in the RTE in the form (2.36) we obtain

$$\frac{d}{d\tau}I_\nu(\tau, \mu, \varphi) = I_\nu(\tau, \mu, \varphi) \tag{11.1}$$

where $d\tau = -k_{\text{ext},\nu}ds$ and s denotes the geometrical distance between P and the top of the atmosphere in viewing direction of the instrument. Equation (11.1) may

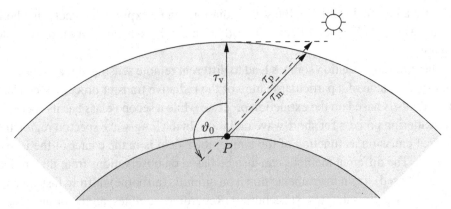

Fig. 11.2 Definition of the optical air mass m for a spherical atmosphere. τ_v: vertical optical depth, $\tau_p = |\mu_0|^{-1}\tau$ and $\tau_m = m\tau$.

be easily integrated yielding

$$I_\nu(\tau, \mu, \varphi) = \delta(\mu - \mu_0)\delta(\varphi - \varphi_0)S_{0,\nu}\exp(-\tau_{\text{obs}}) \tag{11.2}$$

Here, τ_{obs} denotes the optical depth along a straight path between the observation point P of the instrument and the top of the atmosphere in the direction of the Sun. Hence, the radiative flux density registered by the instrument is

$$E_\nu = S_{0,\nu}\exp(-\tau_{\text{obs}}) \tag{11.3}$$

Since the spectral extraterrestrial solar flux density $S_{0,\nu}$ is assumed to be known, the instrument's signal E_ν can be used to derive the extinction optical depth of the entire atmosphere above the observation point

$$\tau_{\text{obs}} = -\ln\left(\frac{E_\nu}{S_{0,\nu}}\right) \tag{11.4}$$

Usually one considers the vertical optical depth τ as defined in (2.51). For a plane–parallel atmosphere we may write

$$\tau_{\text{obs}} = \tau_p = \frac{\tau}{|\mu_0|} \tag{11.5}$$

[1]However, since the atmosphere is spherical in nature, τ_{obs} is smaller than τ_p and is given by

$$\tau_{\text{obs}} = \tau_m = m\tau = \int_z^\infty k_{\text{ext},\nu}(z')\frac{ds}{dz'}\,dz' \tag{11.6}$$

Figure 11.2 depicts the situation.

---

[1] Recall that, according to Figure 2.3, $\mu_0 \leq 0$.

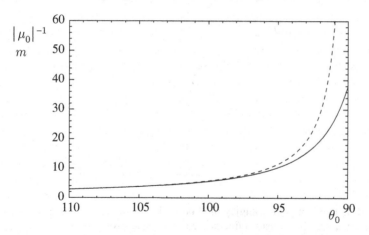

Fig. 11.3 Comparison of the optical air mass of a plane–parallel atmosphere $|\mu_0|^{-1}$ (dashed curve) with the optical air mass m as obtained from (11.7) (solid curve).

In (11.6) we have introduced the *optical air mass, m*. For a perfectly plane–parallel atmosphere we would have $m = |\mu_0|^{-1}$. Kasten and Young (1989) suggested a simple approximation for m as a function of ϑ_0 which is accurate enough for most practical purposes

$$m \approx [|\mu_0| + 0.50572(\vartheta_0 - 83.92)^{-1.6364}]^{-1} \qquad (11.7)$$

This equation applies to a clear and dry atmosphere containing no aerosols, clouds or absorbing gases. In principle the optical air mass is wavelength dependent. Equation (11.7) is to be interpreted in a manner that m multiplied by the total scattering optical thickness of the Rayleigh atmosphere can be used to determine the attenuation of the total solar radiation reaching the surface of the Earth. Figure 11.3 compares the idealized values $|\mu_0|^{-1}$ of a plane–parallel atmosphere with m of the real atmosphere which was calculated by means of (11.7). From the figure it is concluded that the plane–parallel approximation is acceptable for solar zenith angles $\vartheta_0 \geq 100°$. For smaller ϑ_0 values the error becomes increasingly larger. If the Sun is close to the horizon, the neglecting of spherical effects gives rise to big errors. For $\vartheta = 90°$ we have $|\mu_0|^{-1} = \infty$ in contrast to $m = 37.92$.

In an atmosphere consisting of air molecules, aerosol particles and trace gases the optical depth is given by the sum of the three individual optical depths, i.e.

$$\tau = \tau_{air} + \tau_{gas} + \tau_{aer} \qquad (11.8)$$

Since τ_{air} can be computed from standard atmospheric data, the optical depth of the atmospheric aerosol can be directly retrieved if no gas contributions to the

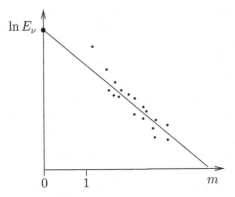

Fig. 11.4 The Bouguer–Langley method to determine $S_{0,\nu}$. For $m = 1$ the Sun is in the local zenith. The solid line is the linear regression.

total optical depth exist. This type of measurement is a common procedure for ground-based observations.

In order to guarantee a high accuracy to determine τ_{aer}, the instrument must be well calibrated with respect to the extraterrestrial solar flux density $S_{0,\nu}$. If this requirement cannot be met the so-called *Bouguer–Langley method* may be applied. This procedure requires a plot of $\ln E_\nu$ versus m by taking measurements for various solar zenith angles. Figure 11.4 depicts an example for a set of measurements. Based on the evaluation of (11.4), from a linear regression of the resulting data points two important quantities can be extracted: (i) the slope of the regression line is equal to τ, and (ii) the intersection of the straight line with the ordinate gives an estimate of $\ln E_\nu$. The Bouguer–Langley method has been employed to determine the spectral solar constant before satellite measurements were available.

Optical thickness observations for trace gases make use of a similar principle. Let us assume two neighboring wave number points (ν, ν') with maximum absorption of a particular gas at ν and very small absorption at ν'. If we then measure E_ν and E'_ν, according to (11.4) and (11.6) we can compute

$$\tau_{gas} - \tau'_{gas} = \frac{1}{m}\left[\ln\left(\frac{S_{0,\nu}}{E_\nu}\right) - \ln\left(\frac{S_{0,\nu'}}{E'_\nu}\right)\right] = \frac{1}{m}\left[\ln\left(\frac{S_{0,\nu}}{S_{0,\nu'}}\right) - \ln\left(\frac{E_\nu}{E'_\nu}\right)\right]$$

$$(11.9)$$

If at ν' the absorption of the gas is small enough, τ'_{gas} may be neglected yielding the optical thickness τ_{gas}. Clearly, the observation is not sensitive to uncertainties in the calibration constant. This relation is strictly valid only if the calibration constants for the wave numbers ν and ν' do not differ from each other.

Example: determination of the aerosol and ozone optical depth

A Sun photometer can be used to determine the total optical depth between the location of the instrument and the top of the atmosphere. From (11.4) and (11.8) the total optical depth is given by

$$\tau = \tau_{air} + \tau_{gas} + \tau_{aer} = -\frac{1}{m} \ln\left(\frac{E_\nu}{S_{0,\nu}}\right) \qquad (11.10)$$

This equation can be used to obtain the optical depth of the aerosol particles τ_{aer}. First we determine the optical depth τ_{air} describing the Rayleigh scattering of the air molecules. For this we need the *scattering cross-section of an air molecule*. Due to Rayleigh (1918) and Cabannes (1929) the scattering cross-section for anisotropic gaseous molecules in random orientation is given by

$$\boxed{\sigma_{sca,air} = \frac{8\pi^3}{3} \frac{(n^2 - 1)^2}{\lambda^4 N^2} \frac{6 + 3\delta}{6 - 7\delta}} \qquad (11.11)$$

where $\lambda = 1/\nu$ is the wavelength, N is the number of air molecules per unit volume, n is the refractive index of the air, and δ the anisotropy factor of the gas molecules. The Earth's atmosphere can be treated as a single gas having $\delta = 0.031$. Equation (11.11) can be used to derive a simple expression for the total Rayleigh optical depth of the atmosphere, see Allen (1963)

$$\boxed{\tau_{air,tot} = 0.008569\lambda^{-4}(1 + 0.0113\lambda^{-2} + 0.00013\lambda^{-4})} \qquad (11.12)$$

where the wavelength λ has to be inserted in units of μm. Note that due to the wavelength dependence of the refractive index $\tau_{air,tot}$ slightly differs from the ideal λ^{-4} law of Rayleigh scattering, see Section 9. For a wavelength of 0.55 μm we obtain $\tau_{air,tot} = 0.0973$. At an arbitrary pressure level p the Rayleigh optical depth as measured from the top of the atmosphere is obtained from

$$\tau_{air}(p) = \frac{p}{p_0} \tau_{air,tot} \quad \text{with} \quad p_0 = 1013.25 \text{ hPa} \qquad (11.13)$$

An important absorbing gas in the UV/VIS spectral region is ozone. By assuming that O_3 is the only trace gas absorber within the atmosphere, for the determination of τ_{aer} in (11.10) we need to know $\tau_{gas} = \tau_{O_3}$. We select the two wavelengths λ and λ' where the ozone absorption is relatively high (λ) and rather low (λ'). The differential absorption technique as described before allows us then to write

$$\tau_{O_3} - \tau'_{O_3} = \frac{1}{m}\left[\ln\left(\frac{S_{0,\nu}}{S_{0,\nu'}}\right) - \ln\left(\frac{E_\nu}{E_{\nu'}}\right)\right] \qquad (11.14)$$

where the spectral solar constants $S_{0,\nu}$, $S_{0,\nu'}$ are known.

Table 11.1 *Examples for Dobson wavelength pairs* (λ, λ')
*and the corresponding differences for the absorption
coefficient of ozone. See also Lenoble (1993).*

Pair	(λ, λ') (nm)	$k_{abs,\nu,O_3} - k_{abs,\nu',O_3}$ (cm atm)$^{-1}$
A	(305.5, 325.4)	4.025
B	(308.8, 325.1)	2.625
C	(311.4, 332.4)	1.842
D	(317.6, 339.8)	0.829
C'	(332.4, 453.6)	0.108

Equation (11.14) follows from (11.10) by assuming that there is no wavelength dependence of τ_{air} and τ_{aer}. To fulfill this requirement as closely as possible, we must choose λ and λ' in close proximity within the spectrum. The aerosol optical depth varies only very weakly with wavelength, and since the optical depth of the air molecules varies approximately with λ^{-4}, an expression such as (11.14) is acceptable. To improve the treatment and make it even more correct, one may add to the right-hand side of (11.14) a small correction term to account for the difference in $\tau_{air} - \tau'_{air}$, and an even smaller correction $\tau_{aer} - \tau'_{aer}$. While the correction for Rayleigh scattering can be carried out with high precision, physically reasonable information from a suitable standard aerosol model is required to obtain the aerosol correction, e.g. Shettle and Fenn (1979), d'Almeida *et al.* (1991), Hess *et al.* (1998).

As a particular example, Table 11.1 presents the generally adopted differences of the ozone absorption coefficient for five different wavelength pairs A, B, C, D, C' which have been selected by the *Dobson Network for total ozone measurements*. Utilizing these data the total ozone column u_{O_3} can be evaluated from

$$u_{O_3} = \frac{\tau_{O_3} - \tau'_{O_3}}{k_{abs,\nu,O_3} - k_{abs,\nu',O_3}} \tag{11.15}$$

In contrast to the total vertical optical depth, it is much more difficult to obtain an optical depth profile as a function of altitude. An important technique to achieve this goal are *spaceborne solar occultation measurements* as shown in Figure 11.5. Here, the instrument views the Sun along a path tangential to an interior atmospheric surface S. This surface is defined by the minimum distance z_T which is known as the *tangent height*. Such observations cannot be performed continually. Only a rather narrow time interval just after sunrise and before sunset, as viewed from the satellite, can be used for the observations. As the satellite moves along its track, the measurements can be performed for different tangent heights. In this manner

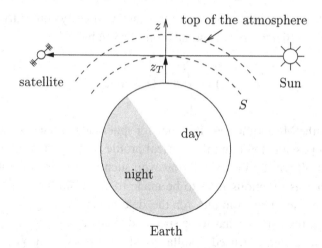

Fig. 11.5 Solar occultation measurements and tangent height z_T.

an optical depth retrieval versus altitude may be achieved. According to (11.6) this can be expressed as

$$\tau(z_T) = 2 \int_{z_T}^{\infty} k_{ext,\nu}(z) \frac{ds}{dz} \, dz \qquad (11.16)$$

where ds is the path element along the viewing direction of the satellite. The above formula further assumes an atmosphere consisting of concentric spherical shells. The factor 2 implies a symmetry in the atmospheric extinction properties with respect to z_T. Utilizing the occultation extinction observations it is possible to find $\tau(z_T)$ so that, in principle, an inversion of (11.16) can be carried out to find the extinction profile $k_{ext,\nu}(z)$.

Let us assume that at a given wavelength only absorption due to an individual gas occurs. In this case the extinction coefficient reduces to

$$k_{abs,\nu}(z) = \sigma_{abs,\nu} N(z) \qquad (11.17)$$

where $\sigma_{abs,\nu}$ is the molecular absorption cross-section in units of m^2, and $N(z)$ is the volume number concentration of the particular gas in question, i.e. the number of gas molecules per unit volume of air located at altitude z. Assuming that the absorption coefficient is known, with the help of the occultation measurements the vertical concentration profile of the gas may be retrieved. In general, a correction for molecular scattering will be necessary.

A further modification of the experiment is possible as will be discussed next. Let us assume that the molecular scattering effects either are exactly known or can be totally neglected, and that the same is true for gaseous absorption. The only extinction that remains is due to a population of Mie scatterers. Assuming spherical particles of radius r and a particle number density concentration $N(z, r)$, the extinction coefficient for this population is given by

$$k_{\text{ext},\nu}(z) = \int\limits_0^\infty \pi r^2 Q_{\text{ext},\nu} N(z, r)\, dr \qquad (11.18)$$

Here $Q_{\text{ext},\nu}$ is the Mie extinction efficiency for spherical particles, see Chapter 9. In principle it is possible to obtain the vertical profile of the particle number density concentration $N(z, r)$ by inversion. However, such an inversion is rather difficult because further assumptions have to be made to constrain the ill-posed problem in a favorable way. For example, for the determination of $Q_{\text{ext},\nu}$ the chemical composition of the particles has to be specified in advance.

If a set of wavelengths is used simultaneously, an inversion of $N(z, r)$ is feasible under certain circumstances. We will not give details and refer to the primary literature, cf. King *et al.* (1978). In connection with the Stratospheric Aerosol and Gas Experiment II (SAGE II) see Wang *et al.* (1989) or Livingston and Russell (1989).

In contrast to the above very complicated situation, an inversion of atmospheric column contents of trace gases can be carried out as follows. Let us assume that some remote sensing experiment provides us with τ. In case of gaseous absorption it is then possible to retrieve the entire column content, $\int_0^\infty N(z)dz$, i.e. the total number of gas molecules per m^2 in an atmospheric column having a 1 m^2 cross-section. A similar approach can be used in case of aerosol particles. This procedure makes it possible to invert a mean particle number density for the entire atmospheric column.

Figures 11.6a–d depict four typical situations of nadir viewing short-wave remote sensing. Figure 11.6a shows the general situation where three contributions must be accounted for describing the instrument's signal: (i) direct solar radiation reflected at the Earth's surface and directly transmitted to the instrument, (ii) solar radiation scattered by air molecules and aerosol particles and then transmitted to the instrument, and (iii) solar radiation being reflected by the cloud tops and transmitted to the satellite. This figure applies to a relatively transparent atmosphere with some embedded clouds lying over a strongly reflecting surface.

Reflection of short-wave radiation at the surface and at the top of the clouds plays an important role for remote sensing performed in an absorption band of an atmospheric trace gas. This situation, shown in Figure 11.6b, is important because the application of Beer–Lambert's extinction law allows the establishing of a direct relation between the transmission of the directly transmitted solar radiation and altitude. If multiple scattering is of relevance, the relationship would turn out to be much more involved and complicated to be exploited.

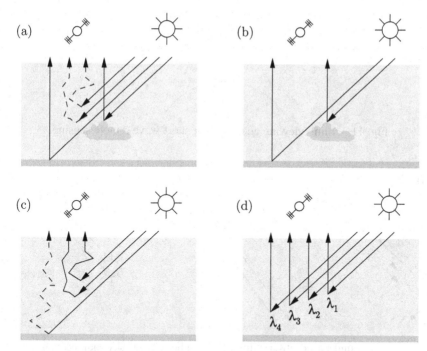

Fig. 11.6 Nadir viewing geometry for different short-wave remote sensing observations. Dashed lines denote contributions of minor importance. See also text.

The case of a turbid atmosphere lying over a dark surface is illustrated in Figure 11.6c. In order to detect a noticeable signal at the instrument, the atmosphere must scatter the incoming solar radiation in the upward direction. Furthermore, the atmosphere itself should not be opaque. Contributions due to ground reflection are of secondary importance, since the ground is assumed to be dark.

Finally, Figure 11.6d depicts the situation that the backscattered solar radiation is observed in the very strong absorption band in the UV-A and UV-B spectral region ($\lambda < 330$ nm). For very short wavelengths ($\lambda < 300$ nm) the ozone absorption optical depth is so large that solar photons entering the top of the atmosphere are absorbed before they reach the surface. Taking advantage of the smoothly varying absorption cross-sections of ozone molecules, the location of the atmospheric backscatter contributions can be varied through the entire altitude range of the Earth's atmosphere. In the figure we have $\lambda_1 < \lambda_2 < \lambda_3 < \lambda_4$. If clouds occur in the pixel observed by the satellite instrument, then it is of key importance to know both the location of the cloud top as well as the cloud cover for the pixel.

The limb viewing geometry is depicted in Figure 11.7. The observing instrument and the Sun are located to the left and to the right of the local vertical with respect

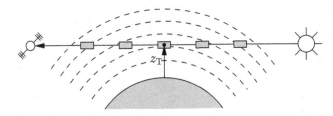

Fig. 11.7 Limb viewing geometry for short-wave remote sensing.

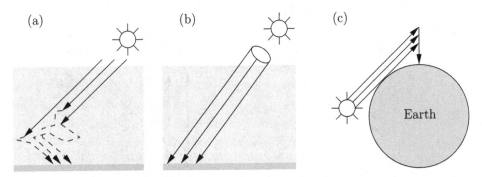

Fig. 11.8 Ground-based observation of sky radiation: (a) sky radiance; (b) solar aureole; (c) zenith radiation during sunset or sunrise.

to the tangent point at height z_T. This point is the lowermost location along the limb viewing path. The instrument's viewing direction is tangential with respect to the Earth's atmosphere.

Note that, in principle, all atmospheric volume elements located along the limb path may deliver a certain contribution to the observed signal. It should be observed, however, that the more distant these volume elements are relative to z_T the less important is their contribution. This is due to the fact that the outer shells of the atmosphere contain less scattering material than the volume elements close to z_T. Therefore, to a very good degree of approximation the major contributions to the signal originate from atmospheric volume elements centered around z_T.

Figure 11.8a–c depict several observation geometries for ground-based remote sensing of solar radiation. Figure 11.8a illustrates the daytime observation of the sky radiance. Due to atmospheric scattering of solar radiation by air molecules, aerosol or cloud particles, the sky radiance exhibits specific features with respect to the viewing zenith and azimuth angle. This information can be exploited to derive physical properties of the scattering particles. More information can be obtained by measuring, modeling and analyzing the polarized radiation field. This can be understood if one realizes that the clear atmosphere (no aerosols and no clouds present) possesses a distinct polarization pattern which is due to the polarizing nature of Rayleigh scatterers. In contrast to this, scattering by aerosol particles or cloud droplets causes a depolarizing effect. This depolarizing nature can be exploited to define specific remote sensing algorithms. We will not treat this subject further and refer to the scientific literature (e.g. Coulson, 1988; Herman *et al.* 1997; Mishchenko and Travis, 1997).

Figure 11.8b depicts the so-called *solar aureole observation*. By this we mean that one observes the immediate neighborhood of the solar disk, i.e. the instrument does not view the solar disk itself but detects diffuse radiation coming from a narrow solid angle interval covering the solar aureole. It is obvious that this type of observation exploits the forward scattering properties of atmospheric aerosol particles or thin or medium thick cirrus clouds. Clearly, such measurements cannot be performed if the sky is overcast because then the radiation field becomes more and more isotropic thus losing any information signature that originates from the single scattering processes. Further information can be found in the literature (e.g. Deirmendjian, 1957; Box and Deepak, 1981; Nakajima *et al.*, 1983; Santer and Herman, 1983).

Another interesting observation geometry is shown in Figure 11.8c. Here one observes zenith radiation during dawn or dusk, i.e. when the Sun is just below the horizon. This observation geometry makes it necessary to take the spherical nature of the atmosphere fully into account. While the Sun moves more and more below the horizon it illuminates only atmospheric layers higher up in the atmosphere. This means that the entire troposphere is located in the darkness zone.

11.2.2 Methods based on thermal emission

We will illustrate the basic principles by making the following assumptions. To a sufficiently high degree of approximation scattering of radiation can be neglected in the long-wave and microwave spectral region yielding

$$k_{\text{ext},\nu} = k_{\text{abs},\nu}, \qquad \omega_{0,\nu} = 0 \tag{11.19}$$

As boundary conditions for the atmospheric radiation field we will assume that at the top of the atmosphere no downwelling radiation exists whereas the upward directed radiation at the Earth's surface consists of the *Planckian emission* of the ground having an emissivity $\varepsilon_{\text{g},\nu}$, plus the reflected part of the downwelling flux density, i.e.

$$I_{-,\nu}(z_t) = 0$$

$$I_{+,\nu}(0) = \varepsilon_{\text{g},\nu} B_\nu(T_\text{g}) + (1 - \varepsilon_{\text{g},\nu}) \frac{E_{-,\nu}(0)}{\pi} \tag{11.20}$$

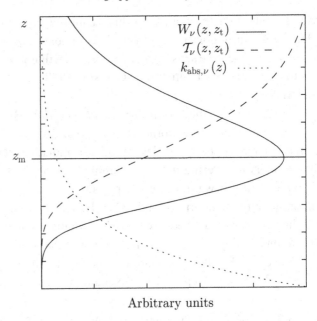

Fig. 11.9 Typical functional shape of the weighting function $W_\nu(z, z_t)$.

In the following we will discuss two arrangements for remote sensing in the long-wave spectral region.

(1) Nadir-looking instrument

In z coordinates the solution of the RTE for the upwelling radiation is given by, (see (2.122))

$$I_{+,\nu}(z_t) = I_{+,\nu}(0)\mathcal{T}_\nu(0, z_t) + \int_0^{z_t} B_\nu(z)\frac{d}{dz}\mathcal{T}_\nu(z, z_t)dz \qquad (11.21)$$

with

$$\mathcal{T}_\nu(z, z_t) = \exp\left(-\int_z^{z_t} k_{abs,\nu}(z')dz'\right) \qquad (11.22)$$

The term

$$\boxed{W_\nu(z, z_t) = \frac{d}{dz}\mathcal{T}_\nu(z, z_t) = k_{abs,\nu}(z)\mathcal{T}_\nu(z, z_t)} \qquad (11.23)$$

is called the *weighting function*.

Usually $k_{abs,\nu}(z)$ decreases with height because for gases like O_2 or CO_2 the number concentration of the gas molecules decreases with increasing z. For such a case Figure 11.9 shows typical vertical profiles of $k_{abs,\nu}$, $\mathcal{T}(z, z_t)$ and the resulting weighting function. In a strong absorption band $\mathcal{T}_\nu(z, z_t)$ increases from 0 to 1 when z increases from $z = 0$ to $z = z_t$. From the figure it will be seen that at a particular altitude $z = z_m$ the weighting function exhibits a maximum. Obviously, with increasing absorption the location of the maximum z_m moves upward. This can be explained by the fact that for a strongly absorbing gas $T_\nu(z, z_t)$ already vanishes at high altitudes. With decreasing absorption z_m moves downward correspondingly.

Let us consider the idealized situation where the weighting function is a δ-function, i.e.

$$W_\nu(z, z_t) = \delta(z - z_m) \tag{11.24}$$

Inserting this equation into (11.21) leads to

$$I_{+,\nu}(z_t) = I_{+,\nu}(0)T_\nu(0, z_t) + B_\nu(z_m) \tag{11.25}$$

Hence, in addition to the contribution coming from the lower boundary, the satellite instrument detects radiation coming directly from the level z_m. If similar weighting functions exist for other wavelengths λ_i and if the corresponding weighting functions W_{ν_i} have their delta function peaks at $z_{m,i}$, then this idealized situation directly provides the atmospheric temperature profile at the discrete set of altitudes $z_{m,i}$.

Unfortunately, atmospheric weighting functions always possess a finite width, and in most cases these weighting functions computed for different wave numbers overlap substantially. Therefore, the inversion of the temperature profile from a set of nadir radiance measurements is a complicated and difficult task.

In the atmospheric window region we have

$$T_\nu \approx 1, \qquad \frac{d}{dz}T_\nu \approx 0 \tag{11.26}$$

Inserting this equation into (11.21) we obtain

$$I_{+,\nu}(z_t) \approx I_{+,\nu}(0) \tag{11.27}$$

i.e. for most practical purposes the instrument registers the signal emitted by the ground. If the emissivity of the ground is equal to 1 it follows from (11.20) that

$$I_{+,\nu}(z_t) = B_\nu(T_g) \tag{11.28}$$

Thus the instrument aboard a satellite measures the *black body emission of the Earth's surface*. This means that the temperature T_g of the ground can be determined by inverting (11.28).

If the observation takes place in a strong absorption band, then the absorption coefficient can be assumed to be large so that

$$T_\nu(0, z_t) \approx 0 \tag{11.29}$$

This statement implies that any ground contribution to the radiation field at the satellite's position can be safely neglected. Now (11.21) reduces to

$$I_{+,\nu}(z_t) = \int_0^{z_t} B_\nu(z) \frac{d}{dz} T_\nu(z, z_t) dz \qquad (11.30)$$

This formula has important applications. For an absorber gas with constant mixing ratio, such as carbon dioxide in the 4.3 and 15 μm bands, or oxygen in the 5 mm microwave band (i.e. the so-called *60 GHz O_2 complex*), the weighting function can be computed very easily, that is the signal observed by the instrument aboard the satellite can be exploited to invert the vertical temperature profile. This makes it necessary to select a specific set of wavelengths so that for each wavelength the weighting function attains its maximum contribution to the observed signal at a particular altitude range.

Note that the determination of the temperature profile heavily relies on the assumption that the mixing ratio of the considered trace gas is constant with height. Otherwise both the concentration profile as well as the temperature profile would be unknown quantities making the inversion much more difficult.

Clouds pose a peculiar problem in atmospheric remote sensing because in the visible and long-wave spectral wavelength region they are not transparent. As a consequence, atmospheric parameters can only be retrieved for the altitude range above the cloud top. However, it becomes possible to invert the cloud top temperature.

So far we have considered only the nadir viewing geometry. However, the previous discussion can be generalized to arbitrary viewing directions using scanning instruments. As seen from the satellite's orbit such instruments provide a larger areal coverage as compared to instruments which mainly look in nadir direction.

(2) Ground-based observations

In the case of ground-based observations of the long-wave sky radiation we assume an isotropic radiation field. Utilizing the boundary conditions (11.20) the downward radiance at ground level expressed in z coordinates follows from (2.122)

$$I_{-,\nu}(0, \mu) = -\int_0^{z_t} B_\nu(z) \frac{d}{dz} T_\nu(0, z, \mu) dz = -\int_0^{z_t} B_\nu(z) W_\nu(0, z, \mu) dz \quad (11.31)$$

with

$$T_\nu(0, z, \mu) = \exp\left(-\frac{1}{\mu} \int_0^z k_{abs,\nu}(z') dz'\right) \qquad (11.32)$$

and where μ is the direction of observation. A similar analysis as in the case of spaceborne observations leads to the following conclusions for ground-based observations: The weighting function typically decreases strongly with increasing altitude. Therefore, the main contributions to the detected signal originate from the lower troposphere thus limiting information extraction to the lowermost parts of the atmosphere. This type of observation geometry in the long-wave spectral region is of much less importance than radiance measurements from satellites. The only exception is the observation of signals in close proximity to the flanks of the strong absorption bands of atmospheric trace gases leading to a complete sounding of the atmosphere from ground level to higher altitudes.

11.3 Inversion of the temperature profile

Using a radiometer in the long-wave electromagnetic spectrum aboard a satellite, one can measure the upwelling radiance in several spectral regions that are called channels. As discussed previously, the information content within an absorption band of a specific trace gas can be exploited to retrieve the atmospheric temperature profile. We will now show how to proceed.

Let us start with the radiative transfer equation of the infrared spectrum, cf. (2.113)

$$\mu \frac{d}{d\tau} I_\nu(\tau, \mu) = I_\nu(\tau, \mu) - B_\nu(\tau) \tag{11.33}$$

The monochromatic differential optical depth is given by

$$d\tau = k_\nu du \quad \text{with} \quad du = -\rho_{abs}(z)dz \tag{11.34}$$

where k_ν is the spectral absorption coefficient, u is the absorber amount and ρ_{abs} is the density of the absorber gas. Employing the hydrostatic equation we can relate optical depth changes to changes in total atmospheric pressure

$$d\tau = k_\nu \frac{q_{abs}}{g} dp \quad \text{with} \quad q_{abs} = \frac{\rho_{abs}}{\rho} \tag{11.35}$$

Here q_{abs} is the mass concentration of the trace gas and ρ is the air density. Substituting (11.35) into (11.33) we obtain for the change in radiance

$$dI_\nu(p, \mu) = \frac{1}{\mu} [I_\nu(p, \mu) - B_\nu(p)] k_\nu \frac{q_{abs}}{g} dp \tag{11.36}$$

For a nadir-looking instrument radiation is observed within a small solid angle element centered around $\mu = 1$ and the *monochromatic transmission* between the top of the atmosphere and the pressure level p is given by

$$\mathcal{T}_\nu(p) = \exp\left(-\frac{1}{g} \int_0^p k_\nu(p') q_{abs}(p') dp'\right) \tag{11.37}$$

In this case the solution to (11.33) can be written as

$$I_\nu(0) = I_\nu(\tau_0)\exp(-\tau_0) + \int_0^{\tau_0} B_\nu(\tau')\exp(-\tau')d\tau' \tag{11.38}$$

where τ_0 is the total optical depth at $z = 0$ corresponding to the surface pressure p_0. Assuming that at τ_0 the surface radiates like a black body with temperature $T(p_0)$ and using the relation

$$\frac{\partial \mathcal{T}_\nu}{\partial p}dp = -\mathcal{T}_\nu d\tau \tag{11.39}$$

we can write the upwelling radiance for $\mu = 1$ at the top of the atmosphere as

$$I_\nu(0) = B_\nu(p_0)\mathcal{T}_\nu(p_0) + \int_{p_0}^0 B_\nu(p)\frac{\partial \mathcal{T}_\nu}{\partial p}dp \tag{11.40}$$

Recall that $\partial\mathcal{T}_\nu(p)/\partial p$ is the weighting function which multiplies the *Planck function* for the upwelling radiation emanating from an elementary atmospheric layer of thickness dp.

Equation (11.40) is valid for *monochromatic radiation* only. In practice, however, a radiation instrument is only capable of detecting radiation within a finite spectral band (ν_1, ν_2) with central frequency $\bar\nu$. The width of the interval and the sensitivity of the radiometer are characterized by the *response function* ϕ_ν. Thus the radiation detected by the radiometer may be expressed by

$$
\begin{aligned}
I_{\bar\nu}(0) &= \frac{\int_{\nu_1}^{\nu_2}\phi_\nu I_\nu(0)d\nu}{\int_{\nu_1}^{\nu_2}\phi_\nu d\nu} \\[2mm]
&= \frac{1}{\int_{\nu_1}^{\nu_2}\phi_\nu d\nu}\left(\int_{\nu_1}^{\nu_2}\phi_\nu B_\nu(p_0)\mathcal{T}_\nu(p_0)d\nu + \int_{\nu_1}^{\nu_2}\int_{p_0}^0 \phi_\nu B_\nu(p)\frac{\partial \mathcal{T}_\nu}{\partial p}dp\,d\nu\right)
\end{aligned}
\tag{11.41}
$$

The *mean transmission* is defined as

$$\mathcal{T}_{\bar\nu}(p) = \frac{\int_{\nu_1}^{\nu_2}\phi_\nu \mathcal{T}_\nu(p)d\nu}{\int_{\nu_1}^{\nu_2}\phi_\nu d\nu} \tag{11.42}$$

so that the average weighting function is given by

$$\frac{\partial \mathcal{T}_{\bar\nu}}{\partial p} = \frac{\int_{\nu_1}^{\nu_2}\phi_\nu \frac{\partial \mathcal{T}_\nu}{\partial p}d\nu}{\int_{\nu_1}^{\nu_2}\phi_\nu d\nu} \tag{11.43}$$

Note that the average weighting function now contains the sensitivity or the efficiency of the instrument. We may apply the approximation that within the given frequency interval the Planck function varies linearly with respect to ν permitting us to replace B_ν by its mean value $B_{\bar{\nu}}$. In this case $B_{\bar{\nu}}$ can be extracted from the frequency integral. Inserting then (11.42) and (11.43) into (11.41) yields

$$I_{\bar{\nu}}(0) = B_{\bar{\nu}}(p_0)\mathcal{T}_{\bar{\nu}}(p_0) + \int_{p_0}^{0} B_{\bar{\nu}}(p)\frac{\partial \mathcal{T}_{\bar{\nu}}}{\partial p}dp \qquad (11.44)$$

where we have also assumed that the frequency integral over the response function is independent of pressure p.

The fundamental principle for deriving the temperature profile of the atmosphere from infrared soundings using satellite instruments is based on (11.44). This is due to the fact that the Planck function contains the temperature information, while the *transmission* of the atmosphere is associated with the absorption coefficient and the vertical profile of the trace gas under consideration. Thus the observed radiation must contain information on the profiles of both the atmospheric temperature as well as the trace gas concentration.

As a particular example let us consider the infrared atmospheric window region where, except for the 9.6 μm ozone band, absorption effects of atmospheric gases are relatively insignificant. Therefore, observations of the upwelling radiance at the top of the atmosphere in the atmospheric window are practically directly related to the Planck radiation emitted by the surface, i.e.

$$I_{\bar{\nu}}(0) \approx B_{\bar{\nu}}(p_0) \qquad (11.45)$$

CO_2 has its main absorption band in the wavelength region stretching from 12–18 μm. With the exception of small-scale local effects, for instance due to biomass burning and other anthropogenic effects, the mixing ratio of CO_2 is essentially constant vertically and horizontally. Presently we may use

$$q_{CO_2} \approx 5.47 \times 10^{-4} \qquad (11.46)$$

being equivalent to a volume mixing ratio of 360 ppmv. The line intensities, the position of all spectral lines and the line half-widths for CO_2 are known with high precision from laboratory or theoretical results. Thus the spectral transmission function and the weighting function for CO_2 can be calculated very accurately as a function of pressure and temperature using model distributions. In many cases the uncertainties in T and p are not essential.

Once a temperature profile has been found by inverting the $B_{\bar{\nu}}$ functions using suitable model values of T and p, we may repeat the transmission function calculations by employing the inverted temperature profile and repeating the calculations to obtain a new temperature profile. We will soon return to this problem.

Given the surface temperature $T(p_0)$ by inversion of (11.45), the vertical temperature profile $T(p)$ can be found by inverting (11.44) for a set of channels in the CO_2

absorption band. Based on the temperature retrieval using CO_2, nadir observation in spectral absorption bands of various other trace gases, such as O_3, H_2O and CH_4, could then be employed to derive total column contents or even profile information for these trace gases. In the following we will limit the discussion to the derivation of the temperature profile based on CO_2.

In the CO_2 absorption band there exist several regions where the transmission function approaches zero, so that the thermal signal emitted by the surface does not reach the top of the atmosphere, i.e.

$$\mathcal{T}_{\bar{\nu}}(p_0) \approx 0 \qquad (11.47)$$

Then (11.44) simplifies to

$$I_{\bar{\nu}}(0) \approx \int_{p_0}^{0} B_{\bar{\nu}}(p) \frac{\partial \mathcal{T}_{\bar{\nu}}}{\partial p} dp \qquad (11.48)$$

For a set of different frequency bands the transmission and weighting functions have to be computed as functions of pressure and temperature. Due to the frequency dependence of the Planck function the average $B_{\bar{\nu}}$ will change from one frequency band to the other. Therefore, it is necessary to eliminate the frequency dependence in (11.48).

Within the CO_2 15 μm band the Planck function can be expressed as a linear function

$$B_{\bar{\nu}}(p) = \alpha_{\bar{\nu}} B_{\bar{\nu}_r}(p) + \beta_{\bar{\nu}} \qquad (11.49)$$

where ν_r is a fixed reference frequency close to the center of the 15 μm band, and $\alpha_{\bar{\nu}}$ and $\beta_{\bar{\nu}}$ are fit constants. Using this linear expression for the Planck function in (11.48) leads to

$$\frac{I_{\bar{\nu}}(0) - \beta_{\bar{\nu}}}{\alpha_{\bar{\nu}}} = \int_{p_0}^{0} B_{\bar{\nu}_r}(p) \frac{\partial \mathcal{T}_{\bar{\nu}}}{\partial p} dp \qquad (11.50)$$

It is customary to introduce the new functions

$$g(\bar{\nu}) = \frac{I_{\bar{\nu}}(0) - \beta_{\bar{\nu}}}{\alpha_{\bar{\nu}}}, \quad f(p) = B_{\bar{\nu}_r}(p), \quad \mathcal{K}(\bar{\nu}, p) = \frac{\partial \mathcal{T}_{\bar{\nu}}}{\partial p} \qquad (11.51)$$

Then (11.50) can be reformulated as

$$g(\bar{\nu}) = \int_{p_0}^{0} f(p) \mathcal{K}(\bar{\nu}, p) dp \qquad (11.52)$$

This is a Fredholm integral equation of the first kind. The weighting function $\mathcal{K}(\bar{\nu}, p)$ is the so-called kernel of the integral equation. $f(p)$ is the function to be determined by employing a set of measurements $g(\bar{v}_i)$, $i = 1, \ldots, M$, where M is the total number of channels provided by the infrared radiometer.

In a first step we will discuss the kernel $\mathcal{K}$ of the integral equation. The spectrally averaged transmission function can be written as

$$\mathcal{T}_{\bar{v}}(p) = \frac{1}{\Delta v} \int_{\Delta v} \exp\left(-\frac{q_{CO_2}}{g} \int_0^p k_v(p')dp'\right) dv \qquad (11.53)$$

with $\Delta v = v_2 - v_1$. For illustrative purposes we have assumed an ideal instrument, i.e. $\phi_v = 1$. For altitudes below about 30–40 km *Lorentz broadening* dominates spectral *line broadening* so that simultaneous *Doppler broadening* can be safely neglected. Inserting the absorption coefficient for the Lorentz line shape, see (7.15), and neglecting the temperature dependence of the Lorentz half-width so that

$$\alpha_L(p, T) = \alpha_{L,0}\frac{p}{p_0}\sqrt{\frac{T_0}{T}} \approx \alpha_{L,0}\frac{p}{p_0} \qquad (11.54)$$

we obtain for the transmission the expression

$$\mathcal{T}_{\bar{v}}(p) = \frac{1}{\Delta v} \int_{\Delta v} \exp\left(-\frac{q_{CO_2}}{g} \int_0^p \sum_{i=1}^L \frac{S_i[T(p')]}{\pi} \frac{\alpha_{L,i}(p')}{(v - v_{0,i})^2 + \alpha_{L,i}(p')^2}dp'\right) dv$$

$$(11.55)$$

where we summed over a total of L individual Lorentz lines contained in the spectral interval Δv.

In principle, the full temperature dependence of the absorption coefficient has to be taken into account. A particular problem for setting up the inversion of (11.44) is that *a priori* we do not even know a rough approximation of the temperature profile that is needed to get the iteration started. As an initial guess one, therefore, assumes a temperature profile which is climatologically representative for the actual position of the satellite.

Wark and Fleming (1966) computed for the US standard atmosphere the kernel function $\mathcal{K}$ for six different rather narrow wave number bands $\bar{v}_i$, $i = 1, \ldots, M = 6$ located within the 15 μm CO_2 band. Figure 11.10 depicts the vertical variation of $\partial \mathcal{T}_{\bar{v}}(p)/\log p$ versus $\log p$.

To make the inversion of (11.44) possible, the individual kernel functions $\mathcal{K}_i(p) = \mathcal{K}(\bar{v}_i, p)$ must possess distinct maxima in different altitude regions. If this is the case for the selected narrow frequency bands, we may extract sufficient information from the upwelling radiation at the instruments level to construct the desired vertical atmospheric profile. As follows from inspection of Figure 11.10, the individual weighting functions overlap considerably which complicates

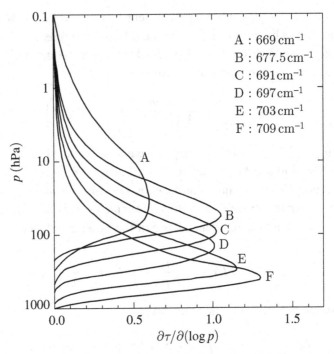

Fig. 11.10 Weighting functions $\partial \mathcal{T}_{\tilde{\nu}}/\partial p$ versus $\log p$ for six reference wave numbers applied to the 15 μm CO_2 band. (Redrawn from Wark and Fleming (1996), with permission from the American Meteorological Society.)

the inversion process. The entire inversion is based on the set of observations $g_i = g(\bar{\nu}_i)$, $i = 1, \ldots, M$ which provide the necessary input information. Owing to the overlapping kernel functions, we have to invert a system of N linear equations for the unknown functions $f(p_j)$, $j = 1, \ldots, N$ where N denotes the total number of pressure levels. In case that the $\mathcal{K}_i$ do not overlap, we would end up with N independent equations which can be inverted directly. Since this is generally not the case, it is advantageous to employ matrix methods to find the proper solution.

11.3.1 Direct linear inversion

Employing the M radiance observations the mathematical problem is stated by a set of M integral equations

$$g_i = \int_{p_0}^{0} f(p)\mathcal{K}_i(p)dp, \qquad i = 1, \ldots, M \tag{11.56}$$

Usually the unknown function $f(p)$ can be expanded in terms of known representation functions $W_j(p)$, $j = 1, \ldots, N$, which may involve Legendre polynomials or trigonometric functions so that

$$f(p) = \sum_{j=1}^{N} f_j W_j(p) = B_{\bar{\nu}_i}(p) \tag{11.57}$$

The f_j appearing in this equation are the unknown expansion coefficients. For (11.56) this leads to

$$g_i = \sum_{j=1}^{N} f_j \int_{p_0}^{0} W_j(p) \mathcal{K}_i(p) dp, \qquad i = 1, \ldots, M \tag{11.58}$$

It is now convenient to introduce a $M \times N$ matrix A whose elements A_{ij} are given by

$$A_{ij} = \int_{p_0}^{0} W_j(p) \mathcal{K}_i(p) dp \tag{11.59}$$

leading to the compact notation

$$g_i = \sum_{j=1}^{N} A_{ij} f_j, \qquad i = 1, \ldots, M \tag{11.60}$$

In vector notation we find

$$\mathbf{g} = A\mathbf{f}, \quad \mathbf{g} = \begin{pmatrix} g_1 \\ g_2 \\ \vdots \\ g_M \end{pmatrix}, \quad \mathbf{f} = \begin{pmatrix} f_1 \\ f_2 \\ \vdots \\ f_N \end{pmatrix} \tag{11.61}$$

For the simple case where A is a square matrix with a nonvanishing determinant, the solution to (11.61) is

$$\mathbf{f} = A^{-1}\mathbf{g} \tag{11.62}$$

Having found $\mathbf{f}$ we can compute the Planck function from (11.57) to find the temperature profile $T(p)$.

Usually for a remote sensing problem the above assumption of an equal number of unknowns and observations is not fulfilled so that the matrix A does not possess an inverse A^{-1}. In most cases the problem is underdetermined ($M < N$) so that we have more unknowns than independent observations. Thus the problem is ill-posed. Even if A is a square matrix, the inverse A^{-1} may still not exist if the determinant of A is close to zero. This type of instability may arise for various reasons such as (i) errors in the computation of the matrix elements A_{ij}, (ii) errors when approximating the Planck function, (iii) round-off errors, or (iv) instrument noise leading to observed radiances which involve random errors. In the Appendix to this chapter we will give an illustrative example of an ill-posed problem.

11.3.2 Linear inversion with constraints

Consider the ill-posed problem

$$g_i = \sum_{j=1}^{N} A_{ij} f_j, \qquad i = 1, \ldots, M \tag{11.63}$$

Now we admit errors in the measurements such that

$$\hat{g}_i = g_i + \varepsilon_i \tag{11.64}$$

Here g_i represents the i-th measurement resulting from an ideal instrument, and ε_i is the measurement error. Let us perform a linear inversion subject to some constraint. For example, this can be expressed by the minimization of the cost function

$$S = \sum_{i=1}^{M} \varepsilon_i^2 + \gamma \sum_{j=1}^{N} (f_j - \bar{f})^2 \tag{11.65}$$

where $\bar{f}$ is the average value of f, and γ is a smoothing parameter which must be prescribed following a mathematically and physically consistent reasoning. It can be seen that S contains as the second term the variance of the f_j. Thus the minimization problem is characterized by the fact that the sum of the squared differences between f_j and $\bar{f}$ is minimized, too. The smoothing parameter γ determines to what extent the discrete values f_j forming the solution are constrained to remain close to the average value $\bar{f}$.

Now we can use (11.64) in (11.63) solving the latter equation for the errors ε_i. Minimization of S then means that the partial derivatives of S with respect to the unknown physical parameters f_k, $k = 1, \ldots, j, \ldots, N$ must vanish. This leads to the expressions

$$\frac{\partial}{\partial f_k} \left[\sum_{i=1}^{M} \left(\sum_{j=1}^{N} A_{ij} f_j - \hat{g}_i \right)^2 + \gamma \sum_{j=1}^{N} (f_j - \bar{f})^2 \right] = 0 \tag{11.66}$$

For the partial derivatives we obtain

$$\sum_{i=1}^{M} \left(\sum_{j=1}^{N} A_{ij} f_j - \hat{g}_i \right) A_{ik} + \gamma (f_k - \bar{f}) = 0 \tag{11.67}$$

Since the average of the f_j is defined as

$$\bar{f} = \frac{1}{N} \sum_{k=1}^{N} f_k \tag{11.68}$$

we find

$$f_k - \bar{f} = -N^{-1} f_1 - N^{-1} f_2 - \cdots + (1 - N^{-1}) f_k - \cdots - N^{-1} f_N \tag{11.69}$$

Inserting this equation into (11.67) leads in matrix notation to

$$A^T A \mathbf{f} - A^T \hat{\mathbf{g}} + \gamma H \mathbf{f} = 0 \tag{11.70}$$

where the $N \times N$ matrix H is given by

$$H = \begin{pmatrix} 1 - N^{-1} & -N^{-1} & \cdots & -N^{-1} \\ -N^{-1} & 1 - N^{-1} & \cdots & -N^{-1} \\ \vdots & \vdots & \cdots & \vdots \\ -N^{-1} & -N^{-1} & \cdots & 1 - N^{-1} \end{pmatrix} \tag{11.71}$$

Inverting (11.70) leads to the final solution

$$\mathbf{f} = (A^T A + \gamma H)^{-1} A^T \hat{\mathbf{g}} \tag{11.72}$$

This solution is due to Phillips (1962) and Twomey (1963).

11.3.3 Chahine's relaxation method

Let us consider the solution to the radiative transfer equation for nadir observation in the M measurement channels at the top of the atmosphere as discussed above, see (11.44)

$$I_i = B_i(p_0) \mathcal{T}_i(p_0) + \int_{p_0}^{0} B_i(p) \frac{\partial \mathcal{T}_i(p)}{\partial \ln p} d \ln p, \quad i = 1, \ldots, M \tag{11.73}$$

The Planck function is given by

$$B_i(p) = B_i [T(p)] = \frac{a v_i^3}{\exp [b v_i / T(p)] - 1} \tag{11.74}$$

We will assume that the weighting functions for the individual channels can be chosen in such a way that each weighting function attains its maximum at a different altitude level or, equivalently, within a pressure layer of thickness $\Delta_i \ln p$. Employing the mean value theorem for integrals, we can approximate the observed radiance $\tilde{I}_i$ by means of

$$\tilde{I}_i \approx B_i(T_0)\mathcal{T}_i(p_0) + B_i(T_i)\frac{\partial \mathcal{T}_i}{\partial \ln p}\bigg|_{p_i} \Delta_i \ln p \qquad (11.75)$$

Note that p_i is the pressure level for which $\partial \mathcal{T}_i/\partial \ln p$ attains its maximum value, and that $\Delta_i \ln p$ can be understood as the effective width of the i-th weighting function in logarithmic pressure coordinates.

Let T_i' stand for an estimate of the temperature at level p_i. Inserting this estimate into (11.75), we obtain an estimate for the observed radiance

$$I_i' \approx B_i(T_0)\mathcal{T}_i'(p_0) + B_i(T_i')\frac{\partial \mathcal{T}_i'}{\partial \ln p}\bigg|_{p_i} \Delta_i \ln p \qquad (11.76)$$

Division of (11.75) by (11.76) leads to

$$\frac{\tilde{I}_i - B_i(T_0)\mathcal{T}_i(p_0)}{I_i' - B_i(T_0)\mathcal{T}_i'(p_0)} \approx \frac{B_i(T_i)}{B_i(T_i')} \frac{\dfrac{\partial \mathcal{T}_i}{\partial \ln p}\bigg|_{p_i}}{\dfrac{\partial \mathcal{T}_i'}{\partial \ln p}\bigg|_{p_i}} \qquad (11.77)$$

Experience indicates that for a temperature change from T to $T + \Delta T$ the Planck function varies much more strongly than the weighting function associated with this particular temperature change. In other words, the change of the transmission versus pressure remains virtually constant with respect to moderate temperature changes. Thus the second factor on the right-hand side involving the ratio of the weighting functions is approximately equal to 1.

Assuming that the surface contribution in (11.77) is negligible we find the simple relaxation equation

$$\boxed{\frac{\tilde{I}_i}{I_i'} \approx \frac{B_i(T_i)}{B_i(T_i')}} \qquad (11.78)$$

which can be used in the following way: insert the measured radiance $\tilde{I}$ in the left-hand side numerator of (11.78). By employing an estimated temperature T_i' the Planck function $B_i(T_i')$ can be evaluated as well as I_i' from (11.76) so that for a fixed channel i the temperature T_i can be found by inverting $B_i(T_i)$.

If necessary, the same form (11.76) of the relaxation formula can be used in case that $T(p_0)$ and $\mathcal{T}_i(p_0) \approx \mathcal{T}_i'(p_0)$ are known to include the surface contribution. In this case $B_i(T_0)\mathcal{T}_i(p_0)$ must be subtracted from $\tilde{I}_i$ in (11.78) and from I_i' in (11.76).

The relaxation equation (11.78) has been originally derived by Chahine (1970). We will now briefly discuss the application of this equation. In the strong absorption regions of the 15 μm band of CO_2 the upwelling radiance measured by the satellite instrument at the top of the atmosphere results from emissions of the upper atmospheric regions. For the weaker absorption bands this radiation originates from progressively lower atmospheric regions. One selects a particular set of M frequency bands which is characterized by overlapping weighting functions covering the altitude range of interest.

The algorithm for Chahine's relaxation method is given by the following steps.

(1) Make use of the radiance observations $\tilde{I}_i$ for all channels $i = 1, \ldots, M$.
(2) Specify the constant mixing ratio of CO_2, the response functions ϕ_{ν_i} of the instrument for all channels, and select the pressure levels p_i for which the weighting functions attain their maximum.
(3) Prescribe an initial estimate $T_i^{(0)}$ ($i = 1, \ldots, M$) for the temperature profile, for example, by using a reasonable profile from a climatology.
(4) For the n-th step, insert $T_i^{(n)}$ into (11.73) and employ an accurate quadrature formula to determine the integral over $d \ln p$.
(5) Make a comparison of the n-th iterate $I_i^{(n)}$ with the measurements $\tilde{I}_i$ for $i = 1, \ldots, M$. If all the residuals $R_i^{(n)} = |\tilde{I}_i - I_i^{(n)}|/\tilde{I}_i$ are smaller than a given bound ε, then $T_i^{(n)}$ is the solution for the temperature profile. If $|\tilde{I}_i - I_i^{(n)}|/\tilde{I}_i > \varepsilon$ apply the relaxation equation (11.78) in the form

$$\frac{\tilde{I}_i}{I_i^{(n)}} \approx \frac{B_i(T_i^{(n+1)})}{B_i(T_i^{(n)})} \tag{11.79}$$

Substituting (11.74) into (11.79) gives the estimate

$$T^{(n+1)}(p_i) = b\nu_i / \ln \left\{ 1 - \left[1 - \exp\left(\frac{b\nu_i}{T_i^{(n)}}\right) \right] \frac{I_i^{(n)}}{\tilde{I}_i} \right\} \tag{11.80}$$

for the iteration step $n + 1$.

(6) Return to step 4 and re-evaluate (11.73) by appropriately interpolating the temperature profile from the set $(p_i, T_i^{(n+1)})$ yielding $I_i^{(n+1)}$. Repeat the above steps until the residuals are smaller than the bound ε.
(7) Use (linear) interpolation to evaluate the temperature profile at pressure levels other than $p_i, i = 1, \ldots, M$.

11.3.4 Smith's iterative inversion method

As an alternative to Chahine's relaxation algorithm we will now discuss an iterative method which was originally introduced by Smith (1970). In this approach (11.73) will be iteratively solved. We start by writing the n-th iteration step of (11.73) as

$$I_i^{(n)} = B_i^{(n)}(p_0) T_i(p_0) + \int_{p_0}^{0} B_i^{(n)}(p) \frac{\partial T_i}{\partial \ln p} d \ln p \tag{11.81}$$

For step $n + 1$ we write analogously

$$\tilde{I}_i = I_i^{(n+1)} = B_i^{(n+1)}(p_0)T_i(p_0) + \int_{p_0}^0 B_i^{(n+1)}(p)\frac{\partial T_i}{\partial \ln p}d \ln p \qquad (11.82)$$

i.e. we use the observations $\tilde{I}_i$ as a first guess in the $(n + 1)$-th iteration step.

The difference between (11.82) and (11.81) is

$$I_i^{(n+1)} - I_i^{(n)} = \left[B_i^{(n+1)}(p_0) - B_i^{(n)}(p_0)\right]T_i(p_0)$$
$$+ \int_{p_0}^0 \left[B_i^{(n+1)}(p) - B_i^{(n)}(p)\right]\frac{\partial T_i}{\partial \ln p}d \ln p \qquad (11.83)$$

The main simplification of this method results from the assumption that for each channel i the difference of the Planck functions under the integral sign does not depend on the entire atmospheric pressure range but only on the temperature at level p. Assuming further that the difference $B_i^{n+1}(p) - B_i^n(p)$ depends only weakly on the pressure level, the integral term in (11.83) can be approximated as

$$\int_{p_0}^0 \left[B_i^{(n+1)}(p) - B_i^{(n)}(p)\right]dT_i(p) \approx \left[B_i^{(n+1)}(p) - B_i^{(n)}(p)\right]T_i(0)$$
$$- \left[B_i^{(n+1)}(p_0) - B_i^{(n)}(p_0)\right]T_i(p_0) \qquad (11.84)$$

Since $T_i(0) = 1$, we obtain for the differences of the nadir radiance between the $(n + 1)$-th and n-th step

$$\tilde{I}_i - I_i^{(n)} = B_i^{(n+1)}(p) - B_i^{(n)}(p) \qquad (11.85)$$

The last expression can be solved for the Planck function in layer i for the $(n + 1)$-th iteration step yielding

$$B_i^{(n+1)}(p) = B_i^{(n)}(p) + \tilde{I}_i - I_i^{(n)} \qquad (11.86)$$

or, equivalently for the temperature, we find

$$T^{(n+1)}(p, v_i) = \frac{bv_i}{\ln\left[1 + av_i^3/B_i^{(n+1)}(p)\right]} \qquad (11.87)$$

For arbitrary p the best approximation for the temperature can be found by weighting all temperature profiles for the individual channels i with their own kernel $W_i(p)$

$$T^{(n+1)}(p) = \frac{\sum_{i=1}^M T^{(n+1)}(p, v_i)W_i(p)}{\sum_{i=1}^M W_i(p)} \qquad (11.88)$$

where we have defined

$$W_i(p) = \begin{cases} dT_i(p), & p < p_0 \\ T_i(p), & p = p_0 \end{cases} \qquad (11.89)$$

Using the above formulas the following steps constitute Smith's iterative retrieval method.

(1) Provide an initial guess for the temperature profile $T^{(n)}(p)$ for $n = 0$.
(2) Compute $B_i^{(n)}(p)$ employing the expression (11.74) for the Planck function. With known Planck function we can proceed to calculate $I_i^{(n)}$ using (11.81).
(3) Compute $B_i^{(n+1)}(p)$ and $T_i^{(n+1)}$ using (11.86) and (11.87). This can be done for arbitrary pressure p.
(4) Using (11.88) a new improved approximation for the temperature profile $T^{(n+1)}(p)$ can be found in the $(n + 1)$-th iteration step.
(5) In the final step we compare the computed nadir radiances $I_i^{(n)}$ and $\tilde{I}_i$. If $R_i^{(n)} = |\tilde{I}_i - I_i^{(n)}|/\tilde{I}_i < \varepsilon$ for all i, then $T^{(n+1)}(p)$ is the solution for the temperature profile. If $R_i^{(n)} > \varepsilon$, then repeat steps 1–5 until the residual term meets the specified bound ε.

Let us now discuss some of the advantages of Smith's iterative method.

(1) For the derivation of the temperature profile there is no prescription of the analytic form of the profile. This is in contrast to Chahine's method where the total number of discrete points of the retrieved temperature profile directly depends on the total number M of radiance observations.
(2) With Smith's method the temperature profile can be evaluated for arbitrary p. Chahine's relaxation method, on the other hand, employs a linear interpolation of the retrieved discrete temperature values $T^{(n)}(p_i)$ $(i = 1, \ldots, M)$.

Figure 11.11 shows a comparison of retrieved temperature profiles using Chahine's relaxation method and Smith's iteration procedure. This figure is due to Liou (1980) who computed the synthetic radiances for the six VTPR (Vertical Temperature Profile Radiometer) channels centered at the wave numbers 669.0, 676.7, 694.7, 708.7, 723.6, and 746.7 cm^{-1} of the 15 μm CO_2 band (solid curve without dots). The VTPR was flown on the NIMBUS 4 satellite in the early 1970s. As initial guess an isothermal temperature profile of 300 K was employed, and the surface temperature was held fixed at 279.5 K. For Chahine's relaxation method the residual was set to $\varepsilon = 10^{-2}$. The final result (solid curve with black dots) was obtained after only four iteration steps. The dots represent the discrete pressure levels p_i. It is seen that the difference between the actual and Chahine's temperature profile amounts to typically 2–3 K with similar over- and underestimations. The result obtained with Smith's iterative method is depicted as the dashed line in the figure. Smith's method converged after about 20 iteration steps whereby the bound

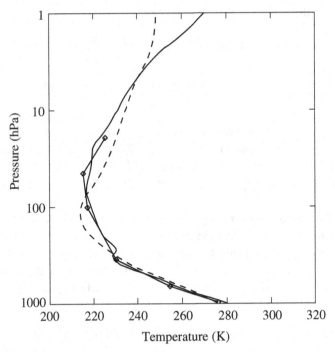

Fig. 11.11 Comparison between the temperature profile retrieval employing Chahine's relaxation and Smith's iteration method for the six VTPR channels. The thick solid line represents the actual temperature profile. (Redrawn from Liou (1980), with permission from Elsevier.)

ε was set to 0.02. It is noteworthy that a reduction of the size of ε did not improve the temperature profile. One can see that the retrieved profiles exhibit less variability than the true profile. Moreover, both methods are not capable of recovering the temperature profile in the upper atmosphere, because the VTPR channels were selected in a manner that the peak for the highest weighting function lies near 30 hPa.

We do not wish to conclude this section without giving special credit to two pioneers in the field of remote sensing of temperature profiles. King (1956) was the first to show in which way the vertical temperature distribution could be inferred from satellite radiance scan measurements. Kaplan (1959) pointed out that the vertical temperature profile of the atmosphere can be inferred from the spectral distribution of the emission spectrum as observed by a satellite. We have already made use of his observation that the upper part of the atmosphere can be seen by probing the radiance of the emitting gas at the band center. By properly selecting spectral regions in the wings of the band we may view the entire atmosphere.

Many papers, too numerous to be listed here, have been written on remote sensing techniques. Moreover, several textbooks on remote sensing are now available for detailed studies, for example Houghton *et al.* (1984) and Stephens (1994).

11.4 Radiative perturbation theory and ozone profile retrieval

We have pointed out already that ozone is one of the most important trace gases in the Earth's atmosphere. It is well known that the concentration of this trace gas strongly depends on altitude. Typically, in the free troposphere the volume concentration is approximately 0.05 ppmv, whereas in the altitude region between about 20 and 40 km ozone has a pronounced maximum of about 10 ppmv.

The radiative effects of ozone can be summarized as follows: it absorbs practically all solar radiation between 240 and 300 nm wavelength. Therefore, the ozone layer in the Earth's atmosphere protects all unicellular organisms and all skin cells of plants, animals and human beings from the dangerous lethal ultraviolet radiation. Nevertheless, solar photons in the wavelength range 300 to 330 nm (UV-B radiation) can penetrate the entire atmosphere and reach the surface. The intensity of the solar radiation is significantly reduced due to the absorbing ozone molecules. Thus UV-B radiation in this spectral region may cause skin cancer for susceptible humans.

The temperature structure and the dynamics of the stratosphere have their origin in the absorption of solar UV radiation. Thus in the stratosphere the temperature increases with increasing altitude while in the troposphere the temperature decreases with increasing altitude. Due to the thermal stability, vertical mixing of trace gases is very slow in the stratosphere. This is in contrast to the situation in the troposphere where vertical convective and turbulent processes cause a rapid mixing of atmospheric gases. In the atmospheric window region ozone possesses a strong absorption band near 9.6 μm and acts here as an important greenhouse gas.

In the stratosphere and lower mesosphere ozone molecules are produced by two steps. First, for wavelengths shorter than 242 nm molecular oxygen is photodissociated into atomic oxygen by the absorption of a photon of energy $h\nu$ (Chapman, 1930a,b)

$$O_2 + h\nu \rightarrow 2\,O \tag{11.90}$$

The subsequent reaction of O and the remaining oxygen molecules is

$$O_2 + O + M \rightarrow O_3 + M \tag{11.91}$$

where M denotes any third atom or molecule. This third atom or molecule is required in order to conserve energy and momentum.

In the mesosphere the concentration of molecular oxygen is so low that the production of atomic oxygen by photodissociation is unimportant. At lower altitudes the concentration of O_2 is higher but the solar photons have been absorbed already by ozone at higher altitudes. Thus there exists an intermediate altitude range characterized by a maximum production of ozone molecules.

In the stratosphere ozone is destroyed due to catalytic reactions (Bates and Nicolet, 1950)

$$O_3 + X \rightarrow O_2 + XO$$
$$O + XO \rightarrow O_2 + X \tag{11.92}$$

leading to a net reaction

$$O + O_3 \rightarrow O_2 + O_2 \tag{11.93}$$

Here X is the catalyst species. Due to the fact that the catalyst is not destroyed in this reaction, only a small amount of X is necessary to destroy a large reservoir of O_3. For this catalytic reaction several gaseous species can be important, e.g. X = H, OH, NO and Cl.

In the troposphere ozone is photochemically produced by the reactions

$$NO_2 + h\nu \rightarrow NO + O$$
$$O + O_2 \rightarrow O_3 \tag{11.94}$$

so that the photodissociation of a NO_2 molecule provides a single oxygen atom. The latter then reacts with molecular oxygen to produce ozone. It should be pointed out that the presence of trace gases like CO, CH_4 and other hydrocarbons allows for a recycling of NO to NO_2 thus making up a catalytic reaction chain. In the lower troposphere a major source of ozone is the *in situ* production via the above reactions. Transport from the stratosphere represents a second source of ozone in the upper and middle troposphere.

In the troposphere ozone plays an important role in atmospheric chemistry. Due to the photodissociation of ozone for wavelengths shorter than 315 nm highly reactive OH (hydroxyl) radicals are produced. These OH radicals can react with practically all atmospheric gases, and, therefore, OH is called an atmospheric detergent. It can be concluded that tropospheric ozone is crucial for the removal of atmospheric pollution. For detailed information on the chemistry of the atmosphere see, for example, Graedel and Crutzen (1994).

A very important field of research is to study the different roles of both tropospheric as well as stratospheric ozone in atmospheric chemistry and climate. For this type of research altitude resolved ozone measurements are an essential input. Ozone profile information on a global scale can only be obtained via satellite remote sensing.

In the following we will discuss the retrieval of ozone profiles from atmospheric spectrometer instruments aboard satellites which observe backscattered solar radiation. The retrieval method is based on radiative transfer modeling involving the solution to the forward and adjoint radiative transfer equation. As will be discussed next, the adjoint problem is used to derive the weighting functions, or Jacobians, via the linear radiative perturbation theory. In general, the Jacobians are the key input for the inversion of atmospheric parameters.

Let us consider the situation where the reflectance $r(\mathbf{x})$ of the atmosphere at the satellite position and in the viewing direction of the instrument is the required information for a given state vector $\mathbf{x}$ of the atmosphere. In the following this state vector represents the vertical profile of the ozone density ρ_{O_3} plus the *Lambertian ground albedo* A_g, i.e. $\mathbf{x} = (\rho_{O_3}, A_g)$. Thus we may use a forward radiative transfer model to simulate $r(\mathbf{x})$. The instrument itself measures the reflectance $\tilde{r}$ in dependence of the unknown atmospheric state. In addition to this, the observation involves the instrument error ε so that

$$\tilde{r} = r(\mathbf{x}) + \varepsilon \tag{11.95}$$

Clearly, the radiative transfer model depends in a nonlinear way on the state vector $\mathbf{x}$. Any retrieval method requires a linearization of the reflectance about a first guess state vector of the atmosphere which we will call $\mathbf{x}_0$. Using a Taylor series expansion, omitting nonlinear terms, we may write

$$r(\mathbf{x}) \approx r(\mathbf{x}_0) + \sum_{k=1}^{K} \frac{\partial r}{\partial x_k}\bigg|_{\mathbf{x}_0} \Delta x_k \quad \text{with} \quad \Delta x_k = (x_k - x_{0,k}) \tag{11.96}$$

where K is the dimension of the state vector $\mathbf{x}$ and x_k is its k-th component. As an example, one may identify this k-th component with the ozone density in the k-th homogeneous atmospheric sublayer.

For the inversion of the forward radiative transfer model we have to find the reflectance r as well as its linearization with respect to the individual components of the state vector. Usually the linearization represents the computational bottleneck of ozone profile retrievals. Therefore, for fast ozone profile retrievals on an operational basis the development of efficient linearized radiative transfer models represents an important task. For the linearization process we will now employ the linear radiative perturbation theory as described previously in Chapter 5 in more detail.

According to (5.5) the radiative transfer equation including Lambertian ground reflection can be formulated in its forward formulation as

$$LI(z, \mathbf{\Omega}) = Q(z, \mathbf{\Omega}) \tag{11.97}$$

where the linear differential operator L including the surface reflection part may be written as (Ustinov, 2001; Landgraf *et al.*, 2002)

$$L = \mu \frac{\partial}{\partial z} + k_{ext}(z) - \frac{k_{sca}(z)}{4\pi}$$

$$\times \int_{4\pi} \left[\mathcal{P}(z, \mathbf{\Omega}' \rightarrow \mathbf{\Omega}) - \frac{A_g}{\pi} \delta(z) U(\mu) |\mu| U(-\mu') |\mu'| \right] \circ d\mathbf{\Omega}' \quad (11.98)$$

The first two terms on the right-hand side represent the extinction of the radiance, the third term describes the scattering process due to air molecules and aerosol particles while the last term denotes the reflection by the Earth's surface with albedo A_g. Again U is the Heaviside step function as defined in (5.50). For simplicity we will not treat the effect of clouds so that only cloud free scenes will be considered in the sequel. For solar radiation the source term Q on the right-hand side of (11.97) is taken from (5.3), that is

$$Q(z, \mathbf{\Omega}) = |\mu_0| S_0 \delta(z - z_t) \delta(\mu - \mu_0) \delta(\varphi - \varphi_0) \quad (11.99)$$

As discussed in Chapter 5, the radiance I has to fulfill *vacuum boundary conditions* for the incoming radiation field

$$\begin{aligned} \text{top of the atmosphere:} \quad & I(z_t, \mu, \varphi) = 0, \quad 0 \le \varphi \le 2\pi, \quad -1 \le \mu < 0 \\ \text{Earth's surface:} \quad & I(0, \mu, \varphi) = 0, \quad 0 \le \varphi \le 2\pi, \quad 0 < \mu \le 1 \end{aligned}$$
$$(11.100)$$

Any *radiative effect* $\mathcal{E}$ is associated with a corresponding response function R

$$\mathcal{E} = \langle R, I \rangle \quad (11.101)$$

where the angular brackets $< \ldots >$ denote the inner product defined by the integration over the full solid angle (4π) and the entire vertical extension of the model atmosphere, see (5.6). In the present situation the radiative effect is the reflectance $r(\mathbf{x})$.

In Chapter 5 we treated specific examples for the response function which yield upward, downward or net flux densities. For satellite applications we have to consider another type of response function. If $\mathbf{\Omega}_v = (\mu_v, \varphi_v)$ is the instrument's viewing direction at the position of the satellite, then the response function

$$R_v = \frac{1}{S_0} \delta(z - z_t) \delta(\mathbf{\Omega} - \mathbf{\Omega}_v) \quad (11.102)$$

extracts from (11.101) the radiance field as seen by an ideal instrument.

Next we have to consider the adjoint radiance field I^+, which is given by the solution of the adjoint radiative transfer equation

$$L^+ I^+(z, \Omega) = Q^+(z, \Omega) \qquad (11.103)$$

where Q^+ is the source of adjoint photons, and the adjoint operator L^+ is defined by

$$L^+ = -\mu \frac{\partial}{\partial z} + k_{\mathrm{ext}}(z) - \frac{k_{\mathrm{sca}}(z)}{4\pi}$$

$$\times \int_{4\pi} \left[\mathcal{P}(z, \Omega' \to \Omega) - \frac{A_g}{\pi} \delta(z) U(-\mu) |\mu| U(\mu') |\mu'| \right] \circ d\Omega' \quad (11.104)$$

Note that in (11.104) as compared to (11.98) the direction μ is inverted. Similarly to this reversion of the zenith angle the meaning of the boundary conditions for the adjoint radiance field has to be changed from the incoming forward radiation field to the outgoing radiance, cf. (5.23)

$$\text{top of the atmosphere:} \quad I^+(z_{\mathrm{t}}, \mu, \varphi) = 0, \quad 0 \le \varphi \le 2\pi, \quad 0 < \mu \le 1$$
$$\text{Earth's surface:} \quad I^+(0, \mu, \varphi) = 0, \quad 0 \le \varphi \le 2\pi, \quad -1 \le \mu < 0$$

$$(11.105)$$

We repeat from Chapter 5 the meaning of the *pseudo-radiance*

$$\Psi(z, \Omega) = I^+(z, -\Omega) \qquad (11.106)$$

by pointing out that $\Psi(z, \mu, \varphi)$ is the solution to the usual forward radiative transfer equation possessing a particular source

$$L\Psi(z, \Omega) = Q^+(z, -\Omega) \qquad (11.107)$$

Again it is noteworthy that for the pseudo-radiance field Ψ the direction of the propagating adjoint photons has to be inverted in the adjoint source Q^+ with respect to the upper and lower 2π hemisphere. Therefore, the transformation (11.106) allows us to use the same standard method of finding the solution for the adjoint radiance field as it is applied for the forward radiance field $I(z, \mu, \varphi)$.

Next we wish to illustrate how the first derivative of the reflectance with respect to the state vector $\mathbf{x}$ is related to the perturbation integral. Let us suppose that we have found the solutions I_0 and I_0^+ for an atmospheric base state (subscript 0) characterized by the state vector $\mathbf{x}_0$. Following the procedure outlined in Chapter 5 and resulting in (5.66), the reflectance for a perturbed atmosphere to first-order accuracy may be approximated by the linear expansion

$$r(\mathbf{x}) \approx r(\mathbf{x}_0) - \sum_{k=1}^{K} \langle I_0^+, \Delta L_k I_0 \rangle \qquad (11.108)$$

Here ΔL_k is the change of the differential operator L if we only perturb the k-th component of the state vector $\mathbf{x}_0$ by Δx_k. By comparing this equation with (11.96) we find for the first derivative of the reflectance r with respect to the physical parameters x_k

$$\frac{\partial r}{\partial x_k} = -\frac{1}{\Delta x_k} \langle I_0^+, \Delta L_k I_0 \rangle \qquad (11.109)$$

Our particular interest lies in perturbing either the ozone molecule concentration $\rho_{O_3,k}$ in units of molecules per m^3 of air in the k-th homogeneous atmospheric sublayer or the Lambertian albedo A_g of the ground. For this we derive the first derivatives of the reflectance with respect to the ozone profile and the surface albedo.

If $\rho_{O_3,k}$ is the mean ozone molecule concentration in the sublayer k and $\rho_{O_3,k,0}$ the corresponding unperturbed value, then the perturbation Δx_k is given by

$$\Delta x_k = \Delta \rho_{O_3,k} = \rho_{O_3,k} - \rho_{O_3,k,0} \qquad (11.110)$$

Because we only perturb $\rho_{O_3,k}$, the change in the absorption optical depth $\Delta \tau_{O_3,k}$ of sublayer k may be written as

$$\Delta \tau_{O_3,k} = \int_{z_k}^{z_{k-1}} (\rho_{O_3,k} - \rho_{O_3,k,0})\sigma_{O_3,k}\, dz = -\sigma_{O_3,k}\Delta\rho_{O_3,k}(z_k - z_{k-1}) \qquad (11.111)$$

where $\sigma_{O_3,k}$ is the absorption cross-section of an ozone molecule and (z_k, z_{k-1}) refer to the lower and upper boundary of the k-th homogeneous sublayer of the model atmosphere. Furthermore, we easily see that, for this particular perturbation, ΔL_k in the perturbation integral (11.108) is

$$\Delta L_k = \begin{cases} \sigma_{O_3,k}\Delta\rho_{O_3,k}, & z_k < z < z_{k-1} \\ 0, & \text{else} \end{cases} \qquad (11.112)$$

Therefore, we directly find the first derivative of r with respect to the mean ozone molecule concentration in sublayer k

$$\frac{\partial r}{\partial \rho_{O_3,k}} = -\frac{1}{\Delta\rho_{O_3,k}} \langle I_0^+, \Delta L_k I_0 \rangle = -\sigma_{O_3,k} \int_{z_{k-1}}^{z_k} \int_{4\pi} \Psi_0(z, -\mathbf{\Omega}) I_0(z, \mathbf{\Omega})\, d\Omega dz \qquad (11.113)$$

where we have replaced I_0^+ by means of (11.106).

For the perturbation of the surface albedo, $\Delta A_g = A_g - A_{g,0}$, where $A_{g,0}$ is the Lambertian albedo for the unperturbed ground, we similarly find from (11.98)

$$\Delta L = -\int_{4\pi} \frac{\Delta A_g}{\pi} \delta(z) U(\mu)|\mu| U(-\mu')|\mu'| \circ d\Omega' \qquad (11.114)$$

Due to the appearance of the delta function $\delta(z)$, the integration over z simplifies to the evaluation of the radiances Ψ_0 and I_0 at ground level $z = 0$. Thus the corresponding first derivative of r with respect to the surface albedo A_g is given by

$$\frac{\partial r}{\partial A_g} = -\frac{1}{\Delta A_g}\langle I_0^+, \Delta L I_0 \rangle = \frac{1}{\pi} E_-(\Psi_0) E_-(I_0) \tag{11.115}$$

Here the quantities $E_-(\Psi_0)$ and $E_-(I_0)$ represent downward flux densities at ground level of the base atmosphere's pseudo-radiance field $\Psi_0(0, \Omega)$ and the forward radiance field $I_0(0, \Omega)$, respectively, i.e.

$$E_-(\Psi_0) = \int_{4\pi} U(-\mu)|\mu|\Psi_0(0, \Omega)d\Omega$$

$$E_-(I_0) = \int_{4\pi} U(-\mu)|\mu|I_0(0, \Omega)d\Omega \tag{11.116}$$

Note that in these expressions the Heaviside step function $U(-\mu)$ just selects all radiation coming from the upper 2π hemisphere.

Substituting (11.113) and (11.115) into (11.96) yields the reflectance $r(\mathbf{x})$ of the actual atmosphere as function of the reflectance $r(\mathbf{x}_0)$ of a given base atmosphere.

Landgraf *et al.* (2001) applied the forward–adjoint perturbation theory to calculate the derivatives of the reflectance with respect to the ozone concentration and the surface albedo. They also compared the results with benchmark calculations which have been performed with the *DISORT* programme package as provided by Stamnes *et al.*, (1988), see also Section 4. Since in the DISORT calculations expressions for the partial derivatives of r with respect to $\rho_{O_3,k}$ and A_g, such as (11.113) and (11.115), are not available, the following discretizations are used in these benchmark calculations

$$\frac{\partial r}{\partial \rho_{O_3,k}} \approx \frac{r(\rho_{O_3,k} + \Delta\rho_{O_3,k}) - r(\rho_{O_3,k})}{\Delta\rho_{O_3,k}}, \qquad \frac{\partial r}{\partial A_g} \approx \frac{r(A_g + \Delta A_g) - r(A_g)}{\Delta A_g}$$

$$\tag{11.117}$$

Figure 11.12a shows the derivatives of r with respect to the ozone concentration. In the left panel of the figure the vertical profile of $\partial r/\partial\rho_{O_3}$ has been computed on the basis of the perturbation theory. These computations have been carried out for the three wavelengths of 299 (solid), 312 (dotted), and 323 nm (dashed). The figure has to be interpreted as follows: in regions where the curves attain large values, the sensitivity of the reflected radiation with respect to a unit ozone perturbation in the associated altitude range is significant. This is the case for all altitudes above 12 km. A negative value for $\partial r/\partial\rho_{O_3}$ means that r decreases if the molecule concentration of ozone is increased in the particular model layer. For similar perturbations of the ozone profile in the middle and lower troposphere this sensitivity decreases with

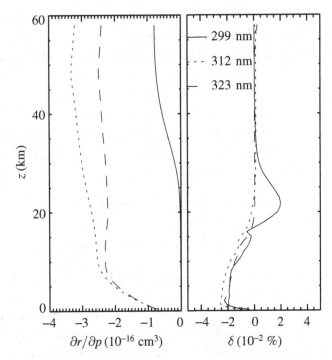

Fig. 11.12 Derivatives $\partial r/\partial \rho$ ($\rho = \rho_{O_3}$) as computed with the forward-adjoint linear perturbation theory. Results are shown for three selected wavelengths $\lambda = 299, 312, 323$ nm. The right part of the figure shows the relative difference δ between the linear perturbation theory and finite difference results obtained with DISORT. After Landgraf *et al.* (2001).

decreasing altitude. For the shorter UV wavelengths the clear-sky atmosphere is optically so thick that photons scattered in the lower part of the atmosphere do not make any significant contribution to the backscattered light observed by the satellite instrument. Hence, we conclude that it is rather difficult to retrieve the ozone profile in the troposphere with satisfactory accuracy.

The right panel of Figure 11.12 depicts the relative difference δ (in $10^{-2}\%$) between the forward–adjoint approach (11.113) and the finite difference approach (11.117) based on DISORT calculations. Again the computations have been performed for clear sky conditions involving only Rayleigh scattering and ozone absorption. The results shown apply to a solar zenith angle of $\vartheta_0 = 45°$ and a viewing zenith angle of $\vartheta_v = 10°$. It can be seen that the linear perturbation approach nearly perfectly reproduces the partial derivatives with relative errors smaller than about 0.03%.

The upper part of Figure 11.13 shows the sensitivity of the reflectance with respect to the surface albedo. The solid curve denotes results with a clear sky

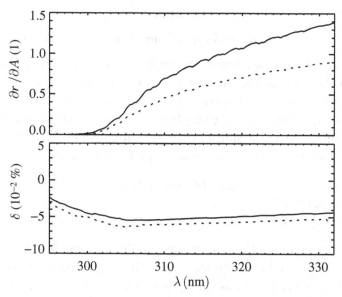

Fig. 11.13 Derivatives $\partial r/\partial A_g$ (upper part with $A = A_g$) and relative differences between the linear perturbation results and the DISORT approach employing finite differencing (lower part). The parameters for these computations were
$\vartheta_0 = 45°$, $\vartheta_v = 10°$, and $A_g = 0.1$. Solid curves: atmosphere without aerosol
particles, dotted curves: atmosphere containing aerosol particles. After Landgraf
et al. (2001).

atmosphere containing no aerosol particles while for the dotted line a typical rural
aerosol scenario has been adopted. The optical parameters of the aerosol particles
were taken from Shettle and Fenn (1979). For the wavelength of 330 nm the total
aerosol optical depth attained a value of 0.85.

As can be seen from the figure, for the smaller wavelengths $\partial r/\partial A_g$ converges
to zero. This can be explained by the fact that solar photons do not reach the
surface because they are already absorbed by the stratospheric ozone. This result
has an important consequence: Any ozone data retrieved with a particular inversion
method will turn out to be insensitive to a possibly unknown value of the Lambertian
surface albedo. Thus, regarding the surface properties it would make no difference
if the retrieval is made over the ocean or over a surface covered by snow or ice.
Figure 11.13 also shows that for increasing wavelength $\partial r/\partial A_g$ increases with λ.
For larger values of λ the wavelength dependence of the sensitivity decreases again.
The lower part of Figure 11.13, depicting the difference between the perturbation
theory and the DISORT approach, reveals that the derivatives of r with respect to
A_g can be accurately computed with the linear radiative perturbation theory. The
relative errors are always smaller than about 0.05%.

11.5 Appendix

11.5.1 Example for an ill-posed inversion problem

In Section 11.3.1 we have pointed out that due to the quasi-singular matrix A the solution to (11.61) may become unstable. Now we will illustrate this feature by virtue of a simple example. Experience with inversion methods tells us that small perturbations in the observations **g** may lead to extreme perturbations in the physical solution **f**. This is the typical situation in case of ill-posedness. Let us briefly discuss the definition of a well-posed mathematical problem, see Hansen (1994).

(1) For each **g** there exists a solution **f** for which $A\mathbf{f} = \mathbf{g}$.
(2) The solution **f** is unique.
(3) Small perturbations in **g** cause only small perturbations for the solution **f**.

If, for a particular problem, one of these conditions is not fulfilled, the problem is said to be ill-posed. Let us illustrate this with a simple example as discussed by Craig and Brown (1986). As weighting function or kernel $\mathcal{K}(x, y)$ of the integral equation we use the Heaviside step function

$$U(y) = \begin{cases} 0, & y \leq 0 \\ 1, & y > 0 \end{cases} \tag{11.118}$$

The problem to be solved is the Volterra integral equation of the first kind for $f(y)$

$$g(x) = \int_a^x f(y)dy = \int_a^b U(x - y)f(y)dy, \quad x \in [a, b] \tag{11.119}$$

The solution to the inversion problem for (11.119) is simply the first derivative of the function g

$$f(y) = \frac{dg}{dx}\bigg|_{x=y} \tag{11.120}$$

This can be easily verified by inserting (11.120) into (11.119).

Figure 11.14 depicts the solutions f_a and f_b which correspond to the two 'measurements'

$$g_a(x) = 1 - \exp(-\alpha x)$$
$$g_b(x) = g_a(x) + \beta \sin(\omega x) \tag{11.121}$$

with $\alpha = 0.8$, $\beta = 0.04$ and $\omega = 20$. Note that the term $\sin(\omega x)$ can be understood as a small perturbation of the measurement g_a having an amplitude β. From (11.120) we find for the solutions of the inversion problem

$$f_a(y) = \alpha \exp(-\alpha y)$$
$$f_b(y) = f_a(y) + \omega\beta \cos(\omega y) \tag{11.122}$$

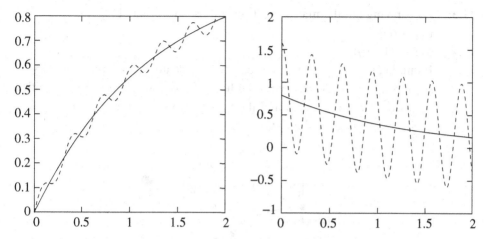

Fig. 11.14 Example for an ill-posed inversion problem. The left panel shows the measurements (solid curve: exact measurement g_a, dashed curve: measurement with perturbations g_b). The right panel depicts the corresponding inversions f_a and f_b.

It is clear that the perturbations in g_b due to the sine term are in the range of a few percent with respect to the measured unperturbed signal g_a. Nevertheless, for $y > 1$ the perturbations in the solution f_b can amount to several 100% of the true solution f_a, see Figure 11.12. Therefore, the solution of the inversion problem (11.119) becomes already unstable for small changes in the observation g.

The typical situation in inverse remote sensing problems is that, unlike to the simple case above, it is not possible to find full analytical solutions of the Fredholm integral equation by virtue of complete function systems (e.g. orthogonal polynomials). Therefore, one formulates the inverse problem in its discretized form and searches for a solution employing methods of linear algebra.

11.6 Problems

11.1: Show that in the Bouguer–Langley method the slope of the regression line is given by

$$\tau_\lambda = \frac{\ln(E_{\lambda,2}) - \ln(E_{\lambda,1})}{m_1 - m_2}$$

11.2: Suppose that the absorption coefficient decreases with height according to $k(z) = k(z = 0)\exp(-z/H)$ where H is a positive constant. Find the height where the weighting function has its maximum. Assume that the transmission function follows an exponential law.

11.3: By carrying out the matrix multiplications, show that (11.70) is equivalent to (11.67).

11.4: Verify (11.80).

11.5: Formula (11.7) is purely empirical. A suitable air mass formula for dry air scattering can be derived with the help of the figure below, the definition $m = (\int_0^\infty \rho\, ds)/(\int_0^\infty \rho\, dz)$ and the basic astronomic formula $(1 + \delta)(R + z)\sin\zeta = const$, see the figure.

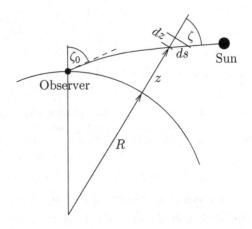

Show that

$$m = \frac{1}{\rho_0 H_0} \int_0^\infty \rho \left[1 - \left(\frac{1 + \delta_0}{1 + \delta} \right)^2 \left(\frac{R}{R + z} \right)^2 \sin^2 \vartheta_0 \right]^{-1/2} dz$$

with $H_0 = p_0/(g\rho_0)$ and $\mathcal{N} = 1 + \delta$ is the index of refraction of the air. The suffix 0 refers to the ground. Usually one assumes that $\delta/\delta_0 = \rho/\rho_0$.

12

Influence of clouds on the climate of the Earth

In the final chapter of this book we will briefly treat the radiative influence of clouds on the climate of the Earth. A brief introduction to this topic has already been presented in the first chapter where we have discussed the radiation budget of the Earth. In the chapters that followed we have studied the radiative transfer theory in some detail and have learned how to calculate the radiances and flux densities in the solar and long-wave spectrum.

We have omitted any discussion of measurement programs such as the satellite experiments ERB and ERBE which were specifically devised to study the global radiation budget. A brief description of some of the sophisticated instrumentation used to measure the reflected solar energy and the outgoing long-wave radiation is given by Lenoble (1993) where many references to this topic can be found.

Globally the planet Earth is in radiative equilibrium implying that the reflected solar radiation and the outgoing long-wave radiation are in balance with the incoming solar radiation. If this balance is disturbed by natural or by anthropogenic processes the global climate will be changed. A detailed study of the radiative impact of clouds on the climate is very difficult and can be carried out only with the help of sophisticated climate models. The reason for this is that many different variables that are responsible for the evolution of the climate interact in a highly nonlinear way. Here we will give a very simple discussion only following the arguments presented by Arking (1991).

12.1 Cloud forcing

The outgoing long-wave radiation $E_{1,t}$ at the top of the atmosphere can be calculated with the help of one of the radiative transfer methods which we have previously discussed. Accepting the insufficiency of any empirical formula, we use the Budyko expression

$$E_{1,t} = 223.0 + 2.2T_0 - (48.0 + 1.6T_0)C \qquad (12.1)$$

to calculate $E_{1,t}$ where T_0 is the air temperature near the ground and C the cloud fraction ranging from 0 to 1. In order to obtain $E_{1,t}$ in units of (W m^{-2}), the temperature T_0 must be expressed in degrees Celsius. The variation of $E_{1,t}$ with cloud fraction C results in the so-called *sensitivity coefficient for long-wave radiation* δ_1 as defined by

$$\delta_1 = \frac{\partial E_{1,t}}{\partial C} = -(48.0 + 1.6T_0) \qquad (12.2)$$

For a constant mean global surface temperature $T_0 = 15°\text{C}$, δ_1 assumes the value of $-72\,\text{W m}^{-2}$.

Analogously, the *sensitivity coefficient for solar radiation* δ_s can be defined by considering the solar flux density $E_{s,t}$ at the top of the atmosphere. If A_C stands for the average albedo of the overcast sky, A_0 for the albedo of the cloud-free atmosphere then $E_{s,t}$ is expressed by

$$E_{s,t} = \frac{S_0}{4}(1 - A) = \frac{S_0}{4}[1 - A_C C - A_0(1 - C)] \qquad (12.3)$$

where $A = A_C C + A_0(1 - C)$ is a weighted average *global albedo* with the cloud fraction as weighting factor. In analogy to (12.2) the sensitivity coefficient for solar radiation δ_s is defined by

$$\delta_s = \frac{\partial E_{s,t}}{\partial C} = -\frac{S_0}{4}(A_C - A_0) \qquad (12.4)$$

Assuming $S_0 = 1368\,\text{W m}^{-2}$, $A_C = 0.5$ and $A_0 = 0.12$ which appear to be reasonable values, the sensitivity coefficient is $\delta_s = -130\,\text{W m}^{-2}$.

Let us now briefly discuss the effect of clouds on the radiation budget at the top of the atmosphere. As follows from (12.3) and simple reasoning, an increase in cloud fraction C increases the reflectance at the top of the atmosphere level thus reducing the energy gain for the system Earth–atmosphere. However, an increase in cloud fraction also reduces the outgoing long-wave radiation at the top of the atmosphere because clouds are colder than the Earth's surface. This fact is also stated by the minus sign of the last term of (12.1).

We will now introduce the *net sensitivity coefficient* which is the difference between the solar and long-wave radiation sensitivity coefficients as expressed by

$$\delta = \delta_s - \delta_1 = \frac{\partial E_{s,t}}{\partial C} - \frac{\partial E_{1,t}}{\partial C} = -\frac{S_0}{4}(A_C - A_0)\left(1 - \frac{\frac{\partial E_{1,t}}{\partial C}}{\frac{\partial E_{s,t}}{\partial C}}\right) \qquad (12.5)$$

Thus, δ amounts to $-58\,\text{W m}^{-2}$. Certainly, this number is no more than a crude estimate, but it gives us an impression of the order of magnitudes which are involved.

We now wish to interpret the net sensitivity coefficient. Holding all other parameters constant, δ is the change in net energy absorbed by the climate system per change in cloud cover fraction. From the above crude estimates of the sensitivity coefficients we conclude that of the two cloud effects, the effect on solar radiation dominates the effect on long-wave radiation emitted to space when looking on the global scale. Moreover, the sensitivity can be determined for any spatial and temporal scale by using instantaneous satellite observations. However, it will have significance only on scales large enough to suppress the 'noise' of day-to-day weather variations.

Inspection shows that (12.5) does not require any explicit cloud fraction information. Since $A_C > A_0$, the sign of δ is determined by the ratio of the sensitivity coefficients appearing in the last factor. If this ratio is greater than 1, net cloud warming is expected. If it is smaller than 1 net cloud cooling should result. Various authors have investigated the effect of clouds on the climate system with the help of satellite data. They concluded that clouds, on the whole, have a very substantial cooling effect on the climate system. For further details we refer to Arking (1991) where an extensive bibliography on the subject can be found.

Let us now consider the difference of flux densities as stated by equations

$$C_{f,l} = E_{l,t}(C = 0) - E_{l,t}(C), \quad C_{f,s} = E_{s,t}(C) - E_{s,t}(C = 0) \qquad (12.6)$$

The quantities $E_{l,t}(C = 0)$ and $E_{s,t}(C = 0)$ refer to cloudless conditions. These differences, now commonly called *cloud forcing*, represent the effect of clouds on the radiative flux densities. The total cloud forcing effect C_f is the sum of the cloud forcing components for solar and long wave radiation, that is

$$C_f = C_{f,l} + C_{f,s} \qquad (12.7)$$

It is a simple exercise to show that cloud forcing and cloud sensitivity coefficients are related by

$$\frac{\partial E_{l,t}}{\partial C} = -\frac{C_{f,l}}{C}, \quad \frac{\partial E_{s,t}}{\partial C} = \frac{C_{f,s}}{C}, \quad \delta = \frac{C_f}{C} \qquad (12.8)$$

Various investigators have attempted to determine the cloud forcing C_f. Ellis (1978), for example, estimated that the mean annual, quasi-global (65° S–65° N) effect of clouds is a cooling of the climate system resulting in $C_f = -20\,\mathrm{W\,m^{-2}}$. The long-wave effect is $C_{f,l} = 22\,\mathrm{W\,m^{-2}}$ and the short-wave $C_{f,s} = -42\,\mathrm{W\,m^{-2}}$. From the ERBE satellite experiments Ramanathan *et al.* (1989) found $C_f = -17\,\mathrm{W\,m^{-2}}$ which is not too different from the value given by Ellis. However, the individual parts of cloud forcing differ substantially with $C_{f,l} = 31\,\mathrm{W\,m^{-2}}$ and $C_{f,s} = -48\,\mathrm{W\,m^{-2}}$.

From climatological data Ellis extracted and computed the individual sensitivity factors and the ratio of sensitivity factors that are independent of cloud amount.

The ratio of the sensitivity factors that can be obtained from the above empirical formulas and the data provided by Ellis and Ramanathan are not too different. However, the ratio proposed by Ohring *et al.* (1981) differs substantially from the values presented by the other authors which may be partly due to the different methods used. It is difficult to decide whose value of the sensitivity coefficient best approximates reality.

12.2 Cloud feedback in climate models

As we know already, the global climate equilibrium is reached if $E = 0$ whereby

$$E = E_{1,t} - \frac{S_0}{4}(1 - A) \qquad (12.9)$$

If $E < 0$ then the system Earth–atmosphere will gain energy and heating will take place. Whenever E differs from zero, primary forcing of the system occurs. This may be due to changes of the solar constant, the global albedo, the upward long-wave radiation or by a combination of these. According to the *Milankovitch theory* a change of S_0 may have occurred over a period of 10^4–10^5 years due to a change of the Earth's orbit. The variation of the solar constant may be responsible for changes of the past climate. Here we will not dwell on this subject. Present measurements show that the solar constant may vary by 0.1% due to changes between a minimum and a maximum of the Sun's activity. Variations of this order of magnitude have been included in climate models to force climate changes. Global albedo changes may be due to natural and anthropogenic causes such as the outburst of volcanoes, biomass burning, agricultural and industrial activities. The observed increase of CO_2 and other trace gases as well as changes in the aerosol content of the air definitely have a non-negligible influence on the radiation budget of the atmosphere.

It stands to reason that changes induced by primary forcing modifies the radiative characteristics of the atmosphere and the Earth's surface thus causing an additional change of the net flux density at the top of the atmosphere. This is known as the *feedback process*. If the feedback increases the primary forcing of the net flux density at the top of the atmosphere, it is called a positive feedback. In contrast, the feedback is called negative if it tends to reduce the primary net flux density change. A detailed study of the feedback process is very complicated partly due to the fact that many open questions in connection with climate modeling still have to be resolved.

However, it is possible to give a very simple linear analysis of the feedback mechanisms which will help us to visualize the overall problem. In the subsequent discussion we will follow the arguments given by Arking (1991) and Lenoble (1993). They assume that without feedback the only response to a variation of the net flux density at the top of the atmosphere level would be a height-independent temperature change which includes a change of the air temperature T_0 near the surface of the Earth. The most simple formulation of this process is given by

$$\Delta T_0 = -k_0 E \qquad (12.10)$$

where k_0 is a positive climate sensitivity parameter. However, if feedback is assumed to take place, this temperature change will be altered. Now we write

$$\Delta T_0' = -k E \qquad (12.11)$$

where k is another positive climate sensitivity parameter. By assuming that all acting feedback processes are linear in T_0 and independent of each other, we may express the modified net flux density at the top of the atmosphere level after feedback has occurred by

$$E' = E - \Delta T_0' \sum_i \alpha_i \qquad (12.12)$$

The sensitivity coefficients α_i have the sign of the feedback. The response to the perturbation E with feedback should be the same as the response to E' without feedback. Thus from

$$\Delta T_0' = -k E = -k_0 E' \qquad (12.13)$$

we find

$$k = \frac{k_0}{1 - k_0 \sum_i \alpha_i} = \frac{k_0}{1 - \sum_i \beta_i} \quad \text{with} \quad \beta_i = k_0 \alpha_i \qquad (12.14)$$

For a particular atmospheric variable y_i the sensitivity coefficient α_i may be expressed by an equation of the form

$$\alpha_i = -\frac{\partial E}{\partial y_i} \frac{dy_i}{dT_0} \qquad (12.15)$$

It should be observed that only the determination of $\partial E / \partial y_i$ is a problem of radiative transfer while the derivative dy_i/dT_0 must be obtained with the help of a climate model. If the variable y_i refers to cloud cover fraction C we have the particular case expressed by

$$\alpha_C = -\frac{\partial E}{\partial C} \frac{dC}{dT_0} \qquad (12.16)$$

12.2.1 Cloud feedback in response to doubling atmospheric CO_2

To give an impression of the complexity of the feedback problem we will briefly discuss the cloud feedback in response to doubling the atmospheric CO_2 and also list several other mechanisms. There are a great number of general circulation models

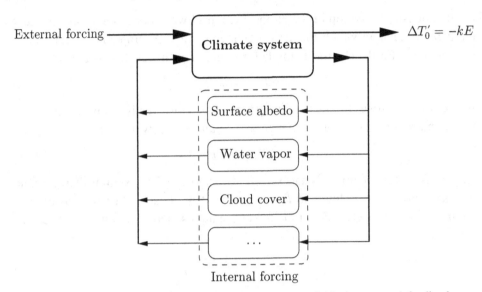

Fig. 12.1 Schematic diagram of the climate system depicting several feedback processes.

(GCM) which can be used to investigate the effect of doubling the atmospheric CO_2 concentrations on the climate. Arking (1991) briefly discusses the results provided by the GISS (Hansen *et al.*, 1984) and the GFDL (Wetherald and Manabe, 1988) GCM models.

Both models have interacting clouds to the extent that the spatial distribution of cloudiness (height and amount) may change as a reaction to climatic change. The optical properties of the clouds remain fixed. In principle, it is possible to include changing optical cloud parameters in response to the climatic change as part of the feedback process. The interpretation of results, however, becomes increasingly difficult due to phase changes and changing configurations of ice particles among other things. Both models yield about the same change of the mean global surface temperature of 4 K in response to doubling the amount of CO_2. The overall conclusions derived from various other models are similar all showing an increase in the surface temperature. However, due to different treatments of the feedbacks, the models produce surface temperature changes ranging from about 1.5–5 K. Below we will briefly dwell on the uncertainties of the feedback effects.

Figure 12.1 shows schematically four internal feedback blocks and their contribution to the climate sensitivity factor k. Let us also inspect Table 12.1 listing the calculated feedback factors due to the modeling of several processes. While the calculated individual feedback factors differ in both models, the sum of these is nearly identical amounting to about 0.7. By substituting 0.7 for the sum over the β_i

Table 12.1 *Feedback factors for several processes calculated with the two GCM models GFDL and GISS*

Feedback process	GFDL	GISS
Surface albedo	0.16	0.09
Water vapor	0.43	0.40
Cloud cover	0.11	0.22
Total feedback factor	0.70	0.71

into (12.14), we find that the reponse is about 3.3 times what it would be without feedback. For further details we refer to the original paper.

12.2.2 Other trace gases

Besides CO_2, the concentrations of several other trace gases are increasing. Because of their strong infrared absorption bands, methane, nitrous oxide and several chlorofluorocarbons are radiatively very important. Although ozone is increasing in the troposphere it is decreasing in the stratosphere.

An increase of other gases such as CO, NO and NO_2, although radiatively not important, cannot be ignored since they may alter the atmospheric chemistry and perturb the radiatively acting gases such as CH_4 and O_3. There seems to be a general agreement that the global effect of all trace gases is of the same order as the effect of the CO_2 increase (Ramanathan *et al.* 1985; Ramanathan, 1987; Wang *et al.* 1986). As the feedback models become more complete we should expect further changes in the results. Deficiencies such as the omission of nonlinear effects and of the complicated chemical processes will eventually be removed.

12.2.3 Liquid water and cloud microphysics feedback

We have previously stated that the optical parameters in the above GCM models were fixed. It has been argued quite early by Paltridge (1980) and later by others that optical parameters are expected to change in response to CO_2-induced atmospheric warming. An increased temperature would increase the liquid water content of clouds thus increasing the cloud optical thickness. Paltridge estimates that the short-wave effect would dominate over the long-wave effect, resulting in a negative feedback for this process alone and reducing the overall sensitivity of the climate to CO_2 changes by about 40%. Subsequent investigations with one-dimensional radiative–convective models also suggest that cloud liquid water would contribute negatively to cloud feedback. It stands to reason, if all else remains the same, the optical thickness of the cloud would be proportional to the vertically integrated liquid water. Additionally, we may argue with Bohren (1985) that changes in the microphysical properties of clouds would be produced due to changes in atmospheric temperature and humidity. If the size of the cloud particles increases along with liquid water content, the magnitude of the negative feedback would decrease. Detailed modeling is required to resolve the remaining questions.

12.2.4 Climatic impact due to aerosols

As is well-known, atmospheric aerosols have a strong impact on the greenhouse effect of the Earth–atmosphere system due to aerosol extinction in the short-wave and long-wave parts of the spectrum. As in case of clouds, the resulting effects depend strongly on the particle concentration, on the optical properties of the particles, and on the vertical as well as on the horizontal aerosol distributions. In general, the effect of increasing concentrations of aerosol particles results in an increase of the planetary albedo. The resulting cooling may counteract the greenhouse warming of CO_2 and other trace gases as pointed out by Hansen and Lacis (1990) and by others.

The single scattering albedo ω_0 of the aerosol particles plays an important role in the radiation budget of the atmosphere since the aerosol effect can switch from cooling to heating when ω_0 crosses a certain critical value. Due to the complex composition of aerosol particles, the complex index of refraction is difficult to measure accurately which is a prerequisite for accurate Mie calculations of the extinction parameters and the phase function.

The relatively small mass of stratospheric aerosols plays an important role in the radiation budget because of their global distribution and their long lifetime. Occasionally, volcanic eruptions increase the stratospheric aerosol concentration over the whole globe for a period of years so that they are an important cause for primary global climate forcing. The stratospheric aerosols, mostly sulfuric acid particles, have a very small imaginary part of the complex index of refraction which results in small values of the absorption coefficient. For an average global albedo of about 30%, they contribute to global cooling. In the stratosphere itself, however, the main effect of the aerosols is an increase of the stratospheric temperature due to the slight absorption of the particles themselves and due to the increase of the photon pathlength in the absorbing gases. This warming was observed after the El Chichon eruption as reported by Labitzke *et al.* (1983). Therefore, the cooling effect of the stratospheric aerosols is limited to the troposphere and to the surface.

Tropospheric aerosols have a much larger optical depth than stratospheric aerosols. However, tropospheric aerosols have a relatively short residence time and vary strongly in type, spatially and temporally. This makes it very difficult to determine their mean radiative characteristics. Of course, they interact with the solar and the infrared radiation field. The influence of tropospheric aerosols on the long-wave radiation budget is small because they are mostly located in the lower tropospheric layers having a temperature not too far removed from the ground temperature. Numerous papers on the radiative effects exist in the literature.

Small aerosols act as cloud condensation nuclei (CCN) to form cloud droplets. An increase of the number density of aerosols increases the number of droplets, and, for a given liquid water content, decreases the droplet size. This effect which is also known as the *Twomey effect* (Twomey, 1977) tends to increase the cloud optical pathlength and cloud reflection. On the other hand, if the CCN contain absorbing constituents they tend to reduce the reflectance. Thus the role of small aerosol particles is quite complex.

12.3 Problems

12.1: To familiarize yourself with the terminology of this chapter, verify equations (12.8).

Answers to problems

Chapter 1

1.1:

$$B_\lambda(T)d\lambda = \frac{2hc^2}{\lambda^5}\left[\exp\left(\frac{hc}{\lambda kT}\right) - 1\right]^{-1} d\lambda$$

(a) $\lambda_{\max}T = const = 2898$ μm K

(b) $\lambda_{\max}(T = 6000) = 0.48$ μm, $\lambda_{\max}(T = 300) = 9.66$ μm

1.2:

(a) $B_\nu d\nu = \dfrac{2\nu^2 kT}{c^2}d\nu$

(b) $B_\nu d\nu = \dfrac{2h\nu^3}{c^2}\exp\left(-\dfrac{h\nu}{kT}\right)d\nu$

1.4:

Lambert's law of photometry: $d\phi = \dfrac{I\,dA_1\,dA_2\cos\alpha_1\cos\alpha_2}{r^2}$

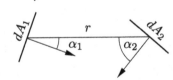

I: radiance from dA_1
$d\phi$: radiant flux received at dA_2

$$E = \frac{\sigma T_c^4}{1 + (z/R)^2}$$

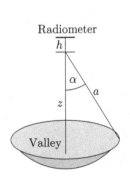

Radiometer

1.5:

(a) $E = \sigma T^4 a^2\dfrac{1 - \cos^2\alpha}{a^2 + h^2 + 2ah\cos\alpha}$

(b) $E = \sigma T^4 a^2\dfrac{1 - \cos^2\alpha}{a^2 + h^2 - 2ah\cos\alpha}$

h is the vertical distance between the radiometer and the center of curvature.

1.6:

(a) $E_r = \dfrac{\pi a^2}{r^2} \displaystyle\int_0^\infty I_\nu d\nu$

(b) $\phi = 4\pi^2 a^2 \displaystyle\int_0^\infty I_\nu d\nu$

(c) $\hat{u}(\mathbf{r}) = \dfrac{2\pi}{c}\left(1 - \sqrt{1 - \dfrac{a^2}{r^2}}\right)\displaystyle\int_0^\infty I_\nu d\nu$

1.7:

$\hat{u}_\nu = \dfrac{S_{0,\nu}}{c}, \qquad \mathbf{E}_{net,\nu} = S_{0,\nu}(\sin\vartheta_0 \cos\varphi_0 \mathbf{i} + \sin\vartheta_0 \sin\varphi_0 \mathbf{j} + \cos\vartheta_0 \mathbf{k}).$

The index 0 refers to the solar radiation and $(\mathbf{i}, \mathbf{j}, \mathbf{k})$ are the unit vectors in the directions of the axes of the Cartesian (x, y, z)-coordinate system.

Chapter 2

2.1:

$C = \dfrac{3}{4}$

2.2:

$$I_+^m = \frac{\omega_0}{4\pi}\mu_0 S_0 \frac{F_1^m(\mu,\mu_0)}{\mu_0+\mu}\left[\exp\left(-\frac{\tau}{\mu_0}\right) - \exp\left(\frac{\tau}{\mu}\right)\exp\left(-\frac{\mu+\mu_0}{\mu\mu_0}\tau_c\right)\right]$$

$$I_-^m = \frac{\omega_0}{4\pi}\mu_0 S_0 \frac{F_2^m(\mu,\mu_0)}{\mu_0-\mu}\left[\exp\left(-\frac{\tau}{\mu_0}\right) - \exp\left(-\frac{\tau}{\mu}\right)\right] \quad \text{with}$$

$$F_1^m(\mu,\mu_0) = \sum_{l=m}^\infty p_l^m(-1)^{l+m} P_l^m(\mu)P_l^m(\mu_0), \quad F_2^m(\mu,\mu_0)=\sum_{l=m}^\infty p_l^m P_l^m(\mu)P_l^m(\mu_0)$$

$$0 \le \mu \le 1$$

2.3:

(a) $I_-^m(\tau,\mu_0) = \dfrac{\omega_0}{4\pi} S_0 F_2^m(\mu_0,\mu_0)\dfrac{\tau}{\mu_0}\exp\left(-\dfrac{\tau}{\mu_0}\right)$

(b) $I_+^m(\tau,0) = \dfrac{\omega_0}{4\pi} S_0 F_1^m(0,\mu_0)\exp\left(-\dfrac{\tau}{\mu_0}\right), \quad I_-^m(\tau,0) = \dfrac{\omega_0}{4\pi} S_0 F_2^m(0,\mu_0)\exp\left(-\dfrac{\tau}{\mu_0}\right)$

(c) $I_+^m(0,0) = \dfrac{\omega_0}{4\pi} S_0 F_1^m(0,\mu_0), \qquad I_-^m(0,0) = 0$

The functions $F_{1,2}^m$ are given in the answer to Problem 2.2.

2.4:

For each integral we find $A(\tau)$.

2.5:

(a) $I_+(\tau, \mu) = B_g$

(b) $\tau = \infty$

(c) with $u = \infty$ you obtain the same result.

2.6:

(a) $E_1(0) = \infty,$ $E_n(0) = \dfrac{1}{n-1},$ $n = 2, 3, \ldots$

(b) $dE_n/dx = -E_{n-1}(x)$

2.8:

$$J(\tau, \mu) = \frac{3(3 - \mu^2)}{8} \frac{1}{2} \int_{-1}^{1} I(\tau, \mu) d\mu + \frac{3(3\mu^2 - 1)}{8} \frac{1}{2} \int_{-1}^{1} I(\tau, \mu) \mu^2 d\mu$$

2.9:

$$I_+(\tau, \mu) = \frac{1}{\mu} \int_{\tau}^{\infty} J(\tau') \exp\left(-\frac{\tau' - \tau}{\mu}\right) d\tau'$$

(a) $I_-(\tau, \mu) = \dfrac{1}{\mu} \displaystyle\int_{0}^{\tau} J(\tau') \exp\left(-\dfrac{\tau - \tau'}{\mu}\right) d\tau'$

(b) $I_n(\tau) = \displaystyle\int_{\tau}^{\infty} J(\tau') E_{n+1}(\tau' - \tau) d\tau' + (-1)^n \int_{0}^{\tau} J(\tau') E_{n+1}(\tau - \tau') d\tau'$

(c) $E_{net} = 2\pi \displaystyle\int_{\tau}^{\infty} J(\tau') E_2(\tau' - \tau) d\tau' - 2\pi \int_{0}^{\tau} J(\tau') E_2(\tau - \tau') d\tau'$

Chapter 3

3.7:

(a) $J(0) = \sqrt{3}/(4\pi) S_0,$ $\Psi = 4/\sqrt{3}$

Chapter 4

4.5:

$I = 0.65860$. The quasi-exact value is $I = 0.65882$.

4.7:

$$\sum_{j=-s}^{s} \left[\left(1 + \frac{\mu_j}{\mu_0}\right) \delta_{ij} - \frac{\omega_0}{2} w_j P(\mu_i, \mu_j)\right] (1 - \delta_{0,j}) Z(\mu_j) = \frac{\omega_0}{4\pi} P(\mu_i, -\mu_0) S_0$$

4.9:

$$(l + 1)\frac{dI_{l+1}^0(\tau)}{d\tau} + l\frac{dI_{l-1}^0(\tau)}{d\tau} + [\omega_0 p_l - (2l + 1)]\, I_l^0(\tau)$$

$$= -\frac{\omega_0}{2\pi} S_0 \exp\left(-\frac{\tau}{\mu_0}\right) p_l^0 P_l^0(-\mu_0) - 2(1 - \omega_o)B(\tau)\delta_{0l}$$

Chapter 6

6.4:

$$E_+(\tau) = \frac{1}{1 + \alpha\tau_c} [E_+(\tau_c)(1 + \alpha\tau) + E_-(0)\alpha(\tau_c - \tau)]$$

$$E_-(\tau) = \frac{1}{1 + \alpha\tau_c}\{E_+(\tau_c)\alpha\tau + E_-(0)[1 + \alpha(\tau_c - \tau)]\}$$

with $\alpha_1 = \alpha_2 = \alpha$.

6.5:

$$C_1 = A\left[E_+(\tau_c) - \gamma_1(B_0 + B_1\tau_c) - \gamma_2 B_1 - \frac{\exp(-\lambda\tau_c)}{\beta_2}(E_-(0) - \gamma_1 B_0 + \gamma_2 B_1)\right]$$

(a) $C_2 = A\left\{\frac{\exp(\lambda\tau_c)}{\beta_2}(E_-(0) - \gamma_1 B_0 + \gamma_2 B_1) - \frac{\beta_1}{\beta_2}[E_+(\tau_c) - \gamma_1(B_0 + B_1\tau_c) - \gamma_2 B_1]\right\}$

$$A = \left[\exp(\lambda\tau_c) - \frac{\beta_1}{\beta_2}\exp(-\lambda\tau_c)\right]^{-1}, \qquad \beta_1 = \frac{\alpha_2}{\alpha_1 + \lambda}, \qquad \beta_2 = \frac{\alpha_2}{\alpha_1 - \lambda}$$

(b) $\dfrac{d}{d\tau}(E_+ - E_-) = C_1(1 - \beta_1)\lambda \exp(\lambda\tau) - C_2(1 - \beta_2)\lambda \exp(-\lambda\tau)$

6.6:

$$\beta_{1j} = \frac{\alpha_{2j}}{\alpha_{1j} + \lambda}, \qquad \beta_{2j} = \frac{\alpha_{2j}}{\alpha_{1j} - \lambda}$$

(i) $\tau = 0$:

$$E_-(0) = C_{11}\beta_{11} + C_{21}\beta_{21} + \gamma_{11}B_{01} - \gamma_{21}B_{11}$$

(ii) $\tau = T_1$:

$$C_{11}\beta_{11}\exp(\lambda_1 T_1) + C_{21}\beta_{21}\exp(-\lambda_1 T_1) + \gamma_{11}(B_{01} + B_{11}T_1) - \gamma_{21}B_{11}$$
$$= C_{12}\beta_{12} + C_{22}\beta_{22} + \gamma_{12}B_{02} - \gamma_{22}B_{12}$$
$$C_{11}\exp(\lambda_1 T_1) + C_{21}\exp(-\lambda_1 T_1) + \gamma_{11}(B_{01} + B_{11}T_1) + \gamma_{21}B_{11}$$
$$= C_{12} + C_{22} + \gamma_{12}B_{02} + \gamma_{22}B_{12}$$

(iii) $\tau = T_1 + T_2$:

$$C_{12}\beta_{12}\exp(\lambda_2 T_2) + C_{22}\beta_{22}\exp(-\lambda_2 T_2) + \gamma_{12}(B_{02} + B_{12}T_2) - \gamma_{22}B_{12}$$
$$= C_{13}\beta_{13} + C_{23}\beta_{23} + \gamma_{13}B_{03} - \gamma_{23}B_{13}$$
$$C_{12}\exp(\lambda_2 T_2) + C_{22}\exp(-\lambda_2 T_2) + \gamma_{12}(B_{02} + B_{12}T_2) + \gamma_{22}B_{12}$$
$$= C_{13} + C_{23} + \gamma_{13}B_{03} + \gamma_{23}B_{13}$$

(iv) $\tau = T_1 + T_2 + T_3$:

$$E_+(T_1 + T_2 + T_3) = C_{13}\exp(\lambda_3 T_3) + C_{23}\exp(-\lambda_3 T_3) + \gamma_{13}(B_{03} + B_{13}T_3) + \gamma_{23}B_{13}$$

Evaluating from these six equations the six unkown quantities C_{ij}, $i = 1, 2$, $j = 1, 2, 3$ and substituting the result into (6.95) solves the problem.

6.7:

$$E_+(\tau) = E_+(\tau_c)2E_3(\tau_c - \tau) + \pi B_0[1 - 2E_3(\tau_c - \tau)]$$
$$E_-(\tau) = E_-(0)2E_3(\tau) + \pi B_0[1 - 2E_3(\tau)]$$

6.9:
$T = -17°C$, effective radiation temperature. The difference of $31°C$ can be explained by the fact that in reality the atmosphere interacts with the radiation field.

6.10:

(a) $T_s^4 = \dfrac{2(1 - A_{tot}) - \bar{A}}{\sigma(2 - \varepsilon)}\dfrac{S_0}{4}$, $T_a^4 = \dfrac{\varepsilon(1 - A_{tot} - \bar{A}) + \bar{A}}{\varepsilon\sigma(2 - \varepsilon)}\dfrac{S_0}{4}$

(b) $T_s = 273.9$ K, $T_a = 248.2$ K.

(c) $T_s = 302.0$ K, $T_a = 254.0$ K.

(d) $\varepsilon = 0.87$, $T_a = 249.8$ K.

Chapter 7

7.2:

(a) 5.7 hPa

(b) 20.3 hPa

7.3:
$T = 25.04\%$

7.4:
$A = 2\pi y L(\bar{u}) + [1 - \exp(-8\bar{u}y^2)] - \sqrt{8\pi\bar{u}y^2}\,\mathrm{erf}(2y\sqrt{2\bar{u}})$ with

$y = \dfrac{\alpha_L}{\Delta\nu}$, $\bar{u} = \dfrac{Su}{2\pi\alpha_L}$ and $\mathrm{erf}(x) = \dfrac{2}{\sqrt{\pi}}\displaystyle\int_0^x \exp(-t^2)dt$ is the error function.

7.6:

$T = 34.17\%$

7.8:

$S_v(p_0) = S_{v,0} \exp\left(-\dfrac{2c_{gas}}{3\mu_0 g} p_0 k_{v,0}\right)$ where c_{gas} is the concentration of the absorbing gas.

$E_{0,\Delta t} = 3600(1.0 - 0.31)2\pi\mu_0 \displaystyle\int_0^\infty S_v(p_0)dv$ J m^{-2}.

7.11:

$T = 81.81\%$

7.12:

$f(k) = \dfrac{S}{\pi k\delta}\left(\dfrac{Sk}{\pi\alpha_L} - k^2\right)^{-1/2}$

7.13:

At 0 km: $\dfrac{\partial T_{rad}}{\partial t} = -2.29$ K day^{-1}, at 1 km: $\dfrac{\partial T_{rad}}{\partial t} = -1.62$ K day^{-1},

at 3 km: $\dfrac{\partial T_{rad}}{\partial t} = -0.57$ K day^{-1}.

7.14:

Lines do not overlap.

$$k_{v,i} = \begin{cases} k_{max}\left(1 - \dfrac{|v-v_i|}{\alpha}\right), & v_i - \alpha \le v \le v_i + \alpha \\ 0 & \text{elsewhere} \end{cases}, \quad i = 1, 2$$

$$W_1 = W_2 = 2\alpha\left\{1 - \dfrac{1}{k_{max}u}[1 - \exp(-k_{max}u)]\right\}, \quad W = W_1 + W_2 = 2W_1$$

$$\lim_{u\to\infty} W_1 = 2\alpha, \quad \lim_{u\to 0} W_1 = 0, \quad \Longrightarrow \quad \lim_{u\to\infty} W = 4\alpha, \quad \lim_{u\to 0} W = 0$$

7.15:

$W = \dfrac{Sc}{g}(p_1 - p_2)$

Chapter 10

10.1:

Left circularly polarized

10.2:

$$\mathbf{E} = \frac{E_0}{\sqrt{2}}(-\mathbf{i}+\mathbf{j})\cos\left[\frac{k}{\sqrt{2}}(x+y)-\omega t\right]$$

10.4:

(a) Left-handed elliptically polarized

(b)

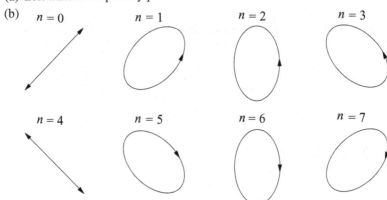

$n=0$ $n=1$ $n=2$ $n=3$

$n=4$ $n=5$ $n=6$ $n=7$

10.10:

$$F_{11} = |A_2|^2, \qquad F_{22} = |A_1|^2, \qquad F_{33} = |A_1||A_2|\cos\delta$$
$$F_{34} = -|A_1||A_2|\sin\delta, \qquad F_{43} = |A_1||A_2|\sin\delta, \qquad F_{44} = |A_1||A_2|\cos\delta$$

The remaining elements are zero.

Chapter 11

11.3:

$z_m = 0$

List of frequently used symbols

In the following we will present a list of the most commonly used symbols of this textbook. The units of all quantities are given in the so-called *Système International d'Unité* (SI) or *MKS system* measuring lengths in **m**eters, mass in **k**ilograms and time in **s**econds. The SI consists of the following seven *fundamental units*:

Physical quantity	Unit name	Unit abbreviation
Length	meter	m
Mass	kilogram	kg
Time	second	s
Current	Ampere	A
Temperature[1]	Kelvin	K
Luminous intensity	candela	cd
Amount of substance	mole	mol

[1] Often temperature is also measured in °C.

In addition to the seven basic units the following *derived units* are used:

Physical quantity	Unit name	Unit abbreviation
Solid angle	steradian	sr
Frequency	Hertz	Hz
Force	Newton	N
Pressure	Pascal	Pa
Energy	Joule	J
Power	Watt	W
Voltage	Volt	V
Magnetic flux density	Tesla	T
Charge	Coulomb	C

459

A :	area		(m^2)
$A(\bar{u})$:	absorption function		
A_o :	average albedo of the cloud-free atmosphere		
A_c :	average albedo of the overcast sky		
A_g :	albedo of the ground		
A_{mn} :	Mie coefficient		
A_n :	Mie coefficient		
$A_{u \to l}$:	Einstein coefficient for spontaneous emission		(s^{-1})
A_ν :	monochromatic absorption		
$\mathbf{A}$:	vector potential		$(V\,m^{-1}\,s)$
$\mathbb{A}$:	coefficient matrix of two-stream method		
$\mathbb{A}$:	transformation matrix for the Stokes vectors		
$\mathbb{A}^m$:	matrix used in spherical harmonics method		
a_1, a_2 :	amplitudes of the electric vector		$(V\,m^{-1})$
a_i :	coefficient to fit the transmission function		
a_m :	expansion coefficient of the wave function		
a_n^s :	expansion coefficient in the Mie theory		
$B_{\nu,g}$:	black body radiation of the ground		$(W\,m^{-2}\,sr^{-1}\,Hz^{-1})$
B_{mn} :	Mie coefficient		
B_n :	Mie coefficient		
$B_{u \to l}$:	Einstein coefficient for induced emission		$(m^3\,J^{-1}\,s^{-2})$
$B_{l \to u}$:	Einstein coefficient for absorption		$(m^3\,J^{-1}\,s^{-2})$
B_λ :	black body radiance		$(W\,m^{-3}\,sr^{-1})$
B_ν :	black body radiance		$(W\,m^{-2}\,sr^{-1}\,Hz^{-1})$
$\mathbf{B}$:	vector for polarized black body radiation		$(W\,m^{-1}\,sr^{-1})$
$\mathbf{B}$:	magnetic field vector		(T)
$\mathbb{B}_\pm$:	matrices for up- and downward radiation		
$\mathbb{B}^m$:	matrix used in spherical harmonics method		
b :	backscattered fraction of radiation		
b_i :	coefficient to fit the transmission function		
b_{ij} :	symbols used in Gaussian quadrature		
b_n^s :	expansion coefficient in the Mie theory		
C_{abs} :	absorption cross-section		(m^2)
C_{ext} :	extinction cross-section		(m^2)
$C_{f,1}$:	cloud forcing for long-wave radiation		$(W\,m^{-2})$
$C_{f,s}$:	cloud forcing for short-wave radiation		$(W\,m^{-2})$
C_i :	cloud fraction of layer i		
c_p :	specific heat at constant pressure		$(m^2\,s^{-2}\,K^{-1})$
C_{sca} :	scattering cross-section		(m^2)
c :	speed of light in vacuum		$(m\,s^{-1})$
D :	dissociation energy		(J)
D_p :	degree of polarization		
$\mathbf{D}$:	displacement vector		$(A\,s\,m^{-2})$
$\mathbb{D}^m$:	matrix used in spherical harmonics method		
$\mathcal{E}$:	radiative effect		
$\mathcal{E}_0$:	radiative effect of the base atmosphere		
$\mathcal{E}_S$:	radiative effect according to the Schwinger functional		
$E_{-,\nu}$:	downward directed radiative flux density		$(W\,m^{-2}\,Hz^{-1})$
$E_{+,\nu}$:	upward directed radiative flux density		$(W\,m^{-2}\,Hz^{-1})$

$\mathbf{E}_{net,\nu}$	:	net radiative flux density vector	$(W\,m^{-2}\,Hz^{-1})$
E_n	:	exponential integral of order n	
$E_{1,r}^{i,s}$	:	components of the electric field vector	$(V\,m^{-1})$
e	:	energy of a photon	(J) or (eV)
e	:	specific internal energy of the air	$(J\,kg^{-1})$
e	:	electric charge	(C)
$\mathbf{E}$	:	electric field vector	$(V\,m^{-1})$
$\mathbf{e}_{\lambda,\varphi,r}$	:	unit vectors in λ,φ,r- direction	
$\mathbb{E}$	:	unit matrix	
f	:	fraction of the forward scattered radiation	
$f(k)$	:	probability density function for absorption coefficient	(m)
$f(t)$	:	time signal of a wave	
$f(\nu-\nu_0)$	:	profile of a spectral line	
f_ν	:	distribution function of photons	$(m^{-3}\,sr^{-1}\,Hz^{-1})$
$G(\omega)$	:	power spectrum	
g	:	asymmetry parameter of the phase function	
$g(k)$	:	cumulative probability function for absorption coefficient k	
$g(\omega)$	:	Fourier transform of $f(t)$	
$g_{l,u}$	:	statistical weights of lower and upper states	
$\mathbb{G}$	:	matrix used in spherical harmonics method	
H	:	Hamilton function	
$H(x)$	:	Hopf function	
$H_n(x)$	:	Hermite polynomial of order n	
$H_n^{1,2}$	:	Hankel functions of the first and second kind and of order n	
$\mathcal{H}$	:	quantum mechanical Hamilton operator	(J)
$\mathcal{H}'$	:	perturbation Hamilton operator	(J)
I_ν	:	radiance in a plane–parallel atmosphere	$(W\,m^{-2}\,sr^{-1}\,Hz^{-1})$
I_ν^m	:	m-th term in the development of I_ν	$(W\,m^{-2}\,sr^{-1}\,Hz^{-1})$
I_ν^+	:	adjoint radiance	
$I_{-,\nu}$	:	downward directed radiance	$(W\,m^{-2}\,sr^{-1}\,Hz^{-1})$
$I_{+,\nu}$	:	upward directed radiance	$(W\,m^{-2}\,sr^{-1}\,Hz^{-1})$
I_n	:	modified Bessel function of first kind and of order n	
I_ν	:	radiance	$(W\,m^{-2}\,sr^{-1}\,Hz^{-1})$
$\mathbf{I}_{\pm,\nu}^m$	:	up- and downward directed radiance vector	$(W\,m^{-2}\,sr^{-1}\,Hz^{-1})$
$i_{1,2}$	:	Mie intensity functions	(sr^{-1})
J	:	rotational quantum number	
J_ν	:	source function for plane–parallel atmosphere	$(W\,m^{-3}\,sr^{-1}\,Hz^{-1})$
J_ν^e	:	source function for true emission	$(W\,m^{-3}\,sr^{-1}\,Hz^{-1})$
J_ν^m	:	m-th term in the development of J_ν	$(W\,m^{-3}\,sr^{-1}\,Hz^{-1})$
J_n	:	Bessel function of first kind and of order n	
j_ν	:	emission coefficient	$(m^{-3}\,s^{-1}\,sr^{-1}\,Hz^{-1})$
$\mathbf{J}$	:	angular momentum	$(kg\,m^2\,s^{-1})$
$\mathbf{J}$	:	current density	$(A\,m^{-2})$
$\mathbf{J}$	:	source function vector	$(W\,m^{-2}\,sr^{-1})$
$\mathbf{J}^q$	:	heat flux	$(J\,m^{-2})$

K :	quantum number	
K :	kinetic energy	(J)
k :	Boltzmann's constant	(J K^{-1})
k :	Hooke's constant	(kg s^{-2})
k :	wave number	(m^{-1})
k_0 :	wave number in vacuum	(m^{-1})
$k_{\text{abs},\nu}$:	absorption coefficient	(m^{-1})
$k_{\text{ext},\nu}$:	extinction coefficient	(m^{-1})
$k_{\nu,\text{D}}$:	absorption coefficient of the Doppler line	(m^{-1})
$k_{\nu,\text{L}}$:	absorption coefficient of the Lorentz line	(m^{-1})
$k_{\nu,\text{V}}$:	absorption coefficient of the Voigt line	(m^{-1})
$k_{\text{sca},\nu}$:	scattering coefficient	(m^{-1})
$\mathbf{k}$:	wave number vector	(m^{-1})
$\mathcal{L}$:	Laplace transform	
L :	Lagrange function	
L :	linear differential operator	
L^+ :	adjoint linear differential operator	
$L(\bar{u})$:	Ladenburg and Reiche function	
M :	magnetic quantum number	
M, m :	mass	(kg)
$\mathbf{m}^{\text{e}}_{\text{o},m,n}$:	wave functions	
$\mathbf{M}$:	magnetization	(A m^{-1})
$\mathbf{M}$:	dipole moment	(A s m)
$\mathbf{M}_\psi$:	vector function to solve the vector wave equation	
$\mathbb{M}$:	diagonal direction matrix	
$\mathbb{M}^m$:	matrix used in spherical harmonics method	
$\mathcal{N}$:	complex index of refraction	
$N_{\text{l,u}}$:	number of systems in lower and upper state	(m^{-3})
N_ν :	total number of photons	
n :	number concentration of molecules	(m^{-3})
$\mathbf{n}$:	normal unit vector	
$\mathbf{N}_\psi$:	vector function to solve the vector wave equation	
$\mathbf{N}$:	Poynting vector	(W m^{-2})
$\mathcal{P}(\cos\Theta)$:	phase function	(sr^{-1})
$\tilde{\mathcal{P}}(\cos\Theta)$:	scattering function	(m^{-1} sr^{-1})
$\mathcal{P}_{\text{HG}}$:	Henyey–Greenstein phase function	(sr^{-1})
P_l :	Legendre polynomial	
P_l^m :	associated Legendre polynomial	
$\bar{P}(\vartheta')$:	probability distribution appearing in Monte Carlo method	
p :	linear momentum	(kg m s^{-1})
p, p_0 :	air pressure	(kg m^{-1} s^{-2})
$p(S)$:	distribution function for line intensities	(m^2)
p_{c} :	number of molecular collisions per unit time	(s^{-1})
p_l, p_l^m :	expansion coefficients of the phase function	
p_m^* :	δ-scaled expansion coefficient	
$\mathbf{P}$:	polarization vector	(A s m^{-2})
$\mathbb{P}$:	phase matrix	
$\tilde{\mathbb{P}}$:	scattering matrix	

Q	:	source function	$(\mathrm{W\,m^{-3}\,sr^{-1}})$
Q_{g}	:	dimensionless surface source function	
Q^+	:	adjoint source function	$(\mathrm{W\,m^{-3}\,sr^{-1}})$
Q_{abs}	:	Mie efficiency factor for absorption	
Q_{ext}	:	Mie efficiency factor for extinction	
Q_{sca}	:	Mie efficiency factor for scattering	
q	:	probability that a molecule does not collide per unit time	
q^i	:	generalized contravariant position coordinate	(m)
q_k	:	normal coordinate	(m)
$\mathbf{q}_i$	:	covariant base vector	
$\mathbf{Q}^m_{\pm}$	:	integral expressions in successive order scattering method	$(\mathrm{W\,m^{-3}\,sr^{-1}})$
$R(z)$	:	response function	
R_0	:	gas constant for dry air	$(\mathrm{m^2\,s^{-2}\,K^{-1}})$
R^m	:	expansion of phase function for m-th component of radiance	
R_{nm}	:	matrix element for transition $n \to m$	
r	:	particle radius	(m)
$\mathbf{r}$	:	position vector	
$\mathbf{r}_{\mathrm{c}}$	:	position vector for center of mass	
$\mathbb{r}$	:	reflectivity matrix	
$\mathbb{r}_{\mathrm{g}}$	:	ground reflection matrix	
$\mathbb{r}^m$	:	discretized reflectivity matrix	
$\mathcal{S}$	:	scattering function	
S	:	line intensity	$(\mathrm{m^{-2}})$
S_0	:	direct sunlight	$(\mathrm{W\,m^{-2}})$
$S_{1,2}$	:	Mie amplitude functions	
S_J	:	part of line intensity that depends only on rotational quantum number J	
S_ν	:	direct solar radiative flux density	$(\mathrm{W\,m^{-2}\,Hz^{-1}})$
$\mathcal{T}$	:	transmission function	
$\mathcal{T}_{\mathrm{f}}$	:	flux transmission function	
T	:	temperature	(K)
$\mathbf{t}$	:	tangential unit vector	
$\mathbb{t}$	:	transmissivity	
$\mathbb{t}^m$	:	discretized transmission matrix	
t	:	time	(s)
U	:	Heavyside step function	
u	:	absorber mass	$(\mathrm{kg\,m^{-2}})$
u	:	real scalar function	
u^*	:	complex scalar function	
$\hat{u}_\nu$	:	energy density	$(\mathrm{J\,m^{-3}})$
$u_n^{s,t}$	:	scalar functions to solve Mie problem	
V	:	potential energy	(J)
V	:	volume	$(\mathrm{m^3})$
$V(r)$	:	Morse function	(J)
v	:	specific volume	$(\mathrm{m^3\,kg^{-1}})$
v	:	vibrational quantum number	
v	:	velocity	$(\mathrm{m\,s^{-1}})$
$\mathbf{v}$	:	velocity vector	$(\mathrm{m\,s^{-1}})$
W	:	equivalent width	$(\mathrm{m^{-1}})$
w	:	absorber mass	$(\mathrm{kg\,m^{-2}})$

w_j	:	weights of the Gaussian quadrature formula	
$\mathbb{W}$	:	diagonal matrix of the Gaussian weights	
$X(\mu)$	:	Chandrasekhar's X-function	
X_{nm}	:	x-component of matrix element of dipole moment for transition $n \to m$	(Asm)
x	:	size parameter	
$\mathbb{X}_j^m$	:	matrix used in spherical harmonics method	
$Y(\mu)$	:	Chandrasekhar's Y-function	
$Y_\pm^m$	:	terms used in successive order of scattering method	
Z	:	partition function	
Z_n	:	cylinder function of order n	
z_n	:	spherical Bessel function of order n	
α_D	:	half-width of the Doppler line	(Hz or m^{-1})
α_L	:	half-width of the Lorentz line	(Hz or m^{-1})
α_V	:	half-width of the Voigt line	(Hz or m^{-1})
α_i	:	sensitivity coefficient for feedback	(W m^{-2} K^{-1})
δ	:	net sensitivity coefficient for feedback	(W m^{-2})
δ	:	phase difference	
$\delta(\mu)$	:	Dirac δ-function	
$\delta_{1,2}$	:	phase angles to describe polarization	
δ_{ij}	:	Kronecker δ	
$\delta_{l,s}$	:	sensitivity coefficients for long-wave and short-wave radiation	(W m^{-2})
ε	:	emissivity	
$\varepsilon, \varepsilon_0$	:	permittivity, permittivity of free space	(C V^{-1} m^{-1})
ε_f	:	flux emissivity function	
ε_g	:	emissivity of the ground	
ε_r	:	relative permittivity, dielectric constant	
$\kappa_{abs,\nu}$	:	mass absorption coefficient	(m^2 kg^{-1})
λ	:	wavelength	(μm)
$\tilde{\lambda}$	:	eigenvalues appearing in two-stream method	
μ	:	reduced mass	
μ, μ_0	:	$\mu = \cos\vartheta$, $\mu_0 = \cos\vartheta_0$	
μ, μ_0	:	permeability, permeability of free space	(V s A^{-1} m^{-1})
μ_r	:	relative permeability	
ν	:	frequency	(Hz)
$\tilde{\nu}$	:	wave number	(m^{-1})
Ψ	:	wave function	
ρ	:	mass density of the dry air	(kg m^{-3})
ρ	:	volume density of electric charge	(A s m^{-3})
ρ_{abs}	:	mass density of an absorber gas	(kg m^{-3})
σ	:	electrical conductivity	(A V^{-1} m^{-1})
σ	:	mean line intensity	(m^{-2})
σ	:	Stefan–Boltzmann constant	(W m^{-2} K^{-4})
σ_c	:	collision cross-section	(m^2)
τ	:	optical depth	
τ^*	:	δ-scaled optical depth	
$\bar{\tau}$	:	average time between two molecular collisions	(s)
Θ	:	scattering angle	

ϑ :	zenith angle indicating direction of radiation	
ϑ_0 :	zenith angle indicating direction of direct solar radiation	
ϕ :	radiative flux	(W)
$\phi(\varphi)$:	wave function	
φ :	azimuthal angle indicating direction of radiation	
φ_0 :	azimuthal angle indicating direction of direct solar radiation	
χ :	electric susceptibility	
Ω :	solid angle	(sr)
$\boldsymbol{\Omega}$:	solid angle vector	(sr)
ω :	angular frequency	(Hz)
ω_0 :	single scattering albedo	
ω_0^* :	δ-scaled single scattering albedo	
$\omega_{a,s}$:	antisymmetric and symmetric vibration	(Hz)

List of constants

c :	speed of light in free space	$(2.997\,924\,58\times10^8 \text{ m s}^{-1})$
$c_{p,0}$:	specific heat at constant pressure, dry air	$(1005 \text{ J kg}^{-1}\text{K}^{-1})$
ε_0 :	permittivity of free space	$(8.8542\times10^{-12} \text{ C V}^{-1}\text{m}^{-1})$
h :	Planck's constant	$(6.626\,196\times10^{-34} \text{ J s})$
k :	Boltzmann constant	$(1.380\,662\times10^{-23} \text{ J K}^{-1})$
μ_0 :	permeability of free space	$(1.2566\times10^{-6} \text{ T m A}^{-1})$
R^* :	universal gas constant	$(8.314\,32 \text{ J mole}^{-1}\text{K}^{-1})$
R_0 :	gas constant of dry air	$(287.05 \text{ J kg}^{-1}\text{K}^{-1})$
σ :	Stefan–Boltzmann constant	$(5.670\,32\times10^{-8} \text{ W m}^{-2}\text{K}^{-4})$
S_0 :	solar constant	(1368 W m^{-2})

References

Abhyankar, K. D., and A. L. Fymat, 1970a: Theory of radiative transfer in inhomogeneous atmospheres. III. Extension of the perturbation method to azimuth-independent terms. *Astrophys. J.*, **159**, 1009–1018.

Abhyankar, K. D., and A. L. Fymat, 1970b: Theory of radiative transfer in inhomogeneous atmospheres. IV. Application of the matrix perturbation method to azimuth-independent terms. *Astrophys. J.*, **159**, 1019–1028.

Abramowitz M., and I. A. Stegun, 1972: *Handbook of Mathematical Functions*. New York: Dover Publications, Inc.

Allen, C. W., 1963: *Astrophysical Quantities*. London: Athlone Press.

Ambartsumian, V., 1936: The effect of the absorption lines on the radiative equilibrium of the outer layers of the stars. *Publ. Obs. Astron. Univ. Leningrad*, **6**, 7–18.

Ambartsumian, V., 1942: Diffusion of light by planetary atmospheres. *Astron. Zh.*, **19**, 30–41.

Anderson, G. P., 1996: *MODTRAN 3 User Instructions*. US Air Force Geophysics Laboratory. [Available from Geophysics Laboratory, US Air Force, Hanscom AFB, MA 01731-5000]

Arking, A. 1991: The radiative effects of clouds and their impact on climate. *Bull. Amer. Meteor. Soc.*, **71**, 795–813.

Asano, S., 1975: On the discrete ordinates method for the radiative transfer. *J. Meteorol. Soc. Japan*, **53**, 92–95.

Atkins, P. W., 1993: *Einführung in die physikalische Chemie*. Weinheim: VCH.

Barker, H. W., and J. A. Davies, 1992: Solar radiative fluxes for broken cloud fields above reflecting surfaces. *J. Atmos. Sci.*, **49**, 749–761.

Barkstrom, B. R., 1976: A finite difference method of solving anisotropic scattering problems. *J. Quant. Spectr. Rad. Transfer*, **16**, 725–739.

Barrow, G. M., 1962: *Introduction to Molecular Spectroscopy*. New York, San Francisco, Toronto, London: McGraw-Hill.

Bates, D. R., and M. Nicolet, 1950: The photochemistry of atmospheric water vapour. *J. Geophys. Res.*, **55**, 301–327.

Bell, G. I., and S. Glasstone, 1970: *Nuclear Reactor Theory*. New York: Van Nostrand Reinhold.

Bohren, C. F., 1985: Comment on "Cloud optical thickness feedbacks in the carbon dioxide climate problem" by R. Somerville and L. A. Remer. *J. Geophys. Res.*, **90**, 5867–5878.

Bohren, C. F., and D. R. Huffman, 1983: *Absorption and Scattering of Light by Small Particles*. New York, Chichester, Brisbane, Toronto, Singapore: John Wiley and Sons.

Born, M., and E. Wolf, 1965: *Principles of Optics: Electromagnetic Theory of Propagation, Interference and Diffraction of Light*. Oxford, London, Edinburgh, New York, Paris, Frankfurt: Pergamon Press.

Bott, A., and W. Zdunkowski, 1983: A fast solar radiation transfer code for application in climate models. *Arch. Met. Geoph. Biocl. B*, **33**, 163–174.

Bott, A., and W. Zdunkowski, 1987: Electromagnetic energy within dielectric spheres. *J. Opt. Soc. Am. A.*, **4**, 1361–1365.

Bott, A., U. Sievers, and W. Zdunkowski, 1990: A radiation fog model with a detailed treatment of the interaction between radiative transfer and fog microphysics. *J. Atmos. Sci.*, **47**, 2153–2166.

Bott, A., T. Trautmann, and W. Zdunkowski, 1996: A numerical model of the cloud-topped planetary boundary-layer: radiation, turbulence and spectral microphysics in marine stratus. *Q. J. R. Meteorol. Soc.*, **122**, 635–667.

Bott, A., 1997: A numerical model of the cloud-topped planetary boundary layer: impact of aerosol particles on radiative forcing of stratiform clouds. *Q. J. R. Meteorol. Soc.*, **123**, 631–656.

Box, M. A., and A. Deepak, 1979: Retrieval of aerosol size distributions by inversion of solar aureole data in the presence of multiple scattering. *Appl. Opt.*, **18**, 1376–1382.

Box, M. A., S. A. W. Gerstl, and C. Simmer, 1988: Application of the adjoint formulation to the calculation of atmospheric radiative effects. *Beitr. Phys. Atmosph.*, **61**, 303–311.

Box, M. A., S. A. W. Gerstl, and C. Simmer, 1989a: Computation of atmospheric radiative effects via perturbation theory. *Beitr. Phys. Atmosph.*, **62**, 193–199.

Box, M. A., B. Croke, S. A. W. Gerstl, and C. Simmer, 1989b: Application of the perturbation theory for atmospheric radiative effects: aerosol scattering atmospheres. *Beitr. Phys. Atmosph.*, **62**, 200–211.

Bronson, R., 1972: *Matrix Methods: An Introduction*. New York: Academic Press.

Brooks, D. L. 1950: A tabular method for the computation of temperature change by infrared radiation in the free atmosphere. *J. Meteor.*, **7**, 313–321.

Bruinenberg, A., 1946: *Een Numerieke Methode voor de Bepaling van Temperatuurs–Veranderingen door Straling in de Vrije Atmosfeer (A Numerical Method for the Calculation of Temperature Changes by Radiation in the Free Atmosphere)*, Koninklijk Nederlands Meteorologisch Instituut, No. 125, Mededelingen en Verhandelingen, Series B, vol. 1, No. 1, 1946.

Buck, R. C., 1965: *Advanced Calculus*. New York, St. Louis, San Francisco, Toronto, London, Sydney: McGraw-Hill Book Company.

Cabannes, J., 1929: *La Diffusion Moléculaire de la Lumilère*. Paris: Les Presses Universitaires de France.

Cahalan, R. F., W. Ridgeway, W. J. Wiscombe, T. L. Bell, and J. B. Snider, 1994: The albedo of fractal stratocumulus clouds. *J. Atmos. Sci.*, **51**, 2434–2455.

Ceballos, J. C., 1988: On two-flux approximations for shortwave radiative transfer in the atmosphere. *Beitr. Phys. Atmosph.*, **61**, 10–22.

Chahine, M. T., 1970: Inverse problems in radiative transfer: determination of atmospheric parameters. *J. Atmos. Sci.*, **27**, 960–967.

Chandrasekhar, S., 1960: *Radiative Transfer*. Oxford: Clarendon Press. Reprinted 1960, New York: Dover Publications, Inc.

Chapman, S., 1930a: A theory of upper atmospheric ozone. *Mem. Roy. Soc.*, **3**, 103–109.

Chapman, S., 1930b: On ozone and atomic oxygen in the upper atmosphere. *Philos. Mag.*, p. 10369.

Chou, M.-D., and A. Arking, 1980: Computation of infrared cooling rates in water vapor bands. *J. Atmos. Sci.*, **37**, 855–867.

Chou, M.-D., and A. Arking, 1981: An efficient method for computing the absorption of solar radiation by water vapor. *J. Atmos. Sci.*, **38**, 798–807.

Chou, M.-D., K.-T. Lee, S.-C. Tsay, and Q. Fu, 1999: Parameterization for cloud longwave scattering for use in atmospheric models. *J. Climate*, **12**, 159–169.

Chou, M.-D., M. Suarez, X.-Z. Liang, and M. M.-H. Yan, 2001: A thermal infrared radiation parameterization for atmospheric studies. *NASA Technical Report Series on Global Modeling and Data Assimilation*, NASA/TM-2001-104606, Volume 19. Greenbelt, MA: Goddard Space Flight Center.

Coulson, K. L., 1975: *Solar and Terrestrial Radiation*. New York: Academic Press Ltd.

Coulson, K. L., 1988: *Polarization and Intensity of Light in the Atmosphere*. Hampton, VA: A. Deepak Publishing.

Coulson, K. L., J. V. Dave, and Z. Sekera, 1960: *Tables Related to Radiation Emerging from a Planetary Atmosphere with Rayleigh Scattering*. Berkeley and Los Angeles: University of California Press.

Courant, R., and D. Hilbert, 1953: *Methods of Mathematical Physics*, Volume 1. New York, London, Sydney: Wiley-Interscience.

Craig, I. J. D., and J. C. Brown, 1986: *Inverse Problems in Astronomy*. Bristol: A. Hilger.

Crisp, D., S. B. Fels, and M. D. Schwarzkopf, 1986: Approximate methods for finding CO_2 15 μm band transmission in planetary atmospheres. *J. Geophys. Res.*, **91**, 11851–11866.

d'Almeida, G. A., P. Koepke, and E. P. Shettle, 1991: *Atmospheric Aerosols: Global Climatology and Radiative Chracteristics*. Hampton, VA: A. Deepak Publishing.

Dave, J. V., 1975: A direct solution of the spherical harmonics approximation to the radiative transfer equation for an arbitrary solar elevation. Part I: Theory. *J. Atmos. Sci.*, **32**, 790–798.

Davis, J. M., S. K. Cox, and T. B. McKee, 1979: The total short wave radiation characteristics of absorbing finite clouds. *J. Atmos. Sci.*, **36**, 508–518.

Debye, P., 1909: Der Lichtdruck auf Kugeln von beliebigem Material. *Ann. Phys.*, **30**, 57–136.

de Haan, J. F., P. B. Bosma, and J. W. Hovenier, 1987: The adding method for multiple scattering calculations of polarized light. *Astron. Astrophys.*, **183**, 371–391.

Deirmendjian, D., 1957: Theory of the solar aureole, part I: scattering and radiative transfer. *Ann. Geophys.*, **13**, 286–306.

Deirmendjian, D., 1959: The role of water particles in the atmospheric transmission of infrared radiation. *Quart. J. Roy. Meteor. Soc.*, **85**, 404–411.

Deirmendjian, D., 1969: *Electromagnetic Scattering on Spherical Polydispersions*. New York: Elsevier.

Demtröder, W., 2003: *Molekülphysik: Theoretische Grundlagen und experimentelle Methoden*. Oldenburg, München, Wien.

Deuze, J. L., C. Devaux, and M. Herman, 1973: Utilisation de la méthode des harmoniques sphériques dans les calculs de transfert radiatif. Extension au cas de couches diffusantes d'absorption variable. *Nouv. Rev. Opt.*, **4**, 307–314.

Eddington, A. S., 1916: On the radiative equilibrium of the stars. *Mon. Not. Roy. Astron. Soc.*, **77**, 16–35.

Ellis, J. S., 1978: Cloudiness, the planetary radiation budget, and climate. Ph.D. Thesis, Fort Collins, CO: Colorado State University.

Elsasser, W. M., 1942: *Heat Transfer by Infrared Radiation in the Atmosphere.* Harvard Meteorological Studies No. 6. Cambridge, MA: Harvard University Press.

Elsasser, W. M., and M. F. Culbertson, 1961: *Atmospheric Radiation Tables.* Meteorological Monographs, Vol. 4, No. 23. Boston, MA: American Meteorological Society.

Emden, R., 1913: Über Strahlungsgleichgewicht und atmosphärische Strahlung: Ein Beitrag zur Theorie der oberen Inversion. *Sitzungsberichte der Königlich Bayerischen Akademie der Wissenschaften, Math.-phys. Klasse*, pp. 55–142.

Eyring, H., J. Walter and G. E. Kimball, 1965: *Quantum Chemistry.* New York, London, Sydney: John Wiley & Sons, Inc.

Fels, S. B., 1979: Simple strategies for inclusion of Voigt effects in infrared cooling rate calculations. *Appl. Opt.*, **18**, 2634–2637.

Flatau, P. J., and G. L. Stephens, 1988: On the fundamental solution of the radiative transfer equation. *J. Geophys. Res.*, **93**, 11037–11050.

Fomichev, V. I., and G. M. Shved, 1985: Parameterization of the radiative divergence in the 9.6 micron O_3 band. *J. Atmos. Terr. Phys.*, **47**, 1037–1049.

Fowles, G. R., 1966: *Analytical Mechanics.* New York, Chicago, San Francisco, Toronto, London: Holt, Rinehart and Winston.

Fowles, G. R., 1967: *Introduction to Modern Optics.* New York, Chicago, San Francisco, Toronto, London: Holt, Rinehart and Winston.

Friedman, B., 1956: *Principles and Techniques of Applied Mathematics.* New York, NY: John Wiley.

Fu, Q., and K. N. Liou, 1992: On the correlated k-distribution method for radiative transfer in nonhomogeneous atmospheres. *J. Atmos. Sci.*, **49**, 2139–2156.

Fu, Q., K. N. Liou, M. C. Cribb, T. P. Charlock, and A. Grossmann, 1997: Multiple scattering parameterization in thermal infrared radiative transfer. *J. Atmos. Sci.*, **52**, 2799–2812.

Fymat, A. L., and K. D. Abhyankar, 1969a: Theory of radiative transfer in inhomogeneous atmospheres. I. Perturbation method. *Astrophys. J.*, **158**, 315–324.

Fymat, A. L., and K. D. Abhyankar, 1969b: Theory of radiative transfer in inhomogeneous atmospheres. II. Application of the perturbation method to a semi-infinite atmosphere. *Astrophys. J.*, **158**, 325–335.

Gantmacher, F. R., 1986: *Matrizentheorie.* Berlin: Springer-Verlag.

Garcia, R. D. M., and C. E. Siewert, 1985: Benchmark results in radiative transfer. *Transp. Theory Statist. Phys.*, **14**, 437–483.

Geleyn, J. F., and A. Hollingsworth, 1979: An economical analytical method for the computation of the interaction between scattering and line absorption of radiation. *Beitr. Phys. Atmosph.*, **52**, 1–16.

Gergen, J. L., 1956: Black ball: a device for measuring atmospheric radiation. *Rev. Scient. Instr.*, **27**, 453.

Gergen, J. L., 1957: Atmospheric infrared radiation over Minneapolis to 30 millibars. *J. Meteorol.*, **14**, 495–504.

Gergen, J. L., 1958: Observations of atmospheric radiation over Mc. Murdo Sound, Antarctica. *Techn. Rep. Atm. Phys. Progr. Navy Contr. Nonr.-710*, **22**.

Gerstl, S. A. W., 1982: Application of the adjoint method in atmospheric radiative transfer calculations. In: Deepak, A., (Ed.), *Atmospheric Aerosols: Their Formation, Optical Properties, and Effects.* Hampton VA: Spectrum Press, pp. 241–254.

Gerstl, S. A. W., and W. M. Stacey, 1973: A class of second order approximation formulations for deep penetration radiation transport problems. *Nucl. Sci. Eng.*, **51**, 339–343.

Gerstl, S. A. W., and A. Zardecki, 1985: Discrete-ordinates finite-element method for atmospheric radiative transfer and remote sensing. *Appl. Opt.*, **24**, 81–93.

Godson, W. L., 1955: The computation of infrared transmission by atmospheric water vapour. *J. Atmos. Sci.*, **12**, 272–284.

Goody, R. M., 1952: A statistical model for water vapour absorption. *Q. J. R. Meteorol. Soc.*, **78**, 165–169.

Goody, R. M., 1964a: *Atmospheric Radiation, I: Theoretical Basis*. Oxford: Clarendon Press.

Goody, R. M., 1964b: The transmission of radiation through an inhomogeneous atmosphere. *J. Atmos. Sci.*, **21**, 575–581.

Goody, R. M., and Y. L. Yung, 1989: *Atmospheric Radiation – Theoretical Basis*. Oxford: Oxford University Press.

Graedel, T. E., and P. J. Crutzen, 1994: *Chemie der Atmosphäre. Bedeutung für Klima und Umwelt*. Heidelberg: Spektrum Akademischer Verlag.

Greiner. W., 1989: *Mechanik, Teil 2*. Frankfurt: Verlag Harri Deutsch.

Hales, J. V., 1951: An atmospheric radiation flux divergence chart and meridional cross-sections of cooling rates. Ph.D. Thesis, Los Angeles, CA: University of California.

Hansen, J. E., 1971: Multiple scattering of polarized light in planetary atmospheres. Part I. The doubling method. *J. Atmos. Sci.*, **28**, 120–125.

Hansen, P. C., 1994: Regularization tools – a Matlab package for analysis and solution of discrete ill-posed problems. *Num. Algor.*, **6**, 1–35.

Hansen, J. E., and A. A. Lacis, 1990: Sun and dust versus greenhouse gases: An assessment of their relative roles in global climate change. *Nature*, **346**, 713–719.

Hansen, J. E., and L. D. Travis, 1974: Light scattering in planetary atmospheres. *Space Sci. Rev.*, **16**, 527–610.

Hansen, J., A. Lacis, D. Rind, *et al.*, 1984: Climate sensitivity: Analysis of feedback mechanisms. In: (J. E. Hansen and T. Takahashi, Eds.) *Climate Processes and Climate Sensitivity* AGU Geophysical Monograph 29. Washington, DC: American Geophysical Union, pp. 130–163.

Hecht, E., 1987: *Optics*. Reading, MA: Addison-Wesley.

Herman, B. M., S. R. Browning, and R. J. Curran, 1971: The effect of atmospheric aerosols on scattered sunlight. *J. Atmos. Sci.*, **28**, 419–428.

Herman, M., J. L. Deuzé, C. Devaux, P. Goloub, F. M. Bréon, and D. Tanré, 1997: Remote sensing of aerosols over land surfaces including polarization measurements and application to POLDER measurements. *J. Geophys. Res.*, **102**, 17039–17049.

Herzberg, G., 1964a: *Molecular Spectra and Molecular Structure. I. Spectra of Diatomic Molecules*. Princeton University, NJ, Toronto, New York, London: D. Van Nostrand Company, Inc.

Herzberg, G., 1964b: *Molecular Spectra and Molecular Structure. II. Infrared and Raman Spectra of Polyatomic Molecules*. Princeton University, NJ. Toronto, New York, London: D. Van Nostrand Company, Inc.

Hess, M., P. Koepke, and I. Schult, 1998: Optical properties of aerosols and clouds: The software package OPAC. *Bull. Amer. Meteorol. Soc.*, **79**, 831–844.

Houghton, H. G., 1985: *Physical Meteorology*. Cambridge, MA: MIT Press.

Houghton, J. T., and S. D. Smith, 1966: *Infra-red Physics*. Oxford: Clarendon Press.

Houghton, J. T., F. W. Taylor, and C. D. Rodgers, 1984: *Remote Sounding of Atmospheres*. Cambridge: Cambridge University Press.

Hovenier, J. W., 1971: Multiple scattering of polarized light in planetary atmospheres. *Astron. Astrophys.*, **13**, 7–29.

Hovenier, J. W., and C. V. M. van der Mee, 1983: Fundamental relationships relevant to the transfer of polarized light in a scattering atmosphere. *Astron. Astrophys.*, **128**, 1–16.

Hunt, G. E., R. Kandel, and A. T. Mecherikunnel, 1986: A history of pre-satellite investigations of the earth's radiation budget. *Rev. Geophys.*, **24**, 351–356.

Jackson, J. D., 1975: *Classical Electrodynamics*. New York: Wiley.

Jahnke, E., and F. Emde, 1945: *Tables of Functions with Formulae and Curves*. New York: Dover Publications.

Johnson, J. C., 1960: *Physical Meteorology*. New York, London: The Technology Press of the Massachusetts Institute of Technology and John Wiley and Sons, Inc.

Joseph, J. H., W. J. Wiscombe, and J. A. Weinman, 1976: The delta-Eddington approximation for radiative flux transfer. *J. Atmos. Sci.*, **33**, 2452–2459.

Kaplan, L. D., 1959: Inference of atmospheric structure from remote radiation measurements. *J. Opt. Soc. Amer.*, **49**, 1004–1007.

Kasten, F., and A. T. Young, 1989: Revised optical air mass tables and approximation formula. *Appl. Opt.*, **28**, 4735–4738.

Keener, J. P., 1988: *Principles of Applied Mathematics: Transformation and Approximation*. Addison-Wesley.

Kiehl, J. T., and K. E. Trenberth, 1997: Earth's annual global mean energy budget. *Bull. Amer. Meteor. Soc.*, **78**, 197–208.

King, J. I. F., 1956: The radiative heat transfer of planet earth. In: *Scientific Use of Earth Satellites*. Ann Arbor, MI: University of Michigan Press, pp. 133–136.

King, M. D., D. M. Byrne, B. M. Herman, and J. A. Reagan, 1978: Aerosol size distribution obtained by inversion of spectral optical depth measurements. *J. Atmos. Sci.*, **35**, 2153–2167.

Kirchhoff, G., 1882: *Gesammelte Abhandlungen*. Leipzig: Barth.

Kneizys, F. X., L. W. Abreu, *et al.*, 1996: The MODTRAN 2/3 Report and Lowtran 7 Model. Hanscom AFB, MA: Phillips Laboratory, Geophysics Directorate. [F19628-91-c-0132]

Kondrat'yev, K. Y., 1965: *Radiative Heat Exchange in the Atmosphere*. New York: Pergamon Press.

Korb, G., and W. Zdunkowski, 1970: Distribution of radiative energy in ground fog. *Tellus*, **22**, 298–320.

Kourganoff, V., 1952: *Basic Methods in Transfer Problems: Radiative Equilibrium and Neutron Diffusion*. Oxford: Clarendon Press.

Kroto, H. W., 1975: *Molecular Rotation Spectra*. London: Wiley.

Kuščer, I., and I. Vidav, 1969: On the spectrum of relaxation lengths of neutron distributions in a moderator. *J. Math. Anal. Appl.*, **25**, 80–92.

Labitzke, K., B. Naujokat, and M. P. McCormick, 1983: Temperature effects on the stratosphere of the April 4, 1982 eruption of El Chichon. *Geophys. Res. Letters*, **10**, 24–26.

Lacis, A. A., and V. Oinas, 1991: A description of the correlated-k distribution method for modeling nongray gaseous absorption, thermal emission, and multiple scattering in vertically inhomogeneous atmospheres. *J. Geophys. Res.*, **96**, 9027–9063.

Lacis, A. A., W. C. Wang, and J. Hansen, 1979: Correlated k-distribution method for radiative transfer in climate models: Application to the effect of cirrus clouds on climate. NASA Conf. Publ. 2076, 309–314.

Ladenburg, R., and F. Reiche, 1913: Über selektive Absorption. *Ann. der Phys.*, **42**, 181–203.

Landgraf, J., O. P. Hasekamp, M. A. Box, and T. Trautmann, 2001: A linearized radiative transfer model for ozone profile retrieval using the analytical forward–adjoint perturbation theory approach. *J. Geophys. Res.*, **106**, 27291–27305.

Landgraf, J., O. P. Hasekamp, and T. Trautmann, 2002: Linearization of radiative transfer with respect to surface properties. *J. Quant. Spectr. Rad. Transfer*, **72**, 327–339.

Lenoble, J. (ed.), 1985: *Radiative Transfer in Scattering and Absorbing Atmospheres: Standard Computational Procedures*. Hampton, VA: A. Deepak Publ.

Lenoble, J., 1993: *Atmospheric Radiative Transfer*. Hampton, VA: A. Deepak Publ.

Levine, I. N., 1975: *Molecular Spectroscopy*. New York: Wiley.

Li, J., and V. Ramaswamy, 1996: Four-stream spherical harmonic expansion approximation for solar radiative transfer. *J. Atmos. Sci.*, **53**, 1174–1186.

Liou, K.-N., 1973: A numerical experiment on Chandrasekhar's discrete-ordinate method for radiative transfer: applications to cloudy and hazy atmosphere. *J. Atmos. Sci.*, **30**, 1303–1326.

Liou, K.-N., 1980: *An Introduction to Atmospheric Radiation*. International Geophys. Series, Vol. 26. New York: Academic Press.

Liou, K.-N., 2002: *An Introduction to Atmospheric Radiation*. Amsterdam: Academic Press.

Livingston, J. M., and P. B. Russell, 1989: Retrieval of aerosol size distribution moments from multiwavelength particulate extinction measurements. *J. Geophys. Res.*, **94**, 8425–8433.

Los, A., M. van Weele, and P. G. Duynkerke, 1997: Actinic flux in broken cloud fields. *J. Geophys. Res.*, **102**, 4257–4266.

Malkmus, W., 1967: Random Lorentz band model with exponential-tailed S^{-1} line-intensity distribution function. *J. Opt. Soc. Am.*, **57**, 323–329.

Manabe, S., and F. Möller, 1961: On the radiative equilibrium and heat balance of the atmosphere. *Mon. Wea. Rev.*, **89**, 503–532.

Manabe, S., and R. F. Strickler, 1964: Thermal equilibrium of the atmosphere with a convective adjustment. *J. Atmos. Sci.*, **21**, 361–385.

Mayer, H., 1947: *Methods of Opacity Calculations*, Los Alamos Report 647. Los Alamos, NM: Los Alamos Scientific Laboratory.

Meador, W. E., and W. R. Weaver, 1980: Two-stream approximations to radiative transfer in planetary atmospheres: A unified description of existing methods and a new improvement. *J. Atmos. Sci.*, **37**, 630–643.

Mie, G., 1908: Beiträge zur Optik trüber Medien speziell kolloidialer Metallösungen. *Ann. Phys.* **25**, 377–445.

Mishchenko, M. I., and L. D. Travis, 1997: Satellite retrieval of aerosol properties over the ocean using polarization as well as intensity of reflected sunlight. *J. Geophys. Res.*, **102**, 16989–17013.

Möller, F., 1941: Die Wärmestrahlung des Wasserdampfes in der Atmosphäre. *Gerlands Beiträge zur Geophysik*, **58**, 11–67.

Möller, F., 1943: Das Strahlungsdiagramm. Reichsamt für Wetterdienst, Luftwaffe.

Möller, F., 1944: Grundlagen eines Diagrammes zur Berechnung langwelliger Strahlungsströme. *Meteorol. Zeitschrift*, **61**, 37–45.

Möller, F., 1951: Long wave radiation. *Compendium of Meteorology*. Boston, MA: American Meteorological Society, pp. 34–49.

Möller F., and S. Manabe, 1961: Über das Strahlungsgleichgewicht der Atmosphäre, *Zeitschrift für Meteorol.*, **15**, 3–8.

Moler, C. B., and C. van Loan, 1978: Nineteen dubious ways to compute the exponential of a matrix. *SIAM Rev.*, **4**, 801–836.

Morse, P. M., and H. Feshbach, 1953: *Methods in Theoretical Physics. Part I*. Boston, MA: McGraw-Hill.

Muldashev, T. Z., A. I. Lyapustin, and U. M. Sultangazin, 1999: Spherical harmonics method in the problem of radiative transfer in the atmosphere–surface system. *J. Quant. Spectr. Rad. Transfer*, **61**, 393–404.

Mügge, R., and F. Möller, 1932: Zur Berechnung von Strahlungsströmen und Temperaturänderungen in Atmosphären von beliebigem Aufbau. *Z. Geophys.*, **8**, 53–64.

Nakajima, T., M. Tanaka, and T. Yamauchi, 1983: Retrieval of the optical properties of aerosols from aureole and extinction data. *Appl. Opt.*, **22**, 2951–2959.

Ohring. G., P. F. Clapp, T. R. Heddinghaus, and A. F. Krueger, 1981: The quasi-global distribution of the sensitivity of the earth–atmosphere radiation budget to clouds. *J. Atmos. Sci.*, **38**, 2539–2541.

O'Hirok, W., and C. Gautier, 1998a: A three-dimensional radiative transfer model to investigate the solar radiation within a cloudy atmosphere. Part I: Spatial effects. *J. Atmos. Sci.*, **55**, 2162–2179.

O'Hirok, W., and C. Gautier, 1998b: A three-dimensional radiative transfer model to investigate the solar radiation within a cloudy atmosphere. Part II: Spectral effects. *J. Atmos. Sci.*, **55**, 3065–3076.

Paltridge, G. W., 1980: Cloud-radiation feedback to climate. *Q. J. R. Meteorol. Soc.*, **106**, 367–380.

Paltridge, G. W., and C. M. R. Platt, 1976: *Radiative Processes in Meteorology and Climatology*. Amsterdam: Elsevier.

Pauling, L., and E. B. Wilson, 1935: *Introduction to Quantum Mechanics*. New York and London: McGraw-Hill Book Company, Inc.

Penner, S. S., 1959: *Quantitative Molecular Spectroscopy and Gas Emissivities*. Reading, MA, London: Addison-Wesley Publishing Company, Inc.

Phillips, D. L., 1962: A technique for the numerical solution of certain integral equations of the first kind. *J. Assoc. Comput. Mach.*, **9**, 84–97.

Plass G. N., 1956: The influence of the 9.6 micron ozone band on the atmospheric infra-red cooling rate. *Q. J. R. Meteorol. Soc.*, **82**, 310–324.

Plass, G. N., G. W. Kattawar, and F. E. Catchings, 1973: Matrix operator theory of radiative transfer. I: Rayleigh scattering. *Appl. Opt.*, **12**, 314–329.

Pomraning, G. C., 1973: *The Equations of Radiation Hydrodynamics*. Oxford, New York: Pergamon Press.

Press, W. H., S. A. Teukolsky, W. T. Vetterling, and B. P. Flannery, 1992: *Numerical Recipes in FORTRAN: The Art of Scientific Computing*. – 2nd edn. Cambridge: Cambridge University Press, pp. 935–1481.

Putzer, E. J., 1966: Avoiding the Jordan canonical form in the discussion of linear systems with constant coefficients. *Amer. Math. Monthly*, **73**, 2.

Ramanathan, V., 1987: The role of earth radiation budget studies in climate and general circulation research. *J. Geophys. Res.*, **92**, 4075–4095.

Ramanathan, V., R. J. Cicerone, H. B. Singh, and J. T. Kiehl, 1985: Trace gas trends and their potential role in climate change. *J. Geophys. Res.*, **90**, 5547–5566.

Ramanathan, V., B. R. Barkstrom, and E. F. Harrison, 1989: Climate and the earth's radiation budget. *Physics Today*, **42**, 22–33.

Rayleigh, Lord (J. W. Strutt), 1871: On the light from the sky, its polarization and colour. *Phil. Mag.*, **41**, 107–120 and 274–279.

Rayleigh, Lord (J. W. Strutt), 1918: On the scattering of light by a cloud of similar, small particles of any shape and oriented at random. *Phil. Mag.*, **35**, 373–381.

Rodgers, C. D., 1967: The use of emissivity in atmospheric radiation calculations. *Q. J. R. Meteorol. Soc.*, **93**, 43–54.

Rothman, L. S., R. R. Gamache, A. Goldman, *et al.*, 1987: The HITRAN database: 1986 edition. *Appl. Opt.*, **26**, 4058–4097.

Rothman, L. S., R. R. Gamache, A. Goldman, *et al.*, 1992: The HITRAN molecular data base. Editions of 1991 and 1992. *J. Quant. Spectr. Rad. Transfer*, **48**, 509–518.

Rozanov, V. V., D. Diebel, R. J. D. Spurr, and J. P. Burrows, 1997: GOMETRAN: A radiative transfer model for the satellite project GOME, the plane-parallel version, *J. Geophys. Res.*, **102**, 16683–16695.

Santer, R., and M. Herman, 1983: Particle size distribution from forward scattering light using the Chahine inversion scheme. *Appl. Opt.*, **22**, 2294–2302.

Schnaidt, F., 1939: Über die Absorption von Wasserdampf und Kohlensäure mit besonderer Berücksichtigung der Druck- und Temperaturabhängigkeit, *Gerl. Beitr. Geophys.*, **54**, 203–234.

Schuster, A., 1905: Radiation through a foggy atmosphere. *Astrophys. J.*, **21**, 1–22.

Schwarzschild, K., 1906: Über das Gleichgewicht der Sonnenatmosphäre. In: *Nachrichten von der Königlichen Gesellschaft der Wissenschaften zu Göttingen. Math.-phys. Klasse*, 195, p. 41–53.

Sekera, Z., 1956: Recent developments in the study of the polarisation of skylight. *Advances in Geophysics*, **3**, 43–104.

Sekera, Z., and A. B. Kahle, 1966: *Scattering Functions for Rayleigh Atmospheres of Arbitrary Thickness*. Santa Monica, CA: RAND Corporation.

Sellers, W. D., 1965: *Physical Climatology*. Chicago, IL: University of Chicago Press.

Shettle, E. P., and J. A. Weinman, 1970: The transfer of solar irradiance through inhomogeneous turbid atmospheres evaluated by Eddington's approximation. *J. Atmos. Sci.*, **27**, 1048–1055.

Shettle, E. P., and R. W. Fenn, 1979: Models for aerosols of the lower atmosphere and the effects of humidity variations on their optical properties. Hanscom Field, Bedford, MA: AFCRL, AFGL-TR-79-0214.

Shurcliffe, W. A., 1966 *Polarized Light: Production and Use*. 2nd Print. Cambridge MA: Harvard University Press.

Siebert, J., A. Bott, and W. Zdunkowski, 1992: Influence of a vegetation-soil model on the simulation of radiation fog. *Beitr. Phys. Atmos.*, **65**, 93–106.

Smirnow, V. I., 1959: *Lehrbuch der Höheren Mathematik*. Teil 3,2. 2. durchgesehene Auflage, Dt. Berlin: Verlag der Wissenschaften.

Smith, W. L., 1970: Iterative solution of the radiative transfer equation for the temperature and absorbing gas profile of an atmosphere. *Appl. Opt.*, **9**, 1993–1999.

Spencer, J. W., 1971: Fourier series representation of the position of the sun, *Search*, **2**, 172.

Spiegel, R. M., 1964: *Schaum's Outline of Theory and Problems of Complex Variables: With an Introduction to Conformal Mapping and Its Applications*. New York: McGraw-Hill.

Stamnes, K., S.-C. Tsay, W. Wiscombe, and K. Jayaweera, 1988: Numerically stable algorithm for discrete-ordinate-method radiative transfer in multiple scattering and emitting layered media. *Appl. Opt.*, **27**, 2502–2509.

Stephens, G. L., 1994: *Remote Sensing of the Lower Atmosphere*. Oxford: Oxford University Press.

Stokes, G. G., 1852: On the composition and resolution of streams of polarized light from different sources. *Trans. Cambridge Philos. Soc.*, **9**, 399–423.

Stratton, J. A., 1941: *Electromagnetic Theory*. New York and London: McGraw-Hill Book Company.

Thomas, G. E., and K. Stamnes, 1999: *Radiative Transfer in the Atmosphere and Ocean*. Cambridge: Cambridge University Press.

Trautmann, T., I. Podgorny, J. Landgraf, and P. J. Crutzen, 1999: Actinic fluxes and photodissociation coefficients in cloud fields embedded in realistic atmospheres. *J. Geophys. Res.*, **104**, 30,173–30,192.

Twomey, S., 1963: On the numerical solution of Fredholm integral equations of the first kind by the inversion of the linear system produced by quadrature. *J. Assoc. Comput. Mach.*, **10**, 97–101.

Twomey, S., 1977: The influence of pollution on the shortwave albedo of clouds. *J. Atmos. Sci.*, **34**, 1149–1152.

Unsöld, A., 1968: *Physik der Sternatmosphären*. Berlin: Springer-Verlag.

Ustinov, E. A., 2001: Adjoint sensitivity analysis of radiative transfer equation: Temperature and gas mixing ratio weighting functions for remote sensing of scattering atmospheres in thermal IR. *J. Quant. Spectr. Rad. Transfer*, **68**, 195–211.

van de Hulst, H. C., 1945: Theory of absorption lines in the atmosphere of the earth. *Ann. Astrophys.*, **8**, 1–11.

van de Hulst, H. C., 1957: *Light Scattering by Small Particles*. New York: John Wiley and Sons, Inc.; London; Chapman and Hall, Ltd.

van de Hulst, H. C., 1963: *A New Look at Multiple Scattering*. Technical Report, Goddard Institute for Space Studies, NASA.

Walshaw, C. D., and C. D. Rodgers, 1963: The effect of the Curtis–Godson approximation on the accuracy of radiative heating-rate computations. *Q. J. R. Meteorol. Soc.*, **89**, 122–130.

Wang, W.-C., D. Wuebbles, W. M. Washington, R. Isaacs, and G. Molnar, 1986: Trace gases and other potential perturbations of global climate. *Rev. Geophys.*, **24**, 110–140.

Wang, P.-H., M. P. McCormick, T. J. Swissler, M. T. Osborn, W. H. Fuller, and G. K. Yue, 1989: Inference of stratospheric aerosol composition and size distribution from SAGE II satellite measurements. *J. Geophys. Res.*, **94**, 8435–8446.

Wark, D. Q., and H. E. Fleming, 1966: Indirect measurements of atmospheric temperature profiles from satellites: 1. Introduction. *Mon. Wea. Rev.*, **94**, 351–362.

Watson, G. N., 1980: *Treatise on the Theory of Bessel Functions*, 2nd edn. Cambridge: Cambridge University Press.

Wedler, G., 2004: *Lehrbuch der Physikalischen Chemie*. Weinheim: Wiley VCH Verlag GmbH.

Welch R., J.-F. Geleyn, W. Zdunkowski, and G. Korb, 1976: Radiative treatment of solar radiation in model clouds. *Beitr. Phys. Atmos.*, **49**, 128–146.

Welch, R. M., and W. G. Zdunkowski, 1981: The effect of cloud shape on radiative characteristics. *Beitr. Phys. Atmosph.*, **54**, 482–491.

Wetherald, R. T., and S. Manabe, 1988: Cloud feedback processes in a general circulation model. *J. Atmos. Sci.*, **45**, 1397–1415.

Whittaker, E. T., and G. N. Watson, 1915: *A Course of Modern Analysis*. 2nd revised edn. Cambridge: Cambridge University Press.

Wilson, E. B. Jr., J. C. Decius, and P. C. Cross, 1955: *Molecular Vibrations. The Theory of Infrared and Raman Vibrational Spectra*. New York, Toronto, London: McGraw-Hill Book Company.

Wiscombe, W. J., 1977: The delta-M method: rapid yet accurate radiative flux calculation for strongly asymmetric phase functions. *J. Atmos. Sci.*, **34**, 1408–1422.

References

Wiscombe, W. J., and J. W. Evans, 1977: Exponential-sum fitting of radiative transmission functions. *J. Comput. Phys.*, **24**, 416–444.

Wiscombe, W. J., and G. W. Grams, 1976: The backscattered fraction in two-stream approximations. *J. Atmos. Sci.*, **33**, 2440–2451.

Wylie, C. R., 1966: *Advanced Engineering Mathematics*. 3rd edn. New York, London: McGraw-Hill.

Yamamoto, G., 1952: On a radiation chart. *Sci. Rep. Tohoku Univ.*, *Ser. 5, Geophys.*, **4**, 9–23.

Yamamoto, G., and G. Onishi, 1953: A chart for the calculation of radiative temperature changes. *Sci. Rep. Tohoku Univ.*, *Ser. 5, Geophys.*, **4**, 108–115.

Zdunkowski, W. G., 1963: Untersuchungen über langwellige Strahlung der Atmosphäre. *Gerl. Beitr. Geoph.*, **72**, 21–47.

Zdunkowski, W., 1974: An alternative derivation of the Ladenburg and Reiche formula for the integrated absorption of a spectral line. *Beitr. Phys. Atmosph.*, **47**, 224–226.

Zdunkowski, W., and A. Bott, 2003: *Dynamics of the Atmosphere. A Course in Theoretical Meteorology*. Cambridge: Cambridge University Press.

Zdunkowski, W., and A. Bott, 2004: *Thermodynamics of the Atmosphere. A Course in Theoretical Meteorology*. Cambridge: Cambridge University Press.

Zdunkowski, W. G., and W. H. Raymond, 1970: Exact and approximate transmission calculations for homogeneous and non-homogeneous atmospheres. *Beitr. Phys. Atmosph.*, **43**, 185–201.

Zdunkowski, W. G., and R. J. Junk, 1974: Solar radiative transfer in clouds using Eddington's approximation. *Tellus*, **26**, 361–368.

Zdunkowski, W. G., and G. Korb, 1974: An approximate method for the determination of short-wave radiative fluxes in scattering and absorbing media. *Beitr. Phys. Atmosph.*, **47**, 129–144.

Zdunkowski, W. G., R. M. Welch, and G. Korb, 1980: An investigation of the structure of typical two-stream-methods for the calculation of solar fluxes and heating rates in clouds. *Beitr. Phys. Atmosph.*, **53**, 147–166.

Zdunkowski, W. G., W.-G. Panhans, R. M. Welch, and G. J. Korb, 1982: A radiation scheme for circulation and climate models. *Beitr. Phys. Atmosph.*, **55**, 215–238.

Zdunkowski, W. G., and G. Korb, 1985: Numerische Methoden zur Lösung der Strahlungsübertragungsgleichung. *PROMET*, **15**, 26–39.

Zdunkowski, W. G., W.-G. Panhans, and T. Trautmann, 1998: Comments on "Four-stream spherical harmonic expansion approximation for solar radiative transfer". *J. Atmos. Sci.*, **55**, 669–672.

Further reading

Anderson, G. P., S. A. Clough, F. X. Kneizys, J. H. Chetwynd, and E. P. Shettle, 1987: AFGL atmospheric constituent profiles (0–120 km). AFGL-TR-86-0110, Air Force Geophysical Laboratory, Hanscom AFB, MA.

Barichello, L., R. Garcia, and C. Siewert, 1996: The Fourier decomposition for a radiative transfer problem with an asymmetrically reflecting ground. *J. Quant. Spectr. Rad. Transfer*, **56**, 363–371.

Box, M. A., and A. Deepak, 1981: Finite sun effect on the interpretation of the solar aureole. *Appl. Opt.*, **20**, 2806–2810.

Duynkerke, P. G., S. R. de Roode, M. C. van Zanten, *et al.*, 2004: Observations and numerical simulations of the diurnal cycle of the EUROCS stratocumulus case. *Q. J. R. Meteorol. Soc.*, **130**, 3269–3296.

Ellis, J. S., T. H. Vonder Haar, S. Levitus, and A. H. Oort, 1978: The annual variation in the global heat balance of the earth. *J. Geophys. Res.*, **83**, 1958–1962.

Godsalve, C., 1996: The inclusion of reflectances with preferred directions in radiative transfer calculations. *J. Quant. Spectr. Rad. Transfer*, **56**, 373–376.

Hansen, J. E., 1951: Multiple scattering of polarized light in planetary atmospheres. Part II. Sunlight reflected by terrestrial water clouds. *J. Atmos. Sci.*, **28**, 1400–1426.

Hasekamp, O., and J. Landgraf, 2001: Ozone profile retrieval from backscattered ultraviolet radiances: The inverse problem solved by regularization. *J. Geophys. Res.*, **106**, 8077–8088.

Hasekamp, O., and J. Landgraf, 2002a: A linearized vector radiative transfer model for atmospheric trace gas retrieval. *J. Quant. Spectr. Rad. Transfer*, **75**, 221–238.

Hasekamp, O., and J. Landgraf, 2002b: Tropospheric ozone information from satellite-based polarization measurements. *J. Geophys. Res.*, 107, No. D17, 4326.

Hasekamp, O., J. Landgraf, and R. van Oss, 2002: The need of polarization modeling for ozone profile retrieval from backscattered sunlight. *J. Geophys. Res.*, 107, No. D23, 4692.

Hecht, E., 1975: *Optics*. Schaum's outline series. New York, St. Louis, Toronto: McGraw Hill Book Company.

Herman, B., and S. Browning, 1965: A numerical solution to the equation of radiative transfer. *J. Atmos. Sci.*, **22**, 559–566.

Houghton, J. T., L. G. Meiro Filho, B. A. Callander, N. Harris, A. Kattenburg, K. Maskell, 1996: *Climate Change 1995, The Science of Climate Change*. Cambridge: Cambridge University Press.

Korn, G. A., and T. M. Korn, 1968: *Mathematical Handbook for Scientists and Engineers. Definitions, Theorems, and Formulas for Reference and Review*. New York: McGraw-Hill.

Liou, K.-N., 1992: *Radiation and Cloud Processes in the Atmosphere: Theory, Observation, and Modeling*. New York: Oxford University Press.

Morse, P. M., 1929: Diatomic molecules according to the wave mechanics. II. Vibrational levels. *Phys. Rev.*, **34**, 57–64.

Sommerfeld, A., 1964: *Partial Differential Equations in Physics*. New York and London: Academic Press.

Strang, G., 1986: *Linear Algebra and its Applications*. San Diego, CA: Harcourt Brace Jovanovich.

Tikhonov, A. N., and A. A. Samarskii, 1959: *Equations of Mathematical Physics*. London: Pergamon Press.

Westphal, W. H., 1959: *Physik*. Berlin: Springer-Verlag.

Zdunkowski, W. G., and R. L. Weichel, 1971: Radiative energy transfer in haze atmospheres. *Beitr. Phys. Atmosph.*, **44**, 53–68.

Index

Printed in the United States
By Bookmasters